SECOND EDITION

# THE APPLIED GENETICS OF
# HUMANS, ANIMALS, PLANTS AND FUNGI

SECOND EDITION

# THE APPLIED GENETICS OF
# HUMANS, ANIMALS, PLANTS AND FUNGI

BERNARD C LAMB

*Imperial College London, London*

Imperial College Press

*Published by*

Imperial College Press
57 Shelton Street
Covent Garden
London WC2H 9HE

*Distributed by*

World Scientific Publishing Co. Pte. Ltd.
5 Toh Tuck Link, Singapore 596224
*USA office:* 27 Warren Street, Suite 401-402, Hackensack, NJ 07601
*UK office:* 57 Shelton Street, Covent Garden, London WC2H 9HE

**British Library Cataloguing-in-Publication Data**
A catalogue record for this book is available from the British Library.

First published 2007 (Hardcover)
Reprinted 2016 (in paperback edition)
ISBN 978-1-911299-68-4

**THE APPLIED GENETICS OF HUMANS, ANIMALS, PLANTS AND FUNGI
(2nd Edition)**

ISBN 13 978-1-86094-610-3
ISBN 10 1-86094-610-0

Typeset by Stallion Press
Email: enquiries@stallionpress.com

This book is dedicated to my wise parents,
for all they have done for me.

# Preface

I was delighted to be asked to prepare an enlarged and updated edition. The title has changed from *The Applied Genetics of Plant, Animals, Humans and Fungi* to *The Applied Genetics of Humans, Animals, Plants and Fungi*, to reflect a different emphasis, but all sections have been expanded and revised.

There have been many advances in all areas of applied genetics since the first edition was published in 2000. There is now a section on genomics, and the large human and medical chapter includes pre-implantation genetic diagnosis, human gene sequencing, mosaics, chimeras and hermaphrodites, stem-cell therapy and cord-blood therapy. The animal examples have been widened to include cats, dogs, guinea pigs, crocodiles, zeedonks, animal cloning, and sex-sorted sperm. Additional plant examples include sugar cane, hops, coca (for cocaine), potatoes and rhododendrons. The fungal section has much more on improving industrial enzyme production and on yeasts. All chapters have updated references and there are many new photographs.

I am particularly grateful to Mrs Jacqueline Lamb and Mrs Brenda Lamb for proofreading. Dr Simon Zwolinski, of the Department of Cytogenetics, Institute of Human Genetics, International Centre for Life, Newcastle Upon Tyne, has kindly supplied fresh human chromosome photographs showing recent advances. In addition to the acknowledgements in the first edition, I wish to thank Rosemary Mates, C. Bedwell, Irving Gottesman, Carrolls' Heritage Potatoes, Groombridge Place Gardens and Richard Williams of Cogent for help with illustrations. Unacknowledged photos are by the author. The breeders who sent me information (e.g., Jack Dunnett) or who talked to me at agricultural shows have been very helpful.

It is a pleasure to thank many Imperial College Biology Department students for their constructive comments, useful literature searches and essays. I particularly thank Basma Bahsoun, Gareth Bennell, Richard Chatwin, Connie Chien, Ju Chin, Mary Connolly, Jennifer Cummins, Calliope Dendrou, Sophia Docherty, Catherine Gapper, Katherine Gregory, Melissa Iacobelli, Malin Johansson, Victoria Judd, Surbjit Kaur, Marvin Lee, Xiaoyun Lee, Wing-Kit Leung, Sally Marsh, Lynsey McInnes, Victoria Neal, Cherlyn Neo, Joanne Roberts, James Rudge, Katherine Scuffham, Daniel Stewart, June Swanston, Mariko Tavernier, Velma Teng, Samantha Westrop, Claire Whittington and Karen Wong.

# Contents

**Chapter 1  Introduction; Aims of Applied Genetics; Revision of Basic Genetic Concepts and Terminology**                                     **1**

1.1    Introduction . . . . . . . . . . . . . . . . . . . . .            1
1.2    The aims of applied genetics in humans, animals, plants
       and fungi; the use of genetic variation; improving the
       harvest index  . . . . . . . . . . . . . . . . . . . .            8
1.3    Revision of basic genetic concepts, definitions and
       symbols  . . . . . . . . . . . . . . . . . . . . . . .           14
       1.3.1    Alleles, genes, loci, wild-types and mutants .          14
       1.3.2    Ploidy . . . . . . . . . . . . . . . . . . . .          17
       1.3.3    Genotype and phenotype; homozygotes and
                heterozygotes; hemizygotes; the time of gene
                expression . . . . . . . . . . . . . . . . . .          18
       1.3.4    Dominance and recessiveness; incomplete
                dominance and additive action; primes;
                overdominance; co-dominance; pure breeding            19
       1.3.5    Additive and multiplicative gene action . . .          22
       1.3.6    Pleiotropy . . . . . . . . . . . . . . . . . .          22
       1.3.7    Mutation . . . . . . . . . . . . . . . . . . .          23
       1.3.8    Recombination; linkage; syntenic and
                non-syntenic loci; coupling and repulsion
                arrangements . . . . . . . . . . . . . . . . .          24
       1.3.9    Allelism and the cis/trans test . . . . . . . .          26
       1.3.10   Heritability . . . . . . . . . . . . . . . . .          29

1.3.11   Selection: natural, sexual, artificial and
         commercial . . . . . . . . . . . . . . . . .   30
1.3.12   Populations . . . . . . . . . . . . . . . . .   31
1.3.13   Polymorphism . . . . . . . . . . . . . .   33
1.3.14   Random mating . . . . . . . . . . . . . .   34
References . . . . . . . . . . . . . . . . . . . . . . .   35

**Chapter 2   The Inheritance and Analysis of Qualitative
and Quantitative Characters**                              **37**

2.1   Single-locus qualitative characters: autosomal loci
      with complete dominance, partial dominance, additive
      action, overdominance; X-linked and holandric loci   .   37
      2.1.1   Qualitative characters . . . . . . . . . . .   37
      2.1.2   Autosomal loci: complete dominance . . . .   38
      2.1.3   Autosomal loci with partial dominance,
              additive action or overdominance . . . . . .   38
      2.1.4   X-linked and holandric loci . . . . . . . .   40
2.2   Multiple-loci qualitative characters: dihybrid ratios and
      gene interactions such as epistasis causing modified
      ratios . . . . . . . . . . . . . . . . . . . . . . .   42
      2.2.1   Standard dihybrid ratios . . . . . . . . . .   42
      2.2.2   Gene interactions such as epistasis causing
              modified dihybrid ratios . . . . . . . . . .   45
2.3   Quantitative characters; quantitative trait loci and
      polygenes; modifiers; threshold characters . . . . . .   49
2.4   Threshold characters . . . . . . . . . . . . . . . .   62
References . . . . . . . . . . . . . . . . . . . . . . .   64

**Chapter 3   Regression, Transgression, Environmental
Effects and Heritability; Correlations Between Characters; Genotype, Phenotype and Breeding Values**         **65**

3.1   Genetic and environmental causes of regression and
      transgression . . . . . . . . . . . . . . . . . . .   65
3.2   Environmental effects on phenotypes . . . . . . . . .   70
      3.2.1   Sex-limited characters . . . . . . . . . . .   70

| | 3.2.2 | Phenocopies, conditional mutants and environmental effects . . . . . . . . . . . | 70 |

3.2.3 Phenotype plasticity . . . . . . . . . . . . . 71

3.2.4 Variable expressivity . . . . . . . . . . . . 72

3.2.5 Incomplete penetrance . . . . . . . . . . . 72

3.3 Narrow and broad sense heritabilities; equations, estimation and use; use of twins; realised heritabilities; correlations between characters . . . . . . . . . . . 74

3.4 Genotype value, phenotype value and breeding value 80

3.5 Estimated Breeding Values (EBV) and Best Linear Unbiased Predictions (BLUP) . . . . . . . . . . . . 83

References . . . . . . . . . . . . . . . . . . . . . . . 84

**Chapter 4 Population Genetics: Allele Frequencies, Genetic Equilibria, Population Mixing, Genetic Drift and Gene Flow** **86**

4.1 Introduction . . . . . . . . . . . . . . . . . . . . . 86

4.2 The Hardy–Weinberg equilibrium for one locus and two loci; linkage disequilibrium, and population mixing 87

4.2.1 One locus, two alleles . . . . . . . . . . . 87

4.2.2 One locus, more than two alleles . . . . . . 89

4.2.3 Allele segregation at two loci and the importance of recombination frequencies . . 89

4.2.4 What happens when two pure-breeding but different populations mix? . . . . . . . . . 91

4.3 Genetic drift, fixation and effects of population size . 97

4.4 Gene flow and population structure . . . . . . . . . 100

4.5 Effects on allele frequencies of selection, mutation, migration and gene conversion; frequency-dependent selection; equilibria between forces in populations . . 105

4.5.1 Selection . . . . . . . . . . . . . . . . . . 105

4.5.2 Migration . . . . . . . . . . . . . . . . . . 109

4.5.3 Mutation . . . . . . . . . . . . . . . . . . 110

References . . . . . . . . . . . . . . . . . . . . . . . 112

**Chapter 5      Types and Uses of Selection**                                    **114**

5.1      Natural, artificial and sexual selection . . . . . . . .      114

5.2      Stabilising selection, towards uniformity . . . . . . .      116

5.3      Directional selection, favouring one extreme . . . . .      118

5.4      Cyclic selection, alternatively favouring different
          extremes . . . . . . . . . . . . . . . . . . . . . . . . . .      119

5.5      Disruptive selection, selecting against the average type      120

5.6      Pedigree selection; breed records, prepotent stud males
          and grading-up . . . . . . . . . . . . . . . . . . . . . .      121

5.7      Progeny testing . . . . . . . . . . . . . . . . . . . . . .      123

5.8      Half-sib and family selection . . . . . . . . . . . . .      125

5.9      Selecting for correlated characters . . . . . . . . . .      126

5.10     Selection for more than one character: tandem
          selection, independent culling levels and index selection      126

          5.10.1      Tandem selection . . . . . . . . . . . . . . .      127

          5.10.2      Independent culling levels . . . . . . . . . .      127

          5.10.3      Index selection, including PTA, PIN and £PLI      128

5.11     Selection intensities and rates of response to selection;
          a key equation for selection responses . . . . . . . .      132

5.12     *In vitro* selection . . . . . . . . . . . . . . . . . . .      134

5.13     Selection at the haploid stage . . . . . . . . . . . . .      135

5.14     Selection for meat characteristics . . . . . . . . . . .      135

References . . . . . . . . . . . . . . . . . . . . . . . . . . . .      138


**Chapter 6      Departures from Random Mating**                                **140**

6.1      Positive and negative assortative mating . . . . . . .      140

6.2      Inbreeding and outbreeding and their consequences;
          inbreeding depression and outbreeding depression . .      143

6.3      Why breeders often use inbreeding . . . . . . . . . .      149

6.4      Wright's inbreeding coefficient, F; Wright's
          equilibrium for genotype frequencies under
          inbreeding; calculation of F from pedigrees . . . . .      151

References . . . . . . . . . . . . . . . . . . . . . . . . . . . .      160

**Chapter 7    Mutation and its Uses**                                    **161**

7.1    Molecular types of mutation and their revertibility;
       mutation frequencies  . . . . . . . . . . . . . . . . .    161
       7.1.1    Revertibility  . . . . . . . . . . . . . . .    161
       7.1.2    Germ-line and somatic mutations  . . . . . .    162
       7.1.3    Base substitutions  . . . . . . . . . . .    163
       7.1.4    Frame shifts  . . . . . . . . . . . . . .    164
       7.1.5    Large deletions . . . . . . . . . . . . . .    164
       7.1.6    Unstable length mutations . . . . . . . . .    165
7.2    Spontaneous and induced mutation; mutagenic agents    166
       7.2.1    Spontaneous mutations; mutation mech-
                anisms; varieties or breeds they have
                produced . . . . . . . . . . . . . . . . .    167
       7.2.2    Induced mutations  . . . . . . . . . . . .    170
7.3    Mutation control and repair systems  . . . . . . . .    174
       7.3.1    Mutagen access  . . . . . . . . . . . . .    174
       7.3.2    Repair systems . . . . . . . . . . . . . .    174
       7.3.3    Suppressor mutations  . . . . . . . . . .    175
       7.3.4    Optimum mutation rates . . . . . . . . . .    176
7.4    How different types of mutation can complicate
       population genetics calculations  . . . . . . . . . .    176
7.5    Using induced mutations  . . . . . . . . . . . . . .    177
References . . . . . . . . . . . . . . . . . . . . . . . . .    180

**Chapter 8    Recombination, Mapping and Genomics**            **181**

8.1    Recombination, genetic distances and the numbers of
       progeny needed to get particular recombinants . . . .    181
8.2    Types of recombination and their effects; meiotic and
       mitotic crossovers; interference and map functions . .    183
8.3    Numbers of types of gamete, and of offspring
       genotypes and phenotypes, for different numbers of
       segregating loci . . . . . . . . . . . . . . . . . .    187
8.4    Calculation of the frequencies of particular genotypes
       and phenotypes . . . . . . . . . . . . . . . . . .    188
8.5    Mapping, including physical mapping  . . . . . . . .    188

8.6      Locating genes on particular chromosomes, including the use of pseudodominance, parasexual methods and hybridisation probes . . . . . . . . . . . . . . . . .  192

    8.6.1      Pseudodominance  . . . . . . . . . . . . . . . . .  192

    8.6.2      Parasexual methods  . . . . . . . . . . . .  194

    8.6.3      Hybridisation probes . . . . . . . . . . . . .  195

8.7      Practical uses of molecular markers in agriculture  . .  196

8.8      Genomics  . . . . . . . . . . . . . . . . . . . . . . . .  198

8.9      Human gene sequencing, including the HapMap Project  201

References . . . . . . . . . . . . . . . . . . . . . . . . . . . . . .  202


**Chapter 9      Structural Chromosome Aberrations: Their Origins, Properties and Uses                                          205**

9.1      Introduction  . . . . . . . . . . . . . . . . . . . . . . .  205

9.2      Deletions   . . . . . . . . . . . . . . . . . . . . . . . .  207

9.3      Inversions, paracentric and pericentric; their effects on fertility . . . . . . . . . . . . . . . . . . . . . . . . . . .  210

9.4      Duplications and the origin of new genes . . . . . . .  213

9.5      Translocations, single and multiple . . . . . . . . . .  215

References . . . . . . . . . . . . . . . . . . . . . . . . . . . . . .  219


**Chapter 10      Changes in Chromosome Number: Their Effects and Uses                                                         220**

10.1      Background . . . . . . . . . . . . . . . . . . . . . . .  220

10.2      Changes in ploidy . . . . . . . . . . . . . . . . . . .  222

10.3      Monoploids and anther culture  . . . . . . . . . . .  224

10.4      Diploids . . . . . . . . . . . . . . . . . . . . . . . . .  228

10.5      Triploids . . . . . . . . . . . . . . . . . . . . . . . . .  228

10.6      Tetraploids . . . . . . . . . . . . . . . . . . . . . . . .  228

10.7      Higher polyploids . . . . . . . . . . . . . . . . . . .  232

10.8      Loss or gain of single chromosomes: aneuploids, monosomics and trisomics  . . . . . . . . . . . . .  232

10.9      Chromosome manipulations and substitutions  . . . .  233

References . . . . . . . . . . . . . . . . . . . . . . . . . . . . . .  236

**Chapter 11    Supernumerary ("B") Chromosomes        238**

11.1    Definition, origins, numbers, size and functions  . . .    238
    11.1.1    Origins . . . . . . . . . . . . . . . . . . . .    238
    11.1.2    Numbers, occurrence and transmission  . . .    238
    11.1.3    Size, sequences and function  . . . . . . .    240
11.2    Effects on phenotype and fertility . . . . . . . . . .    240
References . . . . . . . . . . . . . . . . . . . . . . . .    242

**Chapter 12    Human and Medical Genetics        244**

12.1    Introduction  . . . . . . . . . . . . . . . . . . . .    244
12.2    Finding out how or whether characters are inherited:
    pedigree studies; twin studies; Hardy–Weinberg
    analysis; familial incidence . . . . . . . . . . . . .    246
    12.2.1    Pedigree studies  . . . . . . . . . . . . . .    246
    12.2.2    Twin studies  . . . . . . . . . . . . . . . .    248
    12.2.3    Hardy–Weinberg analysis of populations  . .    253
    12.2.4    Familial incidence . . . . . . . . . . . . .    253
12.3    Single gene characters and disorders, and their
    treatments . . . . . . . . . . . . . . . . . . . . . .    255
    12.3.1    Autosomal genes . . . . . . . . . . . . . .    256
    12.3.2    X-linked genes . . . . . . . . . . . . . . .    265
12.4    Polygenic and multifactorial disorders  . . . . . . .    266
12.5    X-inactivation and Barr bodies  . . . . . . . . . . .    268
12.6    Sex differences in disease susceptibility  . . . . . .    271
    12.6.1    X-linked diseases  . . . . . . . . . . . . .    271
    12.6.2    Non-X-linked diseases . . . . . . . . . . .    273
12.7    Causes of human mutation; cancer . . . . . . . . . .    274
    12.7.1    UV light . . . . . . . . . . . . . . . . . .    277
    12.7.2    Chemicals . . . . . . . . . . . . . . . . .    277
    12.7.3    Ionising radiations . . . . . . . . . . . . .    280
    12.7.4    Temperature . . . . . . . . . . . . . . . .    281
    12.7.5    Infection . . . . . . . . . . . . . . . . . .    282
    12.7.6    Cancer genes . . . . . . . . . . . . . . . .    282
12.8    Chromosome number abnormalities and chromosome
    aberrations . . . . . . . . . . . . . . . . . . . . . .    283
    12.8.1    Human chromosome methods . . . . . . . .    283

12.8.2    Autosomal abnormalities . . . . . . . . . . .    289
12.8.3    Sex chromosome abnormalities  . . . . . . .    296
12.9    Selection before and after birth  . . . . . . . . . . .    297
12.10    Blood groups, especially ABO and Rhesus  . . . . . .    299
12.10.1    Blood groups and transfusions   . . . . . . .    299
12.10.2    The ABO blood group  . . . . . . . . . . .    302
12.10.3    The Rhesus blood group  . . . . . . . . . .    303
12.10.4    Blood groups in farm animals  . . . . . . . .    305
12.11    The major histocompatibility complex  . . . . . . . .    305
12.12    Prenatal, neonatal and adult screening  . . . . . . .    307
12.12.1    Amniocentesis  . . . . . . . . . . . . . . .    307
12.12.2    Early amniocentesis  . . . . . . . . . . . .    308
12.12.3    Chorionic villus sampling  . . . . . . . . .    308
12.12.4    Ultrasound screening . . . . . . . . . . . .    309
12.12.5    Maternal blood sampling   . . . . . . . . .    309
12.12.6    Pre-implantation genetic diagnosis   . . . . .    310
12.12.7    Foetal DNA screening  . . . . . . . . . . .    311
12.12.8    Neonatal screening (birth to one month of age)    312
12.12.9    Adult or adolescent screening  . . . . . . . .    313
12.13    Effects of human inbreeding   . . . . . . . . . . . .    315
12.14    Genetic counselling  . . . . . . . . . . . . . . . . .    319
12.15    Twins and other multiple births  . . . . . . . . . . .    327
12.16    Mosaics, chimeras and hermaphrodites  . . . . . . . .    334
12.17    Gene therapy   . . . . . . . . . . . . . . . . . . . .    338
12.17.1    Introduction   . . . . . . . . . . . . . . . .    338
12.17.2    Methods for somatic gene therapy  . . . . . .    339
12.17.3    Progress   . . . . . . . . . . . . . . . . . .    342
12.18    Stem cell therapy, including the use of cord blood  . .    345
References . . . . . . . . . . . . . . . . . . . . . . . . . . .    348

**Chapter 13    Plant and Animal Breeding Methods and Examples    353**
13.1    Using hybrid vigour . . . . . . . . . . . . . . . . . .    353
13.1.1    Definition of hybrid vigour (heterosis)   . . .    353
13.1.2    Explanations of hybrid vigour  . . . . . . . .    354
13.1.3    Typical F1 hybrid breeding programmes   . .    355
13.1.4    Hybrid maize production   . . . . . . . . . .    358

13.1.5  Hybrid sprout production . . . . . . . . . .  361

13.1.6  Hybrid rice production . . . . . . . . . . .  362

13.1.7  Hybrid animals . . . . . . . . . . . . . . . .  362

13.2  Selection methods for inbreeders . . . . . . . . . .  363

13.2.1  Single line selection . . . . . . . . . . . .  363

13.2.2  Using a mixture of lines, with agricultural mass selection . . . . . . . . . . . . . . . .  364

13.2.3  Bulk population breeding . . . . . . . . .  364

13.2.4  Pedigree breeding . . . . . . . . . . . . .  365

13.3  Selection methods in outbreeders or random-maters .  365

13.4  Recurrent backcrossing for gene transfer . . . . . .  368

13.5  Interspecific and intergeneric hybrids . . . . . . . .  369

13.5.1  Interspecific hybrids . . . . . . . . . . . .  369

13.5.2  Intergeneric hybrids . . . . . . . . . . . .  374

13.6  Making polyploids . . . . . . . . . . . . . . . . . .  376

13.7  Examples of plant and animal breeding programmes .  377

13.7.1  Semi-dwarf wheats and rice — the "Green Revolution" . . . . . . . . . . . . . . . . .  377

13.7.2  Broad beans . . . . . . . . . . . . . . . . .  379

13.7.3  Semi-leafless combining pea . . . . . . . .  379

13.7.4  Coca plants for cocaine . . . . . . . . . .  380

13.7.5  Potato breeding and disease-resistance . . .  380

13.7.6  Lime-tolerant Rhododendrons . . . . . . .  384

13.7.7  Sugar cane . . . . . . . . . . . . . . . . .  385

13.7.8  Hops in Britain — dwarf "hedgerow" hops .  386

13.7.9  Sheep, including cross-breds and Border Leicesters . . . . . . . . . . . . . . . . .  387

13.7.10  Cattle, including Ayrshires (dairy) and Aberdeen-Angus (beef) . . . . . . . . . .  389

13.7.11  Peruvian guinea pigs . . . . . . . . . . . .  396

13.7.12  Chinese crocodiles . . . . . . . . . . . . .  397

13.7.13  Pig breeding to meet the buyer's specifications; meat quality factors . . . . . . . . . . . .  397

13.8  Breeding for shows; breeds and varieties . . . . . . .  399

13.9    Breeding programmes from crosses to selection, to local and national trials, possible commercial release, and approved lists . . . . . . . . . . . . . . . . . . .    402

13.10   Selection in feral animals; feral and farmed animals .    405

13.11   DNA fingerprinting . . . . . . . . . . . . . . . . . .    407

References . . . . . . . . . . . . . . . . . . . . . . . . . . .    408


**Chapter 14    Genetic Engineering in Plants, Animals and Micro-Organisms**    **411**

14.1    Introduction . . . . . . . . . . . . . . . . . . . . . .    411

14.2    Restriction endonucleases and ligases . . . . . . . .    412

14.3    Vectors . . . . . . . . . . . . . . . . . . . . . . . . .    414

14.4    Getting a particular piece of DNA into a vector, and recognising a clone containing it . . . . . . . . . . .    416

14.5    Site-directed mutagenesis . . . . . . . . . . . . . . .    418

14.6    Gene targeting; cosuppression; RNA interference in humans, animals and plants . . . . . . . . . . . . . .    419

14.7    Genetic engineering in plants . . . . . . . . . . . . .    423

14.8    Genetic engineering in animals . . . . . . . . . . . .    428

14.9    Genetic engineering in micro-organisms . . . . . . .    433

14.10   Some dangers of genetic engineering; the amount of genetically engineered crops grown . . . . . . . . . .    434

References . . . . . . . . . . . . . . . . . . . . . . . . . . .    439


**Chapter 15    Genetic Variation in Wild and Agricultural Populations; Genetic Conservation**    **442**

15.1    The forces controlling the amounts of variation in a population . . . . . . . . . . . . . . . . . . . . . . . .    442

    15.1.1    The forces or processes which increase or maintain genetic variation within a population    442

    15.1.2    The forces or processes which reduce genetic variation within a population . . . . . . . .    445

    15.1.3    The interactions of forces or processes affecting the amount of variation within a population . . . . . . . . . . . . . . . . . . .    445

15.2    Using a knowledge of the origins of genetic variation
        to solve a practical problem . . . . . . . . . . . . . .    446
15.3    The maintenance of polymorphism in populations . .    447
15.4    The need for genetic conservation; methods of
        conservation . . . . . . . . . . . . . . . . . . . . . .    450
        15.4.1    The need for genetic conservation and the
                  value of some old varieties . . . . . . . . . .    450
        15.4.2    Conservation programmes and methods of
                  genetic conservation . . . . . . . . . . . .    452
References . . . . . . . . . . . . . . . . . . . . . . . . . .    460

**Chapter 16    Genetic Methods of Insect Pest Control        462**

16.1    Introduction . . . . . . . . . . . . . . . . . . . . . .    462
16.2    The release of sterile insects, or of fertile insects giving
        inviable progeny   . . . . . . . . . . . . . . . . . . .    463
16.3    The breeding of insect-resistant varieties . . . . . . .    466
References . . . . . . . . . . . . . . . . . . . . . . . . . .    469

**Chapter 17    Reproductive Physiology in Plants, Animals
and Humans; Crossing Methods                                 470**

17.1    Plants   . . . . . . . . . . . . . . . . . . . . . . . .    470
        17.1.1    Plant sexual reproduction   . . . . . . . . .    470
        17.1.2    Incompatibility in higher plants and ways of
                  overcoming it   . . . . . . . . . . . . . . .    477
        17.1.3    Crossing methods   . . . . . . . . . . . . .    481
        17.1.4    Getting uniform fruit, seed or bud ripening .    483
        17.1.5    Somaclonal variation and vegetative
                  propagation; grafting and rootstocks . . . . .    484
        17.1.6    Plant protoplast fusion . . . . . . . . . . .    494
        17.1.7    Gene expression and natural and artificial
                  selection at the haploid stage   . . . . . . .    495
17.2    Animals   . . . . . . . . . . . . . . . . . . . . . . .    496
        17.2.1    Sex ratios . . . . . . . . . . . . . . . . . .    496
        17.2.2    Flow cytometry for sexing sperm   . . . . . .    497
        17.2.3    Anatomy, progeny per pregnancy, and
                  temperature effects . . . . . . . . . . . . .    499

17.2.4    Breeding seasons and oestrous cycles  . . . .    501

17.2.5    Sperm; natural and artificial insemination  . .    505

17.2.6    Egg transplantation and embryo freezing  . .    510

17.2.7    Animal cloning . . . . . . . . . . . . . . . . .    511

17.3    Humans  . . . . . . . . . . . . . . . . . . . . . . .    513

References . . . . . . . . . . . . . . . . . . . . . . . . . . . .    516

**Chapter 18    Applied Fungal Genetics**                              **519**

18.1    General fungal genetics: life cycles; wild-types and
mutants; spore types; control of sexual and vegetative
fusions; genomics . . . . . . . . . . . . . . . . . . . .    519

18.1.1    Life cycles  . . . . . . . . . . . . . . . . . .    519

18.1.2    Wild-types and mutants  . . . . . . . . . . .    522

18.1.3    Spores  . . . . . . . . . . . . . . . . . . . .    523

18.1.4    The control of vegetative and sexual fusions    524

18.1.5    Fungal genomics: nuclear, mitochondrial and
plasmid . . . . . . . . . . . . . . . . . . . . .    525

18.2    The commercial importance of fungi . . . . . . . . .    527

18.3    Recombination and sexual mapping  . . . . . . . .    528

18.4    The parasexual cycle and parasexual mapping  . . . .    534

18.5    The induction and isolation of mutants, including
auxotrophs . . . . . . . . . . . . . . . . . . . . . . .    541

18.6    Obtaining improved strains for industry  . . . . . . .    544

18.6.1    Aims and methods  . . . . . . . . . . . . . .    544

18.6.2    Different aims in different yeasts  . . . . . .    547

18.6.3    Improving baker's yeast  . . . . . . . . . . .    550

18.6.4    Improving enzymes in industrial fungi  . . .    554

18.6.5    Penicillin production  . . . . . . . . . . . . .    557

References . . . . . . . . . . . . . . . . . . . . . . . . . . . .    559

**Chapter  19    The  Economics  of  Agricultural  Products
and Breeding Programmes**                                             **562**

19.1    Basic economics: economic systems; price theory;
factors affecting supply and demand; perfect and
imperfect competition; monopolies; inflation . . . . .    562

|  | 19.1.1 | Factors of production; types of economic system | 562 |
|  | 19.1.2 | Price theory | 564 |
|  | 19.1.3 | Types of competition | 570 |
|  | 19.1.4 | Inflation | 574 |
| 19.2 | | Economics applied to agriculture | 575 |
|  | 19.2.1 | Gluts and shortages; how governments intervene in agriculture; European Union policies | 575 |
|  | 19.2.2 | Seasonal and perishable produce | 581 |
|  | 19.2.3 | The value of rarities | 582 |
|  | 19.2.4 | "Health foods" and "organic" products | 585 |
|  | 19.2.5 | Discounted cashflow assessment of breeding programmes | 588 |
|  | 19.2.6 | Breeders' rights | 590 |
|  | 19.2.7 | Breeding or using for niche markets | 592 |
|  | 19.2.8 | Who does the breeding? | 593 |
| References | | | 594 |

# Chapter 1

# Introduction; Aims of Applied Genetics; Revision of Basic Genetic Concepts and Terminology

## 1.1. Introduction

**Genetics** is one of the most exciting subjects today. Almost every week, there are newspaper headlines or TV programmes about the latest advances in medical genetics or plant or animal breeding. Applied genetics is a very stimulating and topical science, and one whose achievements do an enormous amount to feed and clothe us, and to improve our health. **Applied genetics** is the practical application of genetics to medicine, and to plant, animal and microbial breeding.

As the **primary food producers**, plants are basic to agriculture. Only 20 different plant species provide 90% of our food, with wheat, maize, rice and sugar cane providing much more than half of that. According to FAO (Food and Agriculture Organisation of the United Nations), the contributions of various foodstuffs to the world average intake of 2800 kcalories per day in 1999–2001 were: cereals, 48%; vegetable oils and fats, 9%; sugar and its products, 9%; meat and offals, 8%; roots and vegetables, 5%; milk and its products, 4%; others, 17% (FAOSTAT, 2004; http://apps.fao.org/faostat). Rice alone supplies 20% of the world's food energy, is the staple food for more than half the world's population, and provides employment for nearly one billion people in Asia, Africa and the Americas. Unlike some major crops, about 80% of rice is produced by small-scale farmers and consumed locally.

For any crop, one needs varieties to suit **local environments** and **personal tastes**. Thus, different **rice varieties** need to cope with at least four kinds of ecosystem: upland, rain-fed lowland, flood-prone, and irrigated,

1

with different elevations, rainfall patterns, depth of flooding, drainage, nutrient availability/fertiliser use, and local pathogens and pests. The subspecies *Oryza sativa Japonica* has a low amylose content, giving soft sticky rice on cooking, as mainly grown and preferred in Japan, Korea and Northern China. The high amylose *Indica* variety gives the hard fluffy rice, preferred and grown in India. The intermediate *Javanica* variety is grown in Indonesia and is preferred in most of the Western countries. 72% of rice is grown with irrigation, and yield depends heavily on the amount of fertiliser used.

Tables 1.1 and 1.2 illustrate **world production of major plant and animal agricultural products**, and their relative quantitative importance, although they have very different values in money per tonne. In 2001, the percentages by cash value (100% was 412 billion US$) of world agricultural exports were: fruit and vegetables, 16.9; cereals and products, 13.1; "crude materials", 12.7; meat and meat products, 11.1; "others", 9.4; beverages, 8.4; dairy products and eggs, 6.8; coffee, tea, cocoa and spices, 6.1; tobacco, 5.0; animal and vegetable oils, 4.4; sugar and honey, 4.0; live animals, 2.1 (FAO data as above).

There is an estimated worldwide average of 0.3 ha of available **arable land** per person. In 1999, actual harvested land was estimated at 1,401 million hectares for 6 billion people, about 0.23 ha each, equivalent to a strip 23 metres by 100 metres. Especially in the second half of the 20th century, the work of plant, animal and microbial breeders has been essential in raising standards of health and nutrition throughout the world's 6 billion people, through providing better crops, meat and microbial products, including antibiotics. In the "**Green Revolution**", the introduction of high yielding semi-dwarf wheat and rice, with appropriate fertilisers, pesticides and agricultural practices, raised world agricultural output in the period 1972 to 1982 by 25%, with a 33% increase in developing countries and an 18% increase in industrialised countries.

As human populations have increased in the world (6,057 millions in year 2000, estimated by FAO to rise to 9,322 millions by 2050), starvation has often been avoided by the work of **plant and animal breeders**, whether working for private firms or for large institutes such as the former Plant Breeding Institute, Cambridge, England, the Centro Internacional de Mejoramiento de Maiz y Trigo in Mexico (Centre for Improvement of Maize and Wheat), and the International Rice Research Institute in the Philippines.

**Table 1.1.** World production of plant agricultural products in 2003. Figures mainly from FAOSTAT, updated May 2004; accessed October 2004, http://apps. fao.org/faostat.

### Plant products, world annual production in millions of metric tons

| Product | Total | Product | Total | Product | Total |
|---|---|---|---|---|---|
| Alfalfa for forage/ silage | 455.9 | Garlic | 13.70 | Pistachios | 0.54 |
| Almonds | 1.7 | Ginger | 1.00 | Plantains | 33.00 |
| Apples | 48.0 | Gooseberries | 0.17 | Plums | 10.10 |
| Apricots | 2.5 | Grapefruit | 4.70 | Potatoes | 310.80 |
| Asparagus | 6.3 | Grapes | 60.90 | Pulses | 56.50 |
| Bananas | 69.3 | Grasses and legumes, mixed | 1246.80 | Pumpkins, squash, gourds | 19.00 |
| Barley | 141.5 | Green maize | 9.10 | Pyrethrum, dry flowers | 0.01 |
| Beans, dry | 19.0 | Groundnuts | 35.70 | Quinces | 0.40 |
| Beans, green | 5.9 | Hops | 0.10 | Rapeseed | 36.10 |
| Broadbeans, dry | 4.0 | Jute | 2.80 | Raspberries | 0.38 |
| Broadbeans, green | 1.1 | Kiwi | 1.00 | Rice, paddy | 589.10 |
| Buckwheat | 2.0 | Lemon/lime | 12.50 | Rye | 14.90 |
| Cabbage | 66.0 | Lentils | 3.10 | Safflower seed | 0.65 |
| Cantaloupe and melons | 26.7 | Lettuce | 20.80 | Seed cotton | 56.10 |
| Carrots | 23.3 | Linseed | 2.10 | Sesame seed | 2.90 |
| Cashew nuts | 2.0 | Lupins | 1.60 | Sorghum | 59.60 |
| Cassava | 189.1 | Maize | 638.00 | Soya beans | 189.20 |
| Castor beans | 1.1 | Maize for forage/ silage | 366.90 | Spices | 1.90 |
| Cauliflower | 15.9 | Mangoes | 25.60 | Spinach | 11.80 |
| Cherries | 1.9 | Millet | 29.80 | Strawberries | 3.20 |
| Chestnuts | 1.0 | Mushrooms | 3.20 | String beans | 1.60 |
| Chick peas | 7.1 | Mustard seed | 0.63 | Sugar beet | 233.50 |
| Chicory roots | 0.9 | Natural rubber | 7.40 | Sugar cane | 1,333.30 |
| Chillies and peppers | 23.2 | Oats | 26.30 | Sunflower seed | 27.70 |
| Citrus, all | 103.8 | Oil palm | 143.40 | Sweet potatoes | 121.90 |
| Cocoa beans | 3.2 | Okra | 4.90 | Tangerines, mandarins, satsumas | 21.00 |
| Coconuts | 52.9 | Olives | 17.20 | Taro | 8.90 |
| Coffee, green | 7.8 | Onions, dry | 52.50 | Tea | 3.20 |

**Table 1.1.**    (*Continued.*)

| Plant products, world annual production in millions of metric tons | | | | | |
|---|---|---|---|---|---|
| **Product** | **Total** | **Product** | **Total** | **Product** | **Total** |
| Coir fibre from coconuts | 0.63 | Onions, green/ spring | 4.4 | Timber, 3,345 million cubic metres | |
| Cow peas | 3.70 | Oranges | 60.0 | Tobacco leaves | 6.20 |
| Cranberries | 0.34 | Papaya | 6.3 | Tomatoes | 113.30 |
| Cucumbers and gherkins | 39.60 | Peaches and nectarines | 14.8 | Tree nuts | 8.20 |
| Currants | 0.70 | Pears | 17.2 | Triticale | 10.20 |
| Dates | 6.70 | Peas, dry | 10.2 | Vetches | 0.94 |
| Egg plant, aubergines | 29.90 | Peas, green | 8.9 | Walnuts | 1.40 |
| Fibre crops | 24.70 | Pepper, white/ black | 0.37 | Watermelon | 91.80 |
| Figs | 1.10 | Pigeon peas | 3.1 | Wheat | 556.30 |
| Flax fibre | 0.80 | Pineapples | 14.6 | Yams | 39.90 |

**On the medical side**, there has been a revolution in our understanding of many **diseases** caused by single genes or by chromosomal abnormalities, in the treatment of some but not all diseases, and in our ability to detect disorders by prenatal testing, for example, by using gene-specific DNA probes. Plate 1.1 shows a human female karyotype with a complex rearrangement between chromosomes 2, 7 and 8, and a deletion in chromosome 13, where the colouring of individual pairs of chromosomes makes **diagnosis** so much easier than just staining or just banding. **Genetic counselling** is reducing the suffering and lethality caused by certain inherited conditions, such as thalassaemia and Tay–Sachs disease (see Chap. 12).

The massive **gene-sequencing** effort in the **human genome project** should lead to further advances in understanding, albeit at a slow speed. Genomes have now been sequenced from mammals such as man, mouse, fungi as in yeast, diseases such as African sleeping sickness, leishmaniasis and Chagas disease, plants such as maize and rice, from chickens, and the list grows each year. In **dogs**, there is now genome analysis, disease-gene mapping, quantitative trait locus (QTL) mapping, and even gene therapy for inherited metabolic diseases (Ostrander, 2005). Molecular methods are

**Table 1.2.** World production of animal agricultural products in 2003. Figures mainly from FAOSTAT, updated May 2004; accessed October 2004, http://apps.fao.org/faostat.

| Animal product | World production, millions of metric tons per year |
| --- | --- |
| Beef and buffalo meat | 62.1 |
| Chicken meat | 65.0 |
| Deer, reindeer, elk, moose, meat | 0.07 |
| Deer, reindeer, elk, antler velvet | 0.009 |
| Eggs | 60.5 |
| Fish, marine and inland | 130.2 |
| Game meat, total | 1.5 |
| Goat meat | 4.2 |
| Honey | 1.3 |
| Meat total | 253.5 |
| Pigeons and other birds | 0.017 |
| Milk | 600.9 |
| Mutton and lamb | 8.0 |
| Pig meat and products | 98.5 |
| Poultry meat | 75.8 |
| Rabbit | 1.1 |
| Sheep milk | 8.1 |
| Sheepskins | 1.6 |
| Snails | 0.0005 |
| Turkey meat | 5.4 |

increasingly used in animal pedigree determination and in marker-assisted selection methods in plants and animals, where linked molecular markers can be used on very young plants or animals as markers for agricultural traits. DNA markers are replacing blood groups for many purposes. In **sheep**, there is widespread genotyping for scrapie-resistance genes, with breeding and culling policies determined by the 15 possible genotypes (see Sec. 8.8), in the UK.

Although modern genetical knowledge and techniques have led to big improvements in farm animals and plants, great improvements were made by **simple selection and cross-breeding**, and improved husbandry, in the eighteenth century. At the big market in Smithfield, London, average weights in 1710 were 23 kg for calves, 168 kg for beef cattle and 17 kg for sheep. By 1795, these had increased to 60 kg, 363 kg and 36 kg respectively.

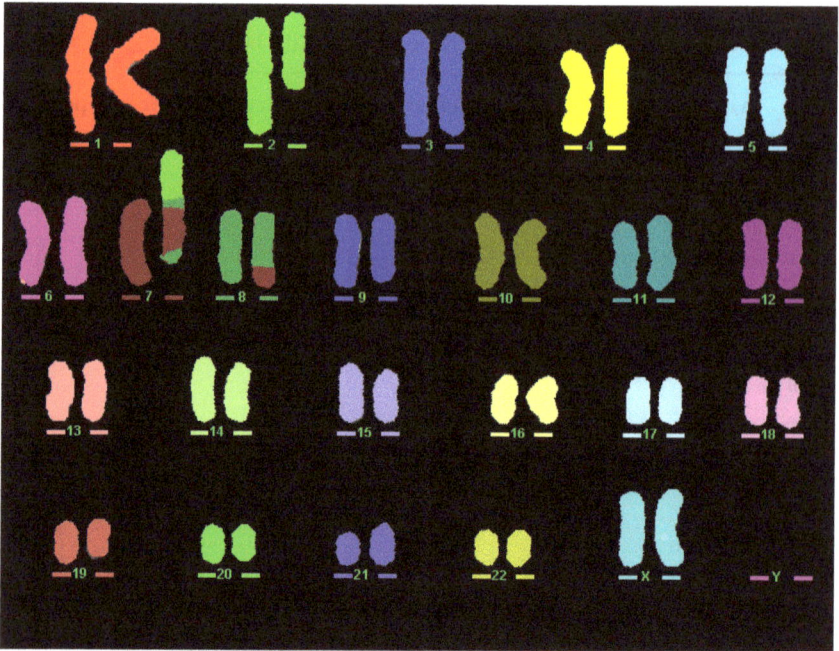

**Plate 1.1.    A fluorescent (*in situ*) hybridisation** (FISH, see Subsec. 12.8.1) **karyotype** from a woman with complex 2;7;8 rearrangements and an interstitial deletion in 13. One chromosome 7 is involved in two separate translocations, the 7p (short) arm with 2q (long arm) and the 7q arm with 8q. The reciprocal distal 7p region on the derivative 2 is too small to show here but does so with subtelomeric probes. The darker green section on the translocated 7 looks as if it is from 8, but is caused by flaring of the fluorochromes and is a technical artefact. The colours are computer-enhanced.

This book is mainly for undergraduates in any biological, medical or agricultural area, but much of it will be understandable by 16–18-year-old biologists, by farmers, medical practitioners, nurses or any layman with interests in biology. It assumes a knowledge of elementary biology and genetics, including meiosis and Mendel's laws, yet revising many of the **basic concepts** and much of the **terminology**. It is essential to include equations and statistics, but I have concentrated on basic principles rather than on mathematical elaboration. Those needing more equations should consult the quoted literature, especially Falconer and Mackay (1996) and Cameron (1997) on quantitative genetics. **Full references** are given at the end of each chapter, with some works quoted in more than one chapter.

Not all works given in the **Suggested reading** sections are mentioned in the text.

While modern **general textbooks** on genetics often have a lot on applied genetics, it is much easier for the student of applied genetics to find all the information collated in one book, instead of having to search many books. The **examples given** are usually from organisms used in medicine, agriculture, horticulture or industry, although classic genetic organisms such as *Drosophila*, *Neurospora* and *Escherichia coli* are referred to, if they provide the best examples of a phenomenon.

This book is based on an **Applied Genetics course** which I have taught for 30 years at Imperial College London to final-year undergraduates in a range of disciplines, including biology, applied biology, microbiology, plant sciences, animal sciences, and ecology; some of them have gone on to read medicine or work in industry or agriculture. This book contains much additional information. It retains the key approach of teaching the **general principles of applied genetics**, e.g., population genetics, quantitative genetics, gene flow, mutation, recombination and chromosome aberrations, which all apply to humans, animals, plants and microbes, with relevant examples from any of those groups. Specialist aspects of each of those groups are also included, such as reproductive physiology (Chap. 17); the microbial area usually excludes bacteria and viruses, as Prokaryotes have such basic differences from Eukaryotes. The section on applied fungal genetics (Chap. 18) is partly included because the application of ideas from fungal genetics helped so much with human genetics, especially for cell fusion and parasexual techniques, and because many of the techniques are applicable to other organisms.

Most topics described here apply to most groups of organisms and are described in the most relevant section. Thus, the effects of different anomalies of sex chromosome number, such as XXY (giving a sterile male), XO (giving a sterile female), and sex mosaics, are described here in the chapter on human genetics (Subsec. 12.8.3 and Sec. 12.16), but exact parallels occur in other mammals such as cats.

The text includes **worked examples of problems** because students find that they greatly improve their understanding of the theory, as well as giving them experience of data analysis and answering practical problems. Readers should **try doing the problems** but should first cover up the answers, which follow immediately.

Readers who are very familiar with genetics can skim over the revision section in Sec. 1.3, checking whether they know all the terms in **bold print**. Understanding the terminology and symbols is essential. This book has extensive cross-referencing between sections and some minor repetitions to help those following their own sequence of topics when reading the book, and because some topics, such as selection, occur in many parts. **Bold print** is used to draw attention to key terms.

When using **genetic symbols**, it is normal to leave no gap between alleles at a locus if the symbols are of one letter only, e.g., *aa*, but to leave a gap when the symbols are of more than one letter, e.g., *vg vg*. In this book, I have left a gap between the two alleles at one locus, whether the symbol is of one or more than one letter, and have put a comma between symbols for different loci, e.g., *a a*, *B b*, *C′ C*, $w^e$ $w^a$. This makes for greater clarity, especially when superscripts and primes are being used. It also helps to distinguish letters used as allele symbols, when spaces are used, and letters used for whole genomes or chromosome regions, when no space is used between them, e.g., *AAAA* for an autotetraploid with four identical genomes.

It is very important that students and users of applied genetics should have the subject placed firmly in a context of the **realities of economics and government regulations**. These are stressed throughout the book, as well as being covered in Chap. 19. Although the prices and laws mentioned will become out of date, they should still be useful for comparison. As the laws in the USA may differ from state to state, most of the references to laws are to the ones in Britain or the EU. The examples of applied genetics are taken from many countries.

## 1.2. The aims of applied genetics in humans, animals, plants and fungi; the use of genetic variation; improving the harvest index

With **humans**, the aims are humanitarian and medical rather than economic, in understanding and finding ways of detecting, curing, treating or preventing disorders, especially hereditary ones, including chromosomal and genetic disorders. As medical expertise, facilities and many drugs are expensive, economic factors cannot be ignored in human applied genetics, e.g., in deciding whether mass screening for a genetic disease is cost-effective.

Chapter 19 deals with basic economics and its application to agriculture. With **animals, plants and fungi**, the **main aims of applied genetics** are to increase the yield and quality of the required product or products, to reduce the costs of production and/or purification, and to reduce vulnerability to disease and to climatic variation. With dairy cows, one wants to increase milk yield and milk nutritional value (protein, fats, vitamins), decrease the cost of feed, shelter, transport, land, labour, especially of feeding and milking, and veterinary expenses, and to minimise yield decreases from unfavourable weather or disease.

Depending on the organism's biology, genetics and economics, the applied geneticist may use **selection** for better phenotypes and genotypes using **pre-existing genetic variation**, or may use **induced mutations** to increase variation, or use **hybridisation**, or **changes in ploidy** or in **individual chromosomes**. The breeder may have **one main product** to consider, such as the drug penicillin from the fungus *Penicillium*, young leaves from tea plants, milk from dairy cattle, meat from beef cattle, or more than one main product. In sheep, the main product is usually meat but wool is important in many breeds, and so are milk and cheese in some breeds. New Zealand has 1.8 million farmed deer, with approximately equal cash value of venison meat and of products used in Chinese medicine, including antler velvet, crushed young antlers, and deer penis (Subsec. 19.2.4).

The **coconut tree** has many products, including coir fibre from the husk, charcoal for filters from the shell, copra (the flesh, used for eating, oil and desiccated coconut), coconut milk from grated flesh plus water, coconut water as a drink, leaves for thatching and wood for building. In coconut plantations, there are also valuable **undercrops** including bananas, coffee, pineapples, spices and limes which could be affected by changes in coconut palm canopy density.

**Camels** are a good example of a multi-product animal. The Arabian camel (*Camelus dromedarius*) principally provides transport, milk and meat, with byproducts such as wool, hair, hides, and use for racing. With **multiple products**, the breeder has to decide which products to try to improve most, taking into account their relative values and whether there are positive or negative correlations between their yields. For example, meat production in sheep might decrease if selection increased the amount of energy devoted to wool production, a negative correlation.

Let us take **beet** as a plant breeding example, with **sugar** as its product. The total yield depends on the amount of root produced and on the sugar content of the root. The sugar content in 1747 was less than 2%, but selection increased this to 6% by the 1820s, and to 9% by the 1830s. By 1858, crosses of *Beta vulgaris* x *B. maritima* gave up to 14% sugar, and modern beets yield over 20% sugar. Triploid varieties are produced by crossing diploids with tetraploids, giving high yields of roots and of sugar. They often have less ash and nitrogen, which is helpful for processing. Yield is more certain as genetic resistance has been incorporated to various diseases, such as *Cercospora* and some viruses. Mechanised production has been simplified by the breeding of plants with "monogerm" seedballs, with only one seed instead of several seeds per fruit, and which therefore do not require thinning after germination.

With a **human disorder**, one would try to find out whether it was caused by infection, by diet or other environmental factors, by heredity from one to many genes, by chromosome number abnormalities or chromosome structural aberrations, and how best to treat the condition and to prevent its occurrence in future births. **Diagnosis** today often includes karyotyping, DNA analysis and biochemical testing, as well as physical examination. One can now treat, but not cure, many hereditary disorders by surgery, physiotherapy, diet or drugs, with gene therapy now being tried (Sec. 12.17).

**Genetic counselling** can make use of advanced cytological techniques (Subsec. 12.8.1) to detect abnormalities and to assess the risk of a parent having an abnormal offspring. For example, fluorescent *in situ* hybridization (FISH) can be used to probe for specific chromosomes in body cells from buccal swabs or white blood cells, or even sperm as shown in Plates 1.2 and 1.3. A man with a translocation between chromosomes 15 and 21 produced sperm with balanced sets (Plate 12.10) or unbalanced sets (Plates 1.2 and 1.3) of chromosomes. This analysis meant that his fertility and risk of passing an unbalanced set to a child could be assessed.

Applied genetics in non-human organisms usually involves the **manipulation of inherited genetic variation** to achieve the aims set out at the beginning of this section. That variation may pre-exist in a population from previous mutations, chromosome aberrations, recombination, or from immigration. See Plate 1.4 for variation in *Sorghum* panicles. Further variation may be introduced by the breeder by mutation, recombination, genetic

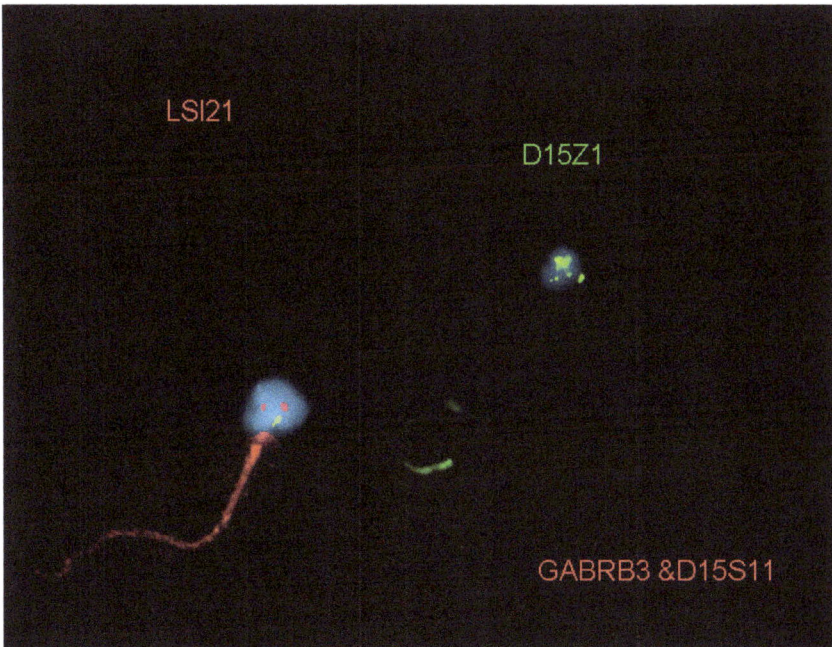

**Plate 1.2.** An **abnormal sperm** (small and misshapen head) from a man with a reciprocal translocation between chromosomes 15 and 21. Four FISH probes have been used, with one green and two red for 15, and one red for 21 (see Plate 12.10 for details). A normal haploid balanced sperm should therefore show one green signal and three red ones. As this sperm has one green signal and only two red ones, it is presumably deficient in chromosome 21 and on fertilisation would have given an inviable zygote with monosomy for 21. This test can be used on males with a balanced rearrangement to estimate their risk of fathering a child with an unbalanced chromosome rearrangement from mis-segregation at meiosis. It can also be used on males who are mosaics (Sec. 12.16).

manipulation, or by bringing in new variants from other populations and even from other species. Where there are heterozygotes, they can be considered as storing genetic variation, with **segregation** from single heterozygotes, e.g., *A a* giving *A A* and *a a* types as well as more *A a*. With double heterozygotes, e.g., *A a, B b*, **recombination** releases variation, giving *A A, B B*; *A A, B b*; *A A, b b*; *A a, B B*; *A a, b b*; *a a, B B*; *a a, B b* and *a a, b b*, as well as more of *A a, B b*.

Over long periods of evolution, species develop genes which work well with each other: **co-adapted gene complexes**. Where different species

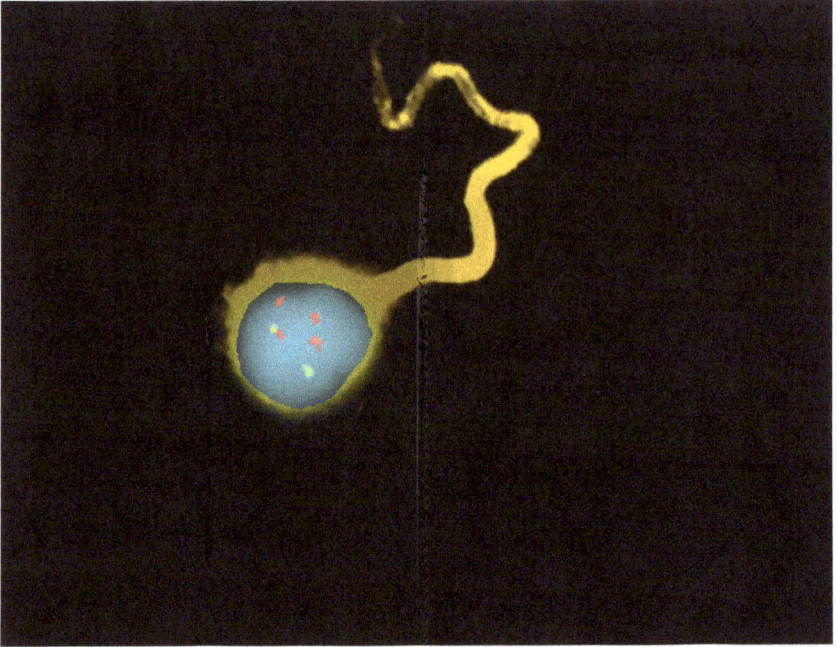

**Plate 1.3.**   Another sperm from the same man as in Plate 1.2, with the same probes for 15 and 21. Because there are two green signals and six red ones (two single and two double), it is deduced that this sperm has two copies of 15 and two of 21, and might be a **diploid unreduced sperm** or haploid with double disomy.

usually grow together, such as clovers and grasses in pasture, one may even get genetic co-adaptation between species. Breaking up well-adapted gene complexes by recombination, mutation or genetic engineering may have adverse effects, and improving the yield of one species in a mixture may have adverse effects on other species. The breeder needs a general biological awareness, as well as specialised knowledge of one organism.

The applied geneticist often works in **multidisciplinary teams**, with people specialising in cytology, molecular biology, biochemistry, human, animal or plant pathology and physiology, soils, fertilisers, agronomy, agricultural engineering, etc. **Economic forecasting** is often crucial to the planning of long-term breeding projects: what will the consumer want by the time a project has been completed? Public and media opinion strongly influences markets, and so do government actions. The current fashion for

**Plate 1.4.  Variation in sorghum panicles**. There is a wide range of colour and form variants available for breeding and selection. Breeders are concerned with grain yield, nutritional value, whether the strain is for feeding animals or humans, and whether the plants are grown for the grain or for grain plus foliage. It is often planted on poor soils, with drought-resistance required in some areas.

vegetarianism, with more than 4 million vegetarians in Britain, reduces the demand for meat products, and government action over bovine spongiform encephalopathy (BSE) has depressed the beef market in the UK and for exports. The present strong consumer opposition in Europe to genetically modified crops (Chap. 14) has seriously affected the economics of firms producing such crops.

**Selection** by breeders may involve measurable characters such as liveweight in meat animals, but care needs to be taken because total body weight is affected by fleece weight in sheep, while in ruminants, 10–25% of liveweight may be gut contents, with large diurnal variations. Selection in animals may directly or indirectly involve economically important **behavioural traits** such as libido in bulls, rams, stallions, boars and cocks, eating persistence, temperament and reactions to handling, and mothering ability in females. The sensitivity of yield of plants, animals and microbes to **environmental variables** must be considered when attempting to select good genotypes from good phenotypes. The selective breeding of humans is ethically unacceptable.

As well as breeding programmes to increase yield, advances are often through **cost-reductions** and increasing the **harvest index** (the percentage of the harvested organism consisting of the most marketable product, e.g., the grain from wheat, or milk from dairy cows). Thus, the very successful **semi-dwarf wheats and rice** have a higher proportion of valuable grain to the less valuable straw, compared with traditional tall varieties. Rice has been selected for less stem, more panicle, more tillers, and a reduced sowing to harvest time, giving three crops a year in many areas. Sorghum was typically about 1.7 m tall, giving about 7% sugar, but short varieties about 1 m tall give about 16% sugar. Dwarf or semi-dwarf **palmyrah palms** (*Borassus flabellifer*) are only 4.5 to 6.1 m high, compared with 30 m for normal palms, and mature in 4 to 5 years instead of 15–20 years. Another example of **breeding for reduction** is selecting clover (*Trifolium* spp.) for less cyanogenic glycosides to reduce toxicity.

In **chickens for egg laying**, the number of eggs per hen has usually reached a plateau in most breeds (about 230 eggs to 500 days of age, with negative correlations between egg number and egg size), but breeding for better feed utilisation has reduced costs after selection for a smaller body size. Similarly, in chickens for meat, breeders have improved the growth rate and the food conversion to meat, with the modern broiler chicken taking only 8 weeks to reach a weight of 1.6 kg from only 3.4 kg of feed. In dairy cows, a cow with a poor milk yield, say 1,450 litres per lactation, might use 44% of food for milk and 56% for keeping the cow alive, while a cow with a good yield of 3,900 litres per lactation might use 65% of food for milk and only 35% for maintenance. Cost-reductions can improve profitability without causing a glut (see Subsec. 19.2.1 for the disastrous effect of gluts on farmers' incomes).

## 1.3. Revision of basic genetic concepts, definitions and symbols

### 1.3.1. *Alleles, genes, loci, wild-types and mutants*

A **locus** is a length of a chromosome occupied by one gene; it is the position of the gene, including all possible alleles, although any one position is occupied by only one allele at a time. Thus, the locus controlling red-green colour blindness in humans is in band q28 on the X chromosome, whatever allele it carries, mutant or wild-type. A **gene** is a unit of hereditary function and information, usually being transcribed into RNA or having controlling functions. Alternative forms of a gene are called **alleles**. In peas, there is a locus of major effect controlling plant height, as investigated by Mendel. At that position (locus) on that chromosome, there may be the $T$ allele, giving tall plants about 183 cm high, or the $t$ allele which as $t\,t$ gives short plants about 30 cm high. Confusingly, people use the term **gene** in the sense of locus, "the gene for height in peas", and in the sense of one allele, "the gene for tallness in peas".

Sometimes there is one allele which is much more common than other forms; it is called the **wild-type allele**, and other forms are regarded as **mutant**, especially if they lack the function coded for by the wild-type allele. Thus, the allele for not having haemophilia A in humans is regarded as the normal or wild-type allele, able to specify the appropriate blood-clotting factor VIII, but alleles giving haemophilia A, an inability to clot blood in wounds, are regarded as mutant. What is regarded as the wild-type may even vary between populations. For example, in Negroid Africans, the allele for brown eyes is the normal one, with very few blue-eyed individuals, but blue-eyed individuals predominate in Scandinavia. In Britain, one could not specify which of the blue or brown eyes was the wild-type, because both are common. In cats, the standard wild-type is taken as the short-haired mackerel-striped tabby as in the European wild cat (*Felix silvestris*) and in most mongrel domesticated cat populations.

One allele may have a similar effect on the products of several related (but not necessarily linked) loci, as in **cat coat colour**. At the "dilute" locus, wild-type allele $D$ gives solid colour throughout each hair. The recessive dilute allele, $d$, when $d\,d$, causes pigment granules to be deposited in clumps, with colourless gaps in between. This dilutes coat colour black to blue, chocolate/brown to lilac, cinnamon to fawn and red to cream, but eye colour is not affected.

At the **DNA level**, one can have several or more different wild-type alleles for a locus, all with complete function but with slightly different base sequences and occasionally with slightly different amino acid sequences, where the differences are not crucial. There are very many

**potential mutants** for a locus, because for any given wild-type sequence, there are three possible different mutant base pairs at any wild-type base pair position, with hundreds or thousands of base pairs in the gene. Many DNA changes from wild-type will have the same phenotype if they give the same loss of function, and may only be distinguishable from each other by DNA analysis or mapping. Thus, different small pea plants may all have the phenotype given by $t\ t$, but may have chemically different $t$ alleles. Some DNA changes may give a wild-type phenotype under some conditions and a mutant phenotype under other conditions. For details of when DNA changes give different phenotypes and when they do not, see Sec. 7.1 and Lamb (1975). Some loci have many known alleles, with over a hundred alleles at some higher plant self-incompatibility loci (Subsec. 17.1.2). At the human *ABO* blood group locus, there are three main alleles, $i^A$, $i^B$ and $i^o$, while **multiple alleles** are very common at the blood group locus *EAB* in cattle, with over 600 alleles, and in the *EAB* locus in sheep, with over 100 alleles.

One often gets **series of different multiple alleles** at a locus where one is the unmutated allele, then a series of different mutant alleles which may give different phenotypes; they are usually recessive to the wild-type allele but may be dominant or partly dominant to alleles lower in the series. This is shown in **cats** at the $C$ locus determining **coat colour expression**, with allele $C$, full colour, completely dominant to all mutant alleles: $c^b$, Burmese; $c^s$, Siamese; $c^a$, blue-eyed albino; $c$, pink-eyed albino, with each mutant allele partly dominant to those shown here to its right. The dominant wild-type allele $C$ gives full colour, which might be tabby ($A\ A, A\ a$), black ($a\ a$), blue tabby ($A\ A$ or $A\ a$ with the dilute allele, $d\ d$), or blue ($a\ a, d\ d$), for example, depending on alleles at unlinked loci. The expression of the mutant alleles at the $C$ locus depends on the alleles at other loci, such as $A$ (agouti, tabby) or $a\ a$ (black). In the presence of $a\ a$, black, Burmese cats are $c^b\ c^b$ (dark Burmese) or $c^b\ c^s$ (light Burmese) or other $c^b$ allele/lower allele heterozygotes for light Burmese. Siamese cats are $c^s\ c^s$ or $c^s\ c^a$ or $c^s\ c$; blue-eyed albinos are $c^a\ c^a$ or $c^a\ c$, and pink-eyed albinos are $c\ c$. If one crossed a full colour black/Siamese heterozygote, $a\ a$, $C\ c^s$ (phenotype full black), with a Burmese/Siamese heterozygote, $a\ a$, $c^b\ c^s$, (phenotype Burmese), one would expect a ratio in the kittens of 2 full black (1 $C\ c^b$; 1 $C\ c^s$), 1 light Burmese ($c^b\ c^s$), and 1 Siamese ($c^s\ c^s$). Mating a light Burmese ($c^b\ c^s$) to another light Burmese ($c^b\ c^s$) would give the typical 1:2:1 ratio for two alleles with incomplete dominance, 1 dark Burmese: 2 light Burmese: 1 Siamese.

There are names for **different types of mutation**. In the fruit fly *Drosophila*, wild-type eye colour is red from dominant allele $W$; the recessive white-eye mutant, $w$, has a total loss of eye-colour function, so is an **amorph**, and there are several mutations with partial loss of function, **hypomorphs**, such as white-eosin, $w^e$, and white-apricot, $w^a$, with some colour left.

## 1.3.2. *Ploidy*

The **ploidy** of a particular stage in the life cycle of an organism is the number of copies of each typical chromosome in the nucleus. Thus, the gamete stage of pea plants and maize (pollen, egg cell), cattle and humans (sperm, eggs) is **haploid**, with one copy of each chromosome per nucleus, while the adult stage is **diploid**, with two copies of each typical chromosome. Many fungi are haploid in the vegetative stage, such as *Neurospora crassa* ($n = 7$). The adult stage in higher organisms is said to have a ploidy of $2n$ (even if polyploid), while the gametic stage is $n$. Yeast can grow and multiply vegetatively by mitosis in both the haploid ($n = 16$) and diploid stages ($2n = 32$), but meiosis only occurs in the diploid. Other examples of diploids (with the diploid, $2n$, 2X, chromosome number in brackets) are man (46), gorilla (48), cattle (60), dog (78), cat (38), mouse (40), chicken (78), frog (26), carp (104), mosquito (6), fruit fly (8), onion (16), barley (14), maize (20), garden pea (14), and pine (24). The Cervidae (deer) show the largest variation in chromosome number found within any mammalian family, from diploid numbers of $2n = 6$ to $2n = 70$. **Polyploids** have more than two copies of each type of chromosome. Examples are commercial bananas, **triploid**, with three of each chromosome; potatoes ($2n = 4X = 48$), **tetraploid**, four copies; bread wheat ($2n = 6X = 42$), **hexaploid**, with six copies; some sugar cane species ($2n = 8X = 80$) and most strawberries ($2n = 8X = 56$), **octaploid**, with eight copies. The leek, *Allium porrum*, is usually tetraploid but its probable ancestor, the wild sand leek, *A. ampeloprasum*, can be diploid ($2n = 16$), tetraploid or rarely hexaploid.

A **euploid** has an exact multiple of the basic chromosome number, e.g., three copies of each type of chromosome, but an **aneuploid** has an inexact multiple of that number, e.g., three copies of six types of chromosomes, but one, two or four of a seventh type. In man, Down syndrome (Subsec. 12.8.2) is an aneuploid with three copies of chromosome 21, $2n = 47$ (Plates 12.15 and 12.16).

### 1.3.3. *Genotype and phenotype; homozygotes and heterozygotes; hemizygotes; the time of gene expression*

The **genotype** is the genetic make-up of an organism, determined by what alleles it carries in its nucleus and in extra-nuclear DNA, while the **phenotype** is the manifested characteristics of the organism, depending on both genotype and environment, and often on age too. Thus, in the diploid pea plant there are three possible genotypes for the $T/t$ locus, and two phenotypes. The $T\,T$ genotype gives a tall phenotype, as does $T\,t$, and $t\,t$ gives a short phenotype.

If the two alleles in a diploid are alike, the genotype is **homozygous**, e.g., $T\,T$ or $t\,t$ in a **homozygote**, while if they are unlike, the genotype is **heterozygous**, e.g., $T\,t$ in a **heterozygote**; one writes the dominant allele before the recessive. The terms are not applicable in haploids, where there is only one allele for each locus, a condition termed **hemizygous** in a **hemizygote**, which also applies in diploids in X Y individuals to those genes on the X sex chromosome which have no counterpart on the Y. Some genes are actually **repeated** many times in one set of chromosomes, or even in one chromosome, such as some genes specifying transfer or ribosomal RNAs, but for simplicity, most genetic definitions relate to single-copy genes. The *Alu* sequence in man occurs more than 500,000 times per genome, involving all chromosomes.

**The time of gene expression:** not all genes are expressed at the time of birth, hatching or germination, so the phenotype shown by the young human, animal or plant may not reflect the organism's genotype. The term "**congenital**" means "present at birth", with some genetic abnormalities that may be present at birth while others may only be present in the genotype, but expressed at a later stage after birth.

In applied genetics, it is most economical to do one's selection as early as possible, to avoid the rearing costs of individuals which will be discarded at selection. In **cucumbers**, the dominant allele *Nb* results in bitter cucurbitacins in all parts of the plant; in adverse cultural conditions, they get into the fruit, causing undesirable bitterness. The recessive allele *nb*, when homozygous, results in non-bitter plants and has been bred into most modern cultivars. Selection is possible at the germinated seed stage by tasting whether the expanded cotyledons are bitter (*Nb Nb* or *Nb nb*) or non-bitter (*nb nb*), so there is no need to grow on the seedlings to the fruiting stage before selection is done. In dogs, **progressive retinal atrophy** is a problem in some breeds, where the young dog sees well and may be used for breeding, but this late-developing genetic disease causes the older dog to become partly or wholly blind.

**Human genetic conditions** may be expressed at almost any stage from the implanting fertilised egg to old age, for example:

- **Polydactyly**, extra fingers and/or toes (Plate 3.1), usually autosomal dominant, expressed in the early embryo.
- **Eye colours** such as blue or brown, expressed a few weeks after birth; the genetics are not fully understood, but brown is usually more dominant than blue. Non-albino babies are all born with bluish eyes due to light-scattering in the iris, and not because of a blue pigment. In people with genes for brown eyes, this blue is masked by brown pigment laid over the light-scattering layer, with the enzyme being produced several weeks after birth.
- **Sex-linked muscular dystrophy**, recessive, giving progressive muscle deterioration, mainly affecting boys. It is expressed from about the age of 4 or 5 years, becoming worse with age so that most sufferers are in a wheelchair by the age of 12 and most are dead by 20.
- **Huntington's chorea**, autosomal dominant, usually expressed between ages 20 and 70, giving jerky movements, madness and death.

### 1.3.4. *Dominance and recessiveness; incomplete dominance and additive action; primes; overdominance; co-dominance; pure breeding*

In a heterozygote, the allele which is expressed in the phenotype is the **dominant allele**, and the one not expressed is the **recessive allele**. In pea plants, $T$ (tall) is dominant over $t$ (short) because the heterozygote $T\,t$ has a tall phenotype. The dominant **allele's symbol** starts with a capital letter, e.g., $T$, $Cn$, $Bw$, but the recessive allele's symbol is all in lower case letters. A particular phenotype may sometimes be dominant and sometimes be recessive, depending on the loci concerned. In blackberry (*Rubus*) for example, thornlessness is dominant in octoploid "Austin Thornless" but recessive in tetraploid "Merton Thornless". In animals, albino is often recessive but sometimes dominant, and the dark melanic form is dominant in some insects, while recessive in others.

Dominance is not always complete. Suppose homozygote $S\,S$ has fruits 20 cm long and homozygote $s\,s$ has fruits 10 cm long. The capital $S$ suggests that long fruit is dominant over short, so the heterozygote $S\,s$ may have fruits 20 cm long. If dominance is not complete at this locus, $S\,s$ might have fruits 18 cm long, so there is **incomplete (partial) dominance** of $S$ to $s$. If $S\,s$ had fruits that are only 10 cm long, short would be dominant over long, but the allele symbols would be wrong. Sometimes, there is **no dominance** with the heterozygote completely intermediate between the two homozygotes, e.g., 15 cm here. This is sometimes called incomplete dominance, but to distinguish it from partial dominance, the term **additive action** is often used. Thus, the basic fruit length in this example is 10 cm, with one allele for increased length adding 5 cm to that, and a second allele for length adding another 5 cm to give 20 cm for the long form. To avoid confusion with symbols for dominant and recessive, a scheme of two identical symbols with an initial capital letter is used for additive action, with the allele giving the increase having a **prime** symbol after it, e.g., $S'$. Thus, $S'\,S'$ gives 20 cm (the basic length of 10 plus two lots of 5 cm), $S'\,S$ gives 15 cm (10+5) and $S\,S$ gives 10 cm. Sometimes a system of lower case superscripts is used for incomplete dominance. Thus, in *Antirrhinum*, white flowers and red flowers are incompletely dominant to each other, giving pink in the heterozygote, with allele symbols $c^w$ and $c^r$ for white and red, with $c$ standing for flower colour. Incomplete dominance for tomato leaf colour is shown in Plate 7.4.

An agricultural example of a gene with **additive effects** is found in Merino sheep, affecting the **ovulation rate**. There is an average of 1.3 ova per oestrous cycle for the homozygous normal ewe, 2.8 for the heterozygote, and 4.3 for the homozygote carrying the Booroola high fecundity allele, $Fec^B$ (the allele symbol does not follow the pattern given here), with the heterozygote exactly intermediate between the two homozygotes. As the average number of lambs born per litter is less than the number of ova, the three genotypes give respectively averages of 1.2, 2.1 and 2.7 lambs per litter, with proportionately more lambs lost per ovum at higher lambing rates than at lower lambing rates. The litter size therefore shows incomplete dominance of the Booroola allele over the normal one, with the heterozygote nearer to the $Fec^B Fec^B$ homozygote than to the normal homozygote.

**Overdominance** is where the heterozygote's phenotype lies outside the range of those of the two homozygotes, e.g., if $S'S$ in the above example had fruits 30 cm long, even longer than those of the 20 cm $S'S'$ long genotype.

**Co-dominance** is a condition where two alleles are both fully expressed in the heterozygote, and so co-dominance differs from incomplete dominance where each allele is only partly expressed in the phenotype. For height, one could not have co-dominance for two alleles, one giving tallness and one giving shortness, as they cannot both be fully expressed simultaneously. In the human ABO blood groups, alleles $i^A$ and $i^B$ are dominant over $i^o$ and show co-dominance with each other. The heterozygote $i^A i^B$ has full expression of blood antigens A and B, and so shows co-dominance for the two alleles. The three allele symbols have $i$ for immunity group and capital letter superscripts for the co-dominant alleles, $i^A$ and $i^B$, and a lower case superscript for the recessive allele, $i^o$.

An organism is **pure breeding** (also known as "true breeding") if it breeds true for a particular character; i.e., on self-fertilising or being crossed with another pure breeding individual with the same genotype, it gives all offspring like each other and like the parent(s) for that character. Heterozygotes will not breed true as they will segregate to give some homozygotes of each genotype and phenotype: $Tt$ selfed or $Tt \times Tt$ gives $TT$, $tt$ and $Tt$ progeny, but $TT$ selfed or $TT \times TT$ gives only $TT$ progeny, so the homozygote is pure breeding. Any crops which are homozygous inbred lines will breed true. In tomatoes, the old variety Ailsa Craig is pure breeding and can be propagated from seed, but most modern varieties are F1 hybrids, taking advantage of hybrid vigour (see Sec. 13.1), and will not breed true from seed.

Some characters will never breed true even if qualitative. An individual having **eyes of different colour**, e.g., one brown, one blue, occurs frequently in some breeds of cats and dogs, less frequently in others, and occasionally in humans. It is a developmental abnormality in a genotype, e.g., for brown eyes, where the pigment-producing cells migrating over the head surface in the embryo from the neural crest reach the eye on one side of the head, but fail to do so on the other, allowing the basic blue colour from light-scattering to show in that eye. Such an animal would breed as a normal brown-eyed one.

## 1.3.5. *Additive and multiplicative gene action*

In some cases, an allele increasing length might **add** a fixed increment, e.g., $S\,S$ is 10 cm long, and with **additive action**, $S'$ might add 5 cm, so that $S'S$ is 15 cm long and $S'\,S'$ is 20 cm long. In other cases, the existing length might be multiplied by each contributing allele. Suppose we now have **multiplicative action**, with $D\,D$ being 10 cm long, and each $D'$ allele **multiplies** the existing length by 2, $D'D$ will then be 20 cm ($10 \times 2$) long and $D'D'$ will be 40 cm long ($10 \times 2 \times 2$). Getting three phenotypes for one locus with lengths of 10 and 40 cm for the two homozygotes and 20 cm for the heterozygote, could also be called incomplete dominance of short over long.

## 1.3.6. *Pleiotropy*

**Pleiotropy** occurs when an allele at one locus affects more than one superficially unrelated phenotypic character. Thus, the dominant allele $W$ for white coat in cats causes all cats with it to have white fur, but it also affects eye colour and hearing, though with incomplete penetrance, causing 70% of cats to be blue eyed, while the remainder are golden eyed or have one eye gold and one eye blue), and about 45% of cats with the $W$ allele are deaf. Those three superficially unrelated characters, coat colour, eye colour and hearing, may have some underlying biochemical link. In the fungus *Neurospora crassa*, the allele $asco^-$ is pleiotropic, causing the sexual ascospores to be colourless instead of black, with very poor germination, and making the fungus lysine-requiring for growth. Supplying lysine in the medium does not restore normal ascospore pigmentation or germination.

## 1.3.7. *Mutation*

The term **mutation** is used both for a process and the result of the process, when a heritable change occurs in a gene. Thus, a change from a wild-type *H* (non-haemophiliac) allele on a human X chromosome to a mutant *h* (haemophilia A) allele is a mutation, involving a change in the DNA base sequence and in the amino acid sequence in the polypeptide. Mutations can occur spontaneously (Plate 1.5) or be induced by man using various mutagens such as X-rays or certain chemicals (Chap. 7 and Plates 7.2, 7.3, 7.4 and 18.1). In yeast, a wild-type *ade-1*$^+$ allele (giving cells not requiring added adenine) might mutate to an *ade-1*$^-$ allele, giving cells requiring the base adenine for growth. A "$^+$" superscript indicates a wild-type allele and a "$^-$" superscript indicates a mutant allele. The symbol *ade-1* stands for adenine locus 1, to distinguish it from other loci controlling other enzymes in the adenine pathway in yeast. Going from the mutant *ade-1*$^-$ back to the wild-type is called **reversion** or **back-mutation**. Some books also use the term **mutation** for a change at the level of a chromosome, not just at the

**Plate 1.5.**   A **spontaneous leaf-shape mutation** in beech (*Fagus sylvatica*). The cut-leaf mutant branch on the left was on the same tree as the normal leaves on the right.

gene level, but it helps to keep the two effects separate if one uses the term **chromosome aberration** for changes on a larger scale than that of a single locus, and **mutation** for changes at the level of a single gene. See Chap. 7 for mutation, Chap. 9 for chromosome aberrations and Chap. 12 for human chromosome aberrations.

### 1.3.8. *Recombination; linkage; syntenic and non-syntenic loci; coupling and repulsion arrangements*

Unlike mutation, where new alleles are produced, **recombination** is the production of **new combinations of existing genes**. For example, a diploid of genotype *A a, B b* formed from the combination of an *A, B* and an *a, b* gamete could produce new recombinant combinations *A, b* and *a, B* in its gametes from meiosis, as well as parental gene combinations *A, B* and *a, b*. The **recombination frequency** is the number of recombinant genotypes $\times 100$, divided by the total number of parental plus recombinant genotypes. Thus, if the above *A a, B b* diploid gave the following gamete numbers, 38 *A, B*; 43 *a, b*; 12 *A, b*; 15 *a, B*, the recombination frequency as a percentage (RF %) is $\frac{(12+15)x100}{38+43+12+15} = \frac{2700}{108} = 25\%$. When measuring recombination frequencies, we want the results from meiosis, which in higher organisms means from the gametes. We can measure recombination frequencies directly from meiotic products in fungi such as *Neurospora crassa* and yeast, where meiosis produces haploid ascospores giving rise to haploid colonies, but in higher organisms, we usually have to cross the gametes and examine diploid progeny, as most genes cannot be scored in gametes.

**Linked genes** are by definition preferentially inherited in the parental arrangements rather than in the recombinant arrangements. As such, genes are linked if they show significantly less than 50% recombination, such as loci *A* and *B* in the above example, in which both are say on chromosome 1. Unlinked genes show about 50% recombination from meiosis. Linkage is thus defined in terms of recombination frequencies. Genes close together on a chromosome show tight linkage with low recombination frequencies, increasing to a maximum of 50% for distant genes.

As genes on non-homologous chromosomes also show 50% recombination, it is helpful to use the term **syntenic** for any genes on the same chromosome (or pair of homologous chromosomes) and **non-syntenic** for genes not on the same chromosomes. If locus $C/c$ is also on chromosome 1 but far from loci $A/a$ and $B/b$, it would show 50% recombination with them, because crossovers and multiple crossovers would be frequent over the long distance. If locus $F/f$ is on a different pair of chromosomes, e.g., chromosome 5, its alleles would show 50% recombination with genes $A$, $B$ and $C$ because they would assort independently at meiosis. Linked loci, with less than 50% RF, e.g., $A/a$ and $B/b$, must be syntenic, but unlinked loci, with 50% RF, may be syntenic but far apart, e.g., $A/a$ and $C/c$, or may be non-syntenic, e.g., $A/a$ and $F/f$.

If two loci are segregating in a cross of pure-breeding (homozygous if diploid) parents, one uses the term **coupling arrangement** (or coupling phase) if all recessive alleles at these two loci are together in one parent and all the dominant alleles are together in the other parent, e.g., $A A$, $B B \times a a$, $b b$ in diploids or $A$, $B \times a$, $b$ in haploids, e.g., in yeast, while in a **repulsion arrangement**, each parent has recessive alleles at one locus and dominant alleles at the other locus, e.g., $A A$, $b b \times a a$, $B B$ in diploids or $A$, $b \times a$, $B$ in haploids.

One can usefully use these terms for a double heterozygote, as genotype $A a$, $B b$ could have come from $A$, $B + a$, $b$ gametes (coupling combination) or from $A$, $b + a$, $B$ gametes (repulsion combination). If the two loci are non-syntenic, the coupling and repulsion double heterozygotes will be identical, with each of the four alleles on a different chromosome. However, if the loci are syntenic, the coupling double heterozygote has one chromosome carrying $A$, $B$ and the homologous chromosome carrying $a$, $b$, while the repulsion double heterozygote has one chromosome carrying $A$, $b$ and the other carrying $a$, $B$. It is essential to register the difference between coupling and repulsion arrangements because they affect the calculation of recombination frequencies for linked loci. Thus, gametes $A$, $B$ and $a$, $b$ are parental when coming from a coupling double heterozygote, but are recombinant when coming from the repulsion double heterozygote. As the $A a$, $B b$ double heterozygote mentioned at the beginning of this section gave significantly more coupling gametes (38 $A$, $B$; 43 $a$, $b$) than repulsion gametes (12 $A$, $b$; 15 $a$, $B$), we can deduce that it is a coupling phase double heterozygote, with $A$ and $B$ on one chromosome and $a$ and $b$ on the homologous chromosome, with linked and syntenic loci.

At meiosis, recombination for non-syntenic loci comes from **independent assortment** of pairs of non-homologous chromosomes (Mendel's second law), while recombination of syntenic loci usually occurs by **reciprocal crossing-over** at pachytene. The non-reciprocal process of **gene conversion** can also recombine genes, but is usually only important for recombining alleles within a locus. At very much lower frequencies than at meiosis, one can also have **mitotic recombination** of syntenic and non-syntenic genes (Secs. 8.1 and 18.4). See Chap. 8 for recombination and mapping.

## 1.3.9. *Allelism and the cis/trans test*

**Alleles** are alternative forms of a gene. If one isolated 60 *Escherichia coli* bacterial strains all requiring the amino acid histidine, some might be mutant at the same locus, say *hisB*, specifying one enzyme (IGP dehydrase) in the histidine pathway, while others might be mutant at *hisC* (specifying IAP transaminase), or *hisA*, or *hisD*, etc., specifying different enzymes. Different loci controlling the same biochemical pathway are distinguished by their final capital letter, e.g., *hisA*, *hisD* (or sometimes by numbers as in *ade-1*, *ade-2*), and different mutants at a locus are distinguished by different final superscript numbers, e.g., $hisB^{1-}$, $hisB^{2-}$, with $hisB^{+}$ representing the wild-type allele. One often needs to find out whether two mutants with similar phenotypes, such as both requiring histidine, are at the same locus (alleles, e.g., $hisB^{1-}$, $hisB^{8-}$) or are at different loci (not alleles, e.g., $hisB^{1-}$, $hisC^{5-}$).

One can map these 60 mutants, where only ones very close together could be alleles, but being close does not prove that they are alleles. In Eukaryotes, two mutations with recombination frequencies more than about 2% are usually too far apart to be alleles. One can also test which precursors of histidine these *E. coli* strains can grow on and what intermediates they accumulate, since alleles will have the same requirements as each other because they are blocked at the same point in the biochemical pathway.

The best test is the *cis/trans* **test for functional allelism**, originally devised by Ed Lewis for *Drosophila*, but extensively used in Prokaryotes and lower and higher Eukaryotes. By crossing (or, depending on the organism, using transformation, transduction, or heterokaryons) one makes two diploids or partial diploids, the *cis* arrangement with the two mutants coming from the same parent and the two wild-type alleles coming from the

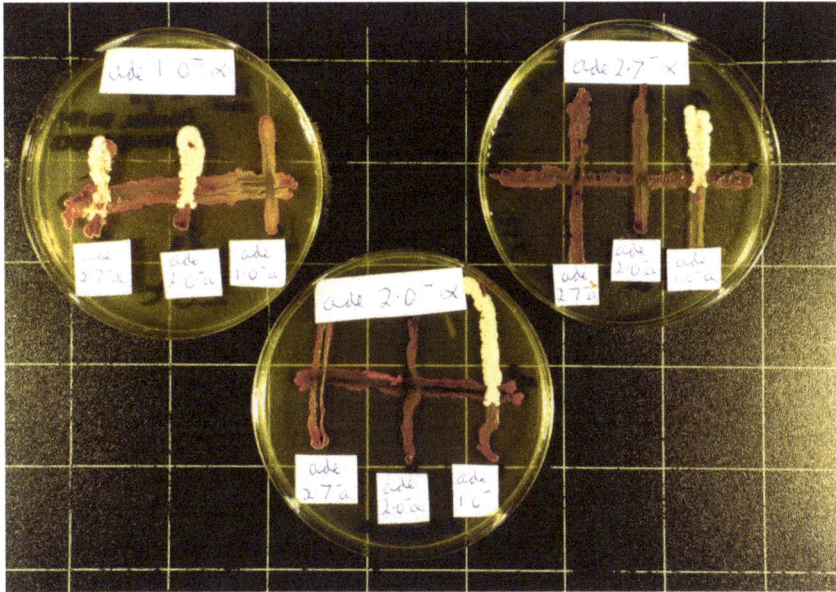

**Plate 1.6.** The *trans* part of a *cis/trans* **test for allelism** between three haploid, red, adenine-requiring mutants in yeast, *ade-1.0⁻*, *ade-2.0⁻* and *ade-2.7⁻*; the wild-type is non-adenine requiring and white. The horizontal streaks are the α (alpha) mating types of these mutations, and the three *a* (a) strains of the same mutations have been streaked upwards through them. The non-mated areas such as below the α strains are controls for spontaneous reversion to wild-type, which would give white colonies. In the mated areas above the α strains, complementation shows as white patches of diploid, while non-complementing mutants give red diploids. The *ade-1* strains complement the *ade-2* strains. No strain complements its own mutation, and the two *ade-2* mutants, *ade-2.0⁻* and *ade-2.7⁻*, do not complement each other because they are at the same locus. The plates are 9 cm in diameter.

other parent, and the *trans* arrangement, with one mutant and one wild-type from one parent, and the other mutant and the other wild-type from the other parent. See specimen genotypes below, and Plate 1.6 for a *cis/trans* test with three adenine-requiring red mutants in yeast. It does not matter whether the two mutants are syntenic or not; if they are non-syntenic or if syntenic but far apart, they cannot alleles.

Let us consider **three** *E. coli* **histidine-requiring single mutations, 1, 2 and 3,** where 1 and 2 are allelic, both at the *hisB* locus, while 3 is not allelic with 1 and 2, being at the *hisC* locus. They will have these genotypes and phenotypes:

Single mutants, all requiring histidine:

mutant 1 is $hisB^{1-}$ $hisB^{2+}$, $hisC^{3+}$ genotype, so is hisB⁻, hisC⁺phenotype, because the *hisB* gene has a defect;

mutant 2 is $hisB^{1+}$ $hisB^{2-}$, $hisC^{3+}$genotype, so is hisB⁻, hisC⁺ phenotype;

mutant 3 is $hisB^{1+}$ $hisB^{2+}$, $hisC^{3-}$ genotype, so is hisB⁺, hisC⁻ phenotype.

In *E. coli*, one can make partial diploids by transduction with bacteriophage. Provided that the two mutations are both recessive to their wild-types, the *cis* arrangement will give a wild-type phenotype both for alleles and for non-alleles. The phenotype difference comes in the *trans* test, where the *trans* partial diploid will be wild-type if the mutations are not alleles, but will be mutant for two alleles. Non-alleles **complement** (complete each other, make good each other's defect) in *trans* because each brings the wild-type allele which the other lacks. In contrast, alleles of the same locus do not complement in *trans* because both mutants are defective at the same locus.

(i) Partial diploids between mutants 1 and 2, **alleles** at the *hisB* locus.

   (a) In *cis*, one genome has $hisB^{1-}$ $hisB^{2-}$, $hisC^{3+}$, specifying bad IGPdH and good IAPt; the other genome has $hisB^{1+}$ $hisB^{2+}$, $hisC^{3+}$, specifying good IGPdH and good IAPt, so the diploid phenotype is **wild-type**, not histidine-requiring, as both enzymes have some good molecules made.

   (b) In *trans*, one genome has $hisB^{1-}$ $hisB^{2+}$, $hisC^{3+}$, specifying bad IGPdH and good IAPt; the other genome has $hisB^{1+}$ $hisB^{2-}$, $hisC^{3+}$, specifying bad IGPdH and good IAPt, so the diploid phenotype is **mutant**, histidine-requiring, as there is no good IGPdH enzyme.

(ii) Partial diploids between mutants 1 and 3, **not alleles**, as they are mutant at different histidine loci, *hisB* and *hisC* respectively.

   (a) In *cis*, one genome has $hisB^{1-}$ $hisB^{2+}$, $hisC^{3-}$, specifying bad IGPdH (the gene has an error at site 1) and bad IAPt; the other genome has $hisB^{1+}$ $hisB^{2+}$, $hisC^{3+}$, specifying good IGPdH and good IAPt, so the diploid phenotype is **wild-type**, not histidine-requiring, as both enzymes have at least some good molecules made.

(b) In *trans*, one genome has $hisB^{1-}$ $hisB^{2+}$, $hisC^{3+}$, specifying bad IGPdH and good IAPt; the other genome has $hisB^{1+}$ $hisB^{2+}$, $hisC^{3-}$, specifying good IGPdH and bad IAPt, so the diploid phenotype is **wild-type**, not histidine-requiring, as there is some good IGPdH enzyme and some good IAPt enzyme.

## 1.3.10. *Heritability*

**Heritability** is the proportion of phenotypic variance ($Vp$) attributable to genetic variation in a population, with the term applying to characters with continuous variation. One distinguishes between narrow-sense heritability, $h^2$, which is the proportion of phenotype variance attributable to genes with additive effects, and broad-sense heritability, $H^2$, which is the proportion of phenotype variance attributable to all types of genetic variation. The variances due to different types of genes are $Va$ for genes with additive effects, $Vd$ for genes with dominance, and $Vi$ for genes with epistasis or gene interactions (Sec. 2.2). $Ve$ represents the amount of variation due to the environment and $Vg/e$ represents variation due to genotype/environment interactions, where different genotypes perform differently in different environments (e.g., pyrethrum production by different varieties at different altitudes, Subsec. 17.1.1). The basic equations are:

$$\text{narrow sense heritability} : h^2 = \frac{Va}{Va + Vd + Vi + Ve + Vg/e};$$

$$\text{broad sense heritability} : H^2 = \frac{Va + Vd + Vi}{Va + Vd + Vi + Ve + Vg/e}.$$

It is extremely important to know how much of the observed phenotypic variation is due to genetic variation and how much is due to environmental variation, when one is contemplating running **selection programmes**. If nearly all the variation is environmental, or if there is very little phenotypic variation, it is probably not worth running a selection programme to choose individuals with better genes. Those interested in the total causes of variation for a character would use broad sense heritabilities, but practical plant and animal breeders often use narrow sense heritabilities because they give a more reliable estimate of the minimum likely gain from selection.

If one considers just one locus, genes with additive action will give three phenotypes, say due to $S'S'$, $S'S$ and $SS$, and selecting the best phenotype, say that due to $S'S'$, will select the best genotype. With complete dominance, however, if one wants the dominant allele, selecting the dominant phenotype selects both for the dominant homozygote $TT$ (wanted) and for the heterozygote, $Tt$, which has an unwanted recessive allele, $t$.

### 1.3.11. *Selection: natural, sexual, artificial and commercial*

**Selection** against a particular phenotype means that that phenotype leaves fewer surviving offspring than corresponding numbers of other phenotypes. Selection acts on phenotypes rather than on genotypes, and may be due to lower viabilities, and/or to lower reproductive success, and/or to producing fewer viable offspring. It can occur at any stage of the life cycle — in gametes, zygotes, embryos, at birth or germination, youth, middle age or old age, although post-reproductive selection is much less effective than earlier selection.

Selection may be **natural** (by the environment), **sexual** (in competition for mates) or **artificial** (by humans, e.g., plant and animal breeders). It is measured by the **selection coefficient**, $s$, which is the **proportional reduction in fitness and/or reproductive success due to selection**. If a phenotype has no selection against it compared to others, $s = 0$ and it is **selectively neutral**. If a phenotype is **lethal or sterile**, leaving no surviving offspring, $s = 1.0$. A **deleterious** but non-lethal phenotype will have an intermediate value for $s$, between 0 and 1.0. Brown versus blue eyes is almost selectively neutral in Britain, but brown eyes have a slight selective advantage in very sunny countries such as Nigeria, giving more protection against strong light, while blue eyes have been favoured in Norway, perhaps allowing better vision in poor light during winter. For males with schizophrenia, $s = 0.51$, so if 100 normal males left 200 surviving offspring, 100 schizoid males would leave $200(1-s)$ offspring $= 98$. For females with schizophrenia, selection is less, $s = 0.22$, so 100 schizoid females would leave $200(1-s) = 156$ surviving offspring, if 100 normal females leave 200 offspring.

Selection coefficients may vary with the **environment**, e.g., penicillin. In a drug-free environment, drug-sensitive bacteria ($pen^S$) usually have a slight selective advantage over drug-resistant bacteria ($pen^R$), because mechanisms of resistance have metabolic costs, but in the presence of the drug, being drug-sensitive is usually lethal. Wild populations of the bacteria are largely drug-sensitive, with occasional spontaneous mutations to drug-resistance, while in someone being treated with penicillin, nearly all surviving bacteria are drug-resistant. The allele frequency of $pen^S$, therefore, drops sharply and that for $pen^R$ increases rapidly when the drug is administered. Once the drug is no longer taken, there is slow selection in favour of drug-sensitive cells, a few of which may have escaped the drug, or they may come from back-mutation, from drug-resistance to drug-sensitivity.

Although it does not appear in genetics textbooks, in an applied genetics context, one could also add **commercial selection** to the usual list of selection types. This selection is between breeds, varieties, cultivars and between the products of different producers. Unlike the other types of selection, commercial selection acts indirectly regarding the survival and mating of specific individuals. If consumers, retailers and wholesalers prefer one type of cabbage to another, that will influence the seed merchants and breeders to multiply up this type rather than a less favoured type. Being readily bought and eaten promotes the breeding of a particular type, although it is not normally the individuals marketed and eaten which are bred from.

There is selection by the wholesalers of what varieties of a plant or animal they purchase from growers; there is also commercial selection by the retailers as to what they will stock and by consumers, as to what they will buy. In many countries, their national supermarket chains have an enormous influence on what is grown. Supermarket buyers are particularly keen on uniform size, attractive appearance, long shelf-life, and suitability for prepacking, as prepacked foods such as vegetables or fruits are sold at higher prices than loose items which have to be bagged, weighed and priced at the time of purchase. Commercial selection at those different levels has two opposite effects such as generally staying with existing varieties because they are familiar and trusted, and less often choosing new ones for novelty value or because of some claimed special advantage.

## 1.3.12. *Populations*

**Populations** are of many different kinds, with no single ideal definition. A simple definition is "a community of sexually interbreeding organisms with a common gene pool". "Community" implies existing at the same time and geographically close to each other. Suppose we have a small population of five organisms, with cross-fertilising males (m) and females (f) numbered 1 to 5, but in generation 1, individual 3 does not reproduce. Figure 1.1 shows some possible matings over two generations, where the numbers in the second and third generations show the numbers of the individuals in generation 1 from which their genes derive. The bottom left individual comes from a brother-sister mating and the bottom right-hand individual comes from a mating of overlapping generations, with a parent (5,f)-offspring (4,5,m) mating. The diagram shows that there is a common gene pool, with genes "flowing" between individuals and generations.

Consider in contrast a community of five bean plants (*Phaseolus vulgaris*) exclusively with self-fertilisation, when there is no gene flow between individuals, as shown in Fig. 1.2. Any individual in generation 3 only contains genes from one individual in generation 1, whereas in Fig. 1.1 an individual in generation 3 could contain genes from two, three or four individuals in generation 1. The self-fertilising group of bean plants satisfies the "community" part of the population definition, but not the "common gene pool" part, as there is no common pool of genes being exchanged between individuals.

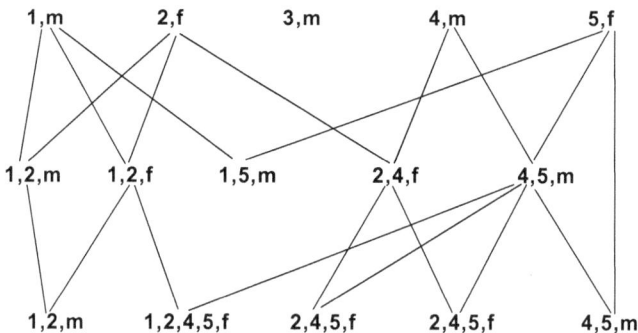

**Fig. 1.1.** A small population with a **common gene pool**. See text. Any individual can leave 0, 1, 2 or more offspring and can mate with any member of the opposite sex. The numbers in the first row label individuals, with *f* for female, *m* for male. In the second and third rows, the numbers indicate which members of the first row they have inherited genes from.

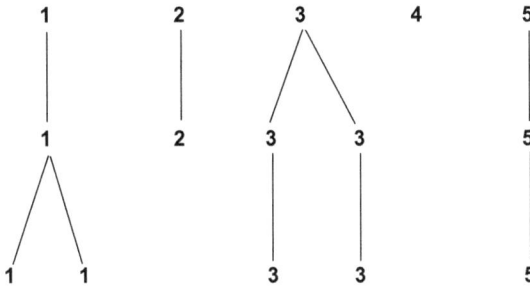

**Fig. 1.2.** A small community of **self-fertilising individuals**, with no common gene pool and no flow of genes between individuals. Individuals may leave 0, 1, 2 or more offspring.

Some populations are **discrete**, separate, like a species of fish in a small pond. Other populations **overlap** each other, with breeding within the population and with neighbouring populations, such as rabbits in different but nearby fields. The amount of **gene flow** between different populations depends on proximity, geography and biology, and on the acceptability of mates from different populations. Most urban humans could be described as being part of several non-discrete populations. Thus, a Catholic undergraduate from Paris studying at a London college is part of that student population, part of the population of one area of London, of Southern England, of France, and of a large religious group. See Sec. 4.4 for more on gene flow and population structure.

## 1.3.13. *Polymorphism*

**Polymorphism** was defined by E. B. Ford as "The occurrence together in the same habitat of two or more discontinuous forms of a species in such proportions that the rarest of them cannot be maintained merely by recurrent mutation." The definition therefore excludes different geographical races and rare, harmful mutations which recur from mutation but do not survive when expressed, such as *albino* in maize, giving white seedlings which cannot photosynthesise and die when seed food reserves have been used up. A locus is sometimes defined as polymorphic if the commonest allele has a frequency of 0.99 or less. Examples of true polymorphisms include human blood groups and eye colours, red, black and intermediate fox fur colour, cat coat colours, and purple and cream flowers in the comfrey plant, *Symphytum officinale*. Although the term is usually used for genes of major effect, it is also used for chromosome aberrations. Polymorphism is part of genetic variation, and therefore can be subjected to natural, sexual and artificial selection. The topic is dealt with in more details in Chaps. 4 and 15.

## 1.3.14. *Random mating*

In **random mating**, mating pairs form independently of genotype and phenotype, and independently of degrees of relatedness between individuals. With **positive assortative mating**, individuals similar with respect to a particular character are more likely to mate with each other than expected by chance. With **negative assortative mating**, similar individuals are less likely to mate with each other than expected by chance. While humans mate approximately at random with respect to blood groups, there is positive assortative mating for height, with tall males tending to marry tall females and short males tending to marry short females. Although it is often said that opposites attract, it is harder to find examples of negative assortative mating, unless one takes sex or self-incompatibility alleles as the character.

While positive and negative assortative mating concern particular characteristics, **inbreeding** and **outbreeding** concern degrees of relatedness, not individual characters. **Inbreeding** is a tendency to mating between individuals more closely related (or less closely related for **outbreeding**) than would occur by chance in that population. Suppose that in a small community the average degree of relationship was fifth cousins, then mating averaging out as being between fifth cousins, with some closer and some less close matings, would be random. If mating was on average between second cousins, that would be inbreeding, while if mating was on average between eighth cousins, that would be outbreeding. In a larger population, where the average relationship between individuals was tenth cousins, then mating on average between eighth cousins would be inbreeding, not outbreeding. **First cousins** have one pair of grandparents in common, **second cousins** have one pair of great-grandparents in common, and so forth. Mating between relatives is not necessarily inbreeding, and of course all members of a species are ultimately related to each other. The terms inbreeding and outbreeding are sometimes used with different meanings in other contexts. See Chap. 6 for departures from random mating, and Sec. 12.13 for the effects of human inbreeding.


## Suggested Reading

Agrawal, R. L., *Fundamentals of Plant Breeding and Hybrid Seed Production.* (1998) Science Publishers, Inc., Enfield, New Hampshire.

ASSINSEL, *Feeding the 8 Billion and Preserving the Planet.* (1997) ASSINSEL (International Association of Plant Breeders), Nyon, Switzerland.

Cameron, N. D., *Selection Indices and Prediction of Genetic Merit in Animal Breeding.* (1997) CAB International, Wallingford.

Connor, J. M. and M. A. Ferguson-Smith, *Essential Medical Genetics*, 5th ed. (1997) Blackwell Scientific, Oxford.

Dalton, D. C., *An Introduction to Practical Animal Breeding.* (1980) Granada, London.

Falconer, D. S. and T. F. C. Mackay, *Introduction to Quantitative Genetics*, 4th ed. (1996) Longman, Harlow.

Fincham, J. R. S., P. R. Day and A. Radford, *Fungal Genetics*, 4th ed. (1979) Blackwell, Oxford.

Gardner, E. J. and D. P. Snustad, *Principles of Genetics*, 9th ed. (1997) Wiley, New York.

Griffiths, A. J. F. *et al.*, *Introduction to Genetic Analysis*, 8th ed. (2004) W. H. Freeman and Co., New York.

Hammond, J., J. C. Bowman and T. J. Robinson, *Hammond's Farm Animals*, 5th ed. (1983) Edward Arnold, London.

Hartl, D. L. and E. W. Jones, *Genetics Analysis of Genes and Genomes*, 6th ed. (2005) Jones and Bartlett, Sudbury, Maryland.

Hayward, M. D., N. O. Bosemark and I. Romagosa (eds.) *Plant Breeding. Principles and Prospects.* (1993) Chapman and Hall, London.

Lamb, B. C., Cryptic mutations: Their predicted biochemical basis, frequencies and effects on gene conversion. *Molec Gen Genet* (1975) 137: 305–314.

Mayo, O., *The Theory of Plant Breeding*, 2nd ed. (1987) Oxford University Press, Oxford.

Ostrander, E. A., *The Dog and Its Genome.* (2005) Cold Spring Harbor Laboratory Press, New York.

Piper, L. and A. Ruvinsky (eds.) *The Genetics of Sheep*. (1997) CAB International, Wallingford.

Poehlman, J. M. and D. A. Sleper, *Breeding Field Crops*, 4th ed. (1995) Iowa State University Press, Iowa.

Simm, G., *Genetic Improvement of Cattle and Sheep*. (1998) Farming Press, Ipswich.

Strachan, T. and A. P. Read, *Human Molecular Genetics*, 3rd ed. (2003) Bios Scientific Publishers, Oxford.

Van Vleck, L. D., E. J. Pollak and E. A. B. Oltenacu, *Genetics for the Animal Sciences*. (1987) W. H. Freeman, New York.

# Chapter 2

# The Inheritance and Analysis of Qualitative and Quantitative Characters

## 2.1. Single-locus qualitative characters: autosomal loci with complete dominance, partial dominance, additive action, overdominance; X-linked and holandric loci

### 2.1.1. *Qualitative characters*

**Qualitative characters** have contrasting phenotypes with a qualitative difference between them. Thus, two or more alleles at a locus give discontinuous variation such as white versus green plants in maize, red versus black fur in foxes, the A, B and O blood groups in man, and adenine-requiring or adenine-non-requiring yeast. The term also applies to measurable characters if there are discontinuous differences, such as between tall and short pea plants. They are usually controlled by two or more alleles at loci of major effect and show typical Mendelian inheritance in diploid Eukaryotes. We will explore genotype and phenotype ratios produced in different circumstances.

Qualitative characters include biochemical polymorphisms which may need immunological or electrophoretic methods to determine which alleles are being expressed, and even DNA markers for which there are qualitative differences, such as the presence or absence of a restriction site (Chaps. 8 and 14). Such polymorphisms are extensively used in sheep and cattle to check pedigrees, e.g., to find out which ram fathered a particular lamb, so that flock pedigree records can be accurate, and breeding values based on progeny-testing reflect the correct father. Both biochemical and morphological markers can be used for population studies, using allele frequencies to work out the origin of particular breeds or varieties or human populations, and the genetic relations between them.

## 2.1.2. *Autosomal loci: complete dominance*

**Autosomal loci** are on chromosomes which are not sex chromosomes. For a locus in diploids, there will be two alleles present which may be alike (homozygous) or unlike (heterozygous). A homozygous individual will be pure breeding, e.g., $T T$ (tall in pea plants) will give all haploid $T$ gametes, so on selfing or on crossing to another $T T$ individual, all offspring will be $T T$, tall, while $t t$ gives all $t$ gametes. Heterozygotes, $T t$, give approximately equal numbers of $T$ and $t$ gametes from Mendel's First Law of Segregation.

Crossing $T t \times T t$ (or selfing $T t$) therefore gives a 1 $T T$: 2 $T t$: 1 $t t$ genotype ratio and a phenotype ratio of 3 $T$ phenotypes (tall) to 1 $t$ phenotype (short), if there is complete dominance.

Crossing $T T$ (gametes all $T$) $\times T t$ (gametes 1 $T$: 1 $t$) gives 1 $T T$ : 1 $T t$, all with tall phenotypes.

Crossing $t t \times T t$ gives 1 $T t$: 1 $t t$, with 1:1 genotype and phenotype (tall: short) ratios.

Crossing $T T \times t t$ gives all $T t$, tall.

Thus, autosomal loci with complete dominance are characterised by **3:1 phenotype ratios** from crossing two heterozygotes, with three of the dominant phenotype and one of the recessive phenotype. See Plate 2.1 for a human example of albinism.

## 2.1.3. *Autosomal loci with partial dominance, additive action or overdominance*

The genotype ratios in these cases are exactly the same as that of complete dominance, but the phenotype ratios are usually different when there are heterozygotes and homozygotes, because incomplete dominance, additive action and overdominance all give heterozygotes which are phenotypically different from both homozygotes. Let $S' S'$ give fruits 20 cm long, $S' S$ give fruits 15 cm long, and $S S$ give fruits 10 cm long, with additive action.

**Plate 2.1.** An autosomal recessive mutation, **albinism**. The parents are unaffected heterozygotes, *A a*, while the **twin boys** are albinos, *a a*, with white hair, pale skins, poor eyesight, and eyes which are pink in direct light, but look blue here. Blue eyes in humans are due to light-scattering, not pigments, and albinos cannot make normal skin, eye or hair melanin pigments. The two unaffected children could be heterozygotes or normal *A A*.

Crossing $S'S \times S'S$ gives a 1 $S'S'$: 2 $S'S$: 1 $S\,S$ genotype ratio, with a one 20 cm: two 15 cm: one 10 cm fruit phenotype ratio. Similarly, in *Antirrhinum*, there is incomplete dominance of red flowers over white flowers. The heterozygote has pink flowers, and selfing it gives a 1 red: 2 pink: 1 white ratio for flower colour. Thus, autosomal loci with partial dominance, additive action or overdominance are characterised by **1:2:1 phenotype ratios** from crossing two heterozygotes.

## 2.1.4. *X-linked and holandric loci*

Where there are sex chromosomes, the one that is present twice in one sex and once in the other sex is called the **X chromosome**; the sex chromosome present as one copy in one sex and none in the other is the **Y**. The parts of the X and Y which can pair in meiosis are called the **homologous regions** and can have crossovers between them; such genes are known as **incompletely sex linked**. The part of the X with no homologue on the Y is known as the **differential region** and the genes are known as **sex linked** or **X-linked**. Genes on the Y with no homologues on the X are in the **holandric region**. As this region's loci occur only in the Y chromosome, they are expressed in the X Y sex, e.g., hairy ears in man. The relative proportions of the three regions vary a lot between organisms, but there are usually few active genes in the holandric region. Genes in the holandric region of Y are **pseudodominant**, usually showing when present, as there are no second copies on another chromosome to dominate them. Thus, hairy ears male × normal female will give hairy eared males and normal females, since the females do not inherit the Y, but the sons must inherit their father's Y.

The X X sex is **homogametic** (producing "alike gametes"), as all gametes carry one X chromosome. X Y is the **heterogametic** (producing "unalike gametes") sex, producing approximately equal numbers of X-bearing and Y-bearing gametes. In mammals, the X X sex is normally female, producing eggs, and X Y is male, producing sperm. In contrast, the homogametic sex is male in birds (including chickens and turkeys), butterflies, moths, and in some fish, reptiles and amphibia, when symbols **Z Z** and **Z W** are usually used for the sex chromosomes. Some plants also have separate sexes, with the male plants producing stamens and the female plants producing ovules. The control is usually by an X/Y chromosome system, with the males XY and the females XX, as in *Silene alba* (white campion) and *Humulus* (hop). In *Asparagus*, maleness (XY) is dominant.

Alleles in the differential region of the X are present twice in females but only once in males, which are **hemizygous** for those loci. As discussed in Sec. 12.6, this causes differences between the sexes in the incidence of X-linked recessive diseases. Let us use **red-green colour blindness in humans** as an example of an **X-linked recessive** condition, when non-colour blindness represents an **X-linked dominant** condition. It is useful to show the chromosomes as well as the alleles, so the recessive allele for colour blindness is shown as $X^{cb}$ and the wild-type dominant allele as $X^{Cb}$, and the Y, with no allele, as Y. Males are either hemizygous normal, $X^{Cb}$ Y, or hemizygous colour-blind, $X^{cb}$ Y, while females are normal, $X^{Cb}X^{Cb}$, or carriers but not sufferers, $X^{Cb}X^{cb}$, or sufferers, $X^{cb}X^{cb}$.

Crosses of a normal homozygous female × a normal hemizygous male give all normal, with a 1:1 sex ratio: $X^{Cb}X^{Cb} \times X^{Cb}$ Y gives 1 $X^{Cb}$ $X^{Cb}$ (female): 1 $X^{Cb}$ Y (male).

Crosses of a colour-blind female × a colour-blind male give all colour-blind offspring: $X^{cb}X^{cb} \times X^{cb}$ Y gives 1 $X^{cb}X^{cb}$: 1 $X^{cb}$ Y.

Crosses of a normal female × a colour-blind male give carrier females (phenotypically normal) and normal sons : $X^{Cb}X^{Cb} \times X^{cb}$ Y gives 1 $X^{Cb}X^{cb}$ : 1 $X^{Cb}$ Y.

Crosses of a colour-blind female to a normal male give carrier females and colour-blind males: $X^{cb}X^{cb} \times X^{Cb}$ Y gives 1 $X^{Cb}X^{cb}$ : 1 $X^{cb}$ Y.

Crosses of a carrier female × a normal male give equal numbers of normal females, carrier females, normal males and colour-blind males: $X^{Cb}$ $X^{cb} \times X^{Cb}$ Y gives 1 $X^{Cb}X^{Cb}$: 1 $X^{Cb}X^{cb}$: 1 $X^{Cb}$ Y: 1 $X^{cb}$ Y.

Crosses of a carrier female to a colour-blind male give equal numbers of carrier females, colour-blind females, normal males and colour-blind males: $X^{Cb}X^{cb} \times X^{cb}$ Y gives 1 $X^{Cb}X^{cb}$ : 1 $X^{cb}X^{cb}$ : 1 $X^{Cb}$ Y : 1 $X^{cb}$ Y.

With incomplete dominance, additive action or overdominance, the female heterozygote has a different phenotype from both female homozygotes.

An easy way to tell whether a locus is X-linked is to make the two **reciprocal crosses** of pure breeding strains, e.g., normal female × sufferer male, and sufferer female × normal male. The results in the next generation would be identical in both crosses for the autosomal loci, with dominant phenotype females and males. For X-linked loci, the first cross gives phenotypically normal carrier females and normal males, but the second cross gives phenotypically normal carrier females and sufferer males (see genotypes above). Differences between reciprocal crosses and the sexes in one cross are found for X-linked genes.

## 2.2. Multiple-loci qualitative characters: dihybrid ratios and gene interactions such as epistasis causing modified ratios

### 2.2.1. *Standard dihybrid ratios*

Crosses with one, two or three segregating loci are called **monohybrid**, **dihybrid** and **trihybrid** crosses respectively. With one segregating locus and complete dominance, selfings of heterozygotes give **3:1 phenotype ratios**, while **test crosses** (heterozygote × homozygous recessive) give **1:1 phenotype ratios** from **Mendel's First Law**, the Law of Segregation. With two segregating loci, both with complete dominance, selfings of double heterozygotes give **9:3:3:1 ratios** for unlinked loci, while test crosses (e.g., *A a, B b* × *a a, b b*) give **1:1:1:1 ratios** from **Mendel's Second Law**, of Independent Assortment, if there is no interaction between the loci's expressions. Linkage gives more parental gametes than recombinant gametes, modifying these ratios in favour of parental combinations.

In cross diagrams and the explanations of what phenotypes correspond to what genotypes, the symbol " - " indicates that either allele could be present at that locus. In any cross diagram where two loci are **segregating** (i.e., two different alleles are present for each of two loci), we must show whether the alleles are in coupling (e.g., *A A, B B* × *a a, b b*) or repulsion phase (e.g., *A A, b b* × *a a, B B*). A cross diagram for a coupling cross with two unlinked and non-interacting segregating loci would look like this for diploid maize, *Zea mays*, where starchy grain (*Su*) is dominant to sugary grain (*su*) and coloured grain (*C*) is dominant to colourless (*c*):

coloured starchy × colourless sugary

| | | |
|---|---|---|
| *C C, Su Su* × *c c, su su* | Parents (P) | |
| *C, Su*    *c, su* | Gametes | |
| *C c, Su su* | First filial generation (F1) | |

coloured starchy grains

On **selfing**, the unlinked loci will assort independently in the F1 meiosis, giving both in eggs cells and pollen a 1 *C, Su*: 1 *C, su*: 1 *c, Su*: 1 *c, su* ratio in the gametes. We can use a **Punnett chequer board** (a multiplying device named after an early Cambridge geneticist) to see the effects of combining the male and female gametes at random. One just multiplies the items in the top and side gamete panels to get the contents of a square; thus, the top left genotype square's contents, 1 *C C, Su Su*, come from combining the male gamete 1 *Cu, Su* at the top with the female gamete 1 *Cu, Su* at the left side. Therefore, we get an F2 (second filial generation) as shown in Table 2.1.

A ratio of 9 coloured starchy is given (*C -, Su -*, where the "-" indicates that either allele could be present at that locus): 3 coloured sugary (*C -, su su*): 3 colourless starchy (*c c, Su -*): 1 colourless sugary (*c c, su su*), which is the standard ratio from a selfing with two unlinked loci with complete dominance at each locus and no locus-interactions, according to Mendel's Second Law, the Law of Independent Assortment. There are 16 F2 genotype squares in Table 2.1, where "1" in a square indicates that 1/16 of the F2 will have that genotype.

If the double heterozygote, the F1 in the above cross, had not been selfed, but had been **test-crossed** (i.e., crossed to a doubly homozygous individual, *c c, su su*, all gametes *c, su*), we would get a **1:1:1:1** phenotype and genotype ratio instead of a **9:3:3:1 ratio**. This is because the double heterozygote has a 1:1:1:1 gamete genotype ratio, and each gamete just combines with a fully recessive *c, su* gamete, giving 1 *C c, Su su*: 1 *C c, su su*: 1 *c c, Su su*: 1 *c c, su su*, i.e., 1 coloured starchy: 1 coloured sugary: 1 colourless starchy: 1 colourless sugary.

If the loci were linked in a coupling phase with a 20% recombination, then the gamete frequencies from the double heterozygote *A a, B b* would not be 1:1:1:1, but would be split into 80% parental gametes (as fractions, 0.4 *A, B*; 0.4 *a, b*) and 20% recombinant gametes (0.1 *A, b*; 0.1 *a, B*), and these would be put into the multiplying grid for gamete mating (Punnett square), instead of the 1:1:1:1 values shown in the maize example. A selfing

**Table 2.1.** **Punnett chequer board** for F2 genotypes and their frequencies, from a dihybrid cross with no linkage.

| Male gametes: | 1 *C Su* | 1 *C su* | 1 *c Su* | 1 *csu* |
|---|---|---|---|---|
| **Female gametes:** | | **F2 genotypes** | | |
| **1 *C Su*** | 1 *C C, Su Su* | 1 *C C, Su su* | 1 *C c, Su Su* | 1 *C c, Su su* |
| **1 *C su*** | 1 *C C, Su su* | 1 *C C, su su* | 1 *C c, Su Su* | 1 *C c, su su* |
| **1 *c Su*** | 1 *C c, Su Su* | 1 *C c, Su su* | 1 *c c, Su Su* | 1 *c c, Su su* |
| **1 *c su*** | 1 *C c, Su su* | 1 *C c, su su* | 1 *c c, Su su* | 1 *c c, su su* |

would then give a 0.66 *A* -, *B* -: 0.09 *A* -, *b b*: 0.09 *a a*, *B* -: 0.16 *a a*, *b b* ratio. A test-cross of the double heterozygote to the *a a*, *b b* double homozygote would give a 0.4 *A a*, *B b*: 0.1 *A a*, *b b*: 0.1 *a a*, *B b*: 0.4 *a a*, *b b* ratio, from which the recombination frequency can be calculated as 20% (0.2 as a fraction). There is a way of calculating recombination frequencies from the F2 of a selfed F1 [see Stephens, 1939, quoted in Strickberger, 1976; or Strickberger, 1976; the table of Z is not given in Strickberger 3rd ed., 1985].

**Linkage modifies** the standard Mendelian dihybrid ratios of 9:3:3:1 and 1:1:1:1, giving more parental combinations and fewer recombinant combinations, with different ratios depending on the particular recombination frequencies. Thus, if selfing or test-crossing a double heterozygote does not give the standard dihybrid ratios, linkage is one explanation to explore where the test-cross is ideal because dominance does not affect the ratios and one can calculate recombination frequencies directly from F2 ratios. Plate 2.2 shows a maize example with complete linkage (0% recombination), with half the grains of each parental type, coloured

**Plate 2.2.** **Complete linkage in maize grain between coloured/colourless, *C*/*c*, and full/shrunken, *Sh*/*sh*.** The parents were coloured full and colourless shrunken, giving 0% recombination in the F2 after test-crossing the F1 (coloured full) to the double recessive, *c c*, *sh sh*. A 1:1:1:1 ratio is expected if there is no linkage.

full and colourless shrunken, and no recombinants, coloured shrunken or colourless full.

## 2.2.2. *Gene interactions such as epistasis causing modified dihybrid ratios*

Let us now examine the effects of **gene interactions** (more accurately known as gene-product interactions) between loci on dihybrid ratios for unlinked loci with complete dominance. By extension, one could also work out expected results with interactions and linkage, or gene interactions and other forms of dominance.

**Epistasis** is a common form of interaction, where the presence of a particular allele at the **epistatic** ("standing above") locus prevents the expression of any alleles at the **hypostatic** ("standing below") second locus. Epistasis can be caused by the dominant allele or by homozygous recessive alleles at the epistatic locus. For example, in maize, the *A* allele gives anthocyan colour while *a* gives colourless to the grains, with *a* showing **recessive epistasis** to a second locus, where the *Pr* allele modifies the red colour from *A* to purple, and the *pr* allele has no effect. If we have a repulsion cross, *A A, pr pr* (red grains) × *a a, Pr Pr* (colourless), the F1 are purple, *A a, Pr pr*. Selfing this double heterozygote with two unlinked loci gives the expected genotype ratio of 9 *A-, Pr -*; 3 *A-, pr pr*; 3 *a a, Pr -*; 1 *a a, pr pr*. The 9 *A -, Pr -* will be purple, with *Pr* modifying the *A* red product to purple. The 3 *A -, pr pr* are red, unmodified, and 3 *a a, Pr -* and 1 *a a, pr pr* are all colourless. There is no red product for the *Pr* allele in *a a, Pr -* to modify, giving a **9:3:4 ratio** of purple: red: colourless, a modified dihybrid ratio alerting us to recessive epistasis. See Plate 2.3.

A possible metabolic pathway to explain the results is to have the epistatic locus controlling an earlier step than does the hypostatic locus, as shown in Fig. 2.1. This shows *A* and *Pr* alleles promoting particular steps, and *a* and *pr* blocking particular steps. It is understood that the recessive allele will only act when homozygous, and only one copy of it is conventionally shown in such diagrams. A horizontal arrow shows the allele promoting the step and a vertical arrow shows the allele blocking that step.

The epistatic allele is dominant in the case of fruit colour in *Cucurbita pepo*, e.g., marrow or pumpkin,where *W*, giving white fruit, is epistatic to

**Plate 2.3.   Recessive epistasis of *a* to the *Pr*/*pr* locus in maize grain**. Selfing the F1 gave a modified dihybrid ratio of 9 purple (*A* -, *Pr* -) : 3 red (*A* -, *pr pr*) : 4 colourless (3 *a a*, *Pr* - : 1 *a a*, *pr pr*). Allele *A* gives red, which *Pr* modifies to purple.

**Fig. 2.1.**   A pathway for **recessive epistasis** for maize grain colour: see text and Plate 2.3.

the unlinked *Y*/*y* locus, with *w w*, *y y* giving green and *w w*, *Y* -giving yellow as in most commercial pumpkins. A repulsion phase cross with selfing of the F1 would be white fruit (*W W*, *y y*) × yellow fruit (*w w*, *Y Y*), giving *W w*, *Y*, *y*, white F1, as *W* shows **dominant epistasis** to *Y*/*y*. Selfing the F1 gives a Mendelian genotype ratio and a modified phenotype ratio, 9 *W* -, *Y* - (white); 3 *W* -, *y y* (white); 3 *w w*, *Y* - (yellow); 1 *w w*, *y y* (green), i.e., a **12:3:1 dominant epistasis phenotype ratio**, quite different from the 9:3:4 for recessive epistasis. The F2 ratio does not depend on whether the cross was in coupling or repulsion, if the loci are unlinked. A possible pathway, with the epistatic allele controlling the earlier step of the pathway, is shown in Fig. 2.2.

white $\xrightarrow[W]{w}$ yellow $\xrightarrow[Y]{y}$ green

**Fig. 2.2.** A pathway for **dominant epistasis** in *Cucurbita* fruit colour: see text.

One can combine the basic dihybrid F2 phenotype ratio, 9:3:3:1, in other ways such as 15:1 (i.e., the double recessive gives one phenotype, and a dominant allele at either or both loci gives a different phenotype), 13:3 (e.g., a dominant epistatic allele inhibits the dominant allele at the hypostatic locus, but has no phenotype of its own), and 9:7. All of these have known examples.

The 9:7 ratio comes from **complementary dominant alleles** at two unlinked loci. In sweet pea, *Lathyrus odoratus*, purple flower colour requires the dominant alleles *R* and *C*. A repulsion phase cross would be *C C, r r* (white because it lacks *R*) × *c c, R R* (white because it lacks *C*), giving purple-flowered F1, *C c, R r*, having both *C* and *R*. Selfing the F1 gives the modified dihybrid ratio of 9 purple (*C* -, *R* -): 7 colourless (3 *C* -, *r r*; 3 *c c, R*-; 1 *c c, r r*). This case can be described as having two complementary dominant genes or as mutual recessive epistasis. As in Fig. 2.3, it can be represented by a pathway, but from the data, we could not tell whether locus *C/c* controls the first step and *R/r* controls the second step, or whether the order is reversed. Plate 2.4 shows this in a maize grain example.

This cross shows several important general points. Firstly, both parents have white flowers with different genotypes and the white phenotype is also given by the double recessive, *c c, r r*, such that the same phenotype can be produced by more than one homozygous genotype, as well as by genotype variations concealed by dominance. Secondly, new phenotypes can appear in the F1 (this cross) and/or the F2 (green fruit in the *Cucurbita* cross). Thirdly, parental phenotypes can differ according to whether the cross is in coupling (*C C, R R*, purple × *c c, r r*, white) or in repulsion (*C C, r r*,

colourless precursor $\xrightarrow[c]{C}$ colourless intermediate $\xrightarrow[r]{R}$ purple

**Fig. 2.3.** A pathway for **complementary dominant genes'** action in sweet pea flower colour: see text.

**Plate 2.4.** **Two unlinked complementary dominant genes in maize grain**, $R$ and $C$, together giving grain colour. Selfing the F1 gave a modified dihybrid ratio of 9 purple ($R$ -, $C$ -) : 7 colourless (3 $R$ -, $c\,c$; 3 $r\,r$, $C$ -; 1 $r\,r$, $c\,c$).

white × $c\,c$, $R\,R$, white), but the F1 genotypes and phenotypes are the same from coupling and repulsion crosses, $C\,c$, $R\,r$, purple.

Alleles at different loci producing the same phenotype, such as $c\,c$ and $r\,r$ both giving colourless grain in maize are sometimes called **mimic genes**. For example, in **cats, Cornish and Devon rexes** (almost hairless cats) are caused by similar mutations at different unlinked loci, $r\,r$ and $re$ $re$ respectively. Crossing a Cornish rex ($r\,r$, $Re\,Re$) with a Devon rex ($R\,R$, $re\,re$) gives normal coat progeny ($R\,r$, $Re\,re$), which on mating among themselves gives the modified dihybrid ratio of 9 normal:7 rex, of which 3 are Cornish, 3 are Devon, and 1 is homozygous for both mutations (Devon-Cornish rex). If the two types of rex had mutations at the same locus, one would not get the 9:7 F2 ratio, as getting wild-types would require a crossover within the locus which is very rare.

Some characters are determined by a very large number of loci with many possible gene interactions such as epistasis between them. In maize, more than 20 loci have qualitative effects on grain colour. In sheep, 50 to

100 loci control wool and hair growth, with the genes often clustered into families on chromosomes.

## 2.3. Quantitative characters; quantitative trait loci and polygenes; modifiers; threshold characters

Many important characters in many organisms are quantitative ones with **continuous variation** such as grain yield, milk yield, meat yield, growth rate of micro-organisms, and human intelligence. Although there are exceptions, quantitative characters typically show greater **effects of the environment** than do qualitative ones. Thus, in maize, the quantitative character of grain yield is greatly affected by light, rain, temperature, soil and disease, while the qualitative character of colourless grain versus purple grain is not affected by such environmental conditions. Environmental variation can make working out the genetics of quantitative inheritance very difficult, so making crosses in controlled environments is recommended if practicable.

The simplest way to understand the inheritance of quantitative characters is to start with simple situations and then build up to more complex ones. Consider **grain colour in wheat**, as studied by Nilsson-Ehle in 1909. He crossed a strain with pure breeding white grain ($A\ A$, $B\ B$) with a strain with pure breeding dark red grain ($A'\ A'$, $B'\ B'$), obtaining an F1 with intermediate light red grain ($A'\ A$, $B'\ B$). On allowing the F1 to self and to cross amongst themselves, he obtained a discontinuous distribution in the F2 ranging from dark red to white. The F2 phenotypes, proportions, numbers of primes and genotypes were respectively:

dark red,   1/16, 4′, $A'\ A'$, $B'\ B'$;
red,        4/16, 3′, $A'\ A'$, $B'\ B$; $A'\ A$, $B'\ B'$;
light red,  6/16, 2′, $A'\ A'$, $B\ B$; $A'\ A$, $B'\ B$; $A\ A$, $B'\ B'$;
pale red,   4/16, 1′, $A'\ A$, $B\ B$; $A\ A$, $B'\ B$;
white,      1/16, 0′, $A\ A$, $B\ B$.

The proportions, in sixteenths, suggest dihybrid inheritance, and getting five phenotypes suggests that there is additive action. We have two unlinked loci with additive action of the alleles at each locus, and cumulative action of the two loci. Thus, one prime (′) adds an equal amount of pigment whichever locus it is at, so the amount of pigment is determined by the total number of primes in the genotype, summed over both loci. This **1:4:6:4:1** is another modified dihybrid ratio. While dominance gives **asymmetrical distributions**, e.g., 3:1, 9:3:3:1, 27:9:9:9:3:3:3:1, for mono-, di- and tri-hybrid crosses respectively, additive action gives **symmetrical distributions** with equal proportions at the two extremes, e.g., 1:2:1 and 1:4:6:4:1. Nilsson-Ehle's results fitted the theory well, with no effects of the environment, but the distribution was discontinuous and qualitative.

Consider a case of height in a diploid controlled by three unlinked loci, with a basic height of 50 cm, and each primed allele adding 2 cm to that. Initially, assuming that the plants are grown in a constant environment with no environmental effects, the cross $A\ A, B\ B, C\ C$ (50 cm) $\times A'\ A', B'\ B', C'\ C'$ (62 cm = 50 + plus the effect of 6 primes, 2 cm each) gives an F1 of height 56 cm [$A'\ A, B'\ B, C'\ C$ has three primes = $50 + (3 \times 2) = 56$ cm]. One can work out the expected F2 distribution for additive and cumulative gene action from a Punnett chequer board, but the easiest way is to use **Pascal's triangle** which gives the coefficients of the binomial expansion. As shown in Fig. 2.4, it starts with the number one, then two ones on the next line, then it expands each successive line by a one at each end, and fills in the intervening numbers by adding the two numbers diagonally above each place, so to get the threes in line four, one adds the ones and twos diagonally above them.

One takes alternate rows, starting with the third row, which gives the F2 proportions for one locus (e.g., $1/4$ red, 2/4 pink, $1/4$ white for flower colour in *Antirrhinum*, where for red, 1 is the figure at the left end of the row, and 4 is the total of the numbers in that row). The fifth row gives the F2 proportions for two loci, 1/16, 4/16, 6/16, 4/16, 1/16, as for wheat grain colour. The case we are currently considering has six segregating alleles, two at each of the three loci, so the seventh line gives us the F2 proportions. These are shown

```
                    1                    (no segregating alleles)
                 1     1
              1    2    1                 (2 segregating alleles, 1 locus)
            1    3    3    1
         1    4    6    4    1            (4 segregating alleles, 2 loci)
       1    5   10   10    5    1
    1    6   15   20   15    6    1       (6 segregating alleles, 3 loci)
```

**Fig. 2.4.** Pascal's triangle: see text.

**Table 2.2.** F2 results for **three unlinked loci** with additive and cumulative action.

| Frequency | Number of primes | Height, cm |
|-----------|------------------|------------|
| 1/64 | 6 | 62 |
| 6/64 | 5 | 60 |
| 15/64 | 4 | 58 |
| 20/64 | 3 | 56 |
| 15/64 | 2 | 54 |
| 6/64 | 1 | 52 |
| 1/64 | 0 | 50 |

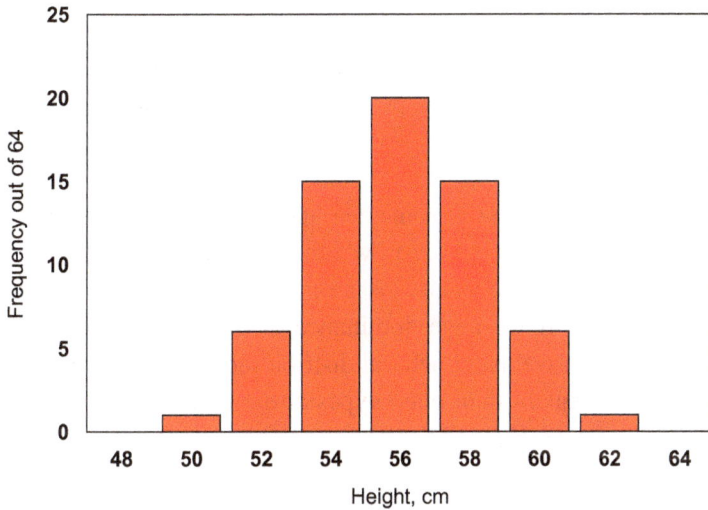

**Fig. 2.5.** **Discontinuous height distribution in the F2** from selfing an F1 with three segregating loci with additive action at each locus and cumulative action between loci, and no environmental variation.

in Table 2.2. This gives a symmetric discontinuous distribution which we can plot as a histogram in Fig. 2.5.

So far, we have been considering a case with no environmental effects and getting a discontinuous distribution with seven separate phenotypes, where each depends entirely on the number of primed alleles at the three controlling loci. Let us now introduce some environmental variation. A genotype with one prime, giving so far only a height of 52 cm, will now give a spread of possible phenotypes, say 50 to 54 cm, as some environmental conditions will increase height by varying amounts while others will decrease it, giving a potentially continuous set of possible phenotypes within that limited range. All seven genotypes will now give a range of possible phenotypes, so what was a discontinuous set of phenotypes now becomes continuous because of the environmental variation. In this case, we will get a symmetric continuous distribution approximating to a Normal (Gaussian) Distribution, as in Fig. 2.6.

The basis of quantitative inheritance with continuous variation is therefore one of **discontinuous variation in genotypes**, but **continuous phenotype distributions** brought about mainly by two factors. The first is the **effects of the environment** and the second is the presence of **segregating polygenes**, each of which gives very small phenotypic effects, hence tending to give continuous variation. A plant in the present case with a height of 51.3 cm might thus have no primes but a favourable environment for increased height, contributing ~1 cm, and a number of height-increasing polygenes, contributing ~0.3 cm, or it might have one prime but a slightly unfavourable environment or some height-decreasing polygenes.

We considered additive action which gives symmetrical distributions and three phenotypes per locus (assuming diploidy and two segregating alleles per locus). If we have dominance, it only gives two phenotypes per locus and can give asymmetrical distributions, with phenotypes at one extreme more frequent than phenotypes at the other extreme. Consider a

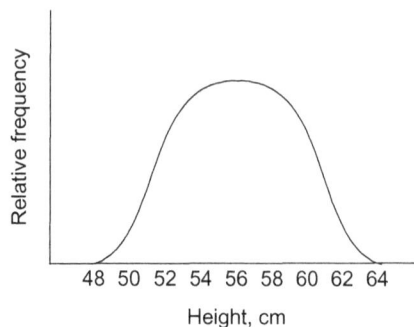

**Fig. 2.6.** **Continuous height distribution** for the conditions of Fig. 2.5, but with some **environmental variation**.

situation where four unlinked loci with complete dominance and cumulative action control height, with all loci having the dominant allele increasing height, e.g., *A*, *B*, *C* and *D* all adding 10 cm to a basic height of 10 cm for *a a*, *b b*, *c c*, *d d*. *A* -, *b b*, *c c*, *d d* is 20 cm and *A A*, *B B*, *C C*, *D D* is 50 cm high. On selfing the quadruple heterozygote, *A a*, *B b*, *C c*, *D d*, one expects a 3:1 phenotype ratio at each locus, over four loci, giving (3/4 dominant phenotype + 1/4 recessive phenotype)$^4$ = 81 (*A* -, *B* -, *C* -, *D* -, 50 cm): 27: 27: 27: 27 (all 40 cm, each with dominants at three loci, e.g., *A* -, *B* -, *C* -, *d d*; *A* -, *B* -, *c c*, *D* -; *A* -, *b b*, *C* -, *D* -, etc.): 9: 9: 9: 9 (all with dominants at two loci, 30 cm, e.g., *a a*, *B* -, *C* -, *d d*): 3: 3: 3: 3 (all with a dominant at one locus, 20 cm, e.g., *a a*, *B* -, *c c*, *d d*): 1 (10 cm, *a a*, *b b*, *c c*, *d d*). With 81 at 50 cm and only 1 at 10 cm, the distribution is clearly skewed in the direction of the dominant phenotypes. The overall phenotype ratio is **81: 108: 36: 12: 1**, with five phenotypes having dominant alleles at 4, 3, 2, 1 or 0 loci respectively.

Where all dominant alleles at different loci give phenotypes all in the same direction for a character, e.g., all increasing or all decreasing height, we have **directional dominance**. If some dominants act in one phenotypic direction and other dominants act in the opposite direction, we have **non-directional dominance**, e.g., *A* and *C* increase height, but *B* and *D* decrease height. Directional dominance tends to give strongly skewed distributions. However, with non-directional dominance, skewing is less, depending partly on how many dominants act in each direction. Another simplifying assumption used thus far is that each major locus has an **equal effect** on the phenotype, which need not be true. These loci which have significant individual amounts of effect on the phenotype are called **quantitative trait loci**, often abbreviated as **QTL**.

In contrast to QTL, we have **polygenes** which are genes whose individual phenotypic effect is very small in relation to the total phenotypic variation for the character. If a plant has heights ranging from 10 to 50 cm, then quantitative trait loci might have individual effects on the height of perhaps several cm, while polygenes might have effects of a few millimetres or less. This explains why segregating polygenes can give continuous phenotype variation when the underlying genotype variation for major genes would, on its own, give a discontinuous set of phenotypes. **In cat coats**, the gradation between ginger and red is continuous, determined by an unknown number of polygenes, all of very small effect, with the X-linked *O* allele (giving yellow, orange, red, while *o* gives black when expressed) being the major gene for that range of colours. Ginger cats can be thought of as having a majority of pigment-reducing alleles and red cats as having a majority of pigment-increasing alleles. The extremes of ginger and of red breed more or less true, with e.g., ginger × ginger giving largely ginger offspring, while ginger × red gives mainly intermediate coat colours.

Polygenes can usually be treated as collectively showing incomplete dominance/additive action, although that may not be true of individual polygenes. They can mutate and recombine, and may show linkage and/or gene interactions. They can be mapped but are difficult to study individually because of their small phenotypic effects which are easily masked by environmental variation. They are generally considered collectively rather than individually.

**Major genes** are ones giving fairly large qualitative phenotypic differences such as black versus red cattle coat colour or 10 versus 15 cm height, so QTL are one type of major gene. Major genes and polygenes are alike at the DNA level, only differing in their extent of effect. Studies in the fungus *Penicillium chrysogenum* gave the interesting finding that for penicillin yield, many major genes with no direct connection with penicillin production act as polygenes with minor effects on penicillin yield. Thus, major gene loci affecting white versus blue-green conidia, or a requirement for methionine or for adenine, acted as polygenes on penicillin yield. Many genes controlling growth rates affect very many other characters.

If polygenes affect the expression of major genes, they are called **modifiers**. Thus, polygenes giving the difference between ginger and red coat for cats expressing the X-linked major gene *O* are modifiers of *O*. Modifiers may act differently on the expression of two alleles of a major gene. Long-haired cats are *l l*, while short-haired ones have *L*. The hair-length modifying polygenes have much greater effects on long-haired cats than on short-haired ones. One could even describe the polygenes controlling human height in non-dwarfs as modifiers of the normal height allele *d*, as non-dwarfs are *d d*. In cattle, cats and some other animals, there exists a dominant major gene causing **white spotting**, but the amount of spotting is often controlled by polygenes. According to Robinson (1991), there are three mechanisms giving spotting in mammals, with the pigment-producing cells in the skin coming from about 34 sites in the developing embryo from which pigment cells migrate to the rest of the body. White spotting occurs in areas from which those cells are absent, (1) because of a reduction in the number of sites, or (2) a slowing down of cell migration, or (3) from a failure of cells to function in a particular area, even though they reach that area.

Failure to realise that modifiers are affecting phenotypes can lead to **misinterpretation of data**. In the house fly, *Musca domestica*, chromosome 2 has a major gene determining the ability to detoxify DDT insecticide (by DDT-dehydrohalogenase) once it enters the fly's body. Allele $DDT^r$ gives resistance to DDT, with an ability to detoxify up to 10 mg per day of DDT. Modifier polygenes at other loci control the uptake of DDT, ranging from 5 mg/day to 15 mg/day. This gives rise to three situations, in which $DDT^r$ may or may not give complete resistance:

- $DDT^r$ present; few DDT-uptake inhibitor modifiers present, so $DDT^r$ gives incomplete DDT resistance, since up to 15 mg/day can be taken up, but only 10 mg/day can be detoxified.
- $DDT^r$ present; many DDT-uptake inhibitor modifiers present, so $DDT^r$ gives complete resistance, since only 5 mg/day of DDT can be taken up, but 10 mg/day can be detoxified.
- $DDT^r$ absent; the fly is sensitive to DDT, with the rate of uptake controlled by modifiers without detoxifying ability.

If organisms are grown in constant environments, we can often estimate the **number of segregating QTL** if each locus has an equal phenotypic effect and additive action. We saw earlier that with one such segregating locus, selfing the heterozygous F1 gave a 1:2:1 genotype and phenotype ratio, two loci gave a 1:4:6:4:1 ratio and three loci gave a 1:6:15:20:15:6:1 ratio. The proportion of the F2 at just one extreme is therefore 1/4 for one locus, 1/16 for two loci and 1/64 for three loci. The proportion of the F2 at just one extreme is therefore $1/(4^n)$, where $n$ is the number of segregating loci. We can use this to analyse data in the following three examples.

**Example 2.1.** In a constant environment, one pure breeding line of diploid rye had a height of 30 cm and was crossed with another pure breeding line of height 60 cm. The F1 had a height of 45 cm and were crossed amongst themselves. In the F2, the 1,000 plants included 16 that were as short as 30 cm and 20 as tall as 60 cm, with the rest made up of a variety of intermediate heights. Analyse these data.

We can see that there was no environmental variation because there was no variation within each parent or within the F1. Two lines of evidence suggest additive action: the F1 were intermediate between the two parental types, and the numbers at both extremes of the F2 were similar (dominance, especially directional dominance, gives skewed distributions). With 16 of the F2 at one extreme and 20 at the other extreme, the average number at one extreme was 18. There were 1,000 F2 plants, so the proportion of the F2 at one extreme was $\frac{18}{1,000} = 0.018$. This is close to 1/64, 0.016, so we can deduce that there were three segregating loci. The formula $1/(4^n)$ gives 1/64 for $n = 3$. We can give the parents' genotypes, $30\,cm = A\,A,\,B\,B,\,C\,C$, and $60\,cm = A'A',\,B'B',\,C'C'$, using primes for additive action alleles. A six primes difference between the parents gives a height difference of 30 cm, so one prime gives 5 cm additional height, with a basic no primes height of 30 cm. We can therefore deduce that F2 heights were 30, 35, 40, 45, 50, 55 and 60 cm, in the ratio of 1:6:15:20:15:6:1 from Pascal's triangle. There were 7 F2 phenotypes, corresponding to 0, 1, 2, 3, 4, 5, and 6 primes, and 27 genotypes, because there are three possible genotypes at each of the three loci, giving $3 \times 3 \times 3$ variations. From the simple data, we draw considerable deductions, providing that our basic assumptions hold, especially that different loci have equal effects and are unlinked.

**Example 2.2.** Yield in barley (a diploid) was studied in a controlled environment. Two inbred lines had yields of 50 and 70 units. When these lines were crossed with each other, the F1 had a yield of 60 units. The F1 were selfed. The 2,000 F2 plants had 8 as low yielding as 40 units and 11 as high yielding as 80 units, the rest being intermediate. Explain these results.

Inbred lines are pure breeding, so each parent was homozygous. The constancy of the parents and F1 shows that there were no environmental effects. We can deduce additive action from the intermediate F1 and the approximately equal numbers at the two ends of the F2 distribution. In this case, however, the F2 extremes were more extreme than either parent. Getting offspring more extreme than the original parents is called **transgression**, with 40 units being lower than the lower yielding parent and 80 units being more extreme than the higher-yielding parent, implying that F2 extremes must have more extreme genotypes than the parents. With

8 units at one extreme and 11 units at the other, there was an average of $\frac{9.5}{2,000}$ at one extreme $= 0.0048$, which is close to $1/256$ ($= 0.004 = 1/4^4$), so we deduce that there are four segregating loci. If the parents are not the most extreme types, we take **the proportion of the F2 at each extreme**, not just the proportion of the F2 as extreme as the parents. Possible pure-breeding genotypes would be $A\,A$, $B\,B$, $C\,C$, $D'\,D'$ for the 50 unit parent and $A'\,A'$, $B'\,B'$, $C'\,C'$, $D\,D$ for the 70 unit parent, with each prime adding 5 units to a basic value of 40 ($A\,A$, $B\,B$, $C\,C$, $D\,D$). The F1 were $A'\,A$, $B'\,B$, $C'\,C$, $D'\,D$; in the F2, the highest yielders, 80 units, had eight primes, and the lowest yielders, 40, had no primes. We can therefore deduce four segregating unlinked loci with additive and cumulative action, each prime adding 5 units of yield, with the parents not having the most extreme genotypes or phenotypes.

> **Example 2.3.** In maize (a diploid), two pure-breeding lines averaged heights of 122 cm (limits 117–127 cm) and 186 cm (limits 179–193 cm). The F1 plants averaged 154 cm (limits 148–160 cm), and were allowed to cross amongst themselves. In 1,000 F2 plants, 8 were shorter than 128 cm, and 10 were taller than 175 cm. What can you deduce?

From the variation within each parent and within the F1, we can deduce that environmental variation affected height. We deduce additive action as the F1 were intermediate between the parents, and the F2 extremes were about equally frequent. The number at one F2 extreme averages $18/2 = 9$, with $\frac{9}{1,000} = 0.009$. This lies between the expected value for three loci ($1/64 = 0.0156$) and that of four loci ($1/256 = 0.0039$). If there are four loci, we have too many at the extremes, and if there are three loci, we have too few at the extremes. Which can we explain by environmental effects, too few or too many at the extremes?

Suppose there were four loci, giving a height difference of 64 cm from 8 primes, or 8 cm a prime. A few of the genotypes and phenotypes are shown in Table 2.3. Looking at the expected limits, one sees that the ranges (limits) of the 0 and 1 prime phenotypes overlap, so that plants of 125–127 cm could have either genotype, so the extreme phenotype of less than 128 cm includes all those with no primes and some, but not all, of those with one prime. We will therefore get more plants at this extreme than the number calculated for no primes alone, and can therefore explain the observed excess over expectation in the data, with a similar explanation at the upper extreme,

**Table 2.3.**    A few of the **genotypes and phenotypes in the F2 from Example 2.3.**

| Number of primes | Average height, cm | Expected limits, cm |
|------------------|--------------------|--------------------|
| 0 | 122 | 117–127 |
| 1 | 130 | 125–135 |
| 2 | 138 | 133–143 |

which will include plants with 7 and 8 primes. We cannot explain having too few plants at the extremes, so the conclusion is that there were four, not three, segregating loci. In general, having environmental effects can blur different genotypic classes and cause more individuals in the extreme classes than one expects in the absence of environmental effects.

These three examples have been of genes with additive action, while others might include dominance, as in Example 2.4.

> **Example 2.4.** When **Aberdeen-Angus beef cattle** (coloured face, no horns, black coat) are crossed with **Hereford cattle** (white face, horns, red coat: see Plate 2.5 for a hornless example), the F1 hybrids show useful hybrid vigour (Sec. 13.1) for beef characters, and have a white face, no horns, and a black coat. When some F1 hybrids were crossed amongst themselves, the following numbers of different types were obtained in the F2: coloured face, horns, black coat, 28; coloured face, horns, red coat, 8; white face, no horns, black coat, 278; white face, no horns, red coat, 85; white face, horns, black coat, 96; white face, horns, red coat, 35; coloured face, no horns, black coat, 89; coloured face, no horns, red coat, 27. Explain these results.

The **dominant characters** are white face, no horns and a black coat, as shown in the F1, which show one or other parental phenotype, but not incomplete dominance. The parents each show a mixture of dominant and recessive traits, so the cross is partly coupling and partly repulsion, e.g., Aberdeen-Angus could be *wf wf, Nh Nh, bc bc* and Hereford could be *Wf Wf, nh nh, Bc Bc*, where the symbols are named for each phenotype, e.g., *Wf* for dominant white face. The F2 show a 278 (same as the F1, showing all three dominants): 96: 89: 85 (all with two dominants showing): 35: 28: 27 (all with one dominant showing): 8 triple recessives, which is clearly

**Plate 2.5.**   **Showing bulls in the Sainsbury Super-Bull Championships for beef bulls under two years of age**, at the South of England Show. **A Charolais** is on the left and **a polled (hornless) Hereford** on the right.

a good fit to the classic Mendelian **27:9:9:9:3:3:3:1 trihybrid ratio** for three unlinked loci with complete dominance at each locus.

In classic **simple quantitative analysis** for traits with additive action, one compares **means, ranges and variances** in the parental, F1 and F2 generations, after crossing pure-breeding parents. The most basic analysis has five assumptions, some of which are tested in the analysis and may not hold. All held for the wheat grain colour considered earlier.

- Each locus affecting the character has an equal effect.
- Each contributing allele has additive effects; there is no dominance between alleles at a locus.
- There are no interactions such as epistasis between loci, just cumulative action.
- There is no linkage between loci.
- Environmental effects are negligible.

Let us analyse two sets of data from Emerson and East (1913) and East (1916), given in Srb, Owen and Edgar (1965).

*Set 1.* **Cob (ear) length in maize**, in cm. The pure-breeding short-eared parent averaged 6.6, limits 4–9, while the long-eared parent averaged 16.8, limits 13–22. The F1 from intercrossing them gave an average of 12.1, limits 8–16. On selfing of the F1, the F2 ranged from 6 to 20 cm. As there was variation within each parent and within the F1 (the F1 are all of uniform genotype), we detect environmental variation. At 12.1 cm, the F1 average is between the parental values of 6.6 and 16.8, but is nearer to the long parent value, suggesting either incomplete dominance of long to short, or dominance for long at some loci and additive action at other loci. The F2 have a wider spread of values than the F1 because the F1 have only one genotype, but the F2 show segregation and recombination for the genes by which the two parents differ and for which the F1 is therefore heterozygous.

Those F2 in the range of 6 to 9 cm overlapped the short-parent range of 4–9 cm, and those in the range of 13 to 20 cm overlapped the long-parent range of 13–22 cm. More than 1/64 of the F2, therefore, overlapped one parental extreme, so that there were three or fewer loci segregating. Possible genotypes which would account for these data are short parent, *A A, B B, c c*, long parent *A' A', B' B', C C*, F1 *A' A, B' B, C c*, and the F2 with a range of different segregants. Postulating one locus with dominance of long and two loci with additive action helps to explain the incomplete dominance in the F1, although other genotypes and explanations are not ruled out.

*Set 2.* **Corolla (flower tube) length in tobacco**, *Nicotiana longiflora*, in mm. The pure-breeding short-flowered parent averaged 40 mm, limits 34–43, while the long-flowered parent averaged 93, limits 86–97. The F1 from intercrossing them gave an average of 61, limits 53–72. On selfing of the F1, the F2 ranged from 51 to 84 mm. As there was variation within each parent and within the F1, we detect the environmental effects. At 61 mm, the F1 average was about midway between the parental values of 40 and 93, so we can deduce no dominance. In the F2, none of the plants had flowers as short as the short parent, and none had flowers as long as the long parent. With none of the F2 at either extreme for 444 plants, we cannot estimate the number of segregating loci accurately. An estimate of more than four loci would be reasonable since the numbers are too small to use in the $1/4^n$ equation, which gives an answer of infinity.

While this kind of analysis can be done by computer, one needs common sense in interpreting the data. Genetic analysis is easiest if environmental influences can be minimised, and if one can deduce that all segregating loci have complete dominance, or all have additive action. If all segregating loci have dominance, one can look at the F2 from selfing a multiple heterozygote, when the phenotype ratio (e.g., 3:1, or 9:3:3:1) should show the number of segregating loci. Note the number of phenotypes as well as the ratio; the rarest phenotype should be that of the multiple recessive. Gene-product interaction and linkage can affect ratios as we have seen, with linkage and recombination frequencies being readily estimated from test crosses (heterozygote × homozygous recessive).

One can have characters controlled both **qualitatively and quantitatively**, with discontinuous and continuous variation. In cattle coat colour, animals may have no spots (*s s*) or some spots (*S -*), with continuous variation for the amount of spots if present, from a series of polygenic loci. In some plants, there may be white flowers, *c c*, or coloured flowers, *C -*, with the amount and type of colour, when present, determined by other loci. Chapter 3 has further information on quantitative characters. For examples of working with QTL, see Georges *et al.* (1995) for milk yield in dairy cattle, and DeVincente and Tanksley (1993) for tomatoes. For much more mathematical treatments of quantitative inheritance, see Cameron (1997), Falconer and Mackay (1996), and Kearsey and Pooney (1995).

In **forest trees**, many commercially important quantitative traits such as height, basal area, stem proportions, and leaf characters such as time of spring bud break, show normal distributions in segregating populations. Their genetics could be polygenic with many loci having very small effects, or there could be a few controlling QTL of large effect, with environmental variation masking discrete phenotypes. Bradshaw and Stettler (1995) studied the parents, F1 and F2 from an interspecific cross of two poplars, *Populus trichocarpa × P. deltoides*. Instead of polygenic control, most of the traits measured had one to five QTL of large effect responsible for much of the genetic variance. For example, the timber trait of stem volume (measured after two years growth in this study) had 45% of the genetic variance and 30% of the phenotypic variance controlled by just two QTL of large effect, even though the F2 showed continuous variation with a roughly normal distribution. There was also hybrid vigour (Sec. 13.1) for this character, with the F1 value ($34 \, dm^3$) clearly higher than the two parental values, $25 \, dm^3$ for *P. trichocarpa* and $3 \, dm^3$ for *P. deltoides*, showing transgression (Sec. 3.1), although none of the F2 had values as high as those of the larger parent or the F1, showing regression (Sec. 3.1) from the F1 value. The QTL were mapped.

For a recent account of QTL in cereals, including genes for disease resistance, stress tolerance, their mapping and use in marker-assisted selection, see Gupta and Varshney (2004). See Sec. 8.5 for more details on mapping QTL and other genes controlling complex quantitative characters such as yield. A typical use of QTL mapping is that of Causse *et al.* (2001); they crossed a cherry (small sweet fruit) tomato line with a large-fruited line to get 144 recombinant inbred lines to study the inheritance of taste, aroma and texture in fresh market tomatoes. Each of these three attributes was controlled by one to five QTL, but positive or negative correlations between characters made selection for individual traits difficult.

## 2.4. Threshold characters

**Threshold characters** have discontinuous qualitative phenotypes, but with an underlying quantitative polygenic distribution, such that the discontinuous phenotypes do not show normal qualitative trait Mendelian ratios. The two phenotypes may not always breed true when they are expected to. The concept of threshold characters is important in human genetics (see Sec. 12.4 on polygenic and multifactorial disorders), plant pathology, animal reproductive physiology, etc.

The classic case is the study by Wright (1934) of **guinea pig hind legs**, where two inbred lines had a discontinuous variation between normal three-toed animals and polydactylous ones with four toes. When the two types were crossed, all F1 had three toes, but F1 × F1 crosses gave 188 three-toed animals to 45 four-toed ones, a poor fit to a classical 3:1 monohybrid ratio, as if three toes were dominant to four toes. However, crossing the F1 to the four-toed line gave 77% four-toed to 23% three-toed animals, not a 1:1 ratio. Wright proposed that there were about 4 pairs of polygenic loci, each with 2 alleles, 1 "polydactyly" and 1 "normal". Guinea pigs with more than 4 polydactyly alleles out of 8 possible alleles were suggested to be four-toed and those with 0 to 4 polydactyly alleles were three-toed, falling below some developmental threshold level for a compound needed for polydactyly.

The normal line had about 0 to 2 or 3 polydactyly alleles and the poly-dactylous line had about 5 to 8 polydactyly alleles, so that the F1 generally had 3 or 4 polydactyly alleles, but the F2, from intercrossing the F1, would have a big range of numbers of polydactyly alleles, giving more three-toed types (0 to 4 polydactyly alleles) than four-toed types (5 to 8 polydactyly alleles). As found, crossing the F1 (3 or 4 polydactyly alleles) to the poly-dactylous (5 to 8 polydactyly alleles) strain would give a majority of four-toed animals. If this model is correct, one would expect that the two parental lines would not always breed true within themselves, and that a small pro-portion of the F1 might be four-toed. The number of polydactyly alleles in gametes from the two lines would depend on homozygosity (most likely in inbred lines) or heterozygosity. Thus, an individual with 2 polydactyly alleles would give gametes with 1 polydactyly allele if they were at the same locus, or 0, 1 or 2 polydactyly alleles if they were heterozygous at two different loci.

Other examples of threshold characters include the **number of ver-tebrae in mice and man**, a number of **multifactorial human disorders** (Sec. 12.4) including club foot, cleft lip and cleft palate, schizophrenia and type 1 diabetes, where the genetics follows that of a threshold character (any individual either has the character/disorder or does not have it), but there may be environmental factors involved too, as in disease-susceptible plants versus disease-resistant ones for a particular disease. The environment plays a much larger part in schizophrenia, developing long after birth, than in cleft palate which develops in a stable womb environment, but environment or chance must also affect cleft lip as one can have identical (monozygotic) twins where one has a cleft lip but the other does not. In human diseases, the underlying continuous genetic variable is often called the **liability** for that disease. In cattle, cows usually have one calf per pregnancy, occasionally twins, but very rarely more than two calves at a time. The factors determin-ing the threshold character of one or two calves are, for dizygotic twins, the levels of circulating blood gonadotrophic hormones, determining the number of eggs shed, and the conditions within the womb affecting embry-onic survival; for monozygotic twins, they are whatever the factors that determine the splitting of the fertilised egg. For a quantitative treatment of threshold characters, including selection and heritability calculations, see Falconer and Mackay (1996). For example, they quote a heritability of 85% for schizophrenia and 35% for congenital heart disease.

## Suggested Reading

Bradshaw, H. D. and R. F. Stettler, Molecular genetics of growth and development in Populus. IV. Mapping QTLs with large effects on growth, form, and phenology traits in a forest tree. *Genetics* (1995) **139**: 963–973.

Cameron, N. D., *Selection Indices and Prediction of Genetic Merit in Animal Breeding*. (1997) CAB International, Wallingford.

Causse, M. *et al.*, Genetic analysis of organoleptic quality in fresh market tomato. 2. Mapping QTLs for sensory attributes. *Theor Appl Genet* (2001) **102**: 273–283.

Dalton, D. C., *An Introduction to Practical Animal Breeding*. (1980) Granada, London.

DeVincente, M. C. and S. D. Tanksley, QTL analysis of transgressive segregation in an interspecific tomato cross. *Genetics* (1993) **134**: 585–596.

Falconer, D. S. and T. F. C. Mackay, *Introduction to Quantitative Genetics*, 4th ed. (1996) Longman, Harlow.

Georges, M. *et al.*, Mapping quantitative trait loci controlling milk production in dairy cattle by exploiting progeny testing. *Genetics* (1995) **139**: 907–920.

Gupta, P. K. and R. K. Varshney, (eds.) *Cereal Genomics*. (2004) Springer, Dordrecht.

Hartl, D. L. and E. W. Jones, *Genetics. Analysis of Genes and Genomes*, 6th ed. (2005) Jones and Bartlett, Sudbury, Maryland.

Kearsey, M. J., Biometrical genetics in breeding, in Hayward, M. D., N. O. Bosemark and I. Romagosa (eds.) *Plant Breeding. Principles and Prospects*. (1993) Chapman and Hall, London, pp. 163–183.

Kearsey, M. J. and H. S. Pooney, *The Genetic Analysis of Quantitative Traits*. (1995) Chapman and Hall, London.

Mayo, O., *The Theory of Plant Breeding*, 2nd ed. (1987) Oxford University Press, Oxford.

Robinson, R., *Genetics for Cat Breeders*, 3rd ed. (1991) Pergamon Press, Oxford.

Srb, A. M., R. D. Owen and R. S. Edgar, *General Genetics*, 2nd ed. (1965) W. H. Freeman and Co., San Francisco.

Strickberger, M. W., *Genetics*, 2nd ed. (1976) Macmillan, New York.

Wright, S., The results of crosses between inbred, strains of guinea pigs, differing in number of digits. *Genetics* (1934) **19**: 537–551.

# Chapter 3

# Regression, Transgression, Environmental Effects and Heritability; Correlations Between Characters; Genotype, Phenotype and Breeding Values

## 3.1. Genetic and environmental causes of regression and transgression

**Regression** (filial regression) is the tendency of the progeny of extreme parents to be less extreme than their parents; they show regression towards the population average value (mean). This applies to quantitative characters and has three causes: dominance, gene interactions such as epistasis, and environmental effects. Dominance causes regression because heterozygotes segregate both dominant and recessive phenotypes.

**Transgression** is the production of offspring (in the F1 or F2) more extreme than their parents for a particular quantitative character. It can be caused by genes with dominance or additive effects, by gene interactions such as epistasis, and by environmental effects.

Regression and transgression are very important in applied genetics. The breeders often want transgression to more extreme types than either parent, e.g., offspring with higher yields or more disease resistance, but sometimes they want regression to an average type, e.g., to eliminate extreme types to get uniformity for sowing-to-ripening time for a mechanically harvested seed crop such as wheat.

The **environment** can produce regression. An excellent year giving a high yield for rice in an area might be followed by an average year giving an average yield. Therefore, there is regression in that high yielding plants have average yielding offspring because of environmental changes. Similarly, an average year with average yields might be followed by a poor year giving low yields, i.e., transgression whereby average parents have a more extreme yielding offspring.

For regression, one needs to distinguish between cases in which there is a general tendency for offspring to be less extreme than their parents, and cases where there is no such general tendency, although some individuals show regression. If the segregating loci all have additive action, then there is no general regression and the offspring average is expected to equal the parental average. In the wheat grain colour example in Sec. 2.3, if a strain with red grains ($A'\,A'$, $B'\,B$, three primes, gametes 1 $A'$, $B'$: 1 $A'$, $B$) is crossed with one with pale red grains ($A'A$, $BB$, one prime, gametes 1 $A'$, $B$: 1 $A$, $B$), we expect a ratio of 1/4 red (3 primes), 1/2 light red (2 primes), and 1/4 pale red (1 prime). The parental average and the offspring average are both two primes, so there is **no overall regression**, even though we have some individuals — light red — less extreme than either parent, showing **individual regression**.

Consider a case with complete and non-directional dominance, with $A$ and $b$ increasing yield and $a$ and $B$ decreasing yield so that $a\,a$, $B$ – has yield 10 units, $a\,a$, $b\,b$ and $A$ –, $B$ – have yield 20 units and $A$ –, $b\,b$ has yield 30 units. In some cases, the offspring average yield equals the parental average yield, but heterozygosity can cause overall regression. In $A\,a$, $b\,b \times A\,a$, $b\,b$, the parental average yield is 30 units and we get offspring in the ratio 3 $A$ –, $b\,b$ (30 units each): 1 $a\,a$, $b\,b$ (20 units). The offspring average, $110/4 = 27.5$, is therefore less than the average of the extreme parents, so there is **overall regression**. Similarly, if we take two plants of lowest yield, with heterozygosity, $a\,a$, $B\,b \times a\,a$, $B\,b$, average 10 units, the offspring are in ratio 3 $a\,a$, $B$ – (10 units each): 1 $a\,a$, $b\,b$ (20 units), then the offspring average yield is $50/4 = 12.5$, showing regression from the extreme low parental average value. **Dominance** therefore give **overall regression** and **individual regression**, while **additive action** can give **individual regression**, but not overall regression. Gene interactions like epistasis can be similar to dominance between alleles of different loci and can cause regression and transgression in a similar way to dominance.

**Transgression** can come from genetic segregation for genes with dominance, additive action or epistasis, or from environmental effects. Thus, in the above case with non-directional dominance and no linkage, $A\,a$, $B\,b$ (20 units) $\times A\,a$, $B\,b$ (20 units) gives 9 $A$ -, $B$ - (20 units each), 3 $A$ -, $b\,b$

(30 units), 3 *a a, B* - (10 units) and 1 *a a, b b* (20 units). Of 16 offspring, 10 will have the same yield as the intermediate yield parents, 3 show transgression in an upward direction and 3 show transgression in a downward direction.

Suppose we have **additive and cumulative action**, with four unlinked loci controlling height, that the breeder wants transgression to increased height and has pure breeding lines available of differing heights. None is of maximum height, which would be $A' A', B' B', C' C', D' D'$. If there is no environmental variation, then the phenotypes reflect the genotypes as far as the number of primes, but give no direct information on which loci carry the primes. If one does not know the **full genotypes**, one does not know the best crosses to make.

Breeders would normally first try a tall line × a different tall line (say both have 6 primes), intercrossing the F1 to get some recombinant homozygotes in the F2, as loci differing between the parents will be heterozygous in the F1. If it turns out that both tall lines have the same genotype, say $A' A', B' B', C' C', D D$, then there will be no recombination, no segregation and no transgression. If the two tall lines have different genotypes, e.g., $A' A', B' B', C' C', D D$ and $A' A', B B, C' C', D' D'$, then the F1 will be heterozygous at the $B$ and $D$ loci. The F2 will be a mixture of some heterozygotes and both parental genotypes (6 primes), but recombination will also give $A' A', B B, C' C', D D$ (4 primes, showing regression) and $A' A', B' B', C' C', D' D'$ (8 primes, the required tallest type from transgression) homozygotes, so the cross will have been a success.

One can easily postulate genotypes from other crosses which will give the desired F2 genotype with eight primes. They include $A' A', B' B', C' C', D D$, (6 primes) × $A A, B B, C' C', D' D'$, (4 primes), $A' A', B B, C' C', D D$ (4 primes) × $A A, B' B', C C, D' D'$, (4 primes) and even $A A, B' B', C' C', D' D'$ (6 primes) × $A' A', B B, C C, D D$ (2 primes). Therefore, not all tall × tall crosses will give increased height, but some will; and so will some tall × medium, some medium × medium and some tall × short. This shows very clearly that getting the desired recombinants with transgression is not just a matter of the height or the number of primes of the strains which might be crossed, but is also very much a matter of **specific genotypes**, of which particular loci carry the primed alleles. If the obvious crosses of tall × tall do not work, this consideration of genotypes shows that trying some tall × medium, medium × medium or even tall × short crosses could give the desired result of transgression. Genetic transgression requires heterozygosity at some stage, either in the parents or the F1, or there can be no segregation or recombination.

The most **extreme transgression** is possible from the most average genotypes, e.g., $A'A'$, $BB$, $C'C'$, $DD$ (4 primes) × $AA$, $B'B'$, $CC$, $D'D'$ (4 primes), giving F2 ranging from 0 primes to all 8 primes. Of the F2 from selfing the F1 in this cross, 186 out of 256 are expected to show transgression, with 93 taller and 93 shorter than the parents. The most extreme regression of the offspring average comes from having dominance at all segregating loci and the maximum amount of heterozygosity.

For a particular quantitative character in a population, regression and transgression will be **opposing forces** with transgression giving more extreme types, especially from average types, and regression giving less extreme types, especially from extreme types. This is shown for size in Fig. 3.1, with R for regression and T for transgression, and showing the R and T effects for the different types of natural cross, assuming random mating. For example, if one has a small × small cross, as with the left-hand group of three arrows, then if these are not the smallest individuals possible, a few offspring will be even smaller, showing transgression from genetic or environmental effects (light arrow, labelled T), while the main tendency (heavy arrow, labelled R) is some regression, but some offspring could show a lot of regression (rightmost light arrow from small × small). Medium × medium will mainly have medium offspring (thick unlabelled arrow), although some will show transgression (light arrows). There will also be large × large, large × small, large × medium and small × medium crosses. In populations where regression and transgression generally balance out, the size distribution in generation 2 will be similar to that of generation 1, as shown in the generation 2 right-hand distribution, unless the environment differs between generations, when the environment might give a population with reduced individual sizes (shown in the left-hand, dashed, distribution) or increased individual sizes.

We can now ask from what kind of parents in the previous generation do extreme individuals come, e.g., very large individuals? A few would have come from regression from extremely large × extremely large, or from very large × very large, while the majority would have come from transgression

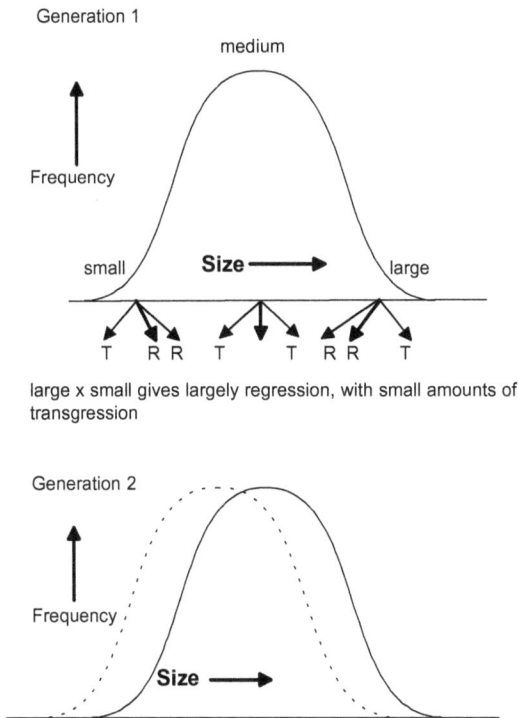

**Fig. 3.1.** **Size distribution in two successive generations** in a random mating population. See text for details. R = regression; T = transgression; thick arrow, average response, e.g., from small × small; thin arrow, extreme response, e.g., from small × small. In generation 2, the thicker curve is the expected one if the environment is similar to that for generation 1, but the thinner one is for a changed environment, giving poorer growth.

from medium × medium, large × small, medium × large, or even a few from small × small. On average, therefore, the parents of large offspring will be less large, and the parents of small offspring will be less small than their offspring.

**Example 3.1.** Compare the average intelligence of university students with that of their parents and that of their own offspring, stating any assumptions which you make.

Assumption 1: university students are on average towards the upper end of the intelligence distribution for their populations. Assumption 2: intelligence is controlled by the environment and by genes with additive action and dominance, and would therefore show regression and transgression. Although some students would have come by regression from even more intelligent parents, the majority would have come from transgression from parents less intelligent on average, as shown in the above discussion about size. As students are near the extreme for intelligence, their own offspring should show regression and on average be less intelligent than their parents. Both assumptions are supported by the available data. The whole answer is framed in terms of averages, and some students would have parents more intelligent than themselves, and some students would have offspring more intelligent than themselves. The answer remains qualitatively the same whether there is positive assortative mating for intelligence, or whether mating is random.

## 3.2. Environmental effects on phenotypes

### 3.2.1. *Sex-limited characters*

Although one usually considers effects of the external environment on a character, there are effects of the **internal environment**, especially hormone levels. In humans, the expression of genes for **hereditary early baldness** depends on the levels of male hormones such as testosterone in the blood. The character is therefore mainly expressed in males, but can be expressed in females with unusually high levels of male hormones, e.g., through glandular malfunction. Characteristics normally only expressed in one sex are called **sex limited**. They must not be confused with sex-linked characters because sex-limited characters are often autosomal. Penis characters in males and ovarian characters in females are obviously sex limited.

### 3.2.2. *Phenocopies, conditional mutants and environmental effects*

The  environment can sometimes change qualitative phenotypes, changing the normal phenotype for one genotype to the phenotype normally associated with a different genotype, causing a **phenocopy**. A woman with genetically dark hair may apply bleach to get fair hair, resembling someone with genetically natural blonde hair. In the fruit fly *Drosophila*, putting silver nitrate in the medium of wild-type larvae gives yellow-bodied adults, resembling the mutant *yellow body*. The genotypes remain unchanged in phenocopies, which will breed as if they had their normal phenotype.

As we have seen, quantitative characters such as grain yield are usually affected by environmental factors such as light, rain, soil, temperature and disease. Whether some qualitative gene mutations are expressed at all may be controlled by the environment. In micro-organisms and higher organisms, one can often obtain such **conditional** mutants. They include **temperature-sensitive** mutants which are expressed (giving the mutant phenotype) at one temperature (**restrictive conditions**) but appear wild-type at another temperature (**permissive conditions**). For example, in the fungus *Neurospora crassa*, the mutant $rib\text{-}1^{ts}$ requires riboflavin for growth at 37° but not at 25°. In the rabbit, $c^h c^h$ gives the Himalayan pattern with off-white body fur but black ears, nose, feet and tail, while $C C$ gives black fur all over. In the fungus and the rabbit, the mutation causes a change in amino acid sequence in an enzyme which is denatured at 37°, but works normally at lower temperatures. The rabbit's extremities are cooler than the body, so the enzyme works in those extremities, giving black pigment there, but is inactivated by the main body temperature. This can be shown by shaving off some body fur, applying a cooling pack for some weeks, and getting black fur in the regrowth. Siamese and some other cat breeds often have dark points on their ears from a similar temperature effect. In maize, the allele "sun-red" gives red colour when parts are exposed to direct sun, but green in shaded parts, while the alternative allele, "not sun-red", gives green in sun and shade.

Environmental effects may be temporary, such as climate one season affecting apple yield; or may be more permanent, such as light intensity during leaf development in some plants, causing leaves to have one of two different types of structure, "sun leaves" or "shade leaves", each anatomically suited to high or low light intensities. Another example of plasticity during development is aquatic plants which have morphologically different submerged and aerial leaves.

### 3.2.3. *Phenotype plasticity*

This is the amount by which different environments can change the expression of a particular phenotypic character in a given genotype. Some quantitative characters are quite plastic like the amount of melanin pigment in a "white" person's skin, often varying reversibly with sun exposure. Other characters may be plastic under fairly extreme environments. Thus, wild-type birch trees might grow to a height of 25 metres in a wide range of environments, but might be stunted, under two metres, if they are grown at high altitudes. The number of noses on a person's face or the number of stamens on an iris flower is constant, so these are non-plastic characters.

Breeders will be very concerned about **phenotype plasticity**, wanting to optimise the environment for the genotypes they are using for their commercial characters. If they are breeding lettuces, for example, their concern will be how different environmental factors affect the yield of leaves. If a genotype reacts unfavourably to a particular local environment, then either the genotype must be appropriately changed, or some other crop should be grown instead. Unusually extreme conditions can be devastating. A high-yielding wheat had been bred for use in Britain, and after being tested for a number of years, it was released for general use. It happened that on a summer that was unusually hot, this variety failed to pollinate properly, giving a very low yield, resulting in farmers ceasing to grow it.

### 3.2.4. *Variable expressivity*

Some alleles are expressed to different extents in different individuals for reasons which are often unknown; they could be genetic, environmental, or both. In humans with the usually dominant allele for **polydactyly**, some individuals have a small extra knob on the outer sides of both hands, showing low expressivity for this disease (Plate 3.1), while others might have two complete extra fingers on each hand and perhaps extra toes, showing high expressivity. The reasons for this difference in expressivity are usually unknown. **Variable expressivity** is a quantitative effect.

### 3.2.5. *Incomplete penetrance*

In contrast to variable expressivity, **incomplete penetrance** is qualitative, when an allele in a particular genotype is sometimes completely expressed and sometimes not expressed at all, like a dagger either penetrating the skin or not penetrating! Most characters have 100% penetrance, but in

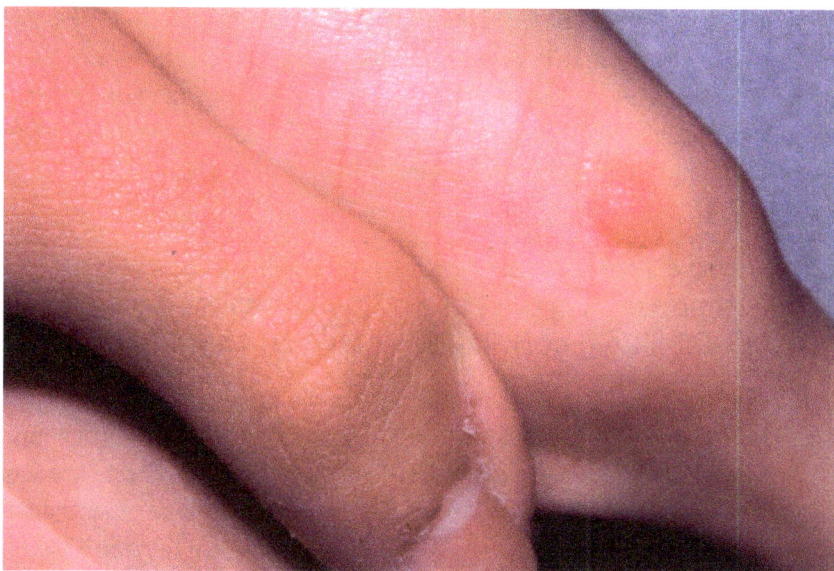

**Plate 3.1.** **Low expressivity human polydactyly**, with only a small knob on one thumb (on right) and an even smaller one on the other thumb, to the left of the top of the vertical crease. Autosomal dominant.

*Drosophila*, the *i i* genotype gives 90% of individuals with fully interrupted wing veins and 10% with the wild-type non-interrupted wing-vein phenotype. The genotypes must be identical in both types of individual, as interrupted × interrupted and non-interrupted × non-interrupted crosses from that *i i* population both give 90% interrupted and 10% non-interrupted progeny. The character is either fully expressed or not expressed at all; perhaps there is some above-threshold concentration of a chemical during development which gives interrupted wings, and those individuals which fail to reach that threshold have normal wings, like those of wild-type genotype lacking that compound completely.

In **man, Huntington's disease** (Subsec. 12.3.1) has 95% penetrance because about 5% of individuals with the dominant allele never develop the symptoms. With **dominant inherited colon cancer**, individuals may sometimes have the gene without showing any symptoms because of incomplete penetrance, but they can pass it on and have affected offspring. Incomplete penetrance is the main reason for a dominant character sometimes "skipping a generation". Multifactorial traits such as schizophrenia can also be considered as showing incomplete penetrance when identical twins are discordant. Gottesman and Bertelsen (1989) showed that for identical twins discordant for schizophrenia, offspring of suffering and of non-suffering twins were equally likely to develop schizophrenia, as both twins of a pair had the same predisposing genes as the other.

## 3.3. Narrow and broad sense heritabilities; equations, estimation and use; use of twins; realised heritabilities; correlations between characters

We saw in Subsec. 1.3.10 that **heritability** is the proportion of phenotypic variance (symbol $Vp$) attributable to genetic variation, with the term applying to characters with continuous variation in a population. For an account of variances, covariances, correlations, linear regressions and other mathematical background, see Cameron (1997). **Narrow sense heritability**, $h^2$, is the proportion of phenotype variance attributable to genes with additive effects, while **broad sense heritability**, $H^2$, is the proportion of phenotype variance attributable to all types of genetic variation. The variances due to different types of genes are $Va$ for genes with additive effects, $Vd$ for genes with dominance, and $Vi$ for genes with epistasis or gene interactions. $Ve$ represents the variance due to the environment and $Vg/e$ represents the variance due to genotype/environment interactions, where different genotypes perform differently in different environments.

$$\textbf{Narrow sense heritability: } h^2 = \frac{Va}{Va + Vd + Vi + Ve + Vg/e}.$$

$$\textbf{Broad sense heritability: } H^2 = \frac{Va + Vd + Vi}{Va + Vd + Vi + Ve + Vg/e}.$$

The equations for heritability therefore contain one or more expressions for genetic variation in the numerator, and for genetic variation and environmental variation in the denominator. A population with more genetic variation than another will therefore tend to have higher heritabilities if its environmental variations are similar. A population with more environmental variation than another will tend to have lower heritabilities if its genetic variation is similar. Of great importance to the breeder is the fact that during selection programmes, heritabilities will decrease as genetic variation is reduced by selecting the better alleles and removing the poorer ones. Genetically uniform populations will have heritabilities of zero for all characters since there is no genetic variation. Heritabilities range from zero, all variation environmental, none genetic, to 1.0, all variation genetic, none environmental. For most populations of most organisms, $Vi$ is relatively low, and $Va$ is larger than $Vd$.

Although one finds published tables of heritabilities for various characters in commercial organisms (e.g., Dalton, 1980), it is essential to remember that while these may be typical values, they may vary widely from herd to herd, depending on how variable the herd is genetically and how variable its environment is. Heritabilities in a herd of pure line sheep or a field of clonally propagated tea (all of one selected genotype) will be much lower than those in a herd of crossbred sheep or a field of seedling tea (which has a wide range of genotypes).

One way to estimate narrow sense heritabilities is from

$$h^2 = \frac{\text{observed correlation between relatives}}{\text{theoretical correlation between relatives}}.$$

The **observed correlation** is the statistical correlation coefficient found for a quantitative trait for a given degree of relationship, often **half-sibs** (which have one parent in common, usually the father, and one parent different). One would take a number of different pairs of half-sibs, measure the character concerned, and work out the correlation coefficient. The **theoretical correlation between relatives** (also known as the coefficient of relationship) is the proportion of genes (alleles) expected to be in common by recent descent. Thus, one parent and an offspring have half their genes in common, as do **full sibs** (full brothers and/or sisters, with both parents in common), while half-sibs have one quarter of their genes in common, as do uncle/niece or aunt/nephew. As long as the parents are not inbred themselves, the theoretical correlation is twice the inbreeding coefficient, F, (see Sec. 6.4) of their offspring. For example, the offspring of half-sibs have F = 1/8, so the theoretical correlation for half-sibs is 0.25.

**Example 3.2.** The correlation between half-sibs for 160-day weight in a herd of pigs is 0.08. What is the heritability?

**Table 3.1.** Coefficients of the variance components in the covariances of relatives.

| Relatives | Estimated proportion of | |
|---|---|---|
| | *Va* | *Vd* |
| **Monozygotic twins** | 1 | 1 |
| **Parent-offspring** | 1/2 | 0 |
| **Full sibs** | 1/2 | 1/4 |
| **Half-sibs** | 1/4 | 0 |
| **Uncle-niece** | 1/4 | 0 |

As half-sibs have 1/4 of their genes in common, the theoretical correlation is 0.25, so $h^2 = \frac{0.08}{0.25} = \mathbf{0.32}$.

There is a series of equations relating the **covariances** for a character between particular types of relatives to estimated proportions of genetic variance due to genes with additive effects, *Va*, to variance due to genes with dominance, *Vd*, and to variance for genes with interactions, *Vi*. For the mathematical background, see Falconer and Mackay (1996), Chap. 9. Assuming that *Vi* and *Vg/e* are relatively small, we get coefficients for the covariances of relatives as shown in Table 3.1.

**Example 3.3.** In a population of chickens, phenotype variance for egg weight, *Vp*, was 16. The covariances between relatives were 1.6 for mother-daughter, 2.8 for sisters, 0.75 for half-sisters. Calculate the heritabilities.

To calculate narrow and broad sense heritabilities, we need to calculate *Va* and *Vd*, and will have to assume that *Vi* and *Vg/e* are negligible so that we can use the values in Table 3.1. We are given the denominator of the heritability equations, the total phenotypic variance, *Vp*. We can calculate *Va* from cases where *Vd* is zero. Parent-offspring (mother-daughter for eggs) covariance = 1/2 *Va* as *Vd* has a coefficient of zero, so 1/2 *Va* = 1.6, and *Va* = 3.2. We can also calculate *Va* from the half-sib data, from which 1/4 *Va* = 0.75, so *Va* = 3.0, in relatively good agreement with the value calculated from parent-offspring. We can take an average of the two *Va* values, 3.1, for use in other calculations. Knowing *Va*, we can use the full sib data to calculate *Vd*, because covariance for full sibs = 1/2 *Va* (when we know $Va$ to be 3.1) + 1/4 $Vd$, so 2.8 = 1/2 (3.1) + 1/4 $Vd$, and $Vd$ = 5.0. Putting these values of $Va$ and $Vd$ into the heritability equations, and assuming that $Vi$ and $Vg/e$ are negligible, we get $h^2 = \frac{3.1}{16} = \mathbf{0.19}$, and $H^2 = \frac{3.1+5.0}{16} = \mathbf{0.51}$. Providing that $Vd$ and/or $Vi$ are not zero, the broad sense heritability is always larger than the narrow sense heritability.

**Twin studies** (Subsec. 12.2.2 and Sec. 12.15) have been used to estimate **human heritabilities**, although there are technical reservations about their accuracy. According to Boomsma *et al.* (2002), if monozygotic (MZ) twins resemble each other more than dizygotic twins (DZ), the heritability ($h^2$) of the phenotype can be estimated as twice the difference between MZ and DZ correlations. For example, the MZ and DZ correlations for depression are about 0.4 and 0.2, so $h^2 = (0.4 - 0.2) \times 2 = 0.4$. For taking up smoking in adolescence, MZ and DZ correlations are 0.9 and 0.7, so $h^2 = 0.4$, but the high correlation between DZ twins shows the importance of a shared environment. The proportion of phenotypic variance due to a **shared environment** is the difference between the twin correlation ($r_{MZ}$ for MZ twins, $r_{DZ}$ DZ twins) and the part explained by heritability, which is $r_{MZ} - h^2$ for MZ twins and $r_{DZ} - h^2/2$ for DZ twins, so is $0.9 - 0.4 = 0.5$ for MZ twins and $0.7 - 0.2 = 0.5$ for DZ twins. Boomsma *et al.* (2002, Fig. 2) give extensive figures for human heritabilities from Dutch twin studies. For females, **estimated percentage heritabilities** include alcohol use 75%, HDL-cholesterol 72%, LDL-cholesterol 70%, sports participation 50%, smoking 35% plus 55% of variance due to shared environmental factors, about 55 to 60% for neuroticism, experience-seeking, anxiety, depression and boredom susceptibility, and values for intelligence being age-dependent, rising from 30% at age 5 to 85% at age 27.

Hartl and Jones (2005) state that the variance for MZ pairs = $Ve$ and the variance for DZ pairs = $Ve + Vg/2$, from which $H^2$ can be estimated. Percentage values for this **human broad sense heritability** which they give include longevity 29%, height 85%, weight 63%, verbal ability 63%, numerical ability 76%, memory 47% and sociability index 66%.

The results of **selection experiments** using observed values of population average, selected parent and offspring average, give **realised heritabilities** (see Sec. 5.11).

The most important **use of heritabilities** is as a guide to whether it is worth running selection programmes to improve a character, bearing in mind the costs of running a **selection programme** and the likely commercial returns from higher yield, better quality or lower costs. For tables of heritabilities, repeatabilities, breeding aims and hybrid vigour in farm animals, see Dalton (1980). Suppose we take some typical narrow sense heritabilities from the 1950s for American dairy cattle, 0.01 for fertility, 0.3 for milk yield and 0.6 for milk butterfat. The actual figures would vary widely between herds, depending on how genetically variable each herd was and on the variability of their environments. We could predict that heritabilities would be lower today, as in 50 years the cattle breeders would have selected better genotypes, getting rid of poorer alleles, thereby reducing genetic variation.

We see that **fertility** had a very low heritability, 0.01, so that only 1% of variation was genetic and 99% was environmental. With so little genetic variation, it would probably not be worth running a major selection programme to improve fertility since there would be so little genetic response, though one would **cull** (kill off or not breed from) any individuals of low fertility. It would be more useful trying to improve the environment (e.g., the feeding regime and disease control) to increase fertility. As high fertility has been encouraged by natural selection before cattle were domesticated and by man since domestication, it is no surprise that heritability for fertility is low.

In dairy cattle, there must have been natural selection in the wild for moderate **milk yields**, enough to feed the calves, but man has selected for increased yields. If milk yield heritability is 0.3, 30% of the variation is genetic; since milk yield is of supreme importance in dairy cattle, it would certainly be worth running a selection programme for increased milk yield with the prospect of a good response, unless the present heritability is much lower than 0.3 in the herd concerned.

There would have been natural selection for a moderate **milk butterfat** content, with selection for increased values since domestication. In many countries, there are legal minimum butterfat levels for particular types of milk. The breed of cattle and the major uses of their milk — for drinking, cream, butter, cheese, yoghurt, etc. — are relevant to the desired butterfat level for commerce. Reduced-fat milks, such as semi-skimmed, are now important. If the heritability is still high, approaching 0.6, one could expect an excellent response to the selection for increased milk butterfat, in grams of fat per litre of milk, as most of the variation is genetic.

We must, however, be conscious that selecting for one character can affect other characters if there are **correlations** between them. If by selection for higher milk butterfat (in grams per litre of milk), one gets more metabolic energy diverted to making butterfat, there would be less energy left for making other milk components. One might therefore predict a negative correlation between milk yield (litres per lactation) and milk butterfat (g/l). One way of getting a response to selection for increased milk butterfat in g/l might be for the cow to produce the same total amount of butterfat per day, while reducing the volume of milk. Breeders would try to select simultaneously for increased butterfat and increased or at least maintained levels of milk yield. If there were to be a negative correlation in spite of dual selection, the breeder would need to work out whether the economic value of the increased butterfat outweighed the reduced value of milk volume.

One cannot always predict **correlations between characters**. Suppose you have successfully selected for **increased leaf length** in lettuce or tobacco, what might happen to **leaf width** and hence to **leaf area**? It would all depend on how the genes controlling leaf width acted. If the segregating genes just affected the number of cell divisions in the leaf meristems in all directions, then selecting for increased length would also select for increased leaf width, giving a positive correlation between length and width, and perhaps a large overall increase in leaf area. If the genes only affect cell divisions in the plane giving length, with no effect on divisions in other planes, then as leaf length increases, there will be no change in leaf width, such that there is no correlation between length and width, and leaf area will increase in proportion to the increase in length. If the genes selected change the number of divisions in different planes, with no effect on the total number of divisions, then as one gets more divisions in the plane giving length, one will get proportionally fewer divisions in the plane giving width, and a negative correlation between the two characters, length and width, perhaps with no overall change in leaf area. There could thus be a positive or a negative correlation, or no correlation, between leaf length and leaf width.

**Correlations between characters** can result from **physiology**, e.g., negative correlations between milk yield and butter fat, or between bean weight and the number of beans per plant, because both characters compete for the same physiological resources. Correlations could also be due to the **linkage** between syntenic loci which are in linkage disequilibrium, but such correlations would be broken in time by recombination. Unbreakable correlations could arise from **pleiotropy**, where two characters, even superficially unrelated ones such as deafness and blue eyes in white cats, are caused by the same allele. Some **examples of correlations** are: in yeast, red colony colour with a requirement for adenine; in cereals, small leaves with many leaves per plant, large grains with few grains per ear, rapid autumn growth with poor winter hardiness; in alfalfa, thick leaves with a high photosynthetic rate; in maize, poor seedling vigour with late maturity. In sheep, there are positive correlations between resistance to fly-strike and to fleece rot, and between food intake and wool production, and a negative correlation between hair follicle density and wool fibre diameter. Cows with the largest udders and biggest milk yield are the most susceptible to disease.

## 3.4. Genotype value, phenotype value and breeding value

For a particular character, an individual's **genotype value** is its value as judged by its genotype; its **phenotype value** is its value as judged by its phenotype, and its **breeding value** is its value as judged by the genes it hands on, which is judged by the mean performance of its progeny, allowing for the merits of the individuals it was crossed with. If the character is entirely controlled by genes with additive action and if there are no environmental effects, then an organism's genotype, phenotype and breeding values are identical.

We can take the case of height from Sec. 2.3, determined by three unlinked loci with additive and cumulative action, a base height of 50 cm and each prime adding 2 cm. If there are no environmental effects, we get a simple linear relation between genotype and phenotype, as shown in Fig. 3.2. Each genotype has only one phenotype, ranging from 50 cm with no primes to 62 cm with six primes. Selecting the tallest individuals, e.g., those of 61 cm or more, is extremely efficient as only individuals with six primes are chosen, which have the best breeding value for height.

If we introduce some environmental effects now, each genotype can then have a range of phenotypes giving continuous variation. This is shown in Fig. 3.3, where in addition to the phenotypes previously shown

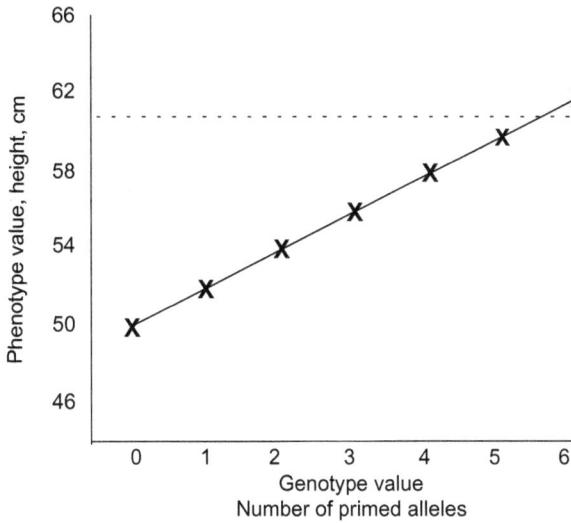

**Fig. 3.2.** **Relation between phenotype value and genotype value with additive action and no environmental effects**. The dashed line indicates the effects of selecting the tallest individuals, of 61 cm or more.

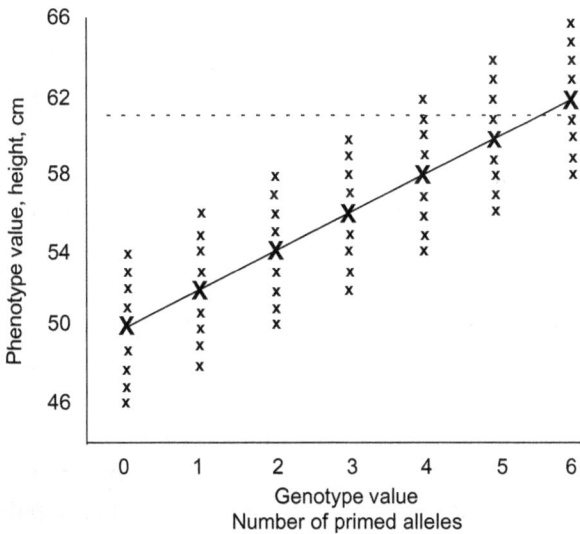

**Fig. 3.3.** As for Fig. 3.2 but **with environmental variation**, so that each genotype now has, in addition to its original phenotype (large X), a range of possible phenotypes (small x's).

(large "X"s), each genotype has other possible phenotypes shown as small "x"s, ranging equally above and below the original points.

If environmental effects are symmetric, giving equal deviations upwards and downwards, then the large X represents the mean phenotype value for each genotype. The **phenotype value**, $P$, for any individual is its genotype value, $G$, plus an **environmental deviation**, $E$, which may be positive, zero or negative. In Fig. 3.3. points on the diagonal line have $E = 0$; the points above the diagonal line have positive environmental deviations, and those below the line have negative deviations. For a whole population, it is usually assumed that the mean environmental deviation is zero, with positive deviations equalling negative ones. For a whole population, the mean phenotype value then equals the mean genotype value and can be used to estimate the latter.

If we consider the **efficiency of selection** in the presence of environmental variation, we see from Fig. 3.3. that selecting the tallest phenotypes, of 61 cm or more, as before, results in selecting individuals for breeding whose genotypes have four, five or six primes, instead of only those with six primes, such that the mean breeding value of the selected individuals will be less than six primes. In addition, Fig. 3.3. shows that we will fail to select some individuals with the ideal genotypes of six primes, as some have phenotypes in the 58 to 60 cm range. As we have seen before, the greater the amounts of environmental variation, the lower the heritabilities, and the less reliable is the selection based on phenotypes. From Fig. 3.3. it can be seen that one could still select only individuals with six primes by taking individuals of heights between 64 and 66 cm, although that is inefficient as it fails to select the majority of individuals with six primes.

Ideally one selects on an individual's breeding value judged by the mean value of the offspring when it (usually males, as this takes too long to do for females) is crossed to a typical range of individuals of the opposite sex. That is **progeny testing** (Sec. 5.7). The tested individual's breeding value is twice the mean deviation of the progeny from the population mean; the deviation is doubled because the individual provides only half the genes in the progeny. While the breeding value could be given in absolute units, it is usual to express it in terms of deviations from the population mean. Over a whole population, the mean breeding value = the mean phenotype value = the mean genotype value.

## 3.5. Estimated Breeding Values (EBV) and Best Linear Unbiased Predictions (BLUP)

In the UK, the Meat and Livestock Commission has a Beefbreeder Service, with a method called **BLUP (Best Linear Unbiased Prediction)** for more accurate estimates of the genetic merit of pedigree cattle, which allows for comparison across herds. BLUP can help in selecting a beef bull to breed replacement heifers with higher maternal performance and higher profitability, possibly worth several thousand pounds over the lifetime of a bull's daughters (based on each bull producing 10 heifers per year by natural service, over four years of breeding).

**BLUP** is used with different models to predict breeding values and to estimate environmental effects. It is most often used to predict sire breeding values from progeny testing (the sire model), or to predict breeding values from animals with repeated records (the repeatability model), or to predict breeding values for all animals in a pedigree (the individual animal model): see Cameron (1997) for equations and examples. In BLUP, the B (Best) refers to maximising the correlation between the true breeding value and the predicted breeding value; the L (Linear) refers to predicted breeding values being linear functions of the observations; the U (Unbiased) refers to the estimates of fixed effects being unbiased and the unknown true breeding values being distributed around the predicted breeding values; the P (Prediction) refers to the procedure predicting the true breeding values.

The home-bred beef replacement heifers should make a direct contribution to the genetic potential of their calves in terms of growth rate, muscling score, etc. They should successfully rear a calf each year through good fertility, ease of calving, and good milk yield. The cow makes as much genetic contribution to the calf as does the bull, as she passes on half her genes. For example, a purebred cow with an **estimated breeding value** (EBV) of +20 kg for 400-day weight will pass on an average of +10 kg weight at 400 days, compared with calves sired by the same bull out of cows with an EBV of 0. The **important direct EBVs** are those for birth weight, 200-day growth, 400-day growth, backfat depth, muscling score and beef value. The BLUP analysis of pedigree Limousin bull "**Grahams Unbeatable**" for Estimated Breeding Values in 2004 was: Gestation Length +1.4 days, Calving Ease −5.0, Birth Weight +3.1 kg, Calving Value

LM-2C, 200-Day Milk Yield 0 kg; 200-Day Growth* +51 kg; 400-Day Growth* +88 kg; Muscling Score +1.5 points, Muscle Depth* +5.7 mm, Fat Depth +0.3 mm, Beef Value LM44. Values marked * are in the top 1% for that breed. He fetched the top price of 35,000 guineas (£36,750) at the ILC Elite Sale in Carlisle in 2004, weighed 761 kg at 400 days, and his semen sold at £70 a straw in 2005. It is more difficult to estimate breeding values in crossbred cows, as an unknown part of their performance will be from hybrid vigour.

The early growth of the suckled calf, and therefore its 200-day weight, depends on its own genetic potential for growth and on the milk yield and milk quality of its mother. In the BLUP scheme, the cow's maternal merit is measured as the cow's ability to rear a calf to 200 days of age; e.g., a cow with a +15 EBV for 200-day milk will wean calves 15 kg heavier on average, than a cow with 0 EBV for this trait.

Some of this information is taken from Simm *et al.* (1993). Their final advice was: select a shortlist of bulls with the highest positive EBVs for 200-day milk, delete bulls with EBVs for 200-day growth of below +10 kg, EBVs for birth weight of above +2 kg, and EBVs for beef value of below £10; visually assess all bulls remaining on the shortlist in terms of physical fitness, rejecting any unsound animals. Selecting for below average birth weight is done to reduce calving difficulties. For estimated breeding values in pigs, ses Subsec. 13.7.13.

Different types of selection and their suitabilities for different heritability ranges are covered in more detail in Chap. 5, including the use of selection indices (Subsec. 5.10.3). Further aspects of selection in plants and animals are given in Chap. 13, in humans in Chap. 12, and in fungi in Chap. 18.

## Suggested Reading

Most of the book references at the end of Chapter 2 also apply here.

Boomsma, D., A. Busjahn and L. Peltonen, Classical twin studies and beyond. *Nat Genet Rev* (2002) **3**: 872–882.

Cameron, N. D., *Selection Indices and Prediction of Genetic Merit in Animal Breeding*. (1997) CAB International, Wallingford.

Dalton, D. C., *An Introduction to Practical Animal Breeding*. (1980) Granada, London.

Falconer, D. S. and T. F. C. Mackay, *Introduction to Quantitative Genetics*, 4th ed. (1996) Longman, Harlow.

Gottesman, I. I. and A. Bertelsen, Confirming unexpressed genotypes for schizophrenia. *Arch Gen Psychiat* (1989) **46**: 867–872.

Hartl, D. L. and E. W. Jones, *Genetics. Analysis of Genes and Genomes*, 6th ed. (2005) Jones and Bartlett, Sudbury, Maryland.

Simm, G., R. E. Crump and B. G. Lowman, *On Choosing a Beef Bull to Breed Replacement Heifers*. SAC Technical Note T336. (1993) Scottish Agricultural College, Edinburgh.

Strickberger, M. W., *Genetics*, 3rd ed. (1985) Collier Macmillan, London.

# Chapter 4

## Population Genetics: Allele Frequencies, Genetic Equilibria, Population Mixing, Genetic Drift and Gene Flow

### 4.1. Introduction

**Population genetics** is extremely important to the understanding of applied genetics, where **plant and animal breeding** often consists of the manipulation of populations (e.g., Pirchner, 1983; Brown *et al.*, 1990). In **human genetic counselling**, a knowledge of allele frequencies is essential in calculating the risks of having affected children for various inherited diseases.

Unless otherwise stated, this chapter refers to diploid Eukaryotes. In Mendelian genetics, one usually considers crosses of pure-breeding individuals, with the F1 being selfed or test-crossed, with F2 ratios being examined as in Chap. 2. In population genetics, we normally consider populations in which several different crosses may be occurring randomly, e.g., $AA \times aa, AA \times AA, AA \times Aa, aa \times aa, Aa \times Aa$ and $Aa \times aa$. Even if we start with only two genotypes, they may not be equally frequent and must be considered as crossing with their own genotype as well as with the other genotype, unless sex or incompatibility barriers prevent that from happening. Mating may be non-random, with inbreeding, outbreeding or assortative mating.

**Allele frequencies** may vary from 0 to 1.0 for any allele in a population, but the sum of the allele frequencies at one locus is always 1.0. In a diploid population of $N$ individuals, the lowest non-zero allele frequency is $\frac{1}{2N}$; in a population of only four individuals, the lowest possible allele frequency greater than zero is therefore 1/8, which is 0.125. **Genotype frequencies** must also sum to one, as must **phenotype frequencies**, and checking that they do so in problems is a useful safeguard against arithmetical errors. While Mendelian genetics is usually concerned with set-piece crosses of different homozygotes over two or three generations and the ratios produced in the F2 or F3, in population genetics, we may be concerned with many generations, with frequencies of alleles, genotypes and phenotypes, with whether mating is random, and with the effects of selection, chance, population size, linkage, mutation, recombination and migration.

## 4.2. The Hardy–Weinberg equilibrium for one locus and two loci; linkage disequilibrium, and population mixing

### 4.2.1. *One locus, two alleles*

The basis of gene frequency analysis is the **Hardy–Weinberg equilibrium**, independently derived in 1908 by Hardy in England and Weinberg in Germany, deduced from Mendelian genetics. If a large population has $p$ as the frequency of one allele at a locus, e.g., $A$, and $q$ as the frequency of the alternative allele, e.g., $a$, so that $p + q = 1$, then the genotypes will reach in one generation and remain at the frequencies $p^2$ **for** $A\,A$, $2pq$ **for** $A\,a$, and $q^2$ **for** $a\,a$. One can obtain this from a simple table, combining the gametes at random, as in Table 4.1.

Obtaining this equilibrium depends on there being only two alleles in the population for the locus concerned: cases of multiple alleles and more than one segregating locus are considered later. The population needs to be large or chance fluctuations in allele frequency may affect genotype frequencies as described later under genetic drift. Achieving Hardy–Weinberg equilibrium also requires that mating be random, not assortative, and without inbreeding

**Table 4.1.** Derivation of the Hardy–Weinberg equilibrium frequencies from random mating.

| Allele frequencies in male gametes: | $p$ of $A$ | $q$ of $a$ |
|---|---|---|
| **Allele frequencies in female gametes** | **Zygotes and their frequencies** | |
| $p$ of $A$ | $A\,A,\ p^2$ | $A\,a,\ pq$ |
| $q$ of $a$ | $A\,a,\ pq$ | $a\,a,\ q^2$ |

or outbreeding; that there are no selection, viability or fertility differences, and that migration and mutation frequencies are negligible.

The genotype frequencies in one generation depend on the previous generation's allele frequencies rather than directly on the latter's genotype frequencies. Thus, if one has a population with 100 $AA$ and 100 $aa$ individuals mating randomly ($AA$ with $AA$ and equally with $aa$, and $aa$ with $aa$ and equally with $AA$), it gives the same genotype frequencies in the next generation as a population of 200 $Aa$ individuals, because both populations have the same allele frequencies, 0.5 for both $A$ and $a$. The genotype frequencies will be $p^2$ for $AA = (0.5)^2 = \mathbf{0.25}$; $2pq$ for $Aa = 2 \times 0.5 \times 0.5 = \mathbf{0.5}$; and $q^2$ for $aa = (0.5)^2 = \mathbf{0.25}$. The genotype frequencies, like the allele frequencies, total 1.0, because they sum alternatives. Hardy–Weinberg analysis enables one to calculate allele and genotype frequencies from phenotype frequencies, even in the presence of complete dominance.

> **Example 4.1.** In maize, the character rust-susceptible, *rp* (to the fungus *Puccinia sorghi*) is recessive to resistance to that fungus, *Rp*. If five maize seedlings are susceptible to rust in a population of 50,000 seedlings, what are the allele and genotype frequencies?

As susceptibility is recessive, the susceptible genotype must be *rp rp*. Let the *rp* allele have frequency $q$, and the *Rp* allele have frequency $p$, with $p + q = 1.0$. If random mating and the other conditions of the Hardy–Weinberg equilibrium are met, then *rp rp* has frequency $q^2$. As we are told that the recessive phenotype has a frequency of 5 in 50,000, $q^2 = \frac{5}{50,000} = 0.0001$, so $q = \sqrt{0.0001} = 0.01$. As $p + q = 1.0$, $p = 1.0 - 0.01 = 0.99$. The allele frequencies are therefore **0.99** for the rust-resistance allele, *Rp*, and **0.01** for the rust-susceptibility allele, *rp*. We know that genotype *rp rp* has frequency **0.0001**. *Rp Rp* has frequency $p^2 = (0.99)^2 = \mathbf{0.9801}$, and *Rp rp* has frequency $2pq = 2 \times 0.99 \times 0.01 = \mathbf{0.0198}$. The three calculated genotype frequencies sum up to 1.0, as expected.

Even though there is complete dominance of resistance to susceptibility, with *Rp Rp* and *Rp rp* having identical phenotypes, just knowing the frequency of the recessive homozygote enables us to calculate the frequencies of these different genotypes. With such a low frequency of the recessive allele, the dominant homozygote is much more frequent (49.5 times) than the heterozygote. As the recessive homozygote is in turn much rarer (198 times) than the heterozygote, we see the very important point that rare recessive alleles are present mainly in the heterozygote, where they are protected by dominance from selection, rather than in the recessive homozygote where they are exposed to selection.

## 4.2.2. *One locus, more than two alleles*

We can easily extend the basic Hardy–Weinberg analysis to **multiple alleles** at a locus. Let the alleles be $a^1$, $a^2$, $a^3$ to $a^n$, with respective allele frequencies $p, q, r$ to $n$, which must add up to 1.0, so $p + q + r + \cdots + n = 1.0$. Many different random matings could occur between homozygotes and heterozygotes. By using a Punnett chequer board as in Chap. 2, Table 2.1, we could show that random mating gives the frequency of any homozygote as the square of its allele frequency, and of any heterozygote as twice the product of the two allele frequencies. Thus, $a^1 a^1$ has equilibrium frequency $p^2$ and $a^3 a^3$ has frequency $r^2$; $a^1 a^3$ has frequency $2pr$ and $a^2 a^3$ has frequency $2qr$.

> **Example 4.2.** A human population has ABO blood group allele frequencies of $i^o = 0.5$, $i^A = 0.4$, $i^B = 0.1$. Alleles $i^A$ and $i^B$ are dominant to $i^o$, and show co-dominance to each other. What are the expected blood group phenotype frequencies?

As $i^o$ is recessive, everyone of blood group O must be $i^o i^o$, with a frequency of $(0.5)^2 = \mathbf{0.25}$.

Blood group A has two genotypes, $i^A i^A = (0.4)^2 = 0.16$, and $i^A i^o = 2 \times 0.4 \times 0.5 = 0.4$, totalling **0.56**.

Blood group B has two genotypes, $i^B i^B = (0.1)^2 = 0.01$, and $i^B i^o = 2 \times 0.1 \times 0.5 = 0.1$, totalling **0.11**.

Blood group AB has one genotype, $i^A i^B$, frequency $2 \times 0.4 \times 0.1 = \mathbf{0.08}$. The four phenotype frequencies sum up to 1.0, as expected.

## 4.2.3. *Allele segregation at two loci and the importance of recombination frequencies*

Although Hardy–Weinberg equilibrium is reached from states of disequilibrium in one generation of random mating for two or more alleles at one locus, there is a gradual approach to equilibrium over many generations for alleles at **two different loci**, because recombination between the loci is involved in decreasing the disequilibrium, and recombination has a maximum frequency of 50%. The maximum reduction in the departure from equilibrium in one generation is therefore 50%, with smaller reductions for linked genes. **Linkage equilibrium** occurs when the alleles of different loci in a population are present in gametes in proportion to the product of the allele frequencies, e.g., the frequency of gamete $A, B$ = the frequency of allele $A \times$ the frequency of allele B, and **linkage disequilibrium** is the amount of departure from this state.

Let $A$ and $a$ be alleles at one locus, with frequencies $p$ and $q$, with $p + q = 1.0$, and let $B$ and $b$ be alleles at a second locus, with frequencies $r$ and $s$, with $r + s = 1.0$. The **equilibrium frequencies for a genotype** are calculated by multiplying the separate genotype frequencies for the two loci, e.g., $A\,A, B\,B = p^2 \times r^2$; $A\,a, B\,b = 2pq \times 2rs$; $A\,A, B\,b = p^2 \times 2rs$. The **equilibrium gamete frequencies** will be $A, B = pr$; $a, b = qs$; $A, b = ps$, and $a, B = qr$. The **recombination fraction** (the recombination frequency as a percentage divided by 100) is $c$, and $d$ is the **departure from equilibrium**, measured by the product of the frequencies of the two coupling allele combinations in the gametes ($A, B$ and $a, b$) minus the product of the frequencies of the two repulsion allele combinations in the gametes ($A, b$ and $a, B$):

$$d = (A, B \times a, b) - (A, b \times a, B).$$

Departure $d$ is zero at equilibrium, when the products of the two types of gametes, coupling and repulsion, are equal; it can have a positive or a negative sign when there is linkage disequilibrium. Any departure from equilibrium is reduced in each generation by a proportion corresponding to the recombination fraction, $c$. Hence, with $d_t$ as the departure from equilibrium at generation $t$ and $d_{t-1}$ as the departure in the previous generation, we get:

$$d_t = d_{t-1} \times (1 - c).$$

For two unlinked loci (50% recombination), $c = 0.5$, so $d_t = d_{t-1} \times (1 - 0.5) = 0.5d_t$, halving the disequilibrium each generation. For two closely linked loci (5% recombination), $c = 0.05$, so $d_t = d_{t-1} \times (1 - 0.05) = 0.95\,d_{t-1}$, with a very small reduction in disequilibrium in each generation.

**Example 4.3.** In part of the USA the frequency of the O blood group allele is 0.62 and that of the Rhesus positive factor is 0.7. What is the expected frequency of people who are blood group O, Rhesus positive?

As blood group O are homozygous recessive, they will be $(0.62)^2 = $ **0.3844**. Rhesus positive is dominant, so Rhesus positives will include $Rh\,Rh = (0.7)^2 = 0.49$ and $Rh\,rh = 2 \times 0.7 \times 0.3 = 0.42$, so the total for Rhesus positive is **0.91**. The equilibrium frequency of people who are group O and Rhesus positive is therefore the product of the two phenotype frequencies, $0.3844 \times 0.91 = \mathbf{0.35}$.

We can multiply phenotype frequencies as well as genotype frequencies for the two loci. The allele frequency of Rhesus negative, *rh*, was calculated as one minus the frequency of *Rh*. As well as calculating the numerical answer, one could add that it depended on all the assumptions of the Hardy–Weinberg equilibrium at each locus, and on the population having reached equilibrium for the two loci. The USA has had much immigration and incomplete mixing of different racial groups, so this population may not have reached equilibrium for the two loci. As the ABO locus is on chromosome 9 and the Rhesus locus is on chromosome 1, the loci are unlinked and any disequilibrium will be halved in each generation.

### 4.2.4. What happens when two pure-breeding but different populations mix?

It helps to clarify many genetical issues, including the ones connected with the equilibrium for two loci, if we consider what happens when two pure-breeding but **different populations mix**, e.g., two small herds of cattle.

**Example 4.4.** Suppose 15 males and 15 females of cattle herd 1, all of genotype *A A, B B*, are mixed with 5 males and 5 females of herd 2, all *a a, b b*, with random mating in the mixed herd, which is counted as generation zero. The herd size then stays constant at 40 animals, with random mating each generation. What happens to the genotype, gamete and allele frequencies, and to the departure from equilibrium, in successive generations?

The **initial allele frequencies** ($p$ for $A$, $q$ for $a$, $r$ for $B$, $s$ for $b$) are therefore 0.75 for $A$ and $B$, and 0.25 for $a$ and $b$, and the initial gamete frequencies in the mixed herd are 0.75 for $A$, $B$ (all from herd 1) and 0.25 for $a$, $b$ (all from herd 2). With no repulsion gametes produced in generation zero, the **initial departure from equilibrium** in their gametes, $d_0 = \{(A, B) \times (a, b)\} - \{(A, b) \times (a, B)\} = (0.75 \times 0.25) - (0.0 \times 0.0) =$ **0.1875**.

I run a **class simulation experiment** on this mixing, using different coloured beads to represent the four gamete genotypes. The role of chance in the population usually causes deviations from the expected results and helps to instruct students in the role of chance. For this book, let us use theoretical results, ignoring chance. With generation 0 having gamete frequencies of 0.75 for $A$, $B$ and 0.25 for $a$, $b$, the expected genotypes frequencies in generation 1 calves and adults for $A\,A$, $B\,B$ are $(0.75)^2 = 0.5625$, for $a\,a$, $b\,b$ are $(0.25)^2 = 0.0625$, and for $A\,a$, $B\,b$ are $2 \times 0.75 \times 0.25 = 0.375$; we can write these particular $A\,a$, $B\,b$ individuals as *AB/ab*, as all these heterozygotes have come from gamete fusions of coupling gametes, $A$, $B + a$, $b$; none of the generation 1 double heterozygotes is from the fusion of repulsion gametes to give *Ab/aB*, as there are no repulsion gametes yet.

When the adults of **generation 1** produce gametes, the homozygotes will only produce one type of allele, so the 0.5625 $A\,A$, $B\,B$ will contribute 0.5625 $A\,B$ gametes to the gamete pool and the 0.0625 $aa$, $bb$ will contribute 0.0625 $ab$ gametes, as we assume no fertility or viability differences. In the first instance, let us assume that the two loci are **unlinked, with 50% recombination**. The **double heterozygotes**, 0.375 $A\,a$, $B\,b$, will produce equal frequencies of the four types of gamete with no linkage, that is $\frac{0.375}{4} = 0.09375$, each of $A$, $B$; $a$, $b$; $A$, $b$; $a$, $B$. The total gamete frequencies from homozygotes and the double heterozygotes will then be 0.65625 of $A$, $B$, 0.15625 of $a$, $b$, 0.09375 of $A$, $b$, and 0.09375 of $a$, $B$, adding up to 1.0. The amount of disequilibrium is now $d_1 = (0.65625 \times 0.15625) - (0.09375 \times 0.09375) = $ **0.09375**, exactly half of $d_0$ because we have 50% recombination, so the disequilibrium has been reduced by proportion $(1 - c) = 0.5$ per generation.

The adults of **generation 2** can be worked out by combining those gametes at random, e.g., using a chequer board. We expect $(0.65625)^2$ of $A\,A$, $B\,B = 0.4307$, etc., however, in addition to homozygotes and coupling

(*AB/ab*) double heterozygotes, we will now get single heterozygotes, e.g., *A a, B B* and *a a, B b*, from combinations of coupling and repulsion gametes, and we can get repulsion (*Ab/aB*) double heterozygotes. The chance of getting a repulsion double heterozygote is twice the frequency of *A, b* gametes times the frequency of *a, B* gametes — the factor of two arises as the double heterozygote could come from sperm *A, b* + egg *a, B* or sperm *a, B* + egg *A, b*. This gives 0.0176 as a fraction, or 0.7 of a cow out of 40. The *a a, B B* class would only be 0.4 individuals out of 40. In a herd of 40 cows, some of the rarer genotypes would probably be absent, as one cannot have viable fractions of a cow.

We can now work out the gamete frequencies produced by generation 2 adults. For double homozygotes and the two double heterozygotes, we can proceed as in generation 1, but now we have some **single heterozygotes**. In these, the recombination frequency is irrelevant; they will segregate at the heterozygous locus in 1:1 ratio, and as the two alleles are identical at the other locus, it does not matter whether there is recombination between the two loci. For example, *A a, B B* will give 1 *A B*: 1 *a B* gametes, whether there is recombination between the two loci or not, and irrespective of the recombination frequency.

We can continue this for many generations, when we will get the calculated equilibrium frequencies, e.g., $A A, B B = p^2 \times r^2 = (0.75)^2 \times (0.75)^2 = 0.3164$. Tables 4.2 and 4.3 show the changes in expected genotype and gamete frequencies in different generations.

Suppose that we repeat the experiment with the same initial numbers of animals, but now let the two loci be **linked, with 5% recombination** at meiosis. There will be no difference in the generation 0 gamete frequencies or in the generation 1 genotype frequencies, but when the double heterozygote *AB/ab* forms gametes, 95% will be parental (0.475 of *A, B* and 0.475 of *a, b*, as fractions) and 5% will be recombinant (0.025 of *A, b* and 0.025 of *a, B*). In generation 2, we will also get some *Ab/aB* double heterozygotes, giving gamete frequencies of 0.025 of *A, B*, 0.025 of *a, b*, 0.475 of *A, b*, 0.475 of *a, B*. If there is no linkage, we get the same gamete frequencies from coupling and repulsion double heterozygotes, but with linkage, it is crucial to record **which type of double heterozygote** is being considered, as the two types have different gamete frequencies. If one compares the cases of 50% and 5% recombination in Tables 4.2 and 4.3, we see large

# Applied Genetics of Humans, Animals, Plants and Fungi

94

**Table 4.2.** The consequences for **genotypes frequencies** of mixing 30 *A A, B B* animals with 10 *a a, b b* animals, from another population, with random mating; (i) with 50% recombination; (ii) with 5% recombination. One would not get fractions of a cow in reality, but these are calculated values.

| Generation | Genotypes and their frequencies | | | | | | | | | |
|---|---|---|---|---|---|---|---|---|---|---|
| | A A, B B | A a, B B | a a, B B | A A, B b | A A, b b | A B/ a b | A b/ a B | A a, b b | a a, B b | a a, b b |
| 0* | 0.75 | | | | | | | | | 0.25 |
| 1* | 0.563 | | | | | 0.375 | | | | 0.063 |
| (i), with 50% recombination | | | | | | | | | | |
| 2 | 0.4307 | 0.1230 | 0.0088 | 0.1230 | 0.0088 | 0.2051 | 0.0176 | 0.0293 | 0.0293 | 0.0244 |
| Numbers out of 40 | 17 | 5 | 0.4 | 5 | 0.4 | 8 | 0.7 | 1 | 1 | 1 |
| (ii), with 5% recombination | | | | | | | | | | |
| 2 | 0.5485 | 0.0139 | 0.0001 | 0.0139 | 0.0001 | 0.3564 | 0.0002 | 0.0045 | 0.0045 | 0.0579 |
| Numbers out of 40 | 22 | 0.6 | 0.004 | 0.6 | 0.004 | 14 | 0.007 | 0.2 | 0.2 | 2 |
| Many* | 0.3164 | 0.2109 | 0.0352 | 0.2109 | 0.0352 | 0.0703 | 0.0703 | 0.0234 | 0.0234 | 0.0039 |
| Numbers out of 40 | 13 | 8 | 1 | 8 | 1 | 3 | 3 | 0.9 | 0.9 | 0.2 |

* The values for generations 0, 1 and "many" apply to both recombination frequencies.

**Table 4.3.** The consequences for **gamete frequencies of mixing 30 *A A, B B* animals with 10 *a a, b b* animals**, from another population, with random mating; (i) with 50% recombination; (ii) with 5% recombination.

| Generation | Gamete frequencies from adults of that generation | | | |
|---|---|---|---|---|
| | A B | a b | A b | a B |
| 0* | 0.75 | 0.25 | 0 | 0 |
| (i), 50% recombination | | | | |
| 1 | 0.6563 | 0.1563 | 0.0938 | 0.0938 |
| (ii), 5% recombination | | | | |
| 1 | 0.7406 | 0.2406 | 0.0094 | 0.0094 |
| (i), 50% recombination | | | | |
| 2 | 0.6094 | 0.1094 | 0.1406 | 0.1406 |
| (i), 5% recombination | | | | |
| 2 | 0.7317 | 0.2317 | 0.0183 | 0.0183 |
| ... | | | | |
| Many* | 0.5625 | 0.0625 | 0.1875 | 0.1875 |

* The values for generations 0 and "many" apply to both recombination frequencies.

differences in generations 1 and 2 for the gamete frequencies, especially for the repulsion gametes, and for particular genotype class such as *Ab/aB*, but no differences for the "many generations". Provided there is some recombination, the actual **recombination frequency** (the fraction *c*) does not affect the **equilibrium genotype or gamete frequencies**, but it affects the **rate of achievement of equilibrium**, as shown in the equation $d_t = d_{t-1} \times (1-c)$. The effect of recombination frequencies on the expected rate of attainment of equilibrium for two loci is shown in Fig. 4.1.

There are a number of useful conclusions from this consideration of the effects of these two populations mixing, followed by random mating.

• When two pure breeding but genetically different populations mix, e.g., *A A*, *B B* and *a a*, *b b*, the next generation has a **limited range of genotypes**, only *A A*, *B B*; *a a*, *b b*; and *A B/a b*, in this case. The production of recombinants from the double heterozygote permits a full range of possible genotypes in generation 2 onwards, but in small populations, some of the rarer genotypes will not be seen as they are too infrequent.

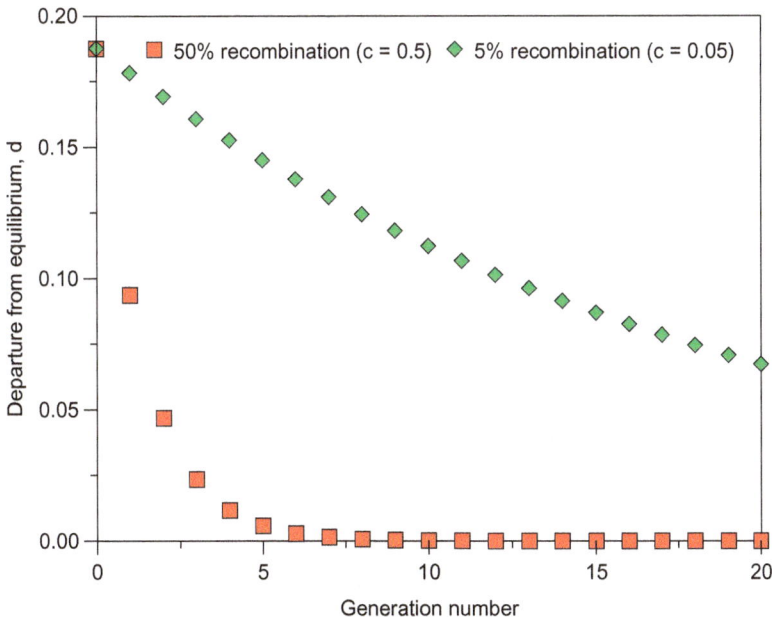

**Fig. 4.1.** The effect of recombination frequency, *c*, on the rate of attainment of equilibrium for two loci.

If one or more allele is rare, there is a very low chance of seeing some rarer genotypes in a small population.

- The departure from linkage equilibrium, $d$, has a **positive sign** if one starts with a coupling combination as above, but a **negative sign** if one starts with a repulsion combination, e.g., $A A$, $b b$ and $a a$, $B B$ mixing. Although disequilibrium populations will tend to go to $d = 0.0$ over many generations, they may overshoot the equilibrium point by chance, so a value of $d$ of $+ 0.001$ in generation 12 in a small population might be followed by $- 0.0003$ in generation 13 and $+ 0.0004$ in generation 14.
- The **equilibrium values** are the same whether one starts with all coupling gametes, all repulsion gametes, or of a mixture of coupling and repulsion gametes, because equilibrium gamete and genotype frequencies depend only on allele frequencies.
- Recombination is only effective in **double heterozygotes**, e.g., *AB/ab*, and not in single heterozygotes, e.g., *AB/aB*. If there is linkage, the two types of double heterozygote, *AB/ab* and *Ab/Ab*, give different gamete frequencies.
- Provided that there is some recombination, the equilibrium genotype and gamete frequencies are unaffected by recombination frequencies, although recombination frequencies determine the **rate of approach to equilibrium** in disequilibrium populations. At linkage equilibrium, the product of the coupling gametes $\{(A, B)(a, b)\}$ is equal to the product of the repulsion gametes $\{(A, b)(a, B)\}$.
- The presence of **linkage disequilibrium** in a population suggests, but does not prove, fairly recent immigration into that population by organisms with different allele frequencies. It could also arise by selection.
- In small populations, it may take many generations for **rare genotypes and phenotypes** to show, even when only two loci are segregating. As several hundred loci may be segregating when two populations mix, it may take hundreds of generations for the rarest types to be present in a small population. The breeders should therefore grow as large a population as practical to find beneficial new types, and should **continue looking** over many generations. Closely linked genes will take much longer than unlinked ones to come into equilibrium, and to produce all possible genotypes in a finite population.

## 4.3. Genetic drift, fixation and effects of population size

**Genetic drift** is a change in allele frequencies by chance and not by selection. It is non-directional, unlike selection, with genetic drift equally likely to increase or decrease the frequency of a particular allele in a population. It may change allele frequencies temporarily, or it may change them permanently if it leads to **fixation**, with one allele being eliminated and another reaching a frequency of 1.0.

Genetic drift has little effect in very large populations (i.e., ones with high numbers of individuals) and is most severe in very small populations. Suppose that a small soil pocket in a rock face has room only for two individuals of an annual grass, with one plant $A A$ and one $A a$, in one generation. By chance, the next generation might have two $A a$ plants, in which case, there has been a **chance change** in the allele frequency in favour of $a$. By chance, the next generation might have two $a a$ plants, with fixation to $a$ and elimination of $A$. The population could only become polymorphic again if a suitable seed from a previous generation germinated to give a mature plant, or if an $a$ allele mutated spontaneously to $A$, or by inward gene flow through the immigration of a seed with $A$, or from ovule fertilisation by pollen from another population which had $A$.

We can get some idea of how effective genetic drift might be in populations of different sizes by using a property of the Normal Distribution, $\sigma$, **the standard deviation of a proportion,** $= \sqrt{\frac{pq}{N}}$, where $p$ and $q$ are fractions adding up to one, such as the frequencies for two alleles at one locus, and $N$ is the total number, e.g., of alleles, which in diploids is twice the number of individuals (parents). Suppose the frequencies of two alleles are 0.75 and 0.25 in a population of two individuals (e.g., one $A A$ and one $A a$), then $\sigma = \sqrt{\frac{0.75 x 0.25}{4}}$, as two parents have 4 alleles in total and $\sigma = 0.217$. A property of the Normal Distribution is that 68% of a population lie within one standard deviation of the mean, and 95% of the population lie within 1.96 standard deviation of the mean. Of all possible daughter populations from this population of two individuals with one $A A$ and one $A a$, 68% should have an allele frequency for $A$ of $p = 0.75 \pm 1\sigma = \mathbf{0.75 \pm 0.217}$, so 68% of possible daughter populations should have $p$ within the range of 0.533 to 0.967, and the remaining 32% will have $p$ values higher than 0.967 or lower than 0.533. Similarly, 95%

of possible daughter populations should have $p$ within 1.96 standard deviations from the mean, which is $0.75 \pm 0.425$, and the remaining 5% will have even more extreme allele frequencies, where possible. **Fixation**, with $p = 1.0$ or $0.0$, is therefore quite likely in such a small population. Fixation is most likely when a population is very small, and/or when one allele is already rare.

If another population had the same allele frequencies of 0.75 and 0.25, but was large with 20,000 individuals, $\sigma = \sqrt{\frac{0.75x0.25}{40,000}} = 0.00216$, so 68% of possible daughter populations will have allele frequencies of $p = $ **0.75 $\pm$ 0.00216**, or 0.748 to 0.752, and 95% will have $p = 0.75\pm0.00423$, or 0.746 to 0.754, so that most daughter populations from this large population will have very small chance changes in the allele frequency. Fixation within one generation is extremely unlikely in the large population, but common in the small one.

The **rate of fixation** (the proportion of populations fixing in one generation) is proportional to 1/(twice the number of parents) and is dependent on the allele frequencies, where the alleles with the lowest frequencies are most likely to be eliminated by drift. The **probability of fixation** to a neutral allele, $A$, in one generation is $p^{2N}$, where $N$ is the number of parents and $p$ is the allele frequency. Thus, in a population of 10 ($2N = 20$), the chance of fixing to $A$ is $(0.9)^{20} = 0.122$ when $p$ is 0.9, 0.012 when $p$ is 0.8, 0.0008 when $p$ is 0.7, and $1 \times 10^{-20}$ when $p$ is 0.1, showing the large effect of allele frequency on the chance of fixation.

Since not every member of a population necessarily reproduces, the genetic drift calculations should be based on the number of parents ($N_{par}$), rather than on the total population size. One needs to be more precise, as not all parents contribute equal numbers of offspring to the next generation, so one uses the **effective population size**, $N_e$, which takes this into account. If a population of 1,000 diploids contains 300 mating couples, all contributing equally to the next generation, $N_e = 600$. **Inequalities between the sexes** and between the individuals reduce the effective population size. Four bulls mated with 400 cows would give $N_e$ of greater than 4 but less than 404, because some individuals are providing many more genes than others to the next generation.

**Sewell Wright** gave the formula:

$$N_e = \frac{4 \times \textbf{number of females} \times \textbf{number of males}}{\textbf{number of males} + \textbf{number of females}},$$

so 4 males mated equally to 400 females would give an effective population size of 15.8, so drift would be as severe as in a population of about 16, even though 404 individuals are reproducing. Different reproductive success of different individuals would further reduce the effective population size. In populations of adult farm animals such as dairy cows and sheep, there are often far more females than males, so the number of males has much more effect on $N_e$ than does the number of females. When one is trying to conserve genetic variability, e.g., in small populations of a rare breed of farm animal (see Chap. 15) or in a pedigree dog or cat line, keeping too few males promotes genetic drift and the loss of variation.

If one has a large population and a few members migrate to colonise a new area, those **founder members** may by chance have a different allele frequency from the original population. The allele frequencies of new populations reflect that of their founder members, so a series of new populations founded from a large one may differ between them in their allele frequencies, and may differ from the parent population (the **founder effect**). For example, **oculopharyngeal muscular dystrophy** is a rare GCG trinucleotide-repeat inherited disease with an unusually high incidence in French-Canadians, where it is traceable to the descendants of a couple who emigrated to Canada in 1634 (Davies *et al.*, 2005).

This founder effect can happen for plants, animals, microbes or humans. If the new populations remain small, they will probably be affected by genetic drift; if fixation occurs, it will reduce their genetic variation and their adaptive capacity. Small populations, e.g., of farm animals or animals in zoos, can easily lose genetic variability and polymorphisms purely through chance genetic drift. Special breeding schemes are now followed in many zoos, to move breeding animals between sites to preserve genetic variation and minimise inbreeding (Sec. 15.3).

In organisms with big **seasonal variations** in number, e.g., with a population size of 1,000,000 in summer, but only 200 individuals successfully overwintering, then drift in winter will have a much larger effect than in summer. "**Bottlenecks**" in population numbers can cause severe drift. Cheetah populations typically have very low amounts of genetic variation, attributed to their having gone through times with extremely low populations sizes when genetic drift reduced genetic variation. When population sizes vary between generations, over $t$ generations $N_e = t/(1/N_1 + 1/N_2 + \cdots + 1/N_t)$, where $N_1, N_2, \ldots, N_t$ are populations sizes in generations 1 to $t$. Two populations with sizes 1,000,000 and 200 therefore have $N_e = \dfrac{2}{\frac{1}{1,000,000} + \frac{1}{200}} = 400$, which is much closer to the small value than to the large one.

Applied geneticists have to consider the numbers, biology and amount of variation in their particular organisms and populations to assess by how much genetic drift will reduce genetic variation. That applies to microbes as well as to plants, animals and humans. For selection, one usually wants a high amount of genetic and phenotypic variation, but the amount of additive genetic variation decreases on average by $\frac{1}{2N_e}$ per generation, by drift.

## 4.4. Gene flow and population structure

**Gene flow** is the physical movement of genes over a distance within or between populations. It depends on an organism's biology with regard to movement and mating arrangements, and on population structure. In Subsec. 1.3.12, we considered the difference between outcrossing individuals and self-fertilising ones for gene flow. Vegetative reproduction (e.g., by tillers in cereal grasses or stolons in strawberries) and various methods of seed- or progeny-production without fertilisation (such as apomixis in plants and parthenogenesis in animals, as in some aphids), involve no exchange of genes between different individuals, although they can result in the movement of unchanged parental genotypes. In contrast, cross-fertilisation is usually much more effective in getting gene flow between individuals and populations, and migration between genetically different populations reduces inbreeding and promotes new recombinants which may be more successful than existing individuals. See Conner and Hartl (2004).

If crops and wild populations of plants are cross-compatible, flower at overlapping periods and grow in close proximity, they will exchange many genes by cross-pollination. This will usually be to the disadvantage of the highly specialised crop varieties and could be to the advantage or disadvantage of the wild individuals. The latter could gain genes for productivity and pest or pathogen resistance, but some of the crop genes would be harmful in the wild, such as the ones for the loss of natural seed dispersal (e.g., brittle rachis in wheat and barley) or the loss of seed dormancy. Burke *et al.* (2002) stated that **crop-weed hybridisation** has produced aggressively weedy crop mimics which share properties of their crop progenitors, making them difficult to control. Ellstrand *et al.* (1999) reported evidence that 12 out of 13 of the world's most important crop plants hybridise with at least one wild relative in at least part of their agricultural cultivation area with the exception of the groundnut, *Arachis hypogaea*. Those authors stated that hybridisation with domesticated species has been implicated in the extinction of various related wild species, so gene flow can clearly have important consequences.

Rieseberg *et al.* (2003) found that three wild annual species of sunflower (*Helianthus annuus*) in extreme habitats (sand dunes, deserts and salt marshes) were all of hybrid origin. **Transgressive segregation**, with individuals more extreme than either parent species, enabled these sunflowers to colonise more extreme habitats. Burke *et al.* (2002) found that about two thirds of cultivated sunflower fields in the USA occurred very close to and flowered at the same time as wild populations, with frequent hybridising between crop and wild populations producing morphologically identifiable hybrids.

Populations may become extremely well adapted to a particular habitat, when outcrossing with another population can produce **outbreeding depression** by breaking up **adaptive gene combinations**, as found by Price and Waser (1979) in *Delphinium nelsoni* crosses between individuals from different distances away. In this herbaceous perennial, seeds drop an average of 11.2 cm from the central stalk and average animal pollen dispersal by hummingbirds and bumble bees is about 1 metre. With hand-pollination, the mean seed set per flower was highest with pollen from plants 1 to 10 metres away, with significantly lower levels from closer or more distant plants. The authors suggested that plants with restricted pollen and seed dispersal may show marked microgeographic genetic differentiation from drift in subpopulations, isolation by distance and/or local ecological adaptation. Very geographically restricted gene flow, with a short outcrossing distance, might then be optimal, giving offspring sufficiently similar to the female parent to be locally adapted, but with more genetic diversity.

Gene flow within and between natural populations or subpopulations can be impeded by various **natural barriers** such as seas, rivers, streams, mountains, glaciers and deserts, and by **man-made barriers** such as towns, roads, factories, farmland, canals, ditches and walls. Su *et al.* (2003) studied the effect of the Juyong-guan part of the Great Wall of China on gene flow in plants. In that section, the wall dates from the Ming Dynasty (1368–1644) and is about 6 metres high and 5.8 m wide. They found that it restricted gene flow between plant populations on either side of the wall, for both insect-pollinated and wind-pollinated plants, permitting more genetic differentiation between subpopulations than occurred between subpopulations just separated by a mountain path. What kinds of barrier affect gene flow obviously depend on an organism's reproductive and general biology, especially on the motility of gametes and individuals at different life stages.

Let us now look at some different **population structures**. Suppose one has a population of snails, which are hermaphrodite but cross-fertilising, spread throughout a large wood, then all snails are potentially able to mate with all other snails. However, because snails move slowly by crawling, two snails that are 100 metres apart have a very low chance of ever meeting and mating, but two snails living one metre apart have a high chance of meeting and mating. We will therefore get gene flow restricted by distance i.e., **isolation by distance**, which is also called **viscous gene flow**; gene flow can happen but is increasingly less likely as distances become greater. We still have one population, not a series of discrete subpopulations.

Most plants are literally rooted to one spot, but their genes can travel in pollen on the wind or on insects, bats or birds, and there are dispersal mechanisms for fruits or seeds, and sometimes for vegetative propagules such as runners or stolons. Animals may crawl, walk, swim, burrow or fly, and humans have many forms of transport, including long-distance flights. The movement of adults, young or gametes can all contribute to gene flow. Gene flow can be **omni-directional or directional**. For example, birds can fly in all directions, but some species have well-defined annual migration patterns. If we have a wind-pollinated plant in a valley with a strong prevailing wind, then plants in the middle of the valley can be pollinated by those upwind of them and can pollinate plants downwind. Directional gene flow also arises from river flows, ocean currents (for fish and coconuts, for example), prevailing winds affecting insect flight, fruits rolling down hills, etc.

One can also have gene flow between discontinuous populations, e.g., by pollen flow between plants in different pockets of soil on a mountainside, or by the hopping of frogs between ponds or streams. For a human example, consider people in Great Britain, Ireland, France and the Channel Islands. Most marriages (or matings) are within each of these four geographical entities, but migration (temporary, such as holidays, or permanent) between them occurs, with marriages or matings occurring between individuals from different communities. On a ship, mice in five different grain holds might mate mainly within their own hold, but migrants might walk to other holds and mate there. This kind of population structure, with mating mainly within discrete populations but occasionally between different populations, is called an **island population structure**.

A variant of this is the **stepping-stone population structure**, where individuals can move between populations, but in a fixed order so that only members of adjacent populations can mate. Imagine a linear chain of islands, in the order A, B, C, D and E, in which a species can move only between adjacent islands since the distances are too great to reach more distant islands by swimming or by wind pollination. The population on island B can receive migrants from A or C, but not directly from D or E. Populations A and E could therefore only indirectly exchange genes if the genes moved one island at a time between them, using each island as a stepping-stone.

In any population structure, gene flow is determined not only by the movement of adults, young or gametes, but also by reproductive considerations. A mouse coming from a different hold on a ship might smell different from mice in the same hold, so the **acceptability of immigrants** for settlement and/or mating is crucial to gene flow. This may be a matter of personal choice and of political decisions in humans, but may be determined by instinct or incompatibility genes in some other organisms. Applied geneticists need to understand gene flow between populations in their organisms, but are often able to control it by fencing or other barriers, especially for animals. Specialist books on population genetics have mathematical models of different population structures.

**The distance pollen can travel** by wind, insects or birds is very important in determining minimal separation distances between crops, where one does not want seed production bastardised by contaminating pollen. It is very relevant to the debate about genetically modified (GM) crops, where producers of pure-line non-GM seed do not want GM pollen from elsewhere contaminating their seed production. The distances over which pollen can fertilise depend on factors such as wind speed, humidity, how long pollen stays viable under those conditions of humidity and temperature, and the effects of UV in sunlight. Under turbulent conditions, pollen can be carried hundreds of miles in a few hours and retain viability. See Sec. 14.10 for the work of Watrud *et al.* (2004) on gene flow from genetically modified glyphosate-resistant (RoundUp-resistant) creeping bentgrass (*Agrostis stolonifera*) to fertilise plants of the same species 24 km away and also a related species 21 km away.

**Seed transport** is often over smaller distances than pollen, but very small seeds may be wind-dispersed over several miles, and transport by water, birds and other animals, as well as by vehicles such as cars, ships and planes, can be over very long distances. Seeds are normally diploid or of a higher ploidy, so seed transport can be more effective in gene flow than pollen transport, since the transported pollen only provides half the genes of any resulting seed. Plants or animals which self-fertilise or circumvent sexual reproduction contribute all the genes in their offspring, while organisms which cross-fertilise others only contribute half the genes of the progeny. Whether there is gene flow by pollination between different generations of plants, depends on whether the organism is annual, biennial or perennial, with long-lived woody populations normally made up of overlapping generations of different ages. Animal biology and life span also determine whether there is cross-fertilisation between different generations.

The **major causes of gene movement** over long and short distances in agricultural animals and plants are the farmers, breeders, seed merchants, wholesale distributors, auctioneers, shippers and shoppers. Much grain is spilled during transport from farms to distributors, markets and processors. Animals from a farm in one country or region may be transported to another country or region and breed there, perhaps being crossed with local animals, and crop seeds grown in one area may be planted hundreds or thousands of miles away. The deliberate transport and accidental spillages anywhere en route apply to normal organisms and GM crops.

**Mitochondrial and chloroplast genes** are usually but not always transmitted via one sex only, largely the female sex. Those organelle genes could still flow over long distances, but would recombine with those of other genomes much less than would nuclear genes transmitted by both sexes and mixing in the progeny. There would still be new combinations produced of nuclear genes with organelle genes.

For a detailed recent account of many aspects of **gene flow in plants**, see the papers from the Royal Society Discussion Meeting, *Mechanisms Regulating Gene Flow in Flowering Plants*, published as *Philosophical Transactions of the Royal Society: Biological Sciences*, (2003), 358, 989–1170. They include the ones on the regulation of mating, mechanisms of gametophytic and sporophytic self-incompatibility, effects of inbreeding on genetic diversity, apomixis, plant hybridisation, transgressive segregation and transgene flow.

## 4.5. Effects on allele frequencies of selection, mutation, migration and gene conversion; frequency-dependent selection; equilibria between forces in populations

**Deviations from the Hardy–Weinberg equilibrium** can be caused by selection, mutation, migration, gene conversion, chance variation in small populations (giving genetic drift), and deviations from random mating. We will consider the first four of those here, and the others elsewhere. The simple two-allele situation usually considered here is over-simplified: see Sec. 7.4 for details.

### 4.5.1. *Selection*

This may be **natural** by environment, **artificial** by plant, animal or microbial breeders, or **sexual**, in the choice of mates. The effectiveness of selection depends strongly on **dominance** considerations. A dominant allele is almost always expressed, even in heterozygotes, unless it is in a hypostatic locus with an epistatic allele present, or unless the developmental stage or environment is unsuitable for its expression, e.g., it is inducible but in non-inducing conditions. Selection against **dominant alleles** can therefore be complete in one generation, e.g., for an expressed lethal dominant allele, all copies of which are eliminated. All the next generation will be homozygous recessives.

Complete selection against **completely recessive alleles** usually takes many generations because they are not expressed in heterozygotes, where they are protected from selection, even if the recessive homozygote is lethal. As we saw in Example 4.1, when recessive alleles are rare, they exist largely in heterozygotes, so even complete selection against the very rare recessive homozygote has little effect on the allele's frequency. Selection against **alleles with incomplete dominance** or additive effects is more effective than against recessives, but not as effective as against dominants. Deleterious recessive alleles can therefore persist for many generations, which can be useful in evolution as they may be beneficial in different environments or in combination with other genes. Thus $a$ might be deleterious in combination with $B$ - and $C$ -, but advantageous in combination with $b\,b$ and $c\,c$.

Let us consider **complete selection against a recessive lethal allele**, $a$, frequency $q$, while $A$ has frequency $p$. We want to know the frequency with which $a\,a$ is produced in each generation, before selection. With complete selection against $a\,a$, all $a\,a$ genotypes must come from $A\,a \times A\,a$ crosses. At Hardy–Weinberg equilibrium, the frequency of these heterozygotes would be $2pq$ out of the total, or $\frac{2pq}{p^2+2pq+q^2}$, but after elimination by selection of the $q^2$ of $a\,a$, this becomes $\frac{2pq}{p^2+2pq} = \frac{p(2q)}{p(p+2q)} = \frac{2q}{1-q+2q}$ (because $p = 1 - q) = \frac{2q}{1+q}$.

The only way to get $a\,a$ in the next generation is from the mating of two heterozygotes, and $1/4$ of the progeny will be $a\,a$, so the frequency of $a\,a$ before selection will be $\frac{1}{4}\left(\frac{2q}{1+q}\right)^2 = \left(\frac{q}{1+q}\right)^2$.

If $q_0^2$ is the frequency of $a\,a$ in generation zero, before any selection, then after one generation of selection, we get from the above equation that $q_1^2 = \left(\frac{q_0}{1+q_0}\right)^2$. The same considerations apply each generation, so after $n$ generations of complete selection against the recessive homozygote, we get $q_n^2 = \left(\frac{q_0}{1+nq_0}\right)^2$. The number of generations, $n$, to reduce $q_0$ to a desired value $q_n$ is given by $n = \frac{1}{q_n} - \frac{1}{q_0}$.

**Example 4.5.** A strain of maize should have unstriped leaves but the recessive homozygous striped form occurs at a frequency of 1% of the plants. With what frequency would the striped form appear (i), after one generation, (ii), after 10 generations, if all plants with striped leaves were destroyed before flowering?

If we assume random mating and other provisos of the Hardy–Weinberg equilibrium, we can say that $q_0^2 = 1\%$ or 0.01, so taking square roots we get $q_0 = 0.1$. We then substitute this in the above formula, $q_n^2 = \left(\frac{q_0}{1+nq_0}\right)^2$, so for (i), we get $q_1^2 = \left(\frac{q_0}{1+1q_0}\right)^2 = \left(\frac{0.1}{1+0.1}\right)^2 = 0.00826$, or about **1 plant in 121**. For (ii), we get $q_{10}^2 = \left(\frac{q_0}{1+10q_0}\right)^2 = \left(\frac{0.1}{1+(10\times0.1)}\right)^2 = 0.0025$, or **1 plant in 400**.

These figures show that selection against a recessive allele, even with complete selection in each generation, only slowly reduces the frequency of the recessive allele because it is present largely in heterozygotes, where it is protected from selection. The rarer the recessive, the slower the progress of selection against it. With complete selection against a recessive, it takes only two generations of selection to reduce its frequency from 0.5 to 0.25 (i.e., by 0.25), but 9,000 generations to reduce its frequency from 0.001 to 0.0001 (i.e., by 0.0009).

We have just considered complete selection, $s = 1.0$. The **selection coefficient**, $s$, is the fraction by which the genotype is reduced in a generation, so if $s = 0.6$ for completely dominant allele $A$, genotypes $A A$ and $A a$ would each leave 40 offspring for every 100 left by the same number of $a a$ individuals. Selection intensity can vary from 0 (selectively neutral) through increasing levels to 1.0 (complete selection, e.g., for a lethal or sterile phenotype). It is common sense that the number of generations required to bring about a particular change in allele frequency is inversely proportional to the selection intensity. Thus, to reduce the frequency of a recessive allele from 0.001 to 0.0001 takes 9,000 generations for $s = 1.0$, but takes 90,023 generations if $s = 0.1$, and 9,002,304 generations, if $s = 0.001$.

It is not necessary to be able to derive all the selection formulae; derivations are given in books such as Strickberger (1985), but the applied geneticist needs to be able to use them. For a **deleterious dominant allele**,

*A*, frequency *p*, selection coefficient *s* against *A A* and *A a*, **Δ*p***, the **change in frequency of *A* in one generation** of selection is:

$$-\frac{sp(1-p)^2}{1-sp(2-p)}$$

As *p* will decrease, the expression has a minus sign, and the expression shows that the rate of progress of selection depends on the allele frequency *p*, as well as on the selection coefficient. We therefore have **frequency-dependent selection**.

For a **deleterious allele *a* with no dominance** (additive effects, so *A'* and *A* symbols could be used), frequency *q*, selection coefficient *s* against *A a* and 2*s* against *a a*, the change in the frequency of *a*, **Δ*q***, in one generation of selection, is:

$$-\frac{sq(1-q)}{1-2sq}.$$

Again, we have frequency dependent selection.

For a **deleterious recessive allele *a*** with selection coefficient *s* against *a a*, the change in the frequency of *a*, **Δ*q***, in one generation of selection, is:

$$-\frac{sq^2(1-q)}{1-sq^2}$$

Again, we have frequency dependent selection.

**This is a very important formula** and will be used to derive other formulae. If selection has already made *q* small, then we can ignore $sq^2$ in the denominator, $(1-q)$ in the numerator will be approximately one, so the change in allele frequency will be approximately $-sq^2$, which we know must be small, because if *q* is small, $q^2$ must be even smaller.

Selection obviously affects **Hardy–Weinberg equilibrium frequencies** because it changes phenotype and therefore genotype frequencies and allele frequencies. Note that selection works directly on phenotypes and only indirectly on genotypes. Since selection can operate at many stages of the life cycle — gamete formation, mating, fertilisation, embryo, germination or birth, and various stages of youth and adulthood — one will get different phenotype frequencies according to whether one scores a population **before or after selection** in that generation. The recessive *albino* mutant of maize gives white seedlings that are unable to photosynthesise. Crossing two heterozygotes gives the Mendelian monohybrid ratio of 3 green: 1 albino at germination, but the albino seedlings die within three weeks, leaving a post-selection ratio of all green: no albino.

If selection occurs in **haploid gametes**, or in a **haploid vegetative stage** as in some fungi, one needs special equations: recessives will usually show fully as they are not hidden in heterozygotes. For a deleterious recessive allele *a* in haploids, frequency *q* and selection coefficient *s* against *a*, the change in the frequency of *a*, **Δ*q***, in one generation of selection is:

$$-\frac{sq(1-q)}{1-sq}.$$

For **X-linked genes**, the formulae also need modifying because the heterogametic sex, X Y, has only one copy of alleles on the X chromosome in the differential region (it is hemizygous). This is considered under human genetics, Subsec. 12.6.1.

## 4.5.2. *Migration*

**Migration** can be natural, but the placing of one group of organisms with another group, e.g., by an animal breeder, is formally equivalent to migration. One has **emigration** from a population and **immigration** into a population. If the emigrants or immigrants have the same allele frequencies as the resident population, migration only changes population size but not allele frequencies. If the migrants have different allele frequencies from the resident population, then the changes in allele frequency in the resident population caused by migration are proportional to both the proportion of migrants to non-migrants, and to the difference in allele frequencies between the resident population and the migrants.

The case of immigrants having different allele frequencies from the resident population is much more common than that of emigrants having different allele frequencies from the non-migrants of the resident population, so the former case will be considered. Let $r_0$ be the frequency of an allele in the resident population before immigration, and let that allele have frequency *i* in the immigrants, with *m* being the proportion of newly introduced genes into the resident population from immigration each generation.

In a generation, the resident population will lose gene frequency $mr_0$ by dilution and gain $mi$ from the migrants. After $n$ generations of immigration,

$$r_n - i = (1 - m)^n (r_0 - i), \quad \text{and} \quad (1 - m)^n = \frac{r_n - i}{r_0 - i}.$$

These equations have been used to work out the proportion of genes ($m$) from "whites" introduced into "blacks" in America, using the present allele frequencies in East African "blacks" ($r_0$), the present allele frequencies in American "whites" ($i$), the present allele frequencies in American "blacks" ($r_n$), and an estimate of 10 generations over the 300 years since African "blacks" were introduced into America and subject to the introduction of genes from "whites". The calculations by Glass and Li (1953) suggested that about 3.6% of genes in the "black" population were introduced from "whites" each generation, with a total of about 30% of the genes in the present American "blacks" having come from "whites".

The plant or animal breeder will usually only be concerned with "migration" over one generation, when the allele frequency will be $r_0(1 - m) + mi$. If one adds 10 animals of genotype $a\,a$ ($i$ for $A = 0.0$) to a resident population of 30 animals of genotype $A\,A$ ($r_0$ for $A = 1.0$), $m$ is $10/40 = 0.25$, then the allele frequency for $A$ in the mixed population is $r_0(1 - m) + mi = 1(0.75) + 0.25(0.0) = \mathbf{0.75}$.

### 4.5.3. Mutation

**Mutation** will affect Hardy–Weinberg equilibrium frequencies only to a very minor amount, unless mutation is unusually frequent or if very rare alleles are being considered, because typical mutation frequencies are about $10^{-4}$ to $10^{-9}$ per locus per generation. Mutation is, however, very important as a source of new genetic variation, and can affect allele frequencies over a long time. Let us take an oversimplified case of allele $A$, frequency $p$, mutating with frequency $u$ to allele $a$, frequency $q$, which backmutates to $A$ with frequency $v$. Initially, we will have

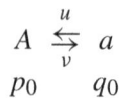

$$A \underset{v}{\overset{u}{\rightleftarrows}} a$$
$$p_0 \qquad q_0$$

The **change in frequency for $a$ in one generation, $\Delta q$**, will be the **gain** from mutation from $A$ to $a$ ($= up_0$) and the **loss** from mutation from $a$

to $A$ ($= vq_0$), so that $\Delta q = up_0 - vq_0$. At mutational equilibrium, there will be no change in allele frequencies, so $up = vq$, and **equilibrium values** (sometimes represented by showing $p$ and $q$ with a ˆ symbol over them) are $\Delta q = 0.0$ and $\hat{p}/\hat{q} = v/u$. If there are no other factors such as selection, drift or migration, the equilibrium frequencies of two alleles are just determined by their relative mutation frequencies, but with typically low mutation frequencies, it can take very many generations to achieve equilibrium.

If $A$ is a functional active allele, it can mutate at many different positions to give a loss of function mutation; if $a$ is such a loss of function mutation, any one $a$ allele needs an extremely specific mutation (a reverse or **back-mutation**), at or near the site of the original mutation, to restore it to the $A$ base sequence. Forward mutations, from $A$ to $a$, will therefore be much more frequent than reverse mutations, $a$ to $A$. Suppose that $u$ is $10^{-5}$ and $v$ is $10^{-7}$, then at equilibrium, the above equation $\hat{p}/\hat{q} = v/u$ suggests that $\hat{p}/\hat{q} = 1/100$, with the mutant allele $a$ 100 times more frequent than the wild-type allele $A$, which is nonsensical, considering the definition of a wild-type allele. The reason that mutant alleles are not much more frequent than wild-type alleles is that mutant alleles are usually selected against and there is an **equilibrium between mutation and selection**.

Consider **selection against a deleterious rare recessive**, $a$, frequency $q$, selection coefficient $s$ against $a\,a$. We saw earlier that the loss by selection of such an allele in one generation was $\frac{sq^2(1-q)}{1-sq^2}$, and as $q$ will be small for a rare allele, this is approximately $sq^2(1 - q)$. As $a$ is rare, we can ignore losses to it from mutation to $A$, and the gain from mutation from $A$ is $pu$ per generation, or $(1-q)u$, as $p+q = 1.0$. At the equilibrium between mutation and selection, the loss from selection $\{sq^2(1 - q)\}$ must equal the gain from mutation $\{(1 - q)u\}$. As $sq^2(1 - q)$ therefore equals $(1 - q)u$, we can cancel the $(1 - q)$, giving $sq^2 = u$, so $q^2 = u/s$ and equilibrium $\hat{q} = \sqrt{\frac{u}{s}}$.

This result pertaining to rare deleterious recessives is important. If we know any two values, we can use this formula to calculate the third value. Thus, if $s = 0.1$ and $\hat{q} = 0.001$, we can calculate the mutation frequency $u$ as 0.0000001 or $1 \times 10^{-7}$. For a **deleterious dominant**, the corresponding equilibrium formula is $\hat{p}(1-\hat{p}) = v/s$.

Although these equations are useful, they do not take account of **gene conversion**, whereby one allele can convert another to being of its own kind at meiosis in a heterozygote. Gene conversion in meiosis can give allele segregation ratios of 3 *A*: 1 *a* or 1 *A*: 3 *a* in tetrads, or in octads (as in some fungi), it can give 8:0, 7:1, 6:2, 5:3, 3:5 (Plate 18.2), 2:6, 1:7 and 0:8 allele ratios for *A*:*a* or wild-type:mutant, from meiosis. The direction of gene conversion is often biased (see Lamb 1998), with more conversions to one allele rather than to another, and in the long term, it can affect allele frequencies in populations. The **strength of the force of gene conversion on allele frequencies** depends on the **frequency of gene conversion** and on the **amount of disparity in conversion** to different alleles. See Lamb and Helmi (1982) for equations taking gene conversion into account in the equilibria involving selection and mutation rates, and Lamb (1998) for references on amounts of gene conversion disparity.

Since most new mutations are harmful, mutations reduce the immediate fitness of a population by an amount called the **mutational load**, although mutations are useful for long-term adaptation and evolution. The mutational load for recessive alleles is equal to the mutation frequency, and for dominant alleles, it is equal to twice the mutation frequency. The **Haldane–Muller principle** is that the amount of harm (loss of fitness) caused by mutation to a population is not dependent on how deleterious the mutations are, but depends solely on the mutation frequency. Over all loci, the mutational load is the mutation frequency multiplied by a factor of between one and two, depending on the proportions of recessive and dominant mutations. Although this principle initially seems illogical, it is sensible when one considers that a very harmful mutation will do a lot of harm while it is present, but will be quickly eliminated by selection, especially if dominant, while a mildly deleterious mutation will do a smaller amount of harm at any time, but over a much longer period as selection against it is much less.

## Suggested Reading

Brown, A. H. D. *et al.* (eds.), *Plant Population Genetics, Breeding, and Genetic Resources*. (1990) Sinauer Associates, Sunderland, Mass.

Burke, J. M., K. A. Gardner and L. H. Rieseberg, The potential for gene flow between cultivated and wild sunflower (*Helianthus annuus*). *Am J Botany* (2002) 89: 1550–1552.

Clarke, B. C. and L. Partridge (eds.), *Frequency-dependent Selection*. (1988) The Royal Society, London.

Conner, J. K. and D. L. Hartl, *A Primer of Ecological Genetics*. (2004) Sinauer Associates, Massachusetts.

Davies, J. E. *et al.*, Doxycycline attenuates and delays toxicity of the oculopharyngeal muscular dystrophy mutation in transgenic mice. *Nat Med* (2005) 11: 672–677.

Ellstrand, N. C., H. C. Prentice and J. E. Hancock, Gene flow and introgression from domesticated plants into their wild relatives. *Annu Rev Ecol Sys* (1999) 30: 539–563.

Falconer, D. S. and T. F. C. Mackay, *Introduction to Quantitative Genetics*, 4th ed. (1996) Longman, Harlow.

Glass, H. B. and C. C. Li, The dynamics of racial intermixture: an analysis based on the American Negro. *Am J Hum Genet* (1953) 5: 1–20.

Griffiths, A. J. F. *et al.*, *Introduction to Genetic Analysis*, 8th ed. (2004) W. H. Freeman and Co., New York.

Hartl, D. L. and E. W. Jones, *Genetics. Analysis of Genes and Genomes*, 6th ed. (2005) Jones and Bartlett, Sudbury, Maryland.

Lamb, B. C., Gene conversion in yeast: its extent, multiple origins and effects on allele frequencies. *Heredity* (1998) 80: 538–552.

Lamb, B. C. and S. Helmi, The extent to which gene conversion can change allele frequencies. *Genet Res* (1982) 29: 199–217.

Pirchner, F., *Population Genetics in Animal Breeding*, 2nd ed. (1983) Plenum, New York.

Price, M. V. and N. M. Waser, Pollen dispersal and optimal outcrossing in *Delphinium nelsoni*. *Nature* (1979) 277: 294–297.

Rieseberg, L. H. *et al.*, The genetic architecture necessary for transgressive segregation is common in both natural and domesticated populations. *Phil Trans Roy Soc Biol Sci* (2003) 358: 1141–1147.

Strickberger, M. W., *Genetics*, 3rd ed. (1985) Collier Macmillan, London.

Su, H. *et al.*, The Great Wall of China: a physical barrier to gene flow? *Heredity* (2003) 90: 212–219.

# Chapter 5

# Types and Uses of Selection

## 5.1. Natural, artificial and sexual selection

**Natural selection** is selection by the environment. For crops and farm animals, this includes the man-influenced environment in which the organisms grow. Natural selective factors include diseases, pests and predators, temperature, rainfall and sunshine, wind and soil. **Artificial selection** is selection by man, e.g., selection for higher yields or for uniform ripening. **Sexual selection** is selection in getting and keeping mates, which may result in different colouring between the sexes, showy crests in some amphibia or showy tails in peacocks. Extreme adaptation in the battle for mates within a species may lead to that species being less competitive against other species, e.g., the long, showy tails of peacocks which attract peahens may reduce the peacocks' fitness relative to competing species. Loud calling by male frogs is selected by females in mating, but it is disadvantageous in leading to increased predation by bats.

The artificial selections imposed on populations by plant, animal or microbial breeders may interact with natural or sexual selection. Human geneticists are aware of natural and sexual selection in mankind, but for ethical reasons, they cannot usually apply artificial selection to humans except in pre-implantation screening (Subsec. 12.12.6). **Medical treatments** for infections and infertility run contrary to natural selection by allowing individuals who would otherwise have died or been sterile to live and reproduce.

Artificial selection may be based on the scorable phenotypic merits of **individuals** if heritabilities are high, but if heritabilities are low so that phenotype is a poor guide to breeding value, it is often better to select on

**whole-family averages**, especially for large families of half-sibs. Other selection methods involve an individual and its ancestors (**pedigree selection**) or an individual and its offspring (**progeny testing**). Some other methods, such as **mass agricultural selection**, are covered in Chap. 13.

There can be unwitting human selection against the best types of plant and animal, if the best types are chosen for consumption instead of for breeding. For example, in Malaysia and Indonesia, local people often pull down the most productive durian trees (*Durio*) in order to collect the fruits for eating; with timber, the best trees are often cut down for wood. In some religions, the best animals are chosen for sacrifice.

*In situ* selection means selection of a new variety well adapted to the local climate, soils and diseases, e.g., with drought-, cold- or salt-tolerance, but those varieties may do poorly when tried in other areas; i.e., there are large **genotype x environment interactions**. It may not be cost-effective to breed many varieties to suit a large number of different niches. *Ex situ* selection is done in a small number of good environments, with the plants or animals then tried in various other areas.

Breeding and selection may sometimes be for a very **specialised type**, e.g., fast-growing, high food-efficiency broiler chickens for factory farms. At other times, one might want a generally hardy and **adaptable type**, such as pack animals for tribes who migrate seasonally between different environments. Different breeds of an animal may be selected for different properties and environments. For example, a Merino sheep bred for wool in Australia is good in extreme environments, provides fine-fibre wool suitable for light to medium weight clothing, and about 80% of the farmer's income will be from the wool. A Southdown sheep (Plate 13.7) in the south of England is suited to fairly rich pasture, provides coarser wool suited to heavyweight clothing, upholstery and interior textiles, and is mainly for meat, with only 10 to 30% of the income generated being from wool. It is used as a sire breed for meat and short wool. The hardy Blackface sheep in the Outer Hebrides islands, off the northwest coast of Scotland, provides wool for the Harris Tweed weaving industry, for heavyweight garments (see Subsec. 13.7.9 for the use of Blackface sheep in crossbreeding). In Scotland, it provides meat and carpet wool.

Selected characters may remain **stable** over long periods, such as resistance to woolly aphid from the apple variety Winter Majetin, which has remained effective for over 150 years. In contrast, resistance to stripe rust in the wheat cultivar Clement was overcome after only one year, by mutation or recombination in the fungus.

The aim of selection is to increase value. Highly selected **pedigree Texel rams** for use in breeding flocks are worth about £5,000 to £32,000 each, but ordinary Texel rams to father lambs for the meat trade are only worth about £550 (figures from G. Crust, Skendleby, Lancashire, *The Daily Telegraph*, 5/6/1999).

See Subsec. 4.5.1 for **selection equations** relating **changes in allele frequencies** to **dominance, selection intensities and allele frequencies**. It is important to realise that selection responses are often **frequency dependent**. For example, selection against a very common deleterious recessive acts fairly quickly as it is often homozygous and expressed, while selection against the same allele works very slowly when it is rare, when it is largely present in the heterozygote and thus not expressed (see Sec. 15.3 for selection for rarity). Alleles for self-incompatibility (Subsec. 17.1.2) are an excellent example of frequency-dependent selection, with selection for rarity. A very common allele would often be present in both pollen and stigma, resulting in an incompatible pollination, while a rare allele is only likely to be present in pollen or stigma, but not in both, in a particular mating, so that it will not be selected against.

## 5.2. Stabilising selection, towards uniformity

In wild populations, natural selection tends towards producing **optimum phenotypes**, ideally suited to their environment. Departures from optimum will usually be less well adapted and thus will be selected against, so that natural selection often stabilises a population around the average type which is ideal for that environment. That is true of height in mice, deer, man, mushrooms, mosses, wheat, etc. Organisms which are much smaller than average have various disadvantages, e.g., small wheat plants being shaded out by taller ones, or small male deer being poor at fighting for mates. Organisms much bigger than average also have disadvantages, e.g., very tall wheat plants being easily knocked down by wind and rain.

Selection in favour of average types and against organisms towards the two extremes is called **stabilising selection**. It often occurs in nature and is used by plant and animal breeders to encourage uniformity and typicalness in a breed. For example, with a mechanically harvested crop such as barley, the farmer wants all plants to ripen their seed simultaneously. When the combine harvester goes through a field, it will harvest all plants, whether ripe, under-ripe or overripe, so **selection for uniformity** of ripening time makes commercial sense, giving the best financial return. Breeders will cull plants or animals not conforming to the desired and recognised form for an animal variety or a plant breed, e.g., maize plants giving brown or purple grains in a strain with yellow grains. Figure 5.1 illustrates stabilising selection for a quantitative character giving a Normal Distribution.

There is evidence for stabilising selection for **human birth weights**. In the data of Karn and Penrose (1951) from a London hospital, the optimum birth weight for survival from birth to one month of age was 3.6 kg, close to the population average of 3.2 kg, with only 1.2% mortality, compared with a mortality of 4.1% averaged over all weights. Very light babies, e.g., below 1.8 kg, had greatly increased mortality, and very heavy babies, e.g., over 4.2 kg, also had a somewhat increased mortality (see Fig. 35-1 in Strickberger, 1985). The optimum weight for minimal mortality in this period just after birth was slightly higher than the average birth weight of 3.2 kg; possibly a lighter weight is optimal earlier, especially at the time of birth when slightly smaller babies may cause fewer complications. In 1999, the annual Confidential Enquiry into Stillbirths and Deaths in Infancy (covering England, Wales and Northern Ireland) reported that heavy babies, those weighing more than 4 kg, had twice the average risk of dying during labour, with one death in 750 births. Out of 669,989 births in 1997, CESDI found that the number of stillbirths, neonatal and postnatal infant deaths was 10,418, a death rate of 1 in 64.

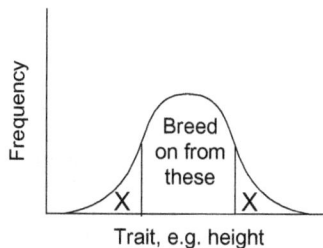

**Fig. 5.1.** **Stabilising selection**, with "x" indicating the organisms to cull (kill, or not to breed from).

## 5.3. Directional selection, favouring one extreme

To improve a breed, breeders often use **directional selection**, breeding from organisms at one end of a distribution, such as dairy cows with the highest milk yield, or rye plants with the most disease resistance. In pigs, breeders reduced back-fat thickness from about 6 cm to about 1 cm at market weight in only a few generations. Directional selection occurs naturally for the highest fertility and fitness, and is illustrated in Fig. 5.2.

The **direction of selection** may vary in different environments. **Light skins** have been selected in humans in less sunny climates where they permit more photosynthesis of vitamin D in the skin, and **dark skins** have been selected in sunnier climates where they provide more protection from damaging and skin-cancer-causing ultraviolet rays, though selection is slow and incomplete.

The breeder may occasionally select in different directions for different purposes. With **tomatoes**, for example, the small, sweet cherry tomatoes have been selected for smallness, whereas large beef tomatoes for stuffing have been selected for large size; other varieties have been selected for medium size. Selection directions often differ for the two main types of tomato: market varieties for selling fresh, and processing tomatoes for tinning, ketchup, soups, paste and sauces. For fresh tomatoes, appearance is important, especially the colour. Processing tomatoes are usually smaller and tougher skinned for mechanical harvesting and crushing-resistance when transported in bulk to factories. Their lower water content reduces the need for water removal in processing, and increases flavour.

One can have selection for **qualitative traits** as well as for quantitative ones. Selection is much simpler when traits are controlled by major genes with Mendelian inheritance. In Sri Lanka, a mutant variety of rice with

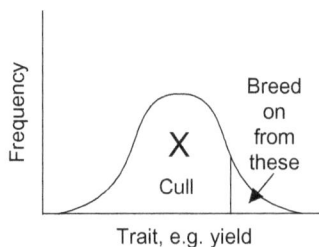

**Fig. 5.2.** **Directional selection**, breeding from one extreme only.

**Plate 5.1.** A **lodging-resistant rice strain** under test at Bentota Regional Research Station, Sri Lanka. This strain can raise the upper stem if beaten down, by bending upwards from the nodes.

altered growth properties was discovered, which could bend upwards at the nodes; if the stem is knocked down by wind or rain, the part of the stem above a node can be raised again. This lodging-resistant strain is shown in Plate 5.1. Animal examples of selectable characters controlled by genes of major effect include the Booroola gene in sheep for increased ovulation rate (Subsec. 2.1.3), a double muscling gene in cattle, and in pigs, two genes associated with meat quality and one which alters the susceptibility to porcine stress syndrome.

## 5.4. Cyclic selection, alternatively favouring different extremes

With **cyclic selection**, one has different phenotypic extremes being favoured alternately. This often occurs through seasonal variation for temperature or light intensity. For **human skin pigmentation** in Britain, there is very mild selection in winter for pale skins which permit more photosynthesis of vitamin D, and in summer, there is very mild selection for dark skins which give more protection from ultraviolet light which can cause malignant skin cancers.

Cyclic selection favours the preservation of genetic variation in a population, by alternately favouring different alleles. Breeders use cyclic selection, usually growing organisms alternately in different habitats or regions, to get **environmental tolerance**. This has been extremely successful in wheat breeding (Subsec. 13.7.1), in getting varieties which do well in a wide range of countries and environments, just needing local adaptation for resistance to local diseases.

Suppose a population of wheat contains some alleles giving specialised success in environment 1 (e.g., sea level), some alleles giving specialised success in environment 2 (e.g., 2,500 metres above sea level) and some alleles giving general environmental tolerance. By growing the plants alternately in the two environments for 10 generations, one selects in every generation for alleles giving general environmental tolerance. The alleles for specialised success in one environment will be selected for in that environment, but be selected against in the alternate environment, so that the alleles most selected for are those for general environmental tolerance.

## 5.5.  Disruptive selection, selecting against the average type

With **disruptive selection**, both extremes are favoured, with selection against the average type. This is mainly a laboratory experimental technique. If breeders want small and large tomatoes for different purposes, they will use directional selection, selecting one population for reduced size and another population for increased size.

Disruptive selection does occur in nature, especially by predation or disease. For example, birds might readily recognise the average type of a butterfly species as being good to eat, but butterflies at either extreme away from the normal type might not be recognised as being of that good food species. In the African butterfly *Pseudacraea eurytus*, colours range from reddish-yellow to blue. The extremes mimic other species not targeted by their normal predators. Intermediate types are therefore more liable to selective predation, so individuals at the extremes are fitter. Diseases might adapt to the commonest form of an organism, and be less well adapted for more unusual types. Sexual selection can lead to increased differentiation between the two sexes, acting as a kind of disruptive selection, selecting one sex in one direction and the other sex in the other direction. Disruptive selection is also likely between different sub-populations, for traits giving different survival in different niches or environments.

## 5.6. Pedigree selection; breed records, prepotent stud males and grading-up

**Pedigree selection** is selecting on an individual's performance and on that of its **recent ancestors**, especially its parents and grandparents, as it should on average have half the genes of each parent and one quarter of the genes of each grandparent. Its genetic basis is that individuals of great phenotypic merit are more likely to possess and to transmit good alleles than individuals of inferior merit. The higher the heritability for a character, the more reliable is the pedigree breeding.

One looks up the **performance records** of ancestors in **herd books**, **stud books** or **farm** or **breeding station records**, and also assesses the performance of the individual being considered for breeding. In a recognised **cattle breed** with registration schemes, one might be able to check the milk yield, milk protein and butterfat content, and calving record, of recent ancestors, and to find out what prizes they won at shows. Since 1800, all **UK foxhounds** have been entered into the Foxhound Kennel Stud Book, so pedigrees can be traced back over very many generations. All British foxhounds are descended from three stallion dogs, Brocklesby Bumper (1743), Menell Stormer (1791) and Gloag Nimrod (1904), with selection for smelling ability, speed, drive, a good "voice", working in a pack, and toughness. In **sheep**, the Border Leicester (Subsec. 13.7.9) was produced by crossing Leicester Longwools with Cheviots around the beginning of the 19th century, with the new breed being recognised in 1869 for shows, and with the herd book starting in 1898.

Pedigree breeding has been used extensively for all major farm animals, for race horses, dogs and cats. At a national cat show, the author asked a successful cat breeder what genetic techniques he used. He replied: "I don't know no genetics. I just breeds champion tom with champion queen!" This policy has been described as "crossing the best with the best and hoping for the best".

With a number of organisms including horses and coconuts, the term **prepotent** is used for individuals and for breeds which are excellent and are known to transmit excellent gene combinations (i.e., have a high breeding value, not just environmentally-determined merit). In horses, the Thoroughbred breed is generally prepotent and has been used to improve other breeds such as Cleveland Bays and Quarter Horses. When individuals of a recognised breed are crossed to other breeds to improve performance, there may be a specified number of backcrosses to the original breed before the offspring can be registered as being of that pure breed. A prepotent Thoroughbred stallion, "Man O' War" (born in 1917), raced for two years and in 21 starts, was first 20 times and second once, establishing five world records and winning much prize money. His immediate progeny were also very successful, showing his high breeding value, with transmissible genetic merit. In 1998, a yearling colt from champion stallion Rainbow Quest × brood mare Silver Lane sold for £2,310,000. The record price for a yearling in Britain was £2,520,000 in the 1980's, with Northern Dancer as a parent, with prices depending on individual and ancestral merit. In Britain, there is a horse National Stud, set up in 1916, and now at Newmarket. The values of stallions range from £400,000 to more than £5 million, costing £3,000 to well over £12,000 a mating. It includes Shaamit, the Derby winner in 1996, who "covers" about 70 mares a year. In 2004, the National Studbook recorded that the Thoroughbred stallion Sadler's Wells made up to £250,000 for siring a foal. In just three months, he was hired out 170 times, proving very valuable to his Irish owners. The pollen of prepotent **oil palms** is very valuable for increasing oil yields.

Pedigree selection is less useful for characters of low heritability. One of its merits is for **selection of opposite-sex characters**. Thus, in dairy cattle, one could select bulls on the milk performance of their female ancestors, such as mothers, grandmothers and great-grandmothers, or select cocks for egg-laying genes on the basis of records of their female ancestors.

With males having much greater reproductive potential than females, and proportionately fewer males than females needing to be kept for breeding each generation, one can have really strong selection of the best males, both on individual phenotypes and on their ancestral pedigrees. Very superior males are used for repeated mating as **stud males**, as in horse, dog and cat breeding, where the stud males have done really well in horse racing or in dog or cat shows. While any breeder may have his (or her) stud males for use with his own females, breeders often use a male in an open stud, even though this is more costly. It helps to reduce inbreeding within a breeder's stocks, introduces genetic variation which he can select on, and may provide an even more superior male than those in his stocks. It is sometimes worthwhile to send even mediocre females to a stud male to "**grade-up**" the whole stock. The breeder should not just choose the most prestigious stud male, but should choose one whose special merits will help to remedy known defects in his own stock. While the superior stud male should improve a breeder's stocks, it is important that those stocks should not lose their distinctive breed characteristics, and selection for them should be made among the offspring.

For this "grading up" to improve a line's average performance, it is best to use a stud male from an inbred line as he should be highly homozygous. Although his phenotypic merits will be diluted in his progeny, as their heritability will be less than 1.0 (and will be diluted by crossing to the inferior stock to be graded up), his high degree of homozygosity should ensure that most gametes contain meritorious genes. A stud male of superior phenotype who is not inbred may have derived some merit from heterozygous advantage which is unlikely to be passed on; he may also owe much of his phenotypic merit to dominant alleles, but be heterozygous for many of those loci, and pass on good and bad alleles to his progeny. Ideally, a prepotent stud male (in animals or as pollen donor in plants such as oil palms) should be selected on the basis of pedigree selection, his individual merits and progeny testing (see next section).

## 5.7. Progeny testing

In **progeny testing**, one judges an individual's merit by its own performance and by that of its **offspring**. It is most useful for individuals capable of having many progeny (e.g., bulls, by artificial insemination, AI; Sec. 17.2), which means males for farm animals, and having a long reproductive life which allows time for producing and assessing the offspring, and then for further breeding from the selected individual.

A typical progeny testing programme for dairy cattle would be to cross a small number of young bulls of very good pedigree to a range of unselected different cows in a common environment. The daughters of each bull are then assessed for their average milk yield and quality. The bull with the best average daughter performance is selected to continue breeding, while the others are culled or just raised for meat. The chosen male can have thousands of daughters by AI. Between 1980 and 2000 in the USA, the average annual milk production by dairy cows increased from about 4,500 kg to 6,800 kg, partly from progeny testing.

A disadvantage is that this testing is expensive and lengthy, keeping bulls of low breeding value alive for several years and producing daughters with poor milk yield (and unwanted sons, unless sex-sorted sperm are used: Subsec. 17.2.2). It is seldom worth progeny-testing females in farm animals, as they have comparatively fewer offspring during their reproductive lives.

Like pedigree breeding, progeny testing is excellent for **opposite-sex character testing**, such as bulls for milk yield and cocks for egg-laying. The method is easily used in plants, using pollen from the plants to be tested on the stigmas of a range of other plants. In annuals such as wheat, one would need to preserve pollen (Sec. 17.1) for use in following seasons once the progeny testing established which plants were prepotent, but there are no such problems with perennials like coconuts and oil palms. The higher the heritabilities, the better progeny testing works, as phenotypic merit is more likely to reflect genotype merit. The reference to Georges *et al.* (1995) in Sec. 2.3 shows the use of progeny testing in QTL mapping.

Traits such as body weight or milk yield can be easily measured on **potential parents**, but many **meat carcase characters** can only be assessed after slaughter, such as the flavour, fat content and marbling, and the killing-out percentage (usable proportion of the carcase, by weight): see Sec. 5.14, and Subsec. 13.7.10 for beef examples. For characters which can only be measured late in life or after killing, generation intervals are rather long, and the rate of gain from selection *per unit time* (years, as opposed to generations), is higher from pedigree selection than from progeny testing or individual selection on an organism's own merits.

"Genus" is a large cattle firm in the UK. Its Genus Sire Improvement Programme (GSIP) is the largest **dairy progeny testing programme** in the UK, and began in 1992 with the intention of using the best genes from anywhere in the world. In 1996, there were more than 2,400 participating farmers, carrying out over 50,000 test inseminations a year. They collect semen from young bulls as early as possible, collecting 1,000 doses between 12 and 14 months. With calving heifers at two years of age, progeny test results are achieved in five years.

## 5.8. Half-sib and family selection

Selection based on individual merits is appropriate for traits with relatively high **heritabilities**, e.g., 0.25 to 1.0, and where important traits can be measured on potential parents, not after killing. When heritabilities are low, e.g., 0.0 to 0.25, the phenotype of an individual is a relatively poor guide to its genotype and breeding values. It is then more efficient to base selection on whole-family averages rather than on individuals, as some of the environmental variation should average out over the family, some having better and some having worse environments. This is **family selection**, often in the form of **half-sib selection**.

For farm animals, it is easy to obtain large families of half-sibs with the same father and different mothers. Half-sibs are more useful than full sibs as the families can be larger, can average out the differences in mothering ability, and can be contemporary, whereas full sibs are usually born over a number of seasons except in litter-bearing species. In litter-bearing animals such as pigs, one can select between different litters, or between a series of litters from the same parents. One takes different families of half-sibs and compares their family averages for particular traits, using the best individuals from the best families as parents in breeding programmes. Comparisons can be made using average index scores for each family. For carcase characters, only some of each family need be slaughtered, leaving others available for breeding. The best family of half-sibs for litter-size in pigs could be used to select males for breeding for this opposite-sex character. One could use information on half-sibs, full sibs and ancestors, but low heritabilities would make records of parental merits inaccurate as regards to breeding values.

The family selection method is thus suitable for characters with low heritability, for opposite sex characters, and for characters scored late in life or after killing. It has been very useful for improving pig carcase quality. Some carcase characters such as mid-back fat depth can now be scored non-destructively, e.g., by the use of echo-sounders.

In family selection, one can use average index scores, or observe which families produce proportionally more superior offspring. In a litter-bearing species such as the pig, one can try different pairings of individuals, e.g., mating male A to females L, M, and N, male B to females O, P and Q to generate different families. This would not only give information on which male was best, averaged over matings to several different females, but would also test which particular combination of parents gave the best offspring average, e.g., A × M. Such good combinations are said to "nick" with each other. In non-litter-bearing species such as cattle, sample sizes from any particular mating would be rather low for that.

## 5.9. Selecting for correlated characters

**Correlations** between characters were considered in Sec. 3.3. Some characters are difficult or expensive to measure, so that it is sometimes better to select for a correlated character rather than the one you are most interested in, once the direction and strength of the correlation have been established. If a character is only scorable late in life or after slaughter, generation intervals can become too long. Selecting for correlated characters which are easier and cheaper to score, or which can be measured earlier in life, is therefore desirable.

For example, in most animals, the growth rate is highly correlated (correlation coefficient, $r$, $= 0.6$ to $0.8$) with **feed conversion efficiency**, which is how efficiently the animal converts food into usable parts of itself, such as meat, wool or leather. Measuring food conversion efficiency on individual animals is expensive, as all feeding has to be monitored individually, while growth rate is easy to measure by weighing the animal at intervals. It is therefore much more convenient to select for increased rate of weight gain rather than for increased feed conversion efficiency.

## 5.10. Selection for more than one character: tandem selection, independent culling levels and index selection

In plants, animals and micro-organisms, the breeder usually wants to **improve several characters**, not just one. In **pigs**, one might want to increase growth rate, food-conversion efficiency, disease-resistance and litter size, and decrease carcase fat. In **beef cattle**, one mainly selects for birth weight, 200 day growth, 400 day growth, backfat depth, muscling score and beef value. In **sheep**, one might select for yields of meat and/or wool, number of lambs born, meat and wool quality, and reduced susceptibility to fly-strike, fleece-rot, dermatophilosis, foot-rot and facial eczema. In **wheat**, one might want to improve grain yield, disease-resistance, heavy-metal- and cold-tolerance, and to reduce straw length and the need for high nitrogen inputs. There are several methods for selecting several characters.

## 5.10.1. *Tandem selection*

With **tandem selection**, one selects for just one character over a number of generations, until the desired improvement has been obtained. One then selects just for a second character for a number of generations, then for a third character, etc. This method is inefficient and is rarely used. It takes too many generations to finish selecting all the desired characters, and any negative correlations between the characters would mean that an initial favourable response for one character might be reversed during later selection for another character.

## 5.10.2. *Independent culling levels*

For **independent culling levels**, one sets separate selection thresholds for each character. In pigs, one might select for breeding only animals above a certain threshold for growth rate (which one wants to increase), and which are simultaneously below a threshold for mid-back fat (which one wants to reduce). Suppose that one needed to keep 1/16th of the herd for breeding, and that growth rate was four times more valuable commercially than mid-back fat, then one could just take the individuals in the best eighth for growth rate and the best half for fat. That would weight the selection for different characters according to their relative value, and would mean taking 1/16th of the herd, and culling the rest.

This is shown in Fig. 5.3, where the individuals' values for the two characters are shown by a single "x". The vertical line is the threshold for fat, taking only animals to the left of this line, with low backfat, and the

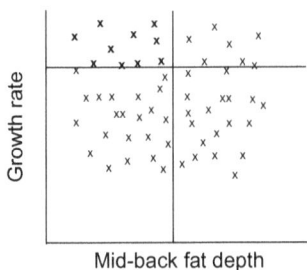

**Fig. 5.3.** Illustration of **independent culling levels for growth rate and mid-back fat in pigs**: see text for details. Each "x" represents one pig, and only pigs in the top left quadrant would be used for breeding, with the best growth rate and the least fat.

horizontal line is the growth-rate threshold, taking only animals above this line. The selected animals for breeding are therefore those in the top left sector, where they are shown in bold type. Independent culling methods are easy to use and are common. One can of course use independent culling levels for a number of different characters, not just for two characters. This will be illustrated in the cat breeding example in the next section on index selection.

## 5.10.3. *Index selection, including PTA, PIN and £PLI*

For just two characters, one might be able to arrive at a single figure combining them, such as the growth rate of the non-fat part of a pig for the above two characters. Usually, however, one constructs **an index** which is based on adding together the merit scores for several key commercial characters, weighting them in terms of relative economic value, heritability, phenotypic variances and genetic correlations between them. It gives an estimate of an individual's overall merit as a single number. See Cameron (1997) for animal selection indices and Baker (1986) for plant indices. Using **index selection** is more complicated than independent culling levels, but it is more efficient.

A **selection index** is a linear function of observable phenotypic values for a number of traits. The observed value for each trait is multiplied by its own **index coefficient**, based on the economic value of the trait, its heritability, its genotypic and phenotypic variances, and covariance with other traits. Suppose it was desired to improve a line of wheat by increasing grain yield, decreasing the days from sowing to maturity, and increasing the protein concentration. A suitable simple index might be: $I = (0.3 \times$ yield in kg/ha) $+ (-1.4 \times$ days to maturity) $+ (4.7 \times$ protein concentration in g protein per 100 g grain). A particular genotype giving 2,000 kg/ha, maturity at 99 days and with a protein content of 11.7% would then have an index score of $(0.3 \times 2,000) + (-1.4 \times 99) + (4.7 \times 11.7) = 516.4$. This genotype could then be compared with others just using the single merit criterion of the index scores.

In the British National Improvement Scheme in 1987, for **pigs**, the average index score for boars was 200, with most boars in the region of 165 to 235, where higher scores indicate more overall merit. Only boars scoring 250 or over were eligible for use in artificial insemination centres, which was just the top four per cent. A rise of 10 points in a boar's index score worked out at an average improvement worth 23 pence per piglet sired.

One can use index selection on subjective character rankings, not just on objective measured characters as in the wheat example above. For example, one might score some character such as cat coat colour for a particular breed on a five-point scale, e.g., 1, very poor; 2, poor; 3, average; 4, good; 5, very good, or on a 10-point scale, again weighting different characters according to their importance. Robinson (1991) gives a theoretical cat example with a 10-point scale for each of the seven characters, with a weighting of 5 for health (H), 4 for coat colour (CC), 3 for coat texture (CT), 3 for body build (BB), 2 for head shape (HS) and 1 each for feet (F) and tail (T), so that head shape is twice as important as feet, coat colour is four times as important as feet, etc., with a maximum score of 190 for a perfect animal for all characters, 10 for each character, weighted. Table 5.1 gives some possible scores for each of the above characters for a number of individual cats, with their total index score.

Examination of Table 5.1 shows that cat A is the best overall, as it scores highly on the highest weighted characters, in spite of being mediocre for the less highly rated characters, and has rather bad feet. Cat B has a slightly lower index score than cat A, but is a better all-rounder, with no feature scoring less than 7. Cat F is quite good overall except for the most important criterion, health. On the selection index, one would choose cats for breeding in the preference order of A, B, C, D, E, and perhaps reject F and G. In contrast, if one used independent culling levels for each character, one might

**Table 5.1.** **Cat character scores and index values**, theoretical values for seven cats. The abbreviations are explained in the text. Adapted from Robinson (1991).

| Cat | Merit score out of 10 for each character, unweighted | | | | | | | Index score; total after weighting |
|---|---|---|---|---|---|---|---|---|
| Weighting | H 5 | CC 4 | CT 3 | BB 3 | HS 2 | F 1 | T 1 | |
| A | 10 | 10 | 10 | 6 | 5 | 3 | 5 | 156 |
| B | 10 | 8 | 7 | 7 | 7 | 8 | 8 | 154 |
| C | 10 | 6 | 7 | 6 | 8 | 9 | 9 | 147 |
| D | 10 | 6 | 6 | 7 | 8 | 5 | 6 | 140 |
| E | 10 | 6 | 6 | 7 | 7 | 6 | 6 | 139 |
| F | 6 | 7 | 7 | 8 | 6 | 7 | 7 | 129 |
| G | 9 | 5 | 4 | 4 | 5 | 5 | 4 | 108 |

reject cat A as it has bad feet, scoring only 3 out of 10 for that. It makes sense to set lower culling levels for less important characters and higher ones for the more important characters. A solution here might be to mate cat A with one having particularly good feet. In animals where performance in the show ring is of the greatest importance, as with pedigree cats or dogs, it makes sense to weight different characters in proportion to their importance in show judging. In agricultural plants and animals, one would weight them according to their commercial value.

In dairy cattle, the **Predicted Transmitting Ability** (Production PTA) is a genetic index for individual production traits, i.e., milk (kg), fat (kg and %) and protein (kg and %), expressing the production potential an animal will pass to its progeny. PTAs for males and females are comparable and in the same format, and form the basis of other indexes such as PIN, £PLI and ITEM. PTAs are calculated from the individual animal model, using the performance of the animal and its relatives. They allow for as many non-genetic variables as possible, by standardising actual production for differences between herds (one may be more high yielding) and within herds (some have larger spreads of values), differences due to the time of year of calving, etc. All PTAs are expressed against the same genetic base, set as the average of all cows born in 1990 being zero. As this was set up in 1995, they were initially PTA95 values and the base is changed every five years, so PTA95 is replaced in 2000 and 2005.

For dairy cattle, the **Profit Index (PIN)** is an economic index (expressed in pounds sterling in the UK), giving the additional profit margin over all feed and quota costs per daughter per lactation, expected from mating with an individual with a PIN value of £0. It weights the PTA values economically. Currently, PIN = $(-0.03 \times$ PTA milk, kg$) + (0.60 \times$ PTA fat, kg$) + (4.04 \times$ PTA protein, kg$)$. Its calculations include the amount of feed energy to produce each milk component, current and expected milk prices, the cost of a milk quota allocation, and the cost of cooling and transporting milk (see Holstein, 1999a and b).

A related index, **ITEM (index of total economic merit)** combines the PIN with a financial value for type, including traits associated with long herd life. Financial values ascribed to key daughter traits for feet and udders are incorporated into the Profit Index to give a single tool for selection of bulls for dairy cattle. For example, Holstein bull Atrius from Italy had a PIN of £80 and an ITEM of £86 in 1995, having given 57 daughters in 49 herds, with milk yield at $+876$ kg, fat at $+20.9$ kg and protein at $+23.3$ kg, and strongly transmitting genes for good udder attachment and teat placement (figures taken from Genus Dairy Directory, 1996).

ITEM was replaced by **£PLI** in February, 1999, also taking into account the contribution of longevity to profitability. To the PIN value, it adds weightings for specific type traits as indirect measures of longevity, then actual longevity information when available. Weightings include $0.043 \times$ foot angle, $0.123 \times$ fore udder attachment, $0.023 \times$ udder depth and $-0.096 \times$ teat length, as the main linear traits associated with survival (see Holstein, 1999a and b). Increased longevity is important to profitability because the age profile increases, with fewer immature non-milking cows, and fewer replacement cows are needed.

**Genetic gains** have been impressive recently, using index selection. For example, the average genetic value of **Holstein Friesian** Society registered females increased by £46 PIN between 1990 and 1997. At £6.44 per £1 PIN, genetic improvements over that period averaged £42 per cow per year, or just under £300 per cow, with the greatest improvement in the top pedigree cows, but even the average cow showed a great improvement: (see Table 5.2).

**Table 5.2.**   Pedigree cow improvement, Holstein Friesians, 1990–1997. Date from Holstein (1999a).

| Group | Average PIN | Average PIN | Annual PIN gain per year | Value of gain, at £6.44 per £1 PIN |
| --- | --- | --- | --- | --- |
| | 1990 | 1997 | 1990–1997 | 1990–1997 |
| **Top 1%** | 44.9 | 99.4 | 7.79 | £351 |
| **Top 5%** | 31.8 | 83.8 | 7.43 | £335 |
| **Top 10%** | 24.8 | 75.4 | 7.23 | £326 |
| **Average** | 0.0 | 45.8 | 6.54 | £295 |

## 5.11. Selection intensities and rates of response to selection; a key equation for selection responses

The **maximum selection intensity** which a breeder can impose on a population depends very much on the organism's biology, especially its reproductive potential. In a microorganism such as yeast, one can have billions of yeast cells in a culture and select hard for rare improved forms, e.g., more alcohol-tolerant wine yeasts. In cattle, one needs to keep half to a third of all female calves just to maintain herd size, so the selection on those females is usually weak. One can select much harder on bulls, where one bull can serve 30,000 cows a year by AI, such that less than about 1 bull in 10,000 needs to be kept. It is easy to monitor weight gain for selecting beef bulls in large herds, but one could not test a large herd of dairy bulls for milk-yield genes by individual progeny testing, as it would require too many cows.

The **generation times** may differ between males and females, depending on the age at which they start and finish useful reproductive life. It is useful to think in terms of genetic gain *per year*, not just *per generation*. The relation between **genetic gain per year**, generation time, selection intensity, phenotype variation and heritability is shown in Fig. 5.4. It is obvious that there is no genetic gain if there is no phenotype variation or if the heritability is zero. For pigs, McPhee (2005) recommends culling sows after their fifth litter and replacing boars before 18 months, giving an average pig generation time of 1.8 years. In typical sheep flocks, rams are genetic gain per year =

$$\frac{1}{2}\left(\frac{\text{selection intensity in male}}{\text{generation time in male}} + \frac{\text{selection intensity in female}}{\text{generation time in female}}\right) \times$$

phenotype variation x narrow sense heritability.

**Fig. 5.4. The equation for genetic gain per year.**

replaced after 1 year and ewes after 3.5 to 4.5 years, giving a generation interval of 2.4 years.

**Simultaneous selection for several characters** reduces the maximum possible selection intensity for any one character. Suppose we were selecting for eight different characters in a plant where we can grow and test 20,000 plants a year, and that 80 plants were needed to give sufficient seed to grow 20,000 plants in the next generation. If we selected the best half of the population of 20,000 plants for each of the eight characters, we would be left with $(0.5)^8 = 0.0039$ of the population, or just 78 plants, when we need 80 to give us 20,000 plants for the next generation. If we selected equally for all eight characters, we could not select harder than the best half for each character, if we needed to maintain the population size. In organisms with a lower reproductive potential, we could select even less hard.

Maintaining and testing 20,000 wheat or barley or maize plants is not too difficult, but maintaining a similar number of pigs, horses, goats, sheep or cattle is more than most individual breeders could manage. It is therefore quite common to have **national or regional breeding schemes**, like the British National Improvement Scheme for Pigs, to help share the load between breeders. In Norway, there is a system of "ram circles" where over 2,000 rams are progeny-tested annually over a number of different flocks, increasing the flock size for selection; in the 1990s, it increased lamb weaning weight by 0.25 kg a year and the number of lambs weaned per ewe by 0.023 a year.

There is a **key equation for quantitative characters**, relating the **offspring yield** ($\bar{Y}_o$) to the **population average** ($\bar{Y}$), the **average of the selected parents** ($\bar{Y}_{par}$), and the **heritability** ($h^2$) for that character. The difference between the selected parent average (also called the "**mid-parent value**") and the population average is called the "**selection**

**differential**".

$$\bar{Y}_o = \bar{Y} + h^2(\bar{Y}_{par} - \bar{Y}).$$

It is obvious from this equation (as well as from common sense) that there will be no gain from the selection if the heritability is zero, and that the gain in offspring yield over the population average value increases as the heritability increases and as the selection differential increases. The better the selected parents are, compared to the population average, the greater the gain, or the greater the reduction if the selected parents have a lower value than the population average, as it might be if selecting for calves which were smaller at birth, to reduce birth trauma, or if selecting for reduced backfat in pigs. This key equation can also be used to calculate the **realised heritability** from observed values of $\bar{Y}_o$, $\bar{Y}$ and $\bar{Y}_{par}$ where these have been determined experimentally. Section 3.3 has various values for human heritabilities.

**Example 5.1.** Calculate the expected offspring average for weight at 160 days for pigs in a population averaging 82 kg, if the heritability is 0.3 and selected parents weighing 95 kg (when they were at that age) are used.

$$\bar{Y}_o = 82 + 0.3(95 - 82) = 85.9 \,\text{kg}.$$

The gain is only 3.9 kg although the selection differential was 13 kg, because $h^2$ was 0.3. If parents of weight 95 kg (when they were at 160 days) were again chosen in the next generation, when the population average was now 85.9 kg, the gain would be less as the selection differential is now only $(95 - 85.9) = 9.1$ kg, and because heritability falls during selection programmes as selection reduces genetic variation.

## 5.12. *In vitro* selection

Although selection is usually carried out on whole animals or plants, it can also be tried in plants on germinating pollen (Subsec. 17.1.7) or cells or tissue culture. For example, resistance to the weedkiller glyphosate was selected in callus tissue from *Petunia*, and persisted into whole plants from which it was transferred to several crop species. Sugar cane tissue cultures were used to isolate resistance to the Fiji virus disease and to the fungus *Helminthosporium sacchari*. Wheat seedlings grown *in vitro* were selected for resistance to *Fusarium culmorum*. See Wenzel and Forough-Wehr (1993). Unfortunately, variation selected in tissue culture often proves not to be inherited or to be unstable. Thus, one can easily select salt-resistant calli, but either the calli fail to regenerate to give whole plants or the whole plants lack the resistance.

## 5.13. Selection at the haploid stage

Section 10.3 describes the uses of monoploid plants for selection by plant breeders, where dominance does not mask the presence of harmful or beneficial recessive alleles. In animals and humans, the diploid genotype of the male parent generally controls the sperm cell's proteins. The **sperm's own haploid genome** is therefore not involved in producing proteins or RNA, so one would not expect the sperms of different genotype to have different powers of fertilisation, with no natural selection at the haploid stage, although many million sperm are released at the same time. The fact that **human chromosome aberrations** are more often passed through the mother than the father is strange, if the sperm genotype (including its karyotype) has no effect on its competitive fertilising ability.

With so many more mobile sperm than eggs, one would not expect competition between the large, non-self-mobile **eggs** for sperm. If there is more than one egg present, there can of course be competition for resources after implantation, as may occur for twins or other multiple births in humans (Sec. 12.15), or in farm animals, especially in litter-bearing species. Competition between eggs is more likely well after fertilisation, when their diploid genomes are expressed, than during fertilisation. Artificial selection amongst **sperm** for altered sex ratios in farm animals, especially to get more females in dairy cattle,  has long been a goal. It was achieved in the late 1990s by flow cytometry (see Subsec. 17.2.2).

## 5.14. Selection for meat characteristics

As **meat characters** can often only be assessed after slaughter, they require special consideration. We saw in Sec. 5.8 that **family selection**, such as using families of half-sibs, with some slaughtered for assessment at the appropriate age and some still available for breeding, is a good method for carcase characters of low heritability. In sheep, Thompson and Ball (1997) concluded that there was genetic variation for body composition within breeds and between breeds, mainly affecting mature size. Lean-growth selection indexes are used for sheep in New Zealand, Australia and the UK, usually weighting growth rate: fat depth: eye-muscle depth at a 3:1:1 ratio of importance, respectively.

While midback fat depth can be assessed by echo-sounding on live animals, a much more detailed study of **fat distribution** within live animals can now be made using **computer-aided tomography**. It is better than echo-sounding, but its high capital and running costs rule out its routine use. **Ultrasonics** can be used for muscle depth as well as fat depth. In sheep, the Texel breed has a low-fat carcase, measured as total body fat. One has to consider both the visible ("dissectible") fat and the intramuscular fat (marbling). In beef, there is a slight correlation of increased marbling fat with increased tenderness. According to Wood and Cameron (1994), selection for decreased backfat in pigs has led to a positively correlated decline in intramuscular fat, and hence to a decline in the eating quality of meat from modern pigs. Fat often has a pleasant flavour in most farm animals, but health advice is usually to cut off any visible fat. As **ostrich meat** is very low in fat, it cooks faster than most other meats and shrinks less during cooking, but low fat meats burn more easily during cooking and can dry out if overcooked. Less fat joints often require more basting with fat during cooking than do leaner meats. See Subsec. 19.2.4 for health aspects of food, and Table 5.3 for a comparison of cholesterol, protein and fat content of different meats.

**Table 5.3.** **Meat composition**, based on 85 g portions. Undated information provided in 1999 from "Ostrich, the Healthy Alternative", Ashdown Foods, Tenterden, Kent.

|  | Beef | Chicken | Lamb | Ostrich | Pork | Turkey |
|---|---|---|---|---|---|---|
| **K-Calories** | 240 | 138 | 205 | 97 | 235 | 135 |
| **Protein, g** | 23 | 27 | 22 | 18 | 23 | 25 |
| **Fat, g** | 15 | 3 | 13 | 2 | 19 | 3 |
| **Cholesterol, mg** | 77 | 72 | 78 | 9 | 82 | 59 |

Important meat characters are the proportions and appearance of muscle, bone and fat in retail cuts, the **yield of usable meat per carcase**, and the **sensory properties** of meat at the time of purchase and when it is eaten. When eaten, which is usually after cooking, meat should have a pleasing appearance with regard to colour, surface texture and the amount of fat. It should smell and taste good, and be tender, juicy and convenient to eat. People prefer large chunks of flesh to small amounts which have to be prised away from the bone, as with some rabbit joints and some very bony fish. See Subsec. 13.7.10 for the critical carcase characters in beef cattle. Subsecs. 13.7.9 to 13.7.13 have meat references, especially Subsec. 13.7.13 on pig breeding.

**Meat colour** generally depends on the amount of myoglobin present, its level of oxidation, and on light-scattering effects which depend on the myofibrillar volume. The colour of meat can change on exposed surfaces with time, so that butchering and meat treatments can influence colour at the time of sale. Oxidation to oxymyoglobin gives a desirable bright red colour. Poultry often has white breast meat and dark leg meat. **Tenderness** depends on the proportion and composition of connective tissue, the amount of intramuscular fat, and on myofibrillar tenderness (see Thompson and Ball, 1997). It can be assessed from the force needed to shear the sample at right angle to the fibre axis, or by tasting panels. **Marbling** of fat amongst the protein is very important too; it can only be measured after slaughter, but it is highly correlated with intramuscular fat which can be assessed on live animals by ultrasound. Reverter *et al.* (2000) compared live-animal ultrasound and carcass traits in cattle, finding a number of useful correlations to guide selection, but meat colour, tenderness, juiciness, flavour, marbling and water loss could only be assessed after slaughter.

In **sheep**, there are three major genes affecting **muscling**. The *callipyge* gene gives greater muscling in the leg, loin and shoulder, lower fat thickness and greater feed efficiency, but the meat is tougher and less juicy. Most meat characters are multifactorial, with low heritabilities and many loci involved. For tables of heritability, repeatability, breeding aims and hybrid vigour in farm animals, see Dalton (1980). For pigs, McPhee (2005) gives these surprisingly high figures for **pig heritabilities** in Australia: backfat thickness, 30–70%; growth rate 20–50%; feed conversion ratio 20–50%; litter size at birth 0–20%; litter size at weaning 0–20%.

Genetic maps of the main farm animals and much sequence data are now available, giving a framework for **marker-assisted selection** (Sec. 8.7), using molecular markers linked to production traits. Gomez-Raya and Klemetsdal (1999) give examples of QTL-marker selection and point out that marker-assisted selection acts on genotypes, not phenotypes, so that they can have faster effects than phenotype selection. One study (Maher *et al.*, 2004) found that the **environment** that an animal is raised in prior to slaughter and the **handling of the carcass** have much greater effects on meat quality than does genetic selection within a breed. In Australia, hot summers and long distances to transport animals from farms to abattoirs mean that pigs expressing the halothane (stress) gene have a higher risk of stress death, thus giving inferior meat.

## Suggested Reading

Abbott, A. J. and R. K. Atkin (eds.) *Improving Vegetatively Propagated Crops*. (1987) Academic Press, London.

Baker, R. J., *Selection Indices in Plant Breeding*. (1986) CRC Press, Boca Raton, Florida.

Cameron, N. D., *Selection Indices and Prediction of Genetic Merit in Animal Breeding*. (1997) CAB International, Wallingford.

Chrispeels, M. J. and D. E. Sadava, *Plants, Genes, and Agriculture*. (1994) Jones and Bartlett, Boston.

Dalton, D. C., *An Introduction to Practical Animal Breeding*. (1980) Granada, London.

Gomez-Raya, L. and G. Klemetsdal, Two-stage selection strategies utilizing marker-quantitative trait locus information and individual performance. *J Anim Sci* (1999) 77: 2008–2018.

Holstein, *Using Genetic Information to Improve Profitability*. (1999a) Holstein UK & Ireland, Rickmansworth.

Holstein, *Holstein Sire Summary 1999*. (1999) Holstein UK & Ireland, Rickmansworth.

Karn, M. N. and L. S. Penrose, Birth weight and gestation time in relation to maternal age, parity, and infant survival. *Annals Eugenics* (1951) 161: 147–164.

Lupton, F. G. H. (ed.) *Wheat Breeding: Its Scientific Basis*. (1987) Chapman and Hall, London.

Maher, S. C. *et al.*, Colour, composition and eating quality of beef from the progeny of two Charolais sires. *Meat Sci* (2004) 67: 73–80.

McPhee, C., Genetic selection in pigs. (2005), *www.dpi.qld.gov.au/pigs/1533.html*

Niks, R. E., P. R. Ellis and J. E. Parlevliet, Resistance to parasites, in Haywood, M. D., N. O. Bosemark and I. Romagosa (eds.) *Plant Breeding. Principles and Prospects.* (1993), pp. 422–447. Chapman and Hall, London.

Piper, L. and A. Ruvinsky (eds.) *The Genetics of Sheep.* (1997) CAB International, Wallingford.

Poehlman, J. M. and D. A. Sleper, *Breeding Field Crops.* 4th ed. (1995) Iowa State University Press, Iowa.

Reverter, A. *et al.*, Genetic analyses of live-animal ultrasound and abattoir carcass traits in Australian Angus and Hereford cattle. *J Anim Sci* (2000) 78: 1786–1795.

Robinson, R., *Genetics for Cat Breeders.* 3rd ed. (1991) Pergamon Press, Oxford.

Strickberger, M. W., *Genetics.* 3rd ed. (1985) Collier Macmillan, London.

Thompson, J. H. and A. J. Ball, Genetics of meat quality, in Piper, L. and A. Ruvinsky (eds.) *The Genetics of Sheep.* (1997), pp. 523–538. CAB International, Wallingford.

Wenzel, G. and B. Forough-Wehr, *In vitro* selection, in Haywood, M. D., N. O. Bosemark and I. Romagosa (eds.) *Plant Breeding. Principles and Prospects.* (1993), pp. 353–370. Chapman and Hall, London.

Wood, J. D. and N. D. Cameron, Genetics of meat quality in pigs. *Proceedings of the 5th World Congress on Genetics Applied to Livestock Production*, Guelph, Ontario, Canada (1994) 19: 458–464.

# Chapter 6

---

# Departures from Random Mating

## 6.1. Positive and negative assortative mating

As mentioned in Subsec. 1.3.14, with **positive assortative mating**, individuals similar with respect to a particular character are more likely to mate with each other than expected by chance. However, with **negative assortative mating**, similar individuals are less likely to mate with each other than expected by chance. Humans are said to mate approximately at random with respect to blood groups, but with positive assortative mating for height, with tall males tending to marry tall females, and short males tending to marry short females, and strong positive assortative mating for intelligence (see Fig. 6.1). **Positive assortative mating for size** has been shown in many groups, including *Drosophila malerkotliana*, the beetle *Diaprepes abbreviatus*, garter snakes and earthworms, *Lumbricus terrestris*. Although it is said that opposites attract, examples of negative assortative mating are harder to find, unless one takes sex or self-incompatibility alleles as the character. **Negative assortative mating** was shown by Houtman and Falls (1994) in the white-throated sparrow, *Zonotrichia albicollis*, where white-striped females preferred tan-striped males, which preferred white-striped females over tan-striped ones. This helps to maintain the polymorphism.

In human genetics, it is useful to consider whether assortative mating might have caused deviations from Hardy–Weinberg equilibrium. Although plants do not make conscious decisions about mating, alike individuals may still mate more often than unlike individuals for some characters, especially for time of flowering. Two early-flowering plants, or two late-flowering plants, are more likely to cross with each other than an early-flowerer with a late-flowerer. Mechanisms allowing self-pollination give the equivalent of **extreme positive assortative mating**, and cross-pollination mechanisms may give **negative assortative mating**, as for pin and thrum flower types in primrose, *Primula vulgaris*, which combines a cross-pollination mechanism with physiological partial self-incompatibility. Plant self-incompatibility alleles give negative assortative mating, with most or all plants heterozygous for such alleles (Subsec. 17.1.2), depending on the incompatibility system and dominance.

Any character causing plants, animals or microbes to occur preferentially in **particular niches** or locations could cause assortative mating because organisms similar for that character will often occur in close proximity to each other within such niches, and away from other types, if they tend to occupy different niches.

Although with farm animals the breeder usually determines which animals mate, there could still be assortative mating taking place within groups of animals, unless AI is used. It is safest to put one sex of one type with the other sex of the other type, so there is no choice of types for mating when it is desired to cross two genetic types.

In **humans, positive assortative mating** occurs for height, intelligence, social status, education level, interests, hair colour and deafness. The closeness of **IQ scores** within 51 American married couples is really striking in the data of Outhit (1933, illustrated in Vogel and Motulsky, 1997, and in Fig. 6.1 below), with very bright individuals usually marrying very bright, and the less bright marrying less bright individuals. For example, of 18 men with IQs between 120 and 140, 15 married women in that IQ range, 1 married a woman of IQ 110–120, and 2 married women of IQ 90–100. The present author has reanalysed Outhit's data, finding a **correlation coefficient for IQ** between members of couples of **+0.77**, significantly different from zero at P = 0.001. The average difference in IQ between members of a couple was 11 points, with 63% of the couples having a difference in IQ of only 0 to 10 points. Only 6% of the couples had an IQ difference greater than 30 points.

Hedrick (1983) gave a table of **correlation coefficients for human assortative mating** for a number of characters from many studies, with fairly poor agreement between studies, and with no indication of which correlation coefficients were significantly greater than zero (positive assortative mating). Typical correlation coefficients were +0.1 to 0.4 for **height**, +0.1 to 0.4 for **weight**, +0.1 to 0.3 for **chest circumference**, +0.1 to 0.4

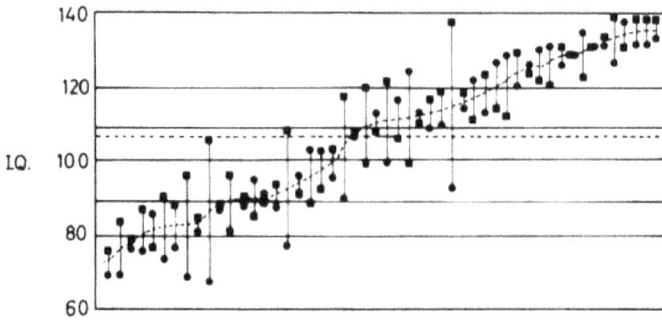

**Fig. 6.1.** **Positive assortative mating for IQ in humans**. Each joined vertical pair of points shows one couple, with the 51 couples arranged from left to right in the order of ascending average IQ. The dotted line shows the average IQ of the population. ■ males; ● females. Data from Outhit (1933) illustrated in Vogel and Motulsky (1997).

for **hair colour**, and a very wide range, from $<0$ to $>0.4$ for **eye colour** in different studies.

In **humans**, **positive assortative mating can arise from** conscious choice, subconscious choice, increased meeting opportunities due to common interests, similar jobs (e.g., one quarter of UK medical doctors are married to other medical doctors) or type of education, or membership of a defined or diffuse subgroup of the population. For example, undergraduates tend to meet other undergraduates more than they meet groups of mentally retarded people. Immigrants from a particular country or region (e.g., Gujaratis in London) or ethnic minorities often marry within their group, as do members of religious groups, castes or sects.

Positive assortative mating for genetically determined traits may increase the phenotype frequency of **recessive traits**. **Red hair** in humans is usually determined by recessive alleles, and there is positive assortative mating between redheads (which could be from choice or from mating within geographic sub-populations, where Scots tend to have red hair more often than in many other groups). Another often recessive trait with strong positive assortative mating is **deafness**, where deaf individuals often form social groups, with special schools and job training, and easier communication and better understanding within the group than with people of normal hearing. The increase in deafness from this nonrandom mating is only slight, as normal alleles at different loci are complementary, so that two deaf parents often have hearing children, with at least 30 different loci with recessive alleles causing deafness.

Inbreeding and outbreeding affect the genotype frequencies for all loci. Assortative mating primarily affects the loci for that trait, such as red hair, but can have **secondary effects on other loci** which share correlations with that character. Thus, if there is positive assortative mating for one Negroid characteristic such as very dark skin, it could cause apparent positive assortative mating for **associated characters** such as hair type, nose shape and certain blood groups.

## 6.2. Inbreeding and outbreeding and their consequences; inbreeding depression and outbreeding depression

**Inbreeding** and **outbreeding** concern degrees of relatedness, not individual characters. **Inbreeding** is a tendency for mating between individuals more closely related (or less closely related for **outbreeding**) than would occur by chance in that population. Mating between relatives is not necessarily inbreeding: all members of a species are ultimately related to each other. The terms "inbreeding" and "outbreeding" are sometimes used with different meanings in other contexts, such as incompatibility studies.

Suppose that in a small community the **average degree of relationship** was fifth cousins, then mating averaging out as being between fifth cousins, with some closer and some less close matings, would be random. If mating were on average between second cousins, that would be inbreeding, while if mating were on average between eighth cousins, that would be outbreeding. In a larger population, where the average relationship between individuals was tenth cousins, then mating on average between eighth cousins would be inbreeding, not outbreeding.

To illustrate the consequences of inbreeding, let us take the most extreme form of inbreeding, **self-fertilisation**, which occurs regularly in many plants and sometimes in insects, trematodes (flukes), cestodes (e.g., tapeworms) and other lower animals. On selfing, each homozygote gives all homozygotes, $A\,A$ giving all $A\,A$ and $a\,a$ giving all $a\,a$. The heterozygotes, however, give half heterozygotes and half homozygotes, with $A\,a$ selfed giving 1 $A\,A$: 2 $A\,a$: 1 $a\,a$. The proportion of heterozygotes is therefore halved each generation, with corresponding and equal increases in both homozygotes.

Let us consider two cases of selfing, one in a population with equal allele frequencies, $A = p = 0.5$, $a = q = 0.5$, and a second in a population with $p = 0.8$ and $q = 0.2$. The genotype frequencies over a number of generations are shown in Table 6.1. The frequency of heterozygotes is halved each generation, going eventually to zero, with corresponding increases in both homozygotes. The allele frequencies are unchanged by inbreeding, and one ends up with the frequencies of the two types of homozygotes exactly reflecting the allele frequencies in that population.

The **phenotype and genotype frequencies** change with inbreeding, but not the allele frequencies. In particular, there is an increase in the frequency of the recessive phenotype, and the rarer the recessive allele is, the greater the proportional increase in the frequency of the recessive phenotype (see Table 6.1), as rare recessives are mainly hidden in heterozygotes, whereas dominant alleles are expressed in homozygotes and heterozygotes. **Less close inbreeding** than selfing, e.g., brother-sister mating or first-cousin mating, has the same ultimate effect of reducing the frequency of heterozygotes and increasing the frequency of the homozygotes, but it takes more generations to have the same effect as selfing, considered at the end of Sec. 6.4.

**Table 6.1.** The effects on genotype frequencies of selfing, which is extreme inbreeding. In generation 0, with no inbreeding, one gets Hardy–Weinberg equilibrium values.

| Generations of selfing | Genotype frequencies in two populations with different allele frequencies | | | | | |
| --- | --- | --- | --- | --- | --- | --- |
| | Allele frequencies $p = q = 0.5$ | | | Allele frequencies $p = 0.8$, $q = 0.2$ | | |
| | $A\,A$ | $A\,a$ | $a\,a$ | $A\,A$ | $A\,a$ | $a\,a$ |
| 0 | 0.2500 | 0.5000 | 0.2500 | 0.6400 | 0.3200 | 0.0400 |
| 1 | 0.3750 | 0.2500 | 0.3750 | 0.7200 | 0.1600 | 0.1200 |
| 2 | 0.4375 | 0.1250 | 0.4375 | 0.7600 | 0.0800 | 0.1600 |
| 3 | 0.4688 | 0.0625 | 0.4688 | 0.7800 | 0.0400 | 0.1800 |
| ... | | | | | | |
| Many | 0.5000 | 0.0000 | 0.5000 | 0.8000 | 0.0000 | 0.2000 |

Virtually all **effects of inbreeding**, good or bad, stem from this reduction in heterozygosity and the increase in homozygosity. If strains or stocks or populations contain deleterious recessives, largely hidden in heterozygotes, then inbreeding will make them homozygous and their deleterious effects will show and can be selected out. Beneficial recessives will also be made homozygous and be exposed. See Sec. 13.1 for the use of inbreeding in making F1 hybrids, and Sec. 13.2 for breeding methods used with inbreeding organisms. Inbreeding will give a deficiency of heterozygotes in Hardy–Weinberg analysis, and outbreeding will give an excess of heterozygotes, compared with random mating. The **amount of inbreeding** in a population is often now estimated from the amount of heterozygosity for codominant molecular markers such as microsatellite loci.

In almost all non-inbred diploid populations, there are many deleterious recessive alleles, largely hidden in heterozygotes. If inbreeding is imposed on such populations, there is usually a loss of fitness and vigour, called **inbreeding depression**. See Charlesworth and Charlesworth (1999) for the genetics of inbreeding depression. In some pine (*Pinus*) species, inbreeding depression was so strong that selfing gave no surviving progeny (Lande *et al.*, 1994). In alfalfa (lucerne, a forage legume, *Medicago sativa*), few inbred lines survived beyond two or three generations of inbreeding, and in **pigs**, an increase in F (Wright's coefficient of inbreeding: Sec. 6.4) of 0.1 decreased average litter size by about one piglet. McPhee (2005) stated that in pigs, an increase in inbreeding of 10% reduced body weight at 160 days by 3 kg and reduced the number of pigs born per litter by 2.5%. He suggested an upper limit of 10% of inbreeding in a herd over 15 years, followed by outcrossing by AI to stock from herds following a similar selection programme. To keep inbreeding in check, he recommended the following minimum percentages of outside matings in relation to herd size: 50 sows, 28%; 100 sows, 12%; 150 sows, 6%; 200 sows, 2%; more than 250 sows, 0%. For the effects of human inbreeding, see Sec. 12.13.

In **rats**, before inbreeding, no matings were nonproductive, average litter size was 7.5 and mortality from birth to four weeks was 3.9%. After about 30 generations of inbreeding (parent/offspring and brother/full sister), 41% of matings were nonproductive, average litter size was 3.2, and mortality from birth to four weeks was 46% (data quoted by Strickberger, 1985).

In **naturally inbreeding populations**, however, natural selection will usually have already eliminated many harmful recessives and selected for beneficial ones. Many cultivated vegetables, cereals, flowers, sheep, pig and cattle breeds are highly inbred whilst still being useful commercially, but except in natural inbreeders, this has only been achieved by intensive selection to overcome inbreeding depression.

Inbreeding depression does not occur in naturally inbreeding species and may cease to occur after some generations of inbreeding in formerly random-mating species, as natural selection will tend to adapt them by eliminating deleterious recessives exposed to selection in homozygotes. The classic example is the **pharaohs** of ancient Egypt, where brother-sister and half-brother/half-sister marriage was practised for many generations. As we shall see in Sec. 12.13, humans carry on average the equivalent of at least three recessive lethal alleles, but usually in the heterozygous form. Presumably, the pharaohs suffered inbreeding depression when the inbreeding started, but must have adapted as they were very successful for thousands of years. An extraordinary case of regular inbreeding by **pre-birth full-sib-mating** is shown by the insect-parasitic mite, *Acarophenax tribolii*. Inside the mother mite, about 15 eggs are produced, giving one male and 14 females. The male fertilises his sisters and dies without ever emerging from his mother, so fertilised females are born ready to have offspring themselves, giving reproductive assurance.

As both dominant and recessive alleles are exposed to selection in inbreeders, selective differences will often lead to inbreeders being less genetically and phenotypically variable than random maters or outbreeders. As **inbreeding restricts gene flow** between individuals (especially with selfing) and populations, favourable new combinations of alleles which exist separately in different individuals are less likely to come together in inbreeders than in random maters, but once they do, they are more likely to be retained together. Suppose that *A A*, *b b*, *C C* would be very favourable and that initially, some individuals have *A A*, *B B*, *c c* and others have *a a*, *b b*, *C C*, then outbreeders are more likely than inbreeders to cross-fertilise and eventually produce *A A*, *b b*, *C C*, but mating of that with the other types could lose it again.

**Evolutionary strategies combining some advantages of inbreeding and outbreeding** include being mainly inbreeding but with some outcrossing, as in wheat, or to protect chunks of genetic material containing favourable gene combinations from recombination by **suppression of crossing over** there, e.g., by localisation of crossovers. Ordinary **linkage**, especially if tight, tends to preserve genes in parental combinations. In the fruit fly, *Drosophila melanogaster*, there is no crossing over in male meiosis, preserving linked combinations, but there is in female meiosis, giving the chance of better (and worse) linked combinations. **Soya bean plants** are bushy herbaceous annual diploids (*Glycine max*, 2n = 40) which are normally self-pollinated by the anthers maturing in the bud and shedding pollen directly on to the stigmas of the same flower. Occasional cross-pollinations are made by bees but have a frequency of less than 0.02% over distances of more than 4.5 metres, so the crop is very strongly inbreeding, with rare outcrossing.

A case of a closed population with restricted mating is shown by the **Chillingham Wild Cattle** (www.chillingham-wildcattle.org.uk). They were the wild cattle of Britain and have been enclosed in a 148 ha Northumberland park since 1270 AD, with no outcrossing at all. The numbers usually vary from 40 to 60, but in the hard winter of 1947, the herd was reduced to 8 cows and 5 bulls, and no youngsters. The population recovered in spite of the mating between close relatives. Inbreeding is increased by their mating system. Usually, there are about 8 bulls, 28 breeding-age females and 14 young, but only the competitively successful "King Bull" mates with the cows, reducing the effective population size to about 4. Although the cattle are small and primitive-looking, they are very hardy and do not appear to suffer from this inbreeding and the population crash, so most deleterious recessives must have been purged a long time ago.

In many but not all **birds and animals**, inbreeding usually gives inbreeding depression. For example, in great tits (*Parus major*) near Oxford in the 1960s and 1970s, nestling mortality was 16% in outbred pairs and 27% in inbred pairs, even when clutch sizes were similar. In brown lemurs, inbreeding had little effect on fitness, while in Sumatran tigers, it gave very low offspring survival. In the Japanese quail, three generations of full-sib mating gave sterility. If inbreeding depression is mainly due to **deleterious recessive alleles** being made homozygous and expressed, then selection should purge the population of those deleterious alleles with continued inbreeding. If, however, the depression is due to **loss of heterozygote**

**advantage**, the loss of fitness is likely to get worse, giving smaller populations with more matings between closer relatives.

**Inbreeding is often associated with the following**: annual plants or short-lived life forms, weeds, the extremes of a species' distribution, marginal habitats and stressful environments. If the organism is genetically well adapted to its environment, inbreeding provides a high proportion of adapted genotypes, while occasional outcrossing allows the production of altered genotypes which might be better adapted if conditions change or permit colonisation of ecologically different areas. In contrast, long-lived and perennial species are often outbreeding or random-mating: the existing genotypes can survive vegetatively for years and can tolerate high variability from sexual reproduction which can give long-term adaptation. Outbreeding is also frequent in short-lived organisms with restricted recombination, which preserves favourable gene combinations.

For a summary of the **effects of inbreeding on genetic diversity**, see Charlesworth (2003). Inbreeders often have less genetic variation than outbreeders, with the former often having more variation between populations than within them. This applies to **organelle genes** as well as nuclear genes, and organelle genes normally have much less variation than nuclear genes in inbreeders and outbreeders. Inbreeding, by promoting homozygosity, reduces any heterozygote advantage present in a population or agricultural stock.

**Inbreeding has some advantages** over random mating or outbreeding. Plants or animals which are self-fertile do not have to find mates to reproduce, a special advantage in species which colonise new areas where there may be no mates. Inbreeding also results in increased transmission of genes already selected to be locally adaptive, whereas outcrossing could introduce genes more suited to other habitats or niches. Hermaphrodite animals such as snails and earthworms cannot usually self-fertilise, but in those particular examples, they have low motility and low dispersion, often mating with near relatives. Some slugs and aquatic snails can self-fertilise, including the tropical freshwater snail *Bulinus truncatus*. In the natural populations of that snail, Viard *et al.* (1997) studied four highly polymorphic microsatellite loci. The average observed heterozygosity was very low, 0.06, compared with the expected value of 0.43 from random mating; the estimated selfing rate was 93%.

**Self-compatible plants** which can also outcross benefit from the advantages of both systems. In some such plants, they are self-incompatible when the flowers are young, but become self-compatible subsequently if they are not cross-fertilised. Some violets (*Viola* spp.) have normal flowers which can cross-fertilise, but near the end of the flowering season, they produce small cleistogamous flowers which are selfed without opening, giving reproductive assurance without the need to attract and reward pollinators with scent and nectar.

**Recombination events** only produce new combinations of alleles if both of two loci are heterozygous (e.g., *A a*, *B b* can give two parental and two recombinant gametes), but not when one or both loci is homozygous (e.g., *A A*, *B b* can only give two parental gametes, *A B* and *A b*). Another consequence of inbreeding is therefore the **loss of effective recombination** as homozygosity increases and heterozygosity decreases, in proportion to F.

For many more **equations and quantitative treatments of nonrandom mating**, see Hedrick (2005) and Falconer and Mackay (1996). **Outbreeding depression** can occur when two well-adapted strains cross to produce a less well adapted strain, especially if they are adapted to different habitats. For example, the Alpine population of mountain ibex (*Capra ibex*) crashed in the late 19th century and attempts were made to restore numbers by crossing them to Turkish and Nubian races. The Alpine population mate in midwinter with offspring born in summer, while the Turkish and Nubian races mate in summer. The hybrids mated in summer, with offspring born in midwinter and dying from the cold.

**Population crashes** give genetic drift in the resulting small populations. The cheetah is deduced to have suffered a big population crash, as it has very low average heterozygosity for allozyme loci, 0.0072, compared with other big cats, such as 0.037 for African lions.

## 6.3. Why breeders often use inbreeding

Breeders usually wish to establish **their own special line**, strain, breed or variety of a plant or animal, so they will want it to be distinctive, with its own particular characteristics in the parents and in all the offspring. It should be recognisably different from other strains and should normally breed true for its main characters. The breeders therefore need to "fix" the line's principal characters, which in practice means making them homozygous, as heterozygous loci will show segregation at meiosis and in offspring. Although uniformity of parental phenotypes might be achieved by strong selection, if some parents are $A\ A$ and some are $A\ a$ for a locus with dominance, then $a\ a$ genotypes with a different phenotype could occur in the offspring from $A\ a \times A\ a$ matings.

**Inbreeding** is the obvious way to fix characters in a homozygous form, especially if a number of different characters are involved. The breeder has to balance a number of factors, such as the effects of inbreeding depression, the time taken in that organism for a given number of generations of inbreeding, the selection intensity permitted by the organism's biology, and the presence of less desirable characters which might also become fixed in that line through inbreeding. The closer the inbreeding, the fewer generations and years it takes to fix a line's distinctive characters, but the worse the inbreeding depression and the fewer generations of selection there are to eliminate any bad characters. The effects of the different methods of inbreeding on the inbreeding coefficient are considered in Sec. 6.4, and shown in Table 6.2. If inbreeding has not yet fixed nearly all characters, then the breeder can continue to select against any poorer features. Initial selection for a line's characters will be followed by further selection during inbreeding.

In animals, one cannot normally self-fertilise an individual, so sib-mating, parent-offspring mating or half-sib mating can be used where intensive inbreeding is wanted. If a **less intense inbreeding scheme** is desired, the **closed stud method** can be used, as is often the case with **pedigree cats** (Robinson, 1991). In any generation, just one or sometimes two outstanding males are used on all the females, with perhaps six females per generation; the smaller the number of males and of females, the stronger the inbreeding and the reduction in heterozygosity. If only one male is used, the offspring in a litter-bearing species such as dogs, cats or pigs will be a mixture of half-sibs and full sibs, but some less related individuals would result from using two males. In dogs, one would want a moderately large number of puppies from each pairing, so that selection intensity in choosing the ones for later breeding can be high. While it is possible to backcross the best daughters to their outstanding father, it is usual to select the best male from the next generation to continue the line, as males in each generation should become progressively more homozygous than those in previous generations. In the closed stud method, one might keep six to ten breeding animals each generation, depending on the organism's biology, and on how many individuals one can rear and select amongst. An alternative, if one could keep say only eight breeding animals per generation, might be to keep four **different full-sib lines** in each of which one carried out full sib-matings, which only need two breeding animals per generation. Selection would be made within lines and between lines, replacing any weak lines by splitting the most successful line into two.

Inbreeding plus selection may sometimes **fail to "fix" characters** and to reduce heterozygosity as quickly as expected, even with high heritabilities. One reason is that the selection for the best, healthiest plants or animals in a generation may choose individuals with heterozygosity at certain loci, if they show heterozygote advantage. Failure to fix characters can also result from environmental effects on them, so that individuals with less good genotypes for that character have good phenotypes for it because of environmental effects. Negative correlations between characters can also be important in preventing success. As mentioned above, inbreeding depression may severely limit progress in inbreeding schemes. For example, in cats and dogs, it may produce small, thin, lethargic animals, reduce litter size, increase birth abnormalities, reduce sex-drive and fertility, and reduce health. In general, pedigree breeds of cats and dogs are less healthy than random-bred mongrels and are more temperamental, because of degrees of inbreeding depression.

## 6.4. Wright's inbreeding coefficient, F; Wright's equilibrium for genotype frequencies under inbreeding; calculation of F from pedigrees

The amount of inbreeding which has taken place is measured by **Wright's coefficient of inbreeding**, **F**, named after Sewell Wright. This is the proportion by which heterozygosity has been reduced by inbreeding. With no inbreeding, one gets $F = 0.0$, with Hardy–Weinberg equilibrium genotype frequencies. With complete inbreeding, one gets $F = 1.0$, with no heterozygotes at all; all individuals are homozygous, with the two homozygote types in proportion to the allele frequencies.

The F value is definitely not an absolute measure of heterozygosity or homozygosity, as those are features of allele frequencies and genetic diversity in a population. Having $F = 0.0$ does not mean that all individuals are heterozygous, just that heterozygotes have frequency $2pq$. If a random-bred sheep ($F = 0.0$) in a particular flock is on average heterozygous at 400 loci, then an individual with an inbreeding coefficient of 0.25 will be heterozygous at 300 loci, as 0.25 of the heterozygosity has been lost through inbreeding. In a more genetically diverse flock, a random-bred sheep might be heterozygous at 4,000 loci, and one with $F = 0.25$ would then be heterozygous at 3,000 loci.

Coefficient F can also be described as the probability that two alleles at a locus in an individual are **identical by recent descent**, that is, they are both derived by replication from the same allele in a common ancestor by inbreeding. A **common ancestor** of two individuals is an individual in some previous generation which is an ancestor of both of them, so that they can have some alleles in common by recent descent from it. When the two individuals mate, they can both provide an identical allele to an offspring.

Suppose we have two individuals and number the individual alleles with superscripts 1 to 3 for $A$, and 4 for the one $a$ allele. If we cross $A^1 a^4 \times A^2 A^3$, then some of the full siblings produced, such as $A^1 A^2$ and $A^1 A^3$, and $A^2 a^4$ and $A^3 a^4$, can be mated as full sib-matings, which is inbreeding. Some possible offspring from those matings are $A^1 A^1$ and $A^1 A^3$ from the former pair, and $A^2 A^3$ and $a^4 a^4$ from the latter pair. In each of $a^4 a^4$ and $A^1 A^1$, we have two alleles identical by descent, for example, the two $A^1$ alleles in $A^1 A^1$ being directly derived from the one $A^1$ allele in the common ancestor (of $A^1 A^2$ and $A^1 A^3$), $A^1 a^4$. In contrast, the two $A$ alleles in $A^1 A^3$ and the two in $A^2 A^3$ are not identical; they are both $A$, but are **similar**, not **identical** by recent common descent. The term **autozygous** is sometimes used of diploids with alleles identical by descent.

Consider a population with two alleles at one locus, $A$, frequency $p$, and $a$, frequency $q$. Without inbreeding, $F = 0$, we expect Hardy–Weinberg equilibrium genotype frequencies, $A A$, $p^2$; $A a$, $2pq$, and $a a$, $q^2$. Alleles in homozygotes will normally be similar, not identical, and the two different alleles in heterozygotes will neither be similar nor identical. With inbreeding, there will then be a chance that alleles in homozygotes are identical, and a correspondingly reduced chance that they are similar. The proportion of similar homozygotes is reduced by the inbreeding coefficient F to $p^2(1 - F)$ for $A\ A$, and to $q^2(1 - F)$ for $a\ a$. We get identical homozygotes in proportion to the allele frequencies, $A\ A$, $p$F, and $a\ a$, $q$F. The heterozygotes, $A\ a$, will by definition be reduced by fraction F to $2pq(1 - F)$.

At **Wright's equilibrium**, we therefore get:

$A\ A = p^2(1 - F)$ for similar alleles $+ p$F for identical alleles,

$a\ a = q^2(1 - F)$ for similar alleles $+ q$F for identical alleles,

$A\ a = 2pq(1 - F)$.

These can be simplified, e.g., $A\ A = p^2(1-F)+p\text{F} = p^2-p^2\text{F}+p\text{F} = p^2 + p\text{F}(1 - p) = p^2 + pq\text{F}$.

After simplifying, the **Wright's equilibrium values** are:

$$A\ A = p^2 + pq\mathbf{F},$$
$$a\ a = q^2 + pq\mathbf{F},$$
$$A\ a = 2pq(1 - \mathbf{F}).$$

These are important formulae and their derivation brings out the difference between similar and identical alleles. A shorter way to derive them is to say that the heterozygote is reduced by inbreeding by fraction F, from $2pq$ to $2pq(1 - F)$, which leaves $2pq$F to be equally distributed between the two homozygotes, so that the $p^2$ for $A\ A$ becomes $p^2 + pq$F and the $q^2$ for $a\ a$ becomes $q^2 + pq$F. If $F = 0.0$, the Wright formulae become the Hardy–Weinberg formulae. One can calculate the **average inbreeding coefficient for a population** from:

$$\mathbf{F} = 1 - \frac{\textbf{Observed frequency of heterozygotes}}{\textbf{Expected frequency of heterozygotes (i.e., } 2pq)}.$$

This will be zero if there is no inbreeding, and 1.0 if all heterozygosity has been eliminated by inbreeding.

**Values of F** vary from zero with no inbreeding (with Hardy–Weinberg levels of heterozygosity), to 1.0, fully inbred (with no heterozygosity). If the parents are not themselves inbred, the maximum F from one generation of inbreeding is $F = 0.5$, from selfing: we saw earlier that selfing reduced heterozygosity by a half, hence the value of 0.5. Values of F for the offspring, if parents are not already inbred, are 0.25 when parents are full sibs (brother and sister) or parent and offspring (father/daughter or mother/son), 0.125 for half-sibs, 0.0625 for first cousins and 0.03125 for second cousins (which are children of first cousins).

> **Example 6.1.** The recessive allele for an inherited recessive disease of sheep has a frequency of 1 in 200. How much is the risk of showing this disease increased if the parents are full sibs, compared to unrelated parents?

The recessive has an allele frequency of $1/200 = 0.005$, so the dominant has a frequency of 0.995. The chance of the disease in a lamb born to unrelated parents is $q^2 = (0.005)^2 = 0.000025$, or **1 in 40,000**. With inbreeding from full sibs, $F = 0.25$, and we have Wright's equilibrium formula, $a\,a = q^2 + pqF$, so the recessive homozygote has a frequency of $0.000025 + (0.995 \times 0.005 \times 0.25) = 0.00127$, or about **1 in 788** of being born showing this disease. The chance of showing the disease has therefore increased by about **51-fold**, compared with random mating, because of the full-sib inbreeding.

We can use Wright's equilibrium frequencies to work out the effects of inbreeding on **quantitative characters** in a population. If we have **additive action**, e.g., with $A\,A$ having height 100 cm, $A'\,A$ 150 cm, and $A'\,A'$ 200 cm, then with inbreeding, some of the heterozygotes (height 150 cm) go equally to the two types of homozygote, the average height of which is also 150 cm. Therefore, inbreeding does not change the average value for a population if there is additive action, with the heterozygote exactly intermediate between the two homozygotes.

If we have **complete dominance**, e.g., with $a\,a$ 100 cm high, $A\,a$ and $A\,A$ 200 cm high, then we could imagine a basic height of 150 cm, to which the dominant allele adds 50 cm, and from which the recessive allele subtracts 50 cm if homozygous; let us call this amount of 50 cm value $v$. For each of the three genotypes, one can work out their frequency, their quantitative value ($+v$ for $A\,A$ and $A\,a$; $-v$ for $a\,a$), and by multiplying their quantitative value by their frequency, their **total quantitative contribution** to the population can be calculated. With no inbreeding, with Hardy–Weinberg genotype frequencies, these total quantitative contributions are $+p^2 v$ for $A\,A$, $+2pqv$ for $A\,a$, and $-q^2 v$ for $a\,a$, totalling $v(p^2 - q^2) + 2pqv = \boldsymbol{v(p - q) + 2pqv}$, because if $p + q = 1$, then $p^2 - q^2 = p - q$. With inbreeding, the genotype frequencies are Wright's equilibrium frequencies and the corresponding total quantitative value becomes $\boldsymbol{v(p - q) + 2pqv - 2pqvF}$. Inbreeding has therefore caused the total quantitative value for the population to be **reduced by $\boldsymbol{2pqvF}$**. This is easy to understand because from inbreeding, we lose $2pqF$ amount of heterozygotes, phenotype value $+v$, with the heterozygotes going equally to $A\,A$, value $+v$, and to $a\,a$, value $-v$, so the $+v$ and $-v$ cancel each other out in the homozygotes produced from the heterozygotes.

As many quantitative traits such as high yield and high fertility are largely determined by dominant alleles, inbreeding, by moving population values towards the phenotype value of the recessives, usually reduces fitness and fertility, giving inbreeding depression.

We can also calculate the inbreeding coefficient F for an individual from **pedigree diagrams**, in which we include any individuals contributing to the inbreeding in that individual. Here, we use the fact that coefficient F can be described as the probability that two alleles at a locus in an individual are identical by recent descent, both derived by replication from the same allele in a common ancestor by inbreeding. We arrange successive generations in descending order in the diagram, oldest at the top, youngest at the bottom, labeling as individual I the individual whose F we wish to determine.

In Fig. 6.2, we have woman A having borne two half-sibs, woman D and man E, by different men, men B and C. The two half-sibs then mate to give child I whose F value we wish to calculate, to find the **F for offspring of half-sibs**. F will be the chance of getting identical alleles in child I. Men B and C can only contribute one allele each, as a maximum, to child I, so they are not common ancestors; only woman A can contribute both alleles to child I, being the only common ancestor. The inbreeding coefficient of I is the chance of it receiving identical alleles, i.e., of receiving $a1$ and $a1$, or $a2$ and $a2$, or $a1$ and $a2$ if those different alleles in woman A are themselves already identical from woman A being somewhat inbred herself. The chance of I getting identical alleles is the same as the chance that egg 3 and sperm 4 contain identical alleles. This will only happen if eggs 1 and 2

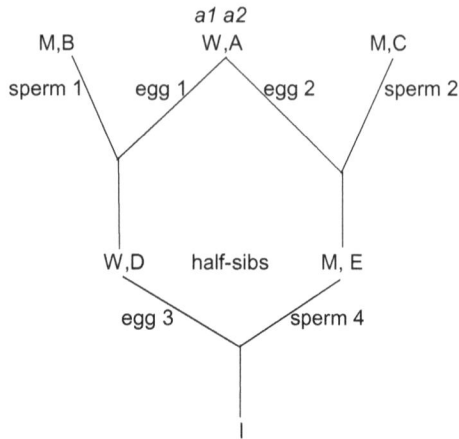

**Fig. 6.2.  Inbreeding diagram for the offspring (individual I) of half-sibs.** M stands for man, and W for woman.

carry identical alleles, and egg 1 and egg 3 have identical alleles, and egg 2 and sperm 4 have identical alleles.

Let us start by working out the chance that woman A's two eggs, 1 and 2, contain **identical alleles**: there is a chance of one quarter for them both being $a1$, a quarter for them both being $a2$, i.e., a chance of a half that the two alleles will be identical for one or other allele, and a half of being one of each allele, $a1$, $a2$. If the woman A is inbred, coefficient $F_A$, then there is a chance $F_A$ of alleles $a1$ and $a2$ being identical by descent, so the chance that eggs 1 and 2 between them contain one allele $a1$ and one allele $a2$ which are identical by previous inbreeding is $1/2\,F_A$. The total chance of eggs 1 and 2 having identical alleles is therefore $1/2$ [= chance of $a1\ a1$ or $a2\ a2$] + $1/2\,F_A$, which gives a total of $1/2\,(1 + F_A)$. The chance of eggs 1 and 3 having identical alleles is $1/2$, because egg 3 has an equal chance of getting its one allele at this locus from egg 1 or from sperm 1. The chance of egg 2 and sperm 4 having identical alleles is $1/2$, as sperm 4 has an equal chance of getting its allele from egg 2 or sperm 2.

The **total chance of child I getting identical alleles** is therefore $1/2\,(1 + F_A) \times 1/2 \times 1/2 = (1/2)^3(1 + F_A)$, or in more general terms, $(1/2)^3(1 + F$ of the common ancestor of that pathway). If we count the number of adults in the inbreeding pathway from the individual I to the common ancestor A and back to I, not counting I, we have three individuals, woman D, woman A (the common ancestor) and man E, and our formula has $1/2$ to the power of three. One can show that the contribution of any common ancestor (for which the symbol A is usually used) is $(1/2)^n(1 + F_A)$, where n is the number of individuals in the inbreeding pathway, excluding individual I, and excluding non-common ancestors such as men B and C in Fig. 6.2. In some pedigrees, there are more than one pathways to individual I, each giving an extra chance of that individual receiving identical alleles, so one has to sum up all the pathways to individual I. We therefore get the **general formula, $F_I = \Sigma \{(1/2)^n(1 + F_A)\}$**, where n is the number of individuals in a particular pathway and $F_A$ is the inbreeding coefficient of the common ancestor in that particular pathway.

One needs to be able to calculate an individual's F value to see how inbred it is, and so how much heterozygosity it might have lost, and for calculations involving Wright's equilibrium frequencies, as in Example 6.1.

**Example 6.2.** Calculate the inbreeding coefficient F for an individual whose parents were first cousins, assuming that the common ancestors are not inbred.

First cousins are descended from full sibs, and thus have one pair of grandparents in common, who will be the two common ancestors of the first cousins' offspring. Although Fig. 6.2 showed the sex of the individuals, that is not usually done in pedigree diagrams, and we do not usually show individuals not contributing to the inbreeding, which in Fig. 6.2 would be men B and C. Our inbreeding diagram for this example is shown in Fig. 6.3, with each grandparent able to contribute both alleles to child I, so that both are common ancestors of I.

There are five individuals in the pathway from I to common ancestor 3 and back again, non-italic numbered pathway, and five individuals from I to common ancestor 3 and back, italic numbered pathway; because the common ancestors are not inbred, $F_A$ is zero. From $F_I = \Sigma\{(1/2)^n(1 + F_A)\}$, we get $F_I = \{(1/2)^5(1+0)\} + \{(1/2)^5(1+0)\} = $ **1/16** or **0.0625** for the F value of the offspring of first cousins.

**Example 6.3.** The pedigree shown in Fig. 6.4 is for beef cattle, listing successive generations in order down the page. Calculate F for I if $F = 1/8$ for B and $1/4$ for C.

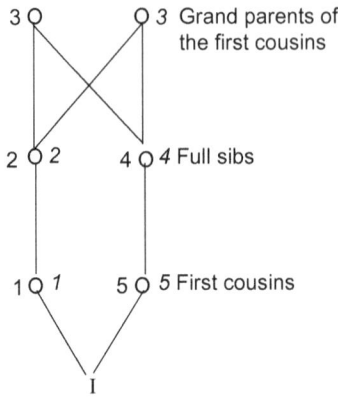

**Fig. 6.3.** **The inbreeding diagram for the offspring of first cousins.** The italic and non-italic numbers represent pathways to different common ancestors.

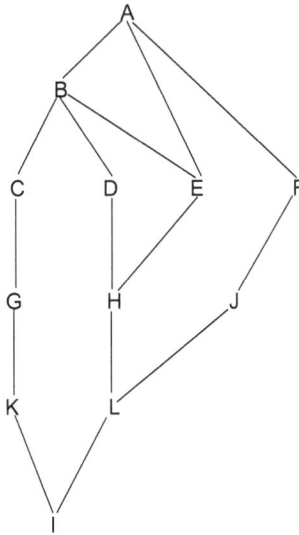

**Fig. 6.4.** **Pedigree diagram for beef cattle**, for calculation of F for individual I.

We see that some individuals, such as G, have only one ancestor shown, because the other parent, such as the individual which mated with C, cannot be a common ancestor or be involved in the inbreeding of I, so it is omitted. We see that A is a parent of B, but mated with B to produce E, a parent-offspring mating. There are five pathways back to common ancestors, which are not all in the same generation. For example, A could have contributed identical alleles to I, by more than one pathway, and so could B.

The various pathways, with their common ancestor underlined and their contribution to $F_I$ from the equation $F_I = \Sigma\{(1/2)^n(1 + F_A)\}$, are:

KGC**B**DHL $= (1/2)^7(1 + 1/8) = 9/1024;$
KGC**B**EHL $= (1/2)^7(1 + 1/8) = 9/1024;$
KGCB**A**EHL $= (1/2)^8(1 + 0) = 1/256;$
KGCB**A**FJL $= (1/2)^8(1 + 0) = 1/256.$

The total value for $F_I$ is the sum of these values, **26/1024**, or **0.0254**. Answers may be given as fractions or decimals.

There are three points to note in such calculations. Firstly, no individual ever occurs twice in any one pathway. Secondly, one may be given irrelevant data, such as F of individuals which are not common ancestors, such as C in this example. In any pathway, the inbreeding coefficients of individuals are not used in the calculation unless they are the common ancestor of the pathway being considered. Thus, in pathway KGCB**A**EHL, one ignores the known F values of C and B (and the unknown values for K, G, E, H and L), as only A is the common ancestor, although one uses F of B when it is the common ancestor, as in KGC**B**DHL. Thirdly, pathways go from I to a common ancestor and back to I without zigzagging, so paths such

**Table 6.2.** The inbreeding coefficient, F, under different systems of close inbreeding.

| Generation | Selfing | Full-sib mating | Half-sib mating |
|:---:|:---:|:---:|:---:|
| 0 | 0.000 | 0.000 | 0.000 |
| 1 | 0.500 | 0.250 | 0.125 |
| 2 | 0.750 | 0.375 | 0.219 |
| 3 | 0.875 | 0.500 | 0.305 |
| 4 | 0.938 | 0.594 | 0.381 |
| 5 | 0.969 | 0.672 | 0.449 |
| 6 | 0.984 | 0.734 | 0.509 |
| 7 | 0.992 | 0.785 | 0.563 |
| 8 | 0.996 | 0.826 | 0.611 |
| 9 | 0.998 | 0.859 | 0.654 |
| 10 | 0.999 | 0.886 | 0.691 |

as KGCBDHE<u>A</u>FJL do not count. It is important to identify all common ancestors and all pathways. One can use pedigree diagrams to work out F for the offspring of matings between relatives, such as father-daughter, aunt-nephew, second cousins, etc., or more complex patterns as in Fig. 6.4.

The increases in F in successive generations under **different degrees of inbreeding** are shown in Table 6.2 so that the effects of selfing, full-sib mating and half-sib mating can be compared. They are shown in graphically by Strickberger (1985) and the recurrence equations are given by Falconer and Mackay (1996), although their Eq. (5.9) for full sib mating has a misprint and should be $F_t = 1/4(1 + 2F_{t-1} + F_{t-2})$ instead.

## Suggested Reading

Most of the books listed in Chapter 4 are suitable.

Charlesworth, B. and D. Charlesworth, The genetic basis of inbreeding depression. *Genet Res* (1999) 74: 329–340.

Charlesworth, D., Effects of inbreeding on genetic diversity of populations. *Philos Trans Royal Soc Lond* B (2003) 358: 1051–1070.

Falconer, D. S. and T. F. C. Mackay, *Introduction to Quantitative Genetics*, 4th ed. (1996). Longman, Harlow.

Hedrick, P. W., *Genetics of Populations*. (1983) Jones and Bartlett, Boston.

Hedrick, P. W., *Genetics of Populations*, 3rd ed. (2005) Jones and Bartlett, Boston.

Houtman, A. M. and J. B. Falls, Negative assortative mating in the white-throated sparrow, *Zonotrichia albicollis*: the role of mate choice and intra-sexual competition. *Anim Behav* (1994) 48: 377–383.

Lande, R., D. W. Schemske and S. T. Schulz, High inbreeding depression, selective interference among loci, and the threshold selfing rate for purging recessive lethal mutations. *Evolution* (1994) 48: 965–978.

McPhee, C., Genetic selection in pigs. (2005), www.dpi.qld.gov.au/pigs/1533.html

Robinson, R., *Genetics for Cat Breeders*, 3rd ed. (1991) Pergamon Press, Oxford.

Strickberger, M. W., *Genetics*, 3rd ed. (1985) Collier Macmillan, London.

Viard, F., F. Justy and P. Jarne, Population dynamics inferred from temporal variation at microsatellite loci in the selfing snail *Bulinus truncatus*. *Genetics* (1997) 146: 973–982.

Vogel, F. and A. G. Motulsky, *Human Genetics. Problems and Approaches*, 3rd ed. (1997) Springer, Berlin.

# Chapter 7

---

# Mutation and its Uses

## 7.1. Molecular types of mutation and their revertibility; mutation frequencies

Mutations can occur in genes in the nucleus, in organelles such as chloroplasts and mitochondria, and in plasmids. A functioning gene typically has hundreds of base pairs, and changes in the base sequence can be of a number of types, with different properties and frequencies. Two different mutations in the same gene may give similar or different phenotypes, or even result in no phenotypic change, as mentioned in Subsec. 1.3.7. It helps one to understand phenotypes, functions and alleles, if one understands the molecular basis of mutation. Section 12.7 gives details of **spontaneous mutation frequencies in human genes** and of the nature of **mutational hotspots** within genes.

### 7.1.1. *Revertibility*

A mutation is **revertible** if it can change back from being mutant to being wild-type, either spontaneously or by induced reversion. For some purposes, it is essential to know if a given mutation is revertible. Suppose you were preparing a live but mutation-weakened strain of polio virus for immunising people against polio. If you used a mutation which was revertible, then people immunised with the attenuated virus would occasionally be infected with the disease, following spontaneous reversion from mutant to wild-type. It would be much safer to use more than one mutation, and to ensure that they were non-reverting. For a gene for yield in rice, e.g., using a revertible mutation would be much less dangerous than for a live vaccine.

**Revertibility** can also be important in **evolution and adaptation**. Suppose that allele $A$ is best in environment 1, but allele $a$ is best in

161

environment 2. An individual from environment 1 could spread to and thrive in environment 2 if there had been an *A* to *a* mutation which had become homozygous, *a a*. If the mutation was a big deletion and was not revertible, then no descendants from the *a a* individuals could produce *A* from reversion, and thus could not recolonise a type 1 environment. A revertible mutant, e.g., a frame shift, could revert to *A* and hence lead to colonisation of a type 1 environment. Within the type 2 environment, the dominant reversions would be unfavourable, imposing a small "reversion load" on fitness compared to non-reverting mutations, but they would confer adaptive flexibility. These two types of mutation (a frame shift and a big deletion) would probably both give a total loss of function and hence have the same selection coefficients, yet differ in their reversion ability.

A **forward mutation** from wild-type to mutant can occur at hundreds of different base pairs, in many different ways, and the spontaneous frequency of forward mutation is typically one in about $10^4$ to one in about $10^8$ or $10^9$ per gene per generation, depending on the organism and the circumstances. Longer genes offer more targets (base pairs) than do shorter genes. As reversion may require a particular change in a particular base pair, spontaneous reversion frequencies are usually much lower than forward mutation frequencies, by factors of 1,000 or more.

## 7.1.2. *Germ-line and somatic mutations*

It is important to distinguish between **germ-line mutations**, which can be passed on to future generations through the gametes, and **somatic mutations** in non-germ-line body tissues. In a fungal hypha, a yeast cell or a bacterium, mutations can affect a non-sexual cell and be passed on to future generations by cell division or by sexual processes. In a higher plant, animal or human, mutations in a leaf, leg or head cell will not be passed on, but a mutation in the egg cell, ovum, pollen or sperm, or in tissues in a young organism which later form gametes, can be inherited. A UV-induced mutation in human skin may be important to that individual if it gives rise to malignant melanoma (a form of skin cancer), but it will not be inherited.

## 7.1.3. *Base substitutions*

If one base pair is replaced by a different base pair in the same place, that is a **base substitution**. Suppose bases 582 to 584 in the coding strand of DNA of a wild-type functional allele are a codon, CCT, then the mRNA will have GGA, coding for the amino acid **glycine**. A base substitution at position 582 of AT for CG gives ACT in DNA, UGA in mRNA, which is a stop codon, terminating polypeptide formation, giving a truncated polypeptide. This kind of base substitution is a **nonsense mutation**, giving no amino acid sense. Such nonsense mutations, unless right at the last-translated end of a gene, usually cause complete loss of function as several or many amino acids are missing.

If the wild-type CCT codon had a different base substitution, CG to GC at position 582, we would get GCT in DNA, CGA in mRNA, giving **arginine**. Such a substitution of one amino acid by another is a **mis-sense mutation**; it gives amino acid sense, but a different amino acid. An amino acid change at a highly specialised part of a enzymatic polypeptide often gives a total loss of function, e.g., for a reactive amino acid at the active site, a fold site, an allosteric regulator-binding point, or a sulphur-containing amino acid at a disulphide bridge point. However, a change in unimportant parts could give a partial loss of function or no change of function, the latter being a **tolerated substitution**. This particular change, from neutral glycine to the basic amino acid arginine, would change the overall charge on the polypeptide which might well affect its function.

If the wild-type CCT codon had a TA to AT change at position 584, we would get CCA in DNA, GGU in mRNA, giving **glycine**. This is a **synonymous base substitution** giving no change in amino acid, and hence no change in function. The change is completely tolerated and if we are going by function, not DNA sequencing, we will not even know that a base change has occurred. Functionally, we would have another wild-type allele although its base sequence, CCA, differed from that of our originally considered wild-type allele, which had CCT there.

Base substitutions therefore have a wide range of phenotypic effects from no effect (a **cryptic mutation**, see Lamb, 1975), to a mild loss of function, to a severe or total loss of function. They are always revertible by the opposite base substitution to the one which caused them, e.g., a

CG to AT substitution is revertible by an AT to CG substitution at exactly the same place in the molecule. These mutations and reversions can occur spontaneously or be induced. Base changes in exons are more likely to have phenotypic effects than those in introns or in "junk DNA".

## 7.1.4. *Frame shifts*

**Frame shifts** are caused by the addition to or deletion from DNA of single base pairs, or of some other number of pairs which is not a multiple of three. If we had three codons in a wild-type gene's DNA, ACA, GGT, TGA with UGU, CCA, ACU in mRNA, we would get cystine, proline, threonine in a polypeptide. If we had a deletion of the first A in the DNA, the mRNA would then have GUC, CAA, CU-, giving valine, proline, and then a series of wrong amino acids to the end of the molecule, or until the wrong sequence of bases gave a stop codon. Unless right at the end of a polypeptide, frame shifts usually give a complete loss of function because many amino acids are changed and stop codons may be created which truncate the polypeptide. Additions and deletions both **change the reading frame** of the triplets of bases in mRNA, hence their name. A deletion or addition of a multiple of three base pairs is not a frame shift. The reading frame remains the same, but there will be missing or additional amino acids which may well affect the function of the product.

Frame shifts are revertible, especially if the number of base pairs added or deleted is low. A base pair addition, e.g., of AT, can be reverted by deletion of that AT base pair, with complete restoration of wild-type function. An addition, e.g., of AT, can sometimes also be reverted, wholly or partly, by the deletion of one base pair near to, but not exactly at the original point of addition. This is because the reading frame is correctly restored, so most amino acids in the polypeptide are correct, and the few amino acids between the point of addition and the point of deletion may not be essential for function. Reversion can be spontaneous or induced.

## 7.1.5. *Large deletions*

**Large deletions** usually cause a complete loss of function as so many amino acids are lost. Loss of 60 base pairs from a coding region will lead to a loss of 20 amino acids. Large deletions which are not multiples of three bases will also be frame shifts, causing further loss of function. Large deletions including "stop" or "start" reading signals will also affect function from that. The deletion of the first 80 bases of a coding sequence can therefore cause a loss of function from a loss of amino acids, from being a frame shift, and from deleting a "start reading" signal.

Large deletions will not revert, spontaneously or by induction, because the chance of getting the right number of bases inserted in the right place in the right order is vanishingly small. Large deletions are therefore extremely useful if one wants a non-revertible total loss of function, as when preparing a live weakened viral vaccine from a wild-type strain.

## 7.1.6. *Unstable length mutations*

**Unstable length mutations** can be intermediate in size between gene mutations and chromosome aberrations. In humans, adult-onset **myotonic dystrophy**, giving progressive muscle weakness, is due to an unstable length mutation where small length mutations cause no or few symptoms, but amplification in successive generations (possibly from unequal crossing-over) gives increasing severity of disease. It is an autosomal dominant where normal individuals have 5 to 37 repeats of a CTG sequence with stable numbers, while 50–99 repeats give mild symptoms, and greater numbers up to 2,000 repeats give severe disease. Larger numbers of repeats are unstable. The condition has a frequency of 1 in 7,500. Affected males have half the offspring affected, half unaffected, but affected females have 50% of offspring unaffected, 29% with late-onset, 12% neonatal deaths, and 9% with severe symptoms at birth, including mental handicap (Connor and Ferguson-Smith, 1997).

The human **fragile-X syndrome** has a frequency of 1 in 3,000 males and with less severity (when heterozygous) in 1 in 5,000 females, causing mental retardation, behavioural problems, physical and physiological abnormalities. It is caused by an unstable length mutation, as is Huntington disease (Sec. 12.3), both with a trinucleotide sequence involved. Fragile-X is the commonest X-linked cause of mental handicap. Under thymidine starvation in culture, X chromosomes with fragile-X often show chromosome or chromatid breaks near the tip of the long arm, at the FRAXA locus at Xq27.3. Normal individuals have 6 to 52 copies of a CGG repeat in the promoter region of the *FMR1* gene (fragile X mental retardation), but small length mutations (premutations) with 60–200 repeats are phenotypically normal. Affected males with the full mutation have 230–1,000 repeats, giving mental handicap, and enlarged testes after puberty. Hypermethylation of the repeating tract and transcriptional silencing of the FMR1 protein are involved in expression of the mutation. Heterozygous females for the full mutation may be normal, but 20–30% have mild mental handicap, and 1% have moderate mental handicap. The premutation exists in 1 in 1,000 males and about 1 in 400 females. The mutation is unstable in inheritance. All daughters of males with the premutation receive a mutation, either an unchanged premutation or an expanded form with the full mutation. Further expansion can happen at mitosis, giving mosaics, or in female meiosis. Mothers heterozygous for the full mutation have half their sons mentally handicapped, while the daughters are carriers or have some mental handicap. Prenatal diagnosis from DNA (Subsec. 12.12.6) is possible in the first three months of pregnancy, using Southern blot or PCR analysis (www.fragilex.org). See Bagni and Greenough, 2005.

## 7.2. Spontaneous and induced mutation; mutagenic agents

**Spontaneous mutations** (Plates 1.5 and 7.1) typically have **frequencies** ranging from about 1 in $10^4$ to about 1 in $10^9$ per gene per generation for forward mutations (to loss of function), and hundreds or thousands of times less than that for backmutations (restoring the lost function for frame shifts or base substitutions). The frequency of spontaneous mutations can increase with age in seeds and men (Subsec. 17.1.1 and Sec. 17.3) and in microbial cultures. **Induced mutations** can have much higher frequencies than spontaneous ones, but very high doses of mutagen can seriously weaken an organism even when the desired mutation is induced, as there are usually many unwanted deleterious mutants.

Neuffer *et al.* (1997) illustrate in colour a wide range of **different maize mutations**. Their pictures show the kinds of mutation one can get in an agricultural plant species, and their book has detailed maps of ordinary and molecular markers in nuclear chromosomes, in mitochondria and in the

**Plate 7.1.** **A spontaneous mutation in the very early development of a red rose**, changing petals to white and anthers to yellow. One could propagate the mutant part but not a half-and-half rose. One meristem cell must give rise to about half of each central petal.

chloroplasts. One cannot assume that an inherited mutation is necessarily nuclear. It may be necessary to study its inheritance to check whether it is in nuclear or organelle DNA.

## 7.2.1. *Spontaneous mutations; mutation mechanisms; varieties or breeds they have produced*

**Spontaneous mutations** are those occurring spontaneously without the help of man, even though they have a range of known causes, sometimes involving natural mutagens. **Mutagens** are chemical or physical agents which cause mutation. **Induced mutations** are the ones induced by man via various means. Spontaneous mutations can occur through **tautomeric shifts** in bases in DNA, where rare tautomers often have different base-pairing properties from the normal tautomer. For example, for cytosine, the normal amino form pairs with guanine, but the rare spontaneous cytosine tautomer in the imino form pairs with adenine. If at replication a cytosine is in the imino form, it will get an adenine opposite it in the new strand.

At the next replication, that adenine will get a thymine pairing with it, so a GC base pair can give rise to an AT base pair, a base substitution.

Spontaneous mutations can arise from non-ionising UV light in sunlight (UV-B, 280 to 320 nm wavelength) and from various ionising radiations, which, in the order of decreasing wavelength and increasing penetrating power, are **X-rays**, **gamma rays** and **cosmic rays**. Their mechanisms of mutation, some of which are not fully understood, are not dealt with in this book.

**UV light** has low penetrating power, being absorbed by water, skin and glass, so while it can cause inherited mutations in microbial cells, it causes superficial somatic mutations rather than germ-line mutations in higher organisms. UV can cause a whole range of mutation types, especially base substitutions and deletions, but including frame shifts and many types of chromosome aberration (Chap. 9). 254 nm is the most mutagenic UV wavelength, being the one most strongly absorbed by DNA.

One gets natural **ionising radiations**, X-rays, neutrons, alpha and beta particles, gamma rays and cosmic rays, from space, rocks, radon gas, and radioactive elements in the environment and within our bodies. These rays have high penetrating power and can therefore cause germ-line and somatic mutations. They cause base substitutions, frame shifts, deletions and chromosome aberrations.

There are also many **chemical mutagens**, some occurring in the environment or in food and drink. **Base analogues** are chemical analogues of DNA bases and get incorporated into DNA at replication. They have a much higher rate of tautomeric shifts than the four normal DNA bases (the **pyrimidines**, thymine and cytosine, and the **purines**, adenine and guanine), giving rise to base substitutions at replication. Natural base analogues include caffeine (present in tea, coffee and some colas), which is a purine analogue.

Some chemicals react directly with DNA, as does nitrous acid, $HNO_2$. This compound replaces an $NH_2$ group with OH, changing cytosine (which pairs with G) to uracil (which pairs with A), and adenine (which pairs with T) to hypoxanthine (which pairs with C). The altered bases paired cause base substitutions at replication. Nitrous acid occurs at low concentrations in animals as a metabolic byproduct. It is a more effective mutagen in Prokaryotes than in Eukaryotes.

**Alkylating agents** add methyl or ethyl groups to bases in DNA, giving altered pairing properties or leading to a loss of purines, especially of guanine, and cause base substitutions and deletions. They can cross-link DNA strands. Like many mutagens, they can be **carcinogenic** as well as mutagenic. They include nitrogen mustards, various nitrosoguanidines such as NMG (N-methyl-N'-nitro-N-nitrosoguanidine), and EMS (ethyl-methanesulphonate). Some of these compounds occur at low concentrations in the body as metabolic byproducts, e.g., from nitrates or nitrites in the diet, sometimes from chemicals used to treat meat.

**Intercalating agents** insert themselves between adjacent purines in DNA, causing base pair additions if inserted into an old strand, or base pair deletions if inserted into new strands; a round of replication is needed before the final mutation is formed. They therefore cause frame shift mutations instead of base substitutions. Intercalating agents include acridine dyes such as proflavin and acridine orange, which were once used as antibiotics for wounds. There are some natural intercalating agents, but most are man-made.

There are many other kinds of natural mutagen, including extremes of temperature and altered pH. There are also mutagenic genetic elements, named "jumping genes", which can replicate themselves from one place in the genome into other places. These **insertion sequences** and **transposons** can transfer themselves to within the coding sequences of other genes, interrupting the polypeptides produced, usually causing complete loss of the gene's function. These transposing elements occur in microbes, plants and animals. It has been estimated that more than half the spontaneous mutations in the fruit fly, *Drosophila melanogaster*, come from such transpositions. The causes of mutation in man are explored further in Sec. 12.7.

**Single spontaneous mutations** have often given rise to **new breeds** of animal or **new varieties** of flowers, vegetables and trees, as in weeping willows, weeping beeches or copper beeches. A classic case is the spontaneous mutation to hairless fruits in peaches to give **nectarines**. In cats, the Cornish **rex mutation** was discovered in 1950, giving a very distinctive short coat, soft and mole-fur-like, apparently with no projecting guard or awn hairs, and often with short and bent whiskers on the face. It is an autosomal recessive, *r*. The Devon rex was discovered in 1960, having the recessive *re* mutation which is unrelated to the Cornish rex mutation. The Devon rex has a short coat similar to that of the Cornish rex, but with a greater tendency to a loss of hair, with bare areas, and for the whiskers to break off. There are now further separate rex mutations and breeds, including the dominant Selkirk rex, *Se*, from Wyoming, and the recessive Oregon rex, *ro*. A German rex mutation proved to be at the same locus as the Cornish rex mutation, with crosses between the Cornish and German rex giving all rex kittens in the limited number produced. In large animals, unlike fungi, it is not practical to test whether the recombination frequency is zero or just very low, as that would require an enormous number of kittens to be tested. The **Manx cat** is another example of a type of cat produced from a single spontaneous mutation, $M$, dominant for tailless, but a recessive prenatal lethal like the human mutation giving achondroplastic dwarfs. The Manx cat owned by the author could walk, run well, climb trees and express her emotions clearly, while having absolutely no tail. As cats have evolved tails with various functions, it was astonishing what little difference having no tail made to this cat. Many Manx cats, however, do have abnormalities of the lower vertebrae and anal region, often having a stiff-legged walk. Possibly through the influence of polygenes, the Manx tailless phenotype can manifest as completely tailless ("rumpy"), or with progressively more very short tail, as "rumpy riser", "stumpy" or "longie".

Other spontaneous harmful mutations occur in cats. They include hairless, ataxic (an inability to coordinate voluntary movements such as walking), hare-lip, cleft-palate, "four-ears" (i.e., a small extra pair of ears, small eyes and an undershot jaw), and urolithiasis (i.e., a tendency to form stones in the urinary system; also known in dogs).

## 7.2.2. *Induced mutations*

**Induced mutations** can be obtained with a much higher frequency than spontaneous mutations, using agents described in Subsec. 7.2.1. Most induced ones are harmful but a few are useful. Higher doses give more harmful mutations as well as more useful ones, and may seriously weaken, kill, or sterilise the organism treated. For any organism, one needs to establish a safe but effective dose for a particular mutagen. One normally wants to induce germ-line mutations, not somatic ones. Dominant mutations are easier to spot than recessive ones, for which selfing or other close inbreeding is needed to make them homozygous to get expression in diploids. As mentioned in Sec. 13.4, recurrent backcrossing of mutant progeny to a non-mutant stock can be used to remove unwanted harmful mutations.

For something as unprotected as yeast cells or bacteria, one can irradiate them in a thin film of water, being careful to remove the Petri dish lid before irradiation with UV, or briefly expose cells to chemical mutagens. See Subsec. 12.7.2 and Plate 7.2 for the chemical induction of histidine-non-requiring backmutations in the **Ames test for mutagenicity**. In animals, one can use mutagens directly on the sperm for AI, or inject

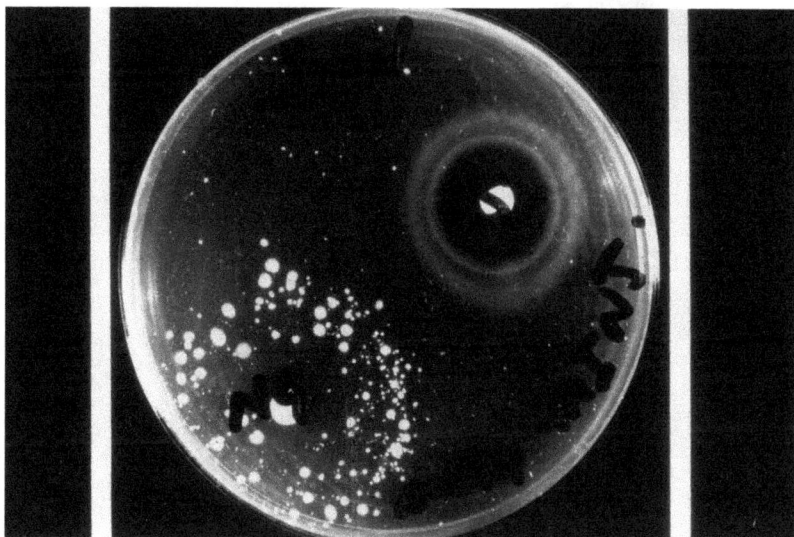

**Plate 7.2.** **The Ames test for mutagenicity.** *Salmonella typhimurium* histidine-requiring bacteria are plated on minimal medium with a trace of histidine, which enables a background lawn to grow slightly, but **revertants** from *his⁻* to *his⁺* can form individual visible colonies. There is a cleared zone of killing close to the disk with NMG (left), but as the concentration of NMG decreases away from the disk, there are lots of individual revertant colonies, showing that this base substitution amber mutation, 881, is reverted by NMG at sub-lethal concentrations. The disk with streptomycin (right) shows a clear halo of growth outside the cleared zone of killing, but no individual revertants. In the halo, misreading of the amber stop codon is induced by streptomycin at low concentration, so that some complete protein is formed, allowing histidine synthesis: that is **phenotype suppression**, not reversion, and the bacteria in the halo would not grow again if subcultured on minimal alone, unlike the revertants from NMG. This plate shows that NMG is mutagenic for base substitutions, but streptomycin is not. There are also some spontaneous revertants away from both disks. The dish is 9 cm in diameter.

chemical mutagens into gonads, or use penetrating radiation such as X-rays or gamma rays on the gonad or whole-body areas. With plants, one can treat pollen, seeds or meristems. Plate 7.3 shows the increasingly damaging effects of 45,000 to 60,000 rads of gamma-ray irradiation on tomato

**Plate 7.3.**   The effects in tomato (*Lycopersicon esculentum*) of **gamma ray irradiation of seeds** by a cobalt-60 source. From the front to the rear, the trays are those with 60,000 rads, 55,000 rads, 45,000 rads, and no irradiation. The seeds were from selfed *Xa-2 xa-2* plants, where there is incomplete dominance of xanthophyllic, *Xa-2*, so one expects a ratio of 1/4 green: 1/2 yellow-green: 1/4 yellow seedlings, where the latter die after germination because they cannot photosynthesise. Irradiation causes stunted growth, morphological deformities and some leaf-colour mutations (Plate 7.4). The seedlings had been planted for five weeks.

seeds. See Sec. 7.5 for more details on the use of induced mutations in agriculture and horticulture.

Some of these treatments will cause a mixture of somatic (Plates 7.3 and 7.4) and germ-line mutations. When seeds are treated, a germ-line mutation will not be visible in the somatic tissue of the plant grown from the seed. One has to cross or self the plants grown up, and look for mutants in the next generation. If one gets a desirable mutation expressed in somatic tissue in a plant, one can often **propagate it vegetatively**, e.g., from buds or callus, to get whole plants and germ-line mutants.

In some organisms, it is more efficient to screen existing populations for spontaneous mutations, but in others, it is better to use induced mutations. Some mutants which are useful in agriculture are disadvantageous in the wild. For example, the wild ancestors of wheat and barley had a brittle ear stem (rachis), aiding grain dispersal by animal contact. Early man must have selected a spontaneous mutant with a **non-brittle rachis**, so the ear held together during the harvest instead of shattering in the field, and it could be threshed back at home. Such a mutation would be disadvantageous in the wild because of poor grain dispersal.

Although induced mutations have been used very extensively in laboratory studies of animal genetics, physiology and development, they have not been used much with farm animals. They have been used extensively with micro-organisms and quite a lot for higher plants, for which gamma rays from a cobalt-60 source are usually used. **Induced mutations in barley** that are commercially used include earlier ripening, shorter stiffer straw, greater diastatic activity (diastase is important in starch breakdown, i.e., in malting barley for beer, gin and whisky), larger grains, and various kinds of disease resistance. Induced mutations for polygenes would be difficult to detect, and for quantitative characters, it is usual to use existing genetic variation. Much of animal breeding is quantitative, with little reference to individual loci.

If generation times are long in plants, as with most trees, one can mutate the buds and then vegetatively propagate them. An advantage of vegetatively propagated mutants is that the favourable gene combinations are not lost through gene reassortment at meiosis. Subsection 17.1.5 covers various propagation methods and **somaclonal variation**, which is the production of altered forms from vegetative reproduction, capable of producing good new cultivars.

## 7.3. Mutation control and repair systems

Organisms can have **partial control** over their mutation frequencies. They can evolve to influence their mutation rates either through **access of mutagens** to the DNA or through the **efficiency of repair mechanisms**, or in specialised cases, through mutations which suppress the expression of other mutations.

### 7.3.1. *Mutagen access*

Many fungi have dark asexual or sexual spores, or fruit bodies, giving some **protection against UV light** (Lamb *et al.*, 1992). Dark melanin skin pigments in humans give some protection against skin mutations which could cause cancer. The effectiveness of the enzyme **nitrite reductase** in the liver and kidneys affects levels of endogenous nitrous acid, which we saw in Subsec. 7.2.1 acted directly on DNA as a mutagen. There can be genetic differences in the amount of uptake of environmental mutagens from the soil by plant roots, or from food by the gut, or in the excretion of mutagens by the liver and kidneys.

   **Mutation rates** are very temperature-sensitive, so in seasonally breeding mammals (Subsec. 17.2.4) such as deer, the time of descent of the testicles from the body cavity to the scrotum affects gonad temperature during spermatogenesis, and hence mutation rates in sperm. Ionising radiations are hard to control because of their penetrating power, although behaviour affecting what kind of rocks an animal lives amongst or burrows under might have minor effects. People living in Cornwall, in the West of England, get exposed to higher levels of radioactive **radon gas** than those in other regions. Thus, migration to other areas could have small effects on the natural radiation dosage (Subsec. 12.7.3).

### 7.3.2. *Repair systems*

An organism can partly control its mutation rates by the efficiency of its **repair systems**, with mutations which decrease their efficiency increasing mutation rates, and mutations increasing their efficiency decreasing mutation rates.

   For UV damage, only a **pre-mutational lesion** is initially produced, mainly thymine-thymine dimers, or cytosine-thymine, and also C6-C4 photoproducts cross-linking adjacent pyrimidines. If not corrected, such within-chain dimers can cause mutations at replication by getting wrong bases inserted opposite them after replication is blocked. Many organisms (micro-organisms, plants, animals and humans) have repair systems which can prevent such pre-mutational lesions from becoming full mutations. One system is **photo-repair**, where a photoreactivating enzyme such as photolyase in *E. coli* uses visible light (especially blue-green, wavelengths 300 to 500 nm, repair peak 384 nm) to break these dimers back to two monomers (for the mechanism, see Kao *et al.*, 2005). **Photolyases** repairing thymine-thymine dimers are found in bacteria, plants and most animals, except mammals which are unusually prone to skin cancers from UV in sunlight. **6-4 photolyases**, repairing C6-C4 photoproducts, occur in plants, insects, reptiles and amphibians, but not in *E. coli*, yeast or mammals.

In **dark-repair**, an endonuclease recognises the dimer, cuts the affected strand, and an exonuclease enzyme erodes the exposed ends, including the dimer. DNA polymerase resynthesises the missing bases, using the other strand as a template, and DNA ligase joins up the ends. Mutations affecting sensitivity to ionising radiations and to UV are known in many organisms. People with the inherited autosomal recessive disease **xeroderma pigmentosum** (which has at least nine subtypes, with mutations in any of at least eight loci) have defective DNA excision repair of UV damage and suffer from multiple skin cancers. They have skin cancers about 1,000 times more frequently than in normal people. The combined frequency of the different types is 1 in 70,000, and sufferers are advised to keep out of bright light, to use strong sun-block creams, and to wear hats and opaque clothing.

Calculations of the expected mutation frequency, on the basis of known frequencies of tautomeric shifts in the DNA bases, show that there must be a **proofreading enzyme system** at replication, which preferentially reduces the incorporation of rare tautomers into the new strands. The efficiency of proofreading and correction systems will obviously affect mutation rates.

## 7.3.3. *Suppressor mutations*

There are mutations which give mutant phenotypes on their own, but which can restore, partly or wholly, a wild-type phenotype to other mutations. The deletion of a base pair gives a frame shift mutation, but the addition of a base pair (which on its own gives a mutation) near to that deletion mutation restores the reading frame, and may give whole or partial restoration of the wild-type phenotype.

One can have mutations which suppress a whole class of mutations. Amber stop codons with UAG in mRNA, whether normal or in a nonsense mutation, can be suppressed by **amber-suppressor mutations**. In these, there has been a mutation in a gene for a transfer RNA (those genes are duplicated, so wild-type genes still remain for that tRNA), changing its anticodon from one binding to an amino acid codon in mRNA, to one binding with the amber stop codon, inserting an amino acid and permitting reading through of the stop codon. Since they suppress normal stop codons as well as mutant ones, such amber-suppressed strains tend to grow slowly, producing some unusually long polypeptides. **Suppressor mutations** are not normally used in commercial organisms to control mutant expression.

### 7.3.4. *Optimum mutation rates*

There is some evidence of **selection for optimum mutation rates** for an organism according to its habitat. Strains of the fungus *Sordaria fimicola* (Plate 18.2) from a stressful exposed environment had higher inherited rates of spontaneous mutation and of induced mutation, than strains from a less stressful environment, as if the production of more genetic variation was useful in adapting to the stressful environment (Lamb *et al.*, 1998). As most mutations are harmful, an ideally adapted organism in a stable environment would do best with a low mutation frequency, but higher mutation frequencies would be appropriate for organisms in unstable environments, or for colonising different environments.

## 7.4. How different types of mutation can complicate population genetics calculations

In population genetics (Chap. 4), we considered an over-simplified system of two alleles, with $A$ mutating with frequency $u$ to $a$, which backmutates to $A$ with frequency $v$. Suppose that $A$ is an essential, functional, wild-type allele of a given base sequence, with no selection against it. It could mutate in many different ways. Some DNA changes such as synonymous mutations can change the DNA sequence without changing the polypeptide or the phenotype. We would probably not detect that the mutation had occurred, thus regarding it as a normal $A$ allele, with $s = 0.0$, but it would mean that there were a number of **different $A$ alleles** which might differ in mutation frequencies and might be able to recombine at meiosis to produce further variants, including mutant phenotypes.

If we had a large deletion, it would almost certainly give a complete loss of function, $s = 1.0$, and it would not revert, so $v = 0.0$. If we had a frame shift or a nonsense mutation, it would probably give a complete loss of function, $s = 1.0$, but could revert spontaneously, so $v$ is not zero. If we had mis-sense mutations, some might give no loss of function, $s = 0.0$, some might give a partial loss of function, e.g., $s = 0.1$, while others, at an active site, might give a complete loss of function, $s = 1.0$. All those mis-sense mutations could revert, but possibly at different frequencies. If all alleles giving a loss of function were called $a$, we would have a whole range of **different $a$ alleles**, differing very much in selection coefficient and in reversion frequency.

It will now be clear that in a very large population, one seldom has just two different alleles at a locus, at the molecular level, even if one can describe the phenotypes as falling into two classes, such as tall and short pea plants. Mutation will eventually produce a whole **range of alleles**, although a new small population may stay genetically uniform for a number of generations if it started with only one type of homozygote for a particular locus.

The simple equations involving only two alleles at a locus are very useful for showing the principles, e.g., of an equilibrium between mutation and selection. If one takes into account that there are many possible alleles with different mutation and reversion frequencies, and different selection coefficients, the equations would become extremely complicated. It is important to realise that we have been using over-simplified equations.

## 7.5. Using induced mutations

As a result of its low penetration, UV has restricted use, mainly to bacteria, fungal spores and hyphae, pollen or thin tissue cultures. X-rays and gamma rays are widely used in plants and animals, as they have good tissue penetration and can be given in precise doses. Chemical mutagens can give high mutation frequencies; the dosage received can be uncertain if tissue penetration to the target tissue is needed, leading to poor reproducibility, especially with variable persistence of the mutagen after treatment. It is difficult to find an **ideal dosage** with a mutagen, since low doses give too few mutations and high doses give too many unwanted mutations, which can seriously weaken an organism's viability and/or fertility, even if it does carry a favourable mutation. All mutagens need strict safety procedures.

The following examples of the **plant part treated**, the mutagen and the dosages given are taken from Micke and Donini (1993). Dry seeds, in seed-propagated plants: oats, rice and maize, gamma rays, 14–28 krad; tomato, EMS, 0.8%, 24 h at 24°; *Triticum durum* (wheat), gamma rays, 10–25 krad, or fast neutrons, 600–800 krad, or EMS, 3.8%; *Triticum durum* pollen, gamma rays, 0.75–3 krad. Vegetatively propagated plants: potato, tubers, EMS, 100–500 ppm, 4 h, 25°, or shoot tips, gamma rays, 2.5–3.5 krad; *Dianthus caryophyllus* (carnation), nodal stems, X rays, 1.5–2 krad; *Malus pumila* (apple), dormant graftwood, gamma rays, 6–7 krad, dormant buds, gamma rays, 2.5–5 krad; *Musa* (banana), shoot tips *in vitro*, gamma rays, 1–2.5 krad; *Citrus sinensis* (orange), ovular callus *in vitro*, gamma rays, 8–16 krad.

As seeds are multicellular, mutation of one cell in them gives a **chimeric M1 generation**, with some parts mutant and most parts wild-type (Plate 7.4). The mutation will only be passed on to progeny if it gets into the germ-line tissue. Gross chromosomal abnormalities may be selected against during the growth and differentiation of meristematic tissues. Selfing the M1 generation is done to get recessive mutations homozygous for expression and selection. Recurrent backcrossing (Sec. 13.4) is often used to get rid of unwanted deleterious mutations, while selecting for the wanted mutation. The use of mutagens on gametes such as pollen is one way of avoiding chimeras. In chimeras, adventitious buds may develop from single cells, and so give genetically uniform (homohistont) plants. Vegetatively propagated plants are often highly heterozygous, with hybrid vigour, so mutating vegetative parts avoids sexual reassortment of the genes but leads to chimeras.

In 1990, a FAO/IAEA symposium recorded 1,363 cultivars produced by mutagenesis, with more than 90% from X rays or gamma rays. The

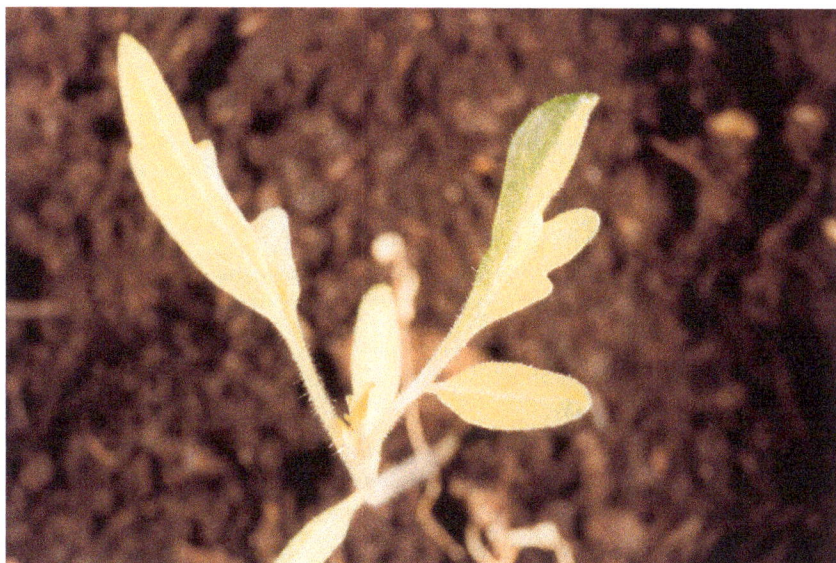

**Plate 7.4.** **A seedling from irradiated tomato seed** (see Plate 7.3 for details). This *Xa xa* yellow-green plant shows some distortion. The upper right leaflet shows a clear **green sector** from an *Xa* to *xa* mutation, giving *xa xa*, and a central **yellow-white sector** from mutation from wild-type *xa* to mutant *Xa*, giving *Xa Xa*. An advantage of using alleles with incomplete dominance is that mutations in either direction in the heterozygote are detectable. One can use induced mutations for **fate-mapping**: this seedling shows what area of leaf is determined by a single mutated cell in the seed.

mutations involved characters such as crop yields, flowering time, flower shape and colour, fruit size and colour, resistance to pests and pathogens, the oil content of sunflowers and soya beans, the fatty acid content of linseed, soya and rapeseed, the protein content, amino acid composition and starch quality of rice, barley, wheat and maize, and fertility restorers and male sterility in maize. In Italy, there was a successful mutation in pasta wheat to lodging resistance, and in Pakistan, there were mutationally improved varieties of cotton, mungbean and chickpea. In vegetatively propagated crops, there were improvements by mutation to apples, cherries, olives and apricots for characters such as compact tree shape, earlier flowering, self-incompatibility, or seedlessness. Ornamental plants had mutations affecting leaves, flowers, temperature requirements, flowering periods, etc. Induced mutations have hardly been used at all in farm animals, although they have been widely employed in lab organisms such as *Drosophila* and mouse, and can be used in human tissue cultures.

## Suggested Reading

Bagni, C. and W. T. Greenhough, From MRNP trafficking to spine dysmorphogenesis: the root of fragile X syndrome. *Nat Rev Neurosci* (2005) 6: 376–387.

Connor, J. M. and M. A. Ferguson-Smith, *Essential Medical Genetics*, 5th ed. (1997) Blackwell Scientific, Oxford.

Friedberg, E. C., DNA damage and repair. *Nature* (2003) 421: 436–440.

Friedberg, E. C., G. C. Walker and W. Siede, *DNA Repair and Mutagenesis*. (1995) Blackwell Science, Oxford.

Kao, Y.-T. *et al.*, Direct observation of thymine dimer repair in DNA by photolyase. *Proceedings of the National Academy of Sciences, USA* (2005) 102: pp. 16128–16132.

Lamb, B. C., Cryptic mutations: their predicted biochemical basis, frequencies and effects on gene conversion. *Mol Gen Genet* (1975) 137: 305–314.

Lamb, B. C. *et al.*, Interactions of UV-sensitivity and photo-reactivation with the type and distribution of ascospore pigmentation in wild-type and mutant strains of *Ascobolus immersus*, *Sordaria brevicollis* and *Sordaria fimicola*. *Genetics (Life Sci Adv)* (1992) 11: 153–160.

Lamb, B. C. *et al.*, Inherited and environmentally induced differences in mutation frequencies between wild strains of *Sordaria fimicola* from "Evolution Canyon". *Genetics* (1998) 149: 87–99.

Micke, A. and B. Donini, Induced mutations, in Haywood, M. D., N. O. Bosemark and I. Romagosa (eds.) *Plant Breeding. Principles and Prospects*. (1993), pp. 152–162, Chapman and Hall, London.

Neuffer, M. G., E. H. Coe and S. R. Wessler, *Mutants of Maize*. (1997) Cold Spring Harbor Laboratory Press, New York.

Venitt, S. and J. M. Parry (eds.) *Mutagenicity Testing: a Practical Approach*. (1984) IRC Press, Oxford.

# Chapter 8

# Recombination, Mapping and Genomics

## 8.1. Recombination, genetic distances and the numbers of progeny needed to get particular recombinants

As we saw in Subsec. 1.3.8, **recombination** is the production of new combinations of existing genes. Recombination frequencies in Eukaryotes are measured from crosses in per cent recombination, which is the same as **centiMorgans** (**cM**, named after the American *Drosophila* geneticist, Thomas Hunt Morgan), where one cM is one per cent recombination. Plate 2.2 shows a maize example with no detected recombination between two very close loci. The dihybrid ratios in Plates 2.3 and 2.4 are modified from 9:3:3:1 (50% RF) by gene interactions between the unlinked loci.

The breeders of microbes, plants and animals will often want to produce new combinations of existing genes. If the relevant loci have been mapped in terms of recombination frequencies, the breeder can then work out the approximate **number of progeny to raise** to be reasonably sure of getting a particular desired recombinant. As shown below, if there is dominance, it may take three generations to get and identify the desired genotypes.

We saw in Sec. 4.2 that recombination for two loci can only occur in double heterozygotes. If one wants the pure-breeding recombinant *a a, B B* from *A A, B B* and *a a, b b*, one must cross them to get the coupling double heterozygote, *AB/ab*, which can then be selfed. If the two loci are unlinked, with 50% recombination, then one quarter of the gametes from *AB/ab* will be *a B* and one sixteenth of the progeny from the selfing will be *a a, B B*. If there is complete dominance at the *B/b* locus, one would need further genetic testing to distinguish *a a, B B* from double that number of *a a, B b* genotypes which share the *a B*-phenotype. If one only reared 16 offspring, then one might or might not find the desired genotype whose expected frequency was 1/16, so it would be safer to rear perhaps 60 offspring, or 120 if one wanted at least one *a a, B B* of each sex. If we were using yeast or another fungus with haploid progeny, 1/4 would be of the desired type, *a, B*, and there would be no complications from dominance in identifying the desired genotype from the phenotype, so fewer progeny need to be raised.

If the two loci were syntenic and closely linked, with 5% recombination instead of 50%, only 2.5% of the gametes would be of the desired type, *a B*, and therefore only 0.0625% of the progeny would be *a a, B B*, or 1 in 1,600. One would have to raise several thousand offspring to be reasonably sure of getting one of the desired recombinants, and in diploids with dominance for *B/b*, one would have to test individuals of the desired phenotype to separate the *a a, B B* types from *a a, B b*.

If one does not know how far apart the loci are, or whether they are syntenic, then one does not know how many progeny to raise to get a desired recombinant. It is therefore very useful to be able to do **genetic mapping**, finding out the recombination frequencies between various pairs of loci, finding which are linked and which are unlinked, and constructing **genetic maps** showing the loci in physical order in different linkage groups. All loci within a linkage group are syntenic, but the more distant ones may be unlinked to each other, with 50% recombination. In a well-mapped organism, linkage groups should be correlated with visually identified chromosomes, with as many linkage groups as there are different types of chromosome.

One can obtain **extremely rare recombinants** quite easily in many micro-organisms, and estimate even very small map distances, if one can use **selection** to identify recombinants. For example, if one wanted to obtain a wild-type recombinant between two auxotrophic mutants within a locus, e.g., $lysA^+$ from $lysA^5$ and $lysA^9$ in a fungus such as yeast or *Neurospora*, one just plates out millions of ascospores (from a repulsion-phase cross) on a series of Petri dishes of minimal medium containing no lysine. The two parental types of spore could not germinate, nor could the double mutant recombinants, because they all require lysine for growth, and only the prototrophic recombinant wild-type ascospores could germinate and grow into colonies. By estimating on minimal medium plus lysine, the number of viable spores plated out, and using the number of wild-type colonies growing up on minimal medium, one can estimate the recombination frequency between the two alleles. Allowing for the fact that the wild-type $lysA^+$ and double mutant $lysA^5 \, lysA^9$ recombinants should be equally frequent, the recombination frequency would be twice the frequency of wild-types. It would be very difficult to identify the double mutant recombinants amongst so many parental single mutants, because they have the same $lysA^-$ phenotype.

**Recombination and chiasma frequencies** are under **genetic and environmental control** in plants, animals and micro-organisms. If a population is segregating for genes affecting recombination frequencies for syntenic loci, selecting for desired recombinants could accidentally select for genotypes with alleles for more recombination.

## 8.2. Types of recombination and their effects; meiotic and mitotic crossovers; interference and map functions

We covered the basic aspects in Sec. 1.3. Non-syntenic loci recombine in meiosis by **independent assortment** because the members of different pairs of homologous chromosomes line up independently of each other at metaphase, giving 50% recombination. Syntenic loci can recombine reciprocally at pachytene of meiosis if a **crossover** occurs between them. The chance of a crossover increases as the distance between loci increases, so close loci show low recombination frequencies and distant loci may show 50% recombination. With reciprocal recombination in a double heterozygote, if one type of recombinant is produced, e.g., *a, B* from an *AB/ab* meiosis, then the reciprocal product is also produced in the same meiosis, *A, b* in this case. Details of the proposed molecular models of recombination are controversial and are beyond the scope of this book, but see Lamb (1996, 2003), Maloisel *et al.* (2004) and Stahl *et al.* (2004) for accounts of crossing-over, gene conversion and interference.

With **gene conversion**, recombination is generally non-reciprocal and non-Mendelian ratios (e.g., 3+:5*m*, Plate 18.2) can be produced in meiotic tetrads or octads. For a heterozygous base substitution mutant, $A/a$, let the *A* chromatid have base pair AT and let the *a* chromatid have a CG base pair at the corresponding position. If one has an *AB/ab* meiosis, and one *A* chromatid forms hybrid DNA by invading an *a* chromatid, then in the formerly *a* chromatid, one might get a mispair such as CT in that chromatid, while the other three chromatids have AT, AT and CG. The CT mispair has three possible fates: it may stay uncorrected, being resolved at the next replication into one AT (allele *A*) chromatid and one CG (allele *a*) chromatid; it may be corrected by enzymes to AT; or it may be corrected to CG. Figure 8.1 shows the consequences for the products of meiosis, and thus how gene conversion can give non-reciprocal recombination.

One can also have **recombination at mitosis**, but with a much lower frequency than at meiosis. At mitosis, there is no regular pairing of homologous chromosomes and no synaptinemal complex. At mitosis, one can have

| Alleles | Base pairs at *A/a*, at the point of the base substitution. | Base pairs at *A/a* after hybrid-DNA formation in one *a* chromatid by invasion by the T-carrying strand of one *A* chromatid. |
|---------|---------|---------|
| *B A* | $\underline{A}$ <br> T | $\underline{A}$ <br> T |
| *B A* | $\underline{A}$ <br> T | $\underline{A}$ <br> T |
| *b a* | $\underline{C}$ <br> G | mispair $\underline{C}$ - can go to $\underline{C}$ or $\underline{A}$ <br>                  T          G   T <br>                        from correction |
| *b a* | $\underline{C}$ <br> G | $\underline{C}$ <br> G |

*B/b* is a segregating locus syntenic to *A/a*, with *B* on the *A* chromatid and *b* on the *a* chromatid.

If mispair C/T at the *A/a* locus is corrected to C/G, it gives an *a* chromatid, with a normal 2*A*:2*a* allele ratio from that meiosis, or 4*A*:4*a* in octads, and no recombination with *B/b*. If it is corrected to A/T, it gives an *A* chromatid, with a 3*A*:1*a*   gene conversion ratio, or 6*A*:2*a* in octads. Correction of the mispair C/T to A/T, an *A* allele, would give a non-reciprocal recombinant, *A, b,* with no corresponding *a, B* recombinant. If it is uncorrected, we will get postmeiotic segregation with a 5*A*:3*a* gene conversion ratio in octads; in tetrads, the meiotic product carrying the mispair would give one *A* and one *a* product after the first mitosis. This can cause a mosaic of two different genotypes; for example, if the post-meiotic segregation occurred in an egg which was fertilised by an *a* sperm, the first zygotic mitotic division would give one cell *A a*, and one cell *a a*.

**Fig. 8.1.  The consequences of gene conversion for allele ratios and recombination.**

**mitotic crossing-over** and **mitotic gene conversion,** but at a frequency reduced by a factor of at least a thousand compared to meiosis. The distribution of crossovers (relative genetic distances in different parts of a chromosome) may also differ between meiosis and mitosis. See Sec. 18.4 for details of mitotic recombination in fungi, where it can be used by breeders in species where sexual reproduction is absent.

**Mitotic recombination** occurs in all Eukaryotes which have two or more copies of each chromosome; in haploids, there is no homologue to cross over with. It was demonstrated in the fruit fly *Drosophila melanogaster* by Stern (1936), long before it was known in fungi. He used mutants *y*, yellow body, and *sn*, singed bristle, which are linked on the X chromosome. Figure 8.2 shows how a single crossover between the centromere and the two loci can give two genetically different nuclei after

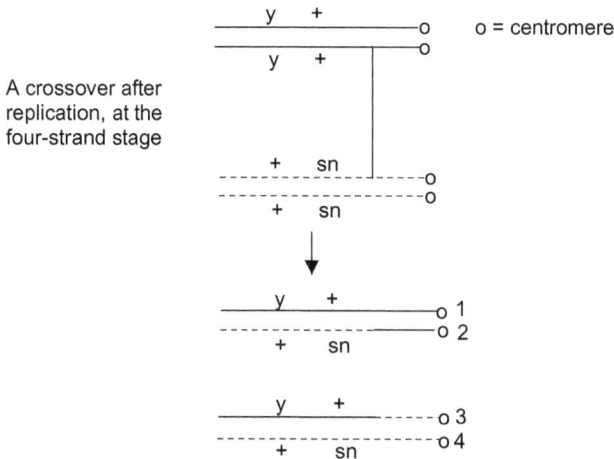

A crossover after replication, at the four-strand stage

If centromeres 1 and 3 go to the same pole, and 2 and 4 go to the other pole, then one nucleus will get y + /y +, giving a yellow, non-singed, phenotype, and the other nucleus will get + sn/+ sn, giving a non-yellow, singed phenotype, which after enough further cell divisions would give visible twin spots. If centromeres 1 and 4 go to the same pole, and 2 and 3 go to the other pole, both nuclei will give wild-type phenotypes.

**Fig. 8.2.** **The origin of twin-spots in *Drosophila* from a mitotic crossover** between the X chromosomes after replication.

mitosis, if the centromere segregation is favourable. If this mitotic recombination occurs early in development, further cell divisions by the daughter cells can give observable twin-spots on the adult fly, where a spot with yellow body and non-singed bristles is next to the one with non-yellow and singed bristles, when the general body appearance is non-yellow and non-singed bristle. **Chromosome aberrations** can also give new combinations of existing alleles (Chap. 9).

In meiosis or mitosis, there can be **interference between crossovers** affecting either the occurrence of double or multiple crossovers, or the distribution of crossovers between strands in double crossovers. See Fincham *et al.* (1979), Foss *et al.* (1993), Stahl *et al.* (2004) and Sec. 18.3. With **chromosome interference** (chiasma position interference) the occurrence of one crossover in an interval may reduce (positive interference), increase (negative interference) or have no effect on the chance of a second crossover in that interval. In **chromatid interference** (strand interference), the occurrence of a crossover between two particular chromatids may affect which chromatids are involved in a second crossover in that interval. If the same strands are involved again less often than one would expect by chance, there is positive chromatid interference, giving a deficiency of two-strand doubles and an excess of four-strand doubles. If the same strands are involved more often in the second crossover than expected by chance, there is negative chromatid interference, with an excess of two-strand doubles and a deficit of four-strand doubles. With no chromatid interference, one expects a ratio of 1 two-strand double crossover: 2 three-strand doubles: 1 four-strand double. Usually there is no chromatid interference, but it requires tetrad analysis to detect it (see Lamb, 1996).

As **undetected crossovers** in double or multiple crossovers reduce the observed genetic map distance in relation to the physical map distance, especially over long distances, it is useful to be able to correct map distances for undetected crossovers. This can be done mathematically by using a **map function** such as **Haldane's mapping function**. That particular function assumes no chromosome or chromatid interference. If $d$ is the true map distance as a fraction (true map units/100), and $\rho =$ the observed recombination distance as fraction (recombination %/100), then $d = -1/2 \ln(1-2\rho)$. From the other form of the equation, $\rho = 1/2(1 - e^{-2d})$, one can see that even very large map distances can only give a maximum recombination frequency of 50% for syntenic loci, because $e^{-2d}$ becomes very small at large $d$ values.

## 8.3. Numbers of types of gamete, and of offspring genotypes and phenotypes, for different numbers of segregating loci

If one has a multiple heterozygote, with two alleles segregating at each of $n$ loci, there are **general formulae for its gametes and offspring**, as follows.

The number of different types of gamete $= 2^n$.
The number of different gamete combinations $= 4^n$.
The number of different offspring genotypes $= 3^n$.
The total number of offspring phenotypes from selfing the F1 are:

(i) with complete dominance, $2^n$;
(ii) with anything giving three phenotypes per locus, e.g., additive action, incomplete dominance or overdominance, $3^n$;
(iii) with $p$ loci with complete dominance and $(n - p)$ loci where the heterozygote is distinguishable, $2^p \times 3^{(n-p)}$.

The offspring phenotype ratio from selfing the F1 with complete dominance at all segregating loci and no linkage is: (3/4 dominant + 1/4 recessive)$^n$; e.g., locus one has starchy grains dominant, sugary grains recessive, and locus two has coloured grains dominant and colourless grains recessive. We expect a Mendelian dihybrid 9:3:3:1 phenotype ratio of (3/4 starchy + 1/4 sugary) (3/4 coloured + 1/4 colourless) = 9/16 starchy coloured + 3/16 starchy colourless +3/16 sugary coloured + 1/16 sugary colourless.

If the heterozygote is distinctive at all segregating loci, as with additive action, we get a phenotype ratio of (1/4 homozygote 1 + 1/2 heterozygote + 1/4 homozygote 2)$^n$. If some loci have complete dominance and some have distinctive heterozygotes, the offspring phenotype ratio is (3/4 dominant + 1/4 recessive)$^p$ (1/4 homozygote 1 + 1/2 heterozygote +1/4 homozygote 2)$^{(n-p)}$. If one has linkage between any two or more loci, these phenotype frequencies will have more of the parental combinations and fewer of the recombinant ones.

## 8.4. Calculation of the frequencies of particular genotypes and phenotypes

To estimate **how many offspring should be reared** to get particular genotypes or phenotypes, one needs to calculate their expected frequencies. The probability of a series of independent events all occurring together is the product of the probabilities of each separate event. If all the loci are unlinked to each other, one just multiplies together the expectations at the different segregating loci.

The **expected frequencies** of different genotypes and phenotypes at any one locus are a simple matter of Mendelian genetics. Thus, if one has *A a* × *A a*, the genotype expectations at that locus in the next generation are $1/4$ *A A*, $1/2$ *A a*, $1/4$ *a a*, and the phenotype expectations with complete dominance are $3/4$ *A* and $1/4$ *a*. For *A a* × *a a*, the genotype expectations are $1/2$ *A a*, $1/2$ *a a*, with phenotypes $1/2$ *A*, $1/2$ *a*.

Suppose we had **seven unlinked segregating loci** in a polyhybrid *A a, B b, C c, D d, E e, F f, G g*, which was selfed, and we wanted the expected frequency of *A A, b b, C -, D -, e e, F f, G g*. Working out the expectations at each locus and multiplying them together, we get $1/4 \times 1/4 \times 3/4 \times 3/4 \times 1/4 \times 1/2 \times 1/2 = 9/4096 = 0.0022$, or about 1 in 455.

Suppose we had *A A, B b, c c, D d, E E, f f* × *A a, B b, C c, D d, E E, F F*; we could again work out the expected frequency for any genotype or phenotype, by working out expectations at each locus and multiplying them together. *A a, b b, C c, D D, E E, F f* would have frequency $1/2 \times 1/4 \times 1/2 \times 1/4 \times 1 \times 1 = 0.015625 = 1$ in 64. **Linkage between any loci** will give an excess of parental types over recombinants. If in this cross *B/b* and *D/d* were linked, with 10% recombination and with *B* and *d* on one chromosome, and *b* and *D* on the homologue, the chance of getting a *b D* gamete from *Bd/bD* is no longer $1/4$, but is 0.45 (because there will be 90% parental gametes), so the chance of getting *b b, D D* is no longer $1/4 \times 1/4$ but is $0.45 \times 0.45$, so the frequency of the desired genotype is 0.050625, or about 1 in 20 instead of 1 in 64.

## 8.5. Mapping, including physical mapping

**Genetic maps** are traditionally based on recombination frequencies from crosses, and can include markers giving phenotypic effects and molecular markers. Much progress is being made with many organisms, microbial, plant, animal and human, with major sequencing programmes under way for whole genomes. **Physical maps** are based on examination of DNA bands on stained gels after electrophoresis. In many organisms, a combination of genetic and physical methods is adding greatly to the detail of maps. For example in 1990, there were only 30 loci on genetic maps of sheep but by 1996, this had increased to about 600 loci and molecular markers. The total number of sheep genes has been estimated at about 100,000. An Australian catalogue of sheep genes and inherited traits can be found at http://www.angis.su.oz.au. There is also a catalogue of inherited traits and disorders in a wide range of animal species, *Mendelian Inheritance in Animals*, at http://morgan.angis.su.oz.au (Nicholas, 1997). For genome mapping in plants, see Paterson (1996).

A physical map is a diagram of a DNA molecule showing the positions of physical landmarks; these commonly include restriction sites and other particular DNA sequences. A **restriction site** is the short base sequence, usually of four to six nucleotides, at which a particular restriction endonuclease enzyme makes a cut. For example, the restriction enzyme *Eco*RI from the bacterium *Escherichia coli* recognises the base sequence 5'-GAATTC-3' and cuts that strand between the G and the A. On the other strand, there must be the complementary sequence 3'-CTTAAG-5', which is also cut between the G and the A, giving **staggered breaks** in the two strands, with **complementary single-strand ends**, which we will see in Chap. 14, are very useful for genetic engineering. Some restriction enzymes cut both strands of DNA at the same point, giving **flush ends**, also known as **blunt ends**.

There are several thousand different restriction enzymes, all with different restriction sites (recognition sequences). They occur in all bacteria and Eukaryotes, with the best known and most used ones being from bacteria. The restriction sites involve **palindromic** sequences, running in opposite directions in the two DNA strands, as in the above example of GAATTC in one strand and CTTAAG in the other.

Much of physical mapping is based on the lengths (judged by migration speed, with smaller fragments migrating faster than long ones) of fragments of DNA after digestion with different restriction enzymes. Pieces of DNA with no, one, two or three *Eco*RI restriction sites would respectively give one, two, three or four fragments of DNA, which could be detected after staining on gels. A piece with one such site near the middle would give two fragments of approximately equal length, while a piece with the site near one end would give two very unequal fragments. Having found out where the *Eco*RI sites are, one could then try a series of different restriction enzymes, e.g., *Hind*III, *Bam*I, and others.

If a mutation changes the base sequence within a restriction site so that the restriction enzyme no longer recognises that sequence, then using that enzyme on the wild-type DNA gives two fragments, but on the mutant DNA, only one fragment is found. Less frequently, mutation can cause restriction sites. Differences in DNA fragment length and number caused by the presence or absence of restriction sites are called **restriction fragment length polymorphisms** (**RFLPs**).

If one has two cleavage sites for a particular restriction enzyme in a given region of DNA, one might expect that DNA treated with that enzyme should produce the same fragment lengths from different individuals. They will produce different numbers of fragments if the individuals (or the two homologous chromosomes in diploids) differ in restriction sites, as mentioned above. Another kind of difference arises when different individuals (or haploid genomes) give different lengths of fragment from the two cleavage sites. A frequent cause is the existence of **tandemly repeated DNA sequences**, repeated e.g., 10 to 300 times, with different numbers of repeats in different individuals. One individual might have the sequence repeated 15 times, while another has it repeated 60 times, giving a different fragment length from the DNA between the two restriction sites which are cut with that enzyme. The general term is a **variable number of tandem repeats**, **VNTR**. See Sec. 9.4 for a general account of duplications.

If the repeat length is just of two bases, one has **simple tandem repeat polymorphisms**, STRPs, where pairs of bases make repeating sequences in tandem, such as 5′-TGTGTGTGTGTG-3′. Different "alleles" (the quotation marks are because these are not really different forms of a functional gene locus) have different numbers of repeats of the pairs of bases, which vary very frequently, so that each STRP site is on average heterozygous in 70% of humans tested. STRPs have been very important in mapping the human genome. Since there is such a high degree of polymorphism between human individuals for STRPs and other VNTRs, they have been extensively used in forensic science for confirming or eliminating suspects on the basis of DNA samples, e.g., comparing sperm DNA in a rape case with DNA from different suspects' hair roots.

One very useful type of marker on the physical map is the **sequence-tagged site, STS**. This is a specific DNA sequence present only once per haploid genome and which can be amplified by the **polymerase chain reaction (PCR)** with a suitable pair of oligonucleotide primers, one at each of its ends. The presence or absence of such sites in a particular piece of DNA is determined after PCR multiplication of that region with those two primers. Many thousands of such markers have been identified in the human genome.

One can compare **physical maps and genetic maps**. This was originally done for the fruit fly, *Drosophila melanogaster*, using physical maps based on DNA bands from salivary gland chromosomes as seen stained under the microscope, and genetic maps from recombination frequencies. The loci on the two kinds of maps were on the same chromosomes and in the same linear order as one would expect, but the relative distances differed in some parts of the maps. This comes from there being relatively more crossovers per unit of physical length in some regions, and fewer crossovers in some other regions. In *Drosophila*, the recombination frequencies have to be obtained from female meiosis, as there is no crossing over in the males, which do however have independent assortment for non-syntenic loci. Humans also have more recombination in female meiosis than in male meiosis, so the total recombination map over 23 pairs of chromosomes in human females is about 4,400 cM long, compared with only 2,700 cM in males. As the human genome is about 3,154 million base pairs long, a million base pairs corresponds very roughly to one cM, or 1% recombination in humans, but this correspondence varies a lot between different groups of organisms.

When trying to map the genes controlling a **complex quantitative character** such as yield, there are complications from environmental effects and genotype by environment effects, as well as the characters usually being controlled by QTL and by polygenes, perhaps with different degrees of dominance and penetrance. For QTL mapping, a cross between inbred lines differing for the character is made and the cosegregation of the alleles at known marker loci and phenotypic characters allows linked markers to be identified. If genotype × environment interactions are large, it may be necessary to use special mapping populations across a range of environments. This has been done for characters such as drought resistance in cotton, growth and yield in rice, and yield in barley (references in Kraakman *et al.*, 2004), enabling them to identify genes affecting yield and yield stability.

Kraakman *et al.* (2004) used a different method, **linkage disequilibrium mapping**, to find markers linked to complex traits of yield, yield adaptability and yield stability in a collection of 146 modern two-row spring barley cultivars. These were all homozygous diploids from inbreeding or doubling haploids. Using 236 AFLP markers, they used linkage disequilibrium to find marker-trait associations. Many of the marker associations were found in chromosome regions where relevant QTL (http://barleyworld.org) had previously been found. Linkage disequilibrium has been found to extend over 3 cM in beet, was very common over distances of less than 10 cM in barley, and extended over longer distances in some *Arabidopsis* populations.

## 8.6. Locating genes on particular chromosomes, including the use of pseudodominance, parasexual methods and hybridisation probes

To complete mapping, one needs to know which linkage group corresponds to which physical chromosome, as observed under the microscope. One expects the number of linkage groups to correspond to the number of different types of chromosome, counting the sex chromosomes as one type, if present. Thus, maize, with 20 chromosomes in the diploid, 10 (one of each type) in the gametes, has 10 linkage groups, and bread wheat, with 42 chromosomes, a hexaploid with 6 copies of each type of chromosome, has 7 linkage groups. As crossing-over is not equally frequent per unit physical distance, there is only a rough correlation between the total length of a chromosome in cM and its physical length in $\mu$m.

### 8.6.1. *Pseudodominance*

One way of correlating visible chromosomes with linkage groups is to use pseudodominance with induced large deletions. In **pseudodominance**, a normally recessive allele is expressed in the phenotype of a diploid because there is only one copy of that locus present; i.e., it is hemizygous because of a deletion opposite it in the other homologue. A maize strain was constructed which was homozygous for a series of linked recessive genes of linkage group II: *virescent, booster* (of *sun-red*), *glossy, liguleless*. This was crossed as female to a pollen parent which carried dominant alleles at all these loci, and had been irradiated to cause chromosome breaks (Chap. 9). Most seedlings were wild-type for all loci, but a few were *virescent*. These proved, on examining their stained chromosomes, to be heterozygous for a terminal deletion on a particular cytologically-mapped chromosome, on examining their stained chromosomes. The experimenters were therefore able to identify which chromosome corresponded to linkage group II and which end of the chromosome corresponded to which end of the linkage group. The irradiation caused chromosome breaks, especially terminal deletions, so the *virescent* seedlings would have been as shown in Fig. 8.3. This kind of chromosome/linkage map correlation was used extensively in many plants. For example, see Khush and Rick (1968) for classical mapping in the tomato, *Lycopersicon esculentum*, before molecular methods were developed. They induced 74 different deficiencies with X-rays and fast neutrons on pollen, using pseudodominance to locate 35 loci on 18 of the 24 arms of the chromosome complement.

Another method of using pseudodominance for mapping in maize involves **reciprocal translocations** between the A (ordinary) and B (supernumerary; see Chap. 11) chromosomes. The second mitotic division in pollen grains after microsporogenesis provides the two generative nuclei, and at that division the B centromeres have frequent non-disjunction, often

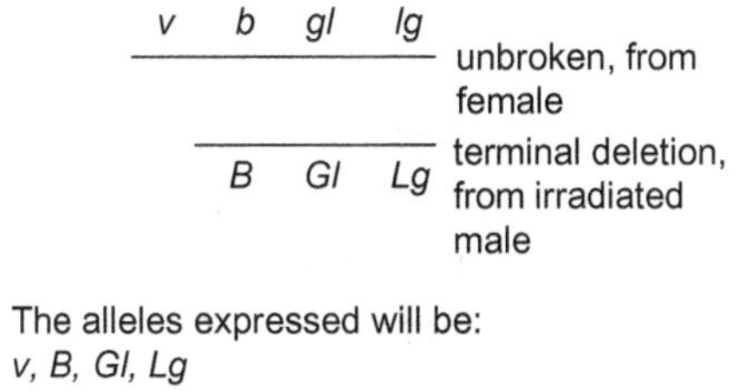

The alleles expressed will be:
*v, B, Gl, Lg*

**Fig. 8.3. Detection of a terminal deletion through pseudodominance** of a recessive allele on the unbroken chromosome carrying recessive alleles in tomato.

with both going to one nucleus and none to the other. A plant to use as male is constructed with one particular A chromosome, a B chromosome, and their $A^B$ and $B^A$ pair of reciprocally translocated chromosomes, where the normal letter indicates the chromosome providing the centromere and the superscript shows the translocated part lacking a centromere. On non-disjunction, one generative nucleus is hyperploid, perhaps getting $A^B$ (lacking the translocated part of A) and two $B^A$s (having two copies of the translocated part of the A chromosome), while the other is hypoploid, perhaps getting one $A^B$ and no $B^A$s (lacking the translocated part of the A, since $A^B$ has part of A missing). When that hypoploid pollen fertilises another maize plant, any recessive genes present in the female and affecting kernals or plants will show pseudodominance, appearing in the phenotype, if they are in the translocated arm of the A chromosome, but not if they are located elsewhere. According to Neuffer *et al.* (1997), there are 89 different B-A reciprocal translocations available in maize, covering 18 out of the 20 chromosome arms. They are very useful for determining in which chromosome and in which chromosome arm recessive genes map.

### 8.6.2. *Parasexual methods*

Other methods for chromosome/linkage group correlations included **parasexual methods**, using hybrid cell cultures rather than crosses. In the 1970's, one could get tissue-culture-adapted mice cells, for example, and fuse these with human cells such as skin fibroblasts or peripheral blood leucocytes, using chemical agents or Sendai virus. For instance, the human cell has 2nH representing its diploid number of 46 human chromosomes, and the mouse cell has 2nM chromosomes, then the fusion of cell membranes and cytoplasm gives a **heterokaryon** (which has unlike nuclei in common cytoplasm) with 2nH + 2nM chromosomes. There can then be rare chance fusion between the two kinds of nucleus to give a **synkaryon**. As the synkaryon proliferates in cell culture to give a hybrid cell population, various chromosomes are lost by **non-disjunction**, a failure to segregate regularly and equally to both daughter cells from a cell division. Different chromosomes, especially human ones, are lost in different cell lines. Typically after 30 generations in the synkaryon proliferation, about 7 out of 46 human chromosomes remain, with a range of about 1 to 20. By correlating the loss of certain genes which are expressible in tissue culture with the loss of visible chromosomes, one can correlate visible chromosomes with linkage maps. It is essential to use tissue-culture-expressed genes or DNA markers, and not ones like eye colour genes.

When a whole chromosome is lost, all genes on it will be lost, so one can find out which loci are syntenic. If only part of a chromosome is lost, e.g., from a spontaneous big terminal deletion, one can correlate genes to a particular end or region of a chromosome. In the 1980's, physical mapping largely took over from other methods in humans, with massive sequencing of all chromosomes in the human genome project from the late 1990's.

One can fuse sheep lymphocytes or fibroblasts to mutant cell lines of mouse or hamster, using polyethylene glycol or Sendai virus, or using rodent cell lines with multiple enzyme deficiencies. During hybrid cell growth, sheep chromosomes are preferentially eliminated. This can be used for mapping by testing hybrid cells for growth in a range of nutrient media, checking which sheep chromosomes can complement particular mouse enzyme deficiencies. For sheep physical maps, see Broad *et al.* (1997); for sheep linkage maps, see Montgomery and Crawford (1997). One can also use cell fusions between individuals of the same species, or between an individual and a reference cell line, to do *cis-trans* **tests for functional allelism** (Subsec. 1.3.9), to see at which locus an individual carries a mutation.

### 8.6.3. *Hybridisation probes*

In molecular genetics, the term **probe** is used for a piece of single-stranded DNA or RNA used in DNA-DNA or DNA-RNA hybridisation assays. It usually has a radioactive label, with hybridisation products being identified through autoradiography, where the radioactivity causes dark spots on sensitive film. Fluorescent or luminescent labels may be used instead, where the probe can be detected under the microscope with the right lighting conditions. The basis of probing is that the single-stranded DNA or RNA probe will, under the right conditions, anneal by base pairing to a complementary sequence in DNA which has been made single-stranded.

If one has cloned a particular gene and wants to find its chromosomal location, it can be grown up in radioactive medium to get it labelled, then it is denatured into single strands by heating or chemical treatment.

A preparation of chromosomes at mitosis is then made, where individual chromosomes can be seen. The chromosomal DNA is denatured to single strands, then the DNA probe is added. Conditions are made to favour annealing of the probe to complementary sequences in the chromosomes, then the excess probe is washed off, so that only base-paired probe remains. For radioactive probes, autoradiography follows, to see where the silver grains from radioactive decay occur on the chromosome. In a diploid with a single-copy gene, one would expect one point to be labelled on two homologous chromosomes, showing where the DNA sequence of the probe is on the chromosomes. If the gene has many copies, as for tRNA genes, many points will be labelled, which helps to identify such multicopy genes and to find whether they are limited to particular chromosomes or regions.

## 8.7. Practical uses of molecular markers in agriculture

Molecular markers can be used for the **indirect selection** of phenotypic agricultural traits, if they are closely linked to that character, and if both markers are segregating in a population. The European Apple Genome Mapping Project (EAGMAP) is led by Horticultural Research International at East Malling, Kent, and Wellesbourne, Warwickshire, developing a reference linkage map for *Malus*. The molecular markers used are isoenzymes and several kinds of DNA markers, including randomly amplified polymorphic DNA (RAPD), restriction fragment length polymorphisms (RFLPs) and microsatellites. The microsatellites are often simple tandem repeat polymorphisms (see Sec. 8.5), which are ubiquitous, highly polymorphic and co-dominant, needing very small amounts of DNA for analysis (King *et al.*, 1996).

The EAGMAP reference population is a cross between Prima, a cultivar susceptible to rosy leaf curling aphid (*Dysaphis devecta*; see Sec. 16.3), and Fiesta (formerly Red Pippin), a resistant cultivar which carries the dominant resistance allele $Sd_1$ from Cox's Orange Pippin which is heterozygous for it. Segregation in the reference population gave 75 resistant plants to 62 susceptible, a good 1:1 allele ratio from $sd_1\ sd_1 \times Sd_1\ sd_1$. Roche *et al.* (1996) also scored segregating RFLP and RAPD markers. Three RFLP markers from Fiesta were very closely linked to $Sd_1$, within 2 cM. Any one of them could therefore be used to screen plants from a segregating

**Plate 8.1.** **Apple seedlings for molecular-marker screening** at Horticultural Research International, East Malling, to decide which are worth growing to fruiting and later determination of commercial characters.

population for the aphid resistance soon after germination (Plate 8.1), before growing plants to maturity. Roche *et al.* (1996) comment, however, that the use of these markers is too laborious for routine screening, and that a PCR-based assay might be more convenient for **marker-assisted selection** (see Plate 8.1 and Sec. 13.3). A visible morphological marker which is closely linked to aphid resistance and which could be scored in seedlings would be even simpler to use. Susceptibility to the aphid can be scored at the seedling stage anyway, but having large populations of pests for testing can be inconvenient.

For a recent account of **cereals** such as wheat, barley and rice, their genetics and genomics, see Gupta and Varshney (2004). The papers in that book include details of marker-assisted selection, QTL, mapping, disease- and stress-resistance, use of molecular markers, cloning, the use of genomics for crop improvement, population structure, adaptation, etc.

There is an **International Chicken Polymorphism Map Consortium** to locate and characterise markers within genes that control traits of interest for chicken breeding, including egg production, muscle mass and growth rate. This requires hundreds of markers and by 2004, nearly three million single nucleotide polymorphisms (SNPs) had been obtained in various breeds. The consortium used Applied Biosystems' SNPlex[TM] Genotyping System assays specifically to target previously mapped QTL for markers of interest and to genotype 48 SNPs in each assay. The chicken genome is described in Sec. 8.8.

## 8.8. Genomics

These are exciting times as **more genomes are being sequenced**, including man, mouse, rat, chickens, rice, maize, yeast and disease organisms causing African sleeping sickness, leishmaniasis and Chagas disease, bacterial pathogens such as *Bacteroides fragilis* and the protozoan parasite giving amoebic dysentery, *Entamoeba histolytica*. In the latter, genome analysis has shown its acquisition from bacteria of genes allowing the use of a wider range of sugars and other energy sources. *Bacteroides fragilis* is a dangerous opportunistic parasite if it escapes from the gut where it is part of the normal flora. Genomic analysis by Cereno-Tarraga *et al.* (2005) has shown that it evades host immune responses by varying its surface coat molecules, by periodically turning surface coat genes on and off. It does this by "**flipping**", turning short fragments of the DNA through 180°.

The **chicken**, *Gallus gallus*, is the first farm animal to be sequenced (International Chicken Genome Sequencing Consortium, 2004). It has about one billion base pairs and an estimated 20,000 to 23,000 genes. The genetic map has about 2,200 loci with a genetic length of about 4,000 cM. Bird chromosomes tend to be of differing lengths, with chickens having small microchromosomes (about 3 Mb and upwards), some of intermediate length, and large macrochromosomes (up to about 200 Mb). The chicken has 2n = 78, with 38 pairs of autosomes, ZW females and ZZ males. The smaller the chromosomes, the higher the recombination rate per unit length, with median values of 6.4 cM per Mb for microchromosomes and 2.8 cM per Mb for macrochromosomes. The chicken has a similar number of genes to humans, but only one third of the number of base pairs. The main difference is that chickens have less than 11% of interspersed segments of short repetitive DNA, compared with 40 to 50% in mammals. Many quantitative loci have been identified in chickens (Muir and Aggey, 2003). See

Andersson and Georges (2004) for more on genomics in domesticated animals, especially in relation to QTL. See Sec. 8.7 for molecular mapping and a cereals genomics reference, Gupta and Varshney (2004).

The **sheep genome** has recently been shown to contain benign "endogenous retroviruses", formerly free living but now integrated into sheep DNA, with about 20 copies related to Jaagsiekte sheep retrovirus which causes an important sheep disease, ovine pulmonary adenocarcinoma. The endogenous form interferes with the replication of the disease form, helping to control it and prevent it from spreading.

There are also many cases of genetic variation in humans affecting **susceptibility to infection and inflammation** (Sec. 12.11). For example, Crohn's disease has been found associated with genetic variants of *NOD2*, a pathogen-associated receptor of the immune system, and *NOD1* with susceptibility to inflammatory bowel disease.

**Genomic analysis** has been used to study **human migrations** and the **origins of particular races or populations**. For example, Hurles *et al.* (2005) studied the origin of the indigenous Malagasy people of Madagascar, 250 miles off the East African coast. Genetic studies of DNA from the Y chromosome (paternally inherited) and the mitochondrial DNA (maternally inherited) showed that two of the four main Malagasy ethnic groups originated from Africa and two from Indonesia. That fits with the Malagasy language sharing 90% of its basic vocabulary with Maanyan, spoken in Southern Borneo, 4,500 miles away, and there are borrowings from the Bantu languages in East Africa. There is archaeological evidence supporting such a migration.

Genome analysis has also shown that the **domestic pig**, *Sus*, originated in the islands of South East Asia then spread to the west, with repeated domestication by humans in the Far East, Near East, and further westward. The pig is interesting in that domestic pigs often have surviving wild pig relatives in the same country, with some genetic exchange through unplanned matings between wild and domestic varieties.

The **pig genome** contains 30,000–40,000 genes, with 2n = 38. The Pig Improvement Company UK has a database of millions of samples and production records, to test for associations between animal performance and particular single nucleotide polymorphisms (SNPs). When a new SNP is found, it is genotyped across the DNA samples and compared with performance records. Many markers have been found in different genes which affect growth, carcase composition, meat pH (see Subsec. 13.7.13), coat colour and disease susceptibility.

**Microarray technology** can analyse the expression of many thousands of genes in a single experiment. Minute spots of DNA, each from a different gene, are placed by robots on glass slides ("chips") in arrays, so that the position of each gene spot is known. With 15,000 genes on one slide, the whole pig genome would need only two or three slides. Usually, all known copy DNAs from a genome are spotted on, then exposed to sets of fluorescently labelled probes, e.g., of all mRNA molecules from a given cell type for **gene expression determination**. Binding of the probes to particular DNAs is automatically monitored by laser-beam microscopy. The microarrays can be used to compare gene expression between physiologically extreme samples, such as disease susceptible and disease resistant, or low meat drip loss and high drip loss (see Subsec. 13.7.13) in pigs. DNA chips can also be used to study fluorescently tagged DNA-binding proteins. There are now **customised gene chips** available, for example, for rapid scans of tumour samples for specific DNA changes for prognosis in the cases of neuroblastoma, a common form of cancer in children. By comparing peripheral blood DNA with DNA from the tumour, a particular deletion in chromosome 11 which is associated with aggressive neuroblastoma can be identified, as can other neuroblastoma changes. See Sec. 13.11 for DNA fingerprinting.

The **National Scrapie Plan for Great Britain** involves very extensive **genotyping of sheep** for all three codons associated with scrapie susceptibility. Scrapie is a neurodegenerative disease of sheep and goats caused by changes in brain prion proteins, leading to death. It is one of the **transmissible spongiform encephalopathies**, as is BSE in cattle. Genotyping does not detect the disease, but shows the sheep's susceptibility if exposed to the scrapie agent. There are five different allele combinations, ARR, AHQ, ARH, ARQ and VRQ, where letters represent amino acids, giving 15 different diploid genotypes. Type 1, ARR/ARR, is the most scrapie-resistant; type 2, ARR/AHQ, ARR/ARH and ARR/ARQ, is resistant but could have susceptible offspring; type 3 has little resistance; type 4, ARR/VRQ, is susceptible and should not be used for breeding; neither should type 5, which is highly susceptible.

If scrapie is found in a flock, the whole flock is culled or genotyped followed by selective culling of susceptible genotypes. Compensation is paid for culled sheep, e.g., adult sheep £90 each, lambs £50, embryos £150, ova £5 (2005 rates). Only sheep of certain genotypes may be retained for breeding. Sheep breeders try to cross rams with ewes to produce only or mainly type 1 or type 2 lambs. In its fourth year, the National Scrapie Plan has involved more than 1.3 million blood samples and resulted in an increase in the percentage of types 1 and 2 from 69% in 2002 to 83% in 2004, averaged over all breeds. That is a very **rapid genotype change** in only three seasons. CBS Technologies claim to give results of tests within three weeks of sampling. See www.defra.gov.uk/nsp.

**Databases** in 2005 include: www.yeastgenome.org; www.flybase.org; mouse — www.informatics.jax.org; rat — rgd.mcw.edu; rice, www.rice. genomics.purdue.edu/. Various fungal genome sequences and information can be found at www.broad.mit.edu/annotation/fungi/fgi, including *Rhizopus*, *Chaetomium*, *Saccharomyces cerevisiae RM11*, and three *Candida* species. For recent advances in **rice genomics**, see International Rice Genome Sequencing Project (2005) and *Plant Molecular Biology*, special issue, 59, (1), 2005. For **fungal genomics**, see Subsec. 18.1.5.

## 8.9. Human gene sequencing, including the HapMap Project

The **sequencing of the human genome** was a major step forward in understanding human genetics. Since then, much faster high-throughput whole-genome genotyping has been possible, so that a few hundred thousand single nucleotide polymorphisms (SNPs) can be analysed in major collaborative studies such as the HapMap Project (see below). Human evolution can be much better understood now that the draft sequence of the chimpanzee has become available.

Feuk *et al.* (2005) **compared human and chimpanzee DNA sequences**, finding that **inversions** are many times more frequent in primates than previously thought. The two species diverged about six million years ago, retaining about 98% identity. This study identified **1,576 presumed inversions** between the species, 33 of which were larger than 100 kb, when the average human gene is about 60 kb long. Three out of 23 experimentally confirmed inversions were polymorphic in humans, with some humans heterozygous, with one human and one chimp version, which could cause fertility problems in meiosis (Sec. 9.3). About 10% of the inversions contained a complete gene or caused a breakpoint within a gene, which could affect its function. The largest inversion was a 4.3 megabase one at human chromosome 7p14. There are karyotypically visible pericentric inversions on chromosomes 1 and 16. The other inversions include pericentric and paracentric examples.

In human genetics, **association studies** are used to find loci contributing to disease susceptibility, by comparing genetic patterns in those with and without a disease. According to McVean *et al.* (2005), meiotic recombination in humans is mainly concentrated into short hotspots about 1 to 2 kb long, occurring every 100 to 200 kb. Alleles at loci close to each other, especially in the same recombination "cold-spot", often show **linkage disequilibrium** (Sec. 4.2).

In 2002, the **International HapMap Project** was started to map the structure of allelic associations across the human genome, initially to genotype one single nucleotide polymorphism (SNP) every 5 kb in the human genome across 270 individuals from four "panels". These consist of Yoruba in Nigeria, Han Chinese in Beijing, Japanese from Tokyo and Americans in Utah.

Each single chromosome's genes constitutes a **haplotype**. Within regions of less than 500 kb, one can find combinations of particular SNPs in multiple unrelated individuals, the regions often separated from each other by the recombination hotspots. As one might expect, certain haplotypes only occur in one of the four panels and others have different frequencies in different panels, and some have medical associations, e.g., with genes for lactose intolerance or malaria-resistance. The HapMap Project is also providing extensive data on human genetic diversity, recombination, selection, evolution and linkage disequilibrium.

## Suggested Readings

Andersson, L. and M. Georges, domestic-animal genomics: deciphering the genetics of complex traits, *Nat Rev Genet* (2004) 5: 202–212.

Broad, T. E., H. Hayes and S. E. Long, Cytogenetics: physical chromosome maps, in Piper, L. and A. Ruvinsky (eds.) *The Genetics of Sheep.* (1997), pp. 241–295, CAB International, Wallingford.

Cerdeno-Tarraga, A. M. *et al.*, Extensive DNA inversions in the *B. fragilis* genome control variable gene expression. *Science* (2005) 307: 1463–1465.

Feuk, L. *et al.*, Discovery of human inversion polymorphisms by comparative analysis of human and chimpanzee DNA sequence assemblies. *PLoS* (2005) 1: 489–498.

Fincham, J. R. S., P. R. Day and A. Radford, *Fungal Genetics*, 4th ed. (1979) Blackwell, Oxford.

Foss, E. *et al.*, Chiasma interference as a function of gene distance. *Genetics* (1993) 133: 681–691.

Griffiths, A. J. F. *et al.*, *Introduction to Genetic Analysis*, 8th ed. (2004) W. H. Freeman and Co., New York.

Gupta, P. K. and R. K. Varshney (eds.) *Cereal Genomics.* (2004) Springer, Dordrecht.

Hartl, D. L. and E. W. Jones, *Genetics. Analysis of Genes and Genomes*, 6th ed. (2005) Jones and Bartlett, Sudbury, Maryland.

Hurles, M. E. *et al.*, The dual origin of the Malagasy in Island Southeast Asia and East Africa: evidence from maternal and paternal lineages. *Am J Hum Genet* (2005) 76: 894–901.

International Chicken Genome Sequencing Consortium, Sequence and comparative analysis of the chicken genome provide unique perspectives on vertebrate evolution. *Nature* (2004) 432: 695–716.

International Rice Genome Sequencing Project, The map based sequence of the rice genome. *Nature* (2005) 436: 739–800.

Khush, G. S. and C. M. Rick, Cytogenetic analysis of the tomato genome by means of induced deficiencies. *Chromosoma* (1968) 23: 452–484.

King, G., N. Periam and C. Ryder, Development of microsatellite markers in the Rosaceae. *Annual Report* 1995–1996. Horticulture Research International, Wellesbourne, Warwickshire, pp. 64.

Korol, A. B. and I. A. Preygel, *Recombination Variability and Evolution.* (1994) Chapman and Hall, London.

Kraakman, A. T. W. *et al.*, Linkage disequilibrium mapping of yield and yield stability in modern spring barley cultivars. *Genetics* (2004) 168: 435–466.

Lamb, B. C., Ascomycete genetics: the part played by ascus segregation phenomena in our understanding of the mechanisms of recombination. *Mycol Res* (1996) 100: 1025–1059.

Lamb, B. C., Meiotic recombination in fungi: mechanisms and controls of crossing-over and gene conversion, in Arora, D. K. and G. C. Khachatourians (eds.) *Applied Mycology and Biotechnology, Vol. 3, Fungal Genomics.* (2003) Elsevier, Amsterdam.

Maloisel, L., J. Bhargava and G. S. Roeder, A role for DNA polymerase $\delta$ in gene conversion and crossing over during meiosis in *Saccharomyces cerevisiae*, *Genetics* (2004) 167: 1133–1142.

McVean, G., C. C. A. Spencer and R. Chaix, Perspectives on human genetic variation from the HapMap project, *PLoS Genetics* (2005) 1: 413–418.

Montgomery, G. W. and A. M. Crawford, The sheep linkage map, in Piper, L. and A. Ruvinsky (eds.) *The Genetics of Sheep*. (1997), pp. 297–351, CAB International, Wallingford.

Muir, W. M. and S. E. Aggrey (eds.) *Industrial Perspectives on Problems and Issues Associated with Poultry Breeding*. (2003) CAB International, Wallingford.

Neuffer, M. G., E. H. Coe and S. R. Wessler, *Mutants of Maize* (1997). Cold Spring Harbor Laboratory Press, New York.

Nicholas, F. W., Genetics of morphological traits and inherited disorders, in Piper, L. and A. Ruvinsky (eds.) *The Genetics of Sheep*. (1997), pp. 87–132, CAB International, Wallingford.

Paterson, A. H. (ed.) *Genome Mapping in Plants*. (1996) Academic Press, London.

*Plant Molecular Biology*. (2005) 59: 1, September. Special complete issue on rice functional and comparative genomics.

Roche, P. *et al.*, Molecular markers linked to aphid resistance. *Annual Report* 1995–1996, Horticulture Research International, Wellesbourne, Warwickshire, pp. 64–65.

Stahl, F. W. *et al.*, Does crossover interference count in *Saccharomyces cerevisiae*? *Genetics* (2004) 168: 35–48.

# Chapter 9

# Structural Chromosome Aberrations: Their Origins, Properties and Uses

## 9.1. Introduction

In Sec. 1.3, we saw that **chromosome aberrations** were changes in the chromosomes on a scale larger than a single locus, with changes within a locus being called mutations. They may involve changes to just one chromosome, with the loss of a region of the chromosome (**deletion**), reversal of the order of loci in part of the chromosome (**inversion**), or duplication of parts (**duplication**). They may also involve two non-homologous chromosomes as in **translocations**, which may be reciprocal or non-reciprocal. Some chromosome aberrations are usable by plant and animal breeders, while others cause infertility or other problems in micro-organisms, plants, animals and humans. Male gametes tend to be more sensitive than female gametes to the adverse effects of chromosome aberrations which are therefore often transmitted more through females than through males. This chapter is concerned with changes in chromosome structure; changes in chromosome number are covered in Chap. 10. Human chromosome aberrations are also covered in Sec. 12.8, and see Gersen and Keagle (2005). Section 8.9 dealt with the many and often large inversion differences between humans and chimpanzees, with heterozygosity for them in some humans (Feuk *et al.*, 2005).

All the kinds of aberration described here occur in fungi, plants, animals and humans. In man, the translocation form of Down syndrome (an unbalanced aberration) can be caused by a translocated chromosome 21 on chromosome 13 or 14, in addition to two normal 21s, or by a 21 on 21 translocation plus one normal 21 (Subsec. 12.8.2). Aberrations may be **balanced**, with no change in the total amount of genetic material, as in a translocation of part of chromosome 1 onto chromosome 12 (Plates 12.6

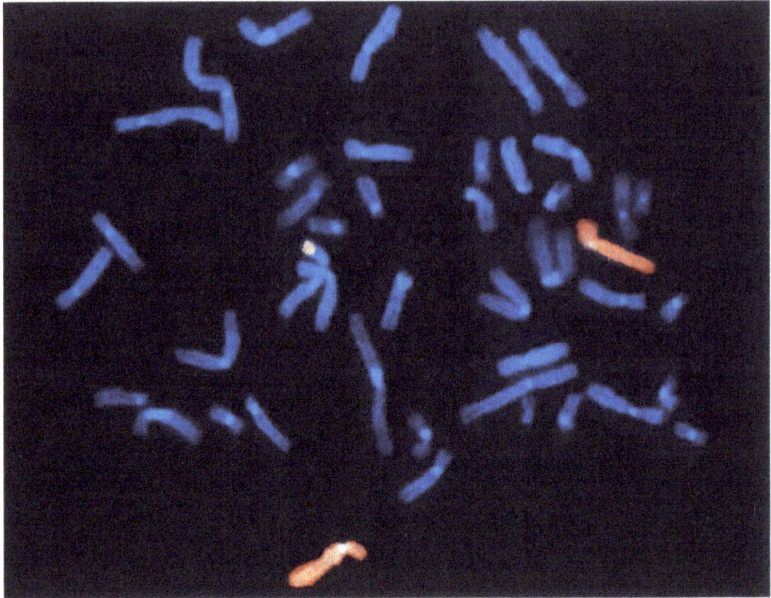

Case: B99-904   Slide: wcp   Cell: 5d   Patient:

**Plate 9.1.** A human cell at metaphase, with a **"whole chromosome 5" FISH paint**. There are two normal chromosome 5's, showing pink, and a 21 with part of 5, showing pink for its left end, with the translocated part of 5 almost as long as the 21, which does not physically join the two longer chromosomes below it. This is **an unbalanced non-reciprocal translocation**.

and 12.7), or **unbalanced**, as in the addition of part of chromosome 5 to chromosome 21, in the presence of two normal chromosome 5's (Plate 9.1). Unbalanced aberrations usually have **harmful phenotypic effects** from gene-dosage imbalance, while balanced ones may be neutral or harmful phenotypically, and often affect fertility.

The production of chromosome aberrations involves spontaneous or induced **double-strand breaks** of DNA (see Sec. 7.2 on which types of mutagen cause these). Aberrations may also involve the rejoining of the broken ends of chromosomes; this occurs frequently because the broken ends act as if they were "sticky". In the following diagrams, the wild-type chromosome is usually considered to have a sequence of regions (not loci, but longer stretches) *ABCDoEFGH*, where *o* represents the centromere. The regions involved in the aberration will be shown in **bold typeface**.

Any **acentric** fragment (without a centromere) cannot segregate properly at mitosis, and will eventually be lost, with the **centric** fragment being retained and usually segregating normally.

Some aberrations arise from **illegitimate crossing-over**, where supposedly non-homologous regions cross-over, e.g., a crossover starting between *B* and *C* in one homologue of *ABCDoEFGH*, and ending between *F* and *G* in the other homologue (see Fig. 9.3.). Some small genetic elements, such as transposons and insertion sequences, are present many times in some genomes and can thus provide points of homology in the otherwise non-homologous regions within and between chromosomes, as can repeated gene sequences; they may promote illegitimate crossovers (i.e., between generally non-homologous regions), as shown in Fig. 9.3.

Chromosome aberrations change map distances, or even whether two loci are syntenic or non-syntenic. Some balanced aberrations may have phenotypic effects just because they move genes into different parts of the chromosome (**position effects**). Thus, moving a gene from the **euchromatic** middle of a chromosome arm to near the centromere, into a **heterochromatic region** (one with a different time of DNA condensation in the cell cycle from normal euchromatic DNA), may affect function and therefore the phenotype, even if there is no change in what genes are present or how many times they are present. Aberrations changing the **gene dosage** (the number of times a gene is present) can change the phenotype in ways which are difficult to predict. Aberrations are normally only microscopically visible if extending over at least 4 Mb, although probes can detect much smaller aberrations.

Chromosome ends need **telomeres** and telomerase enzyme to prevent shortening of chromosomes at replication, as the lagging strand would otherwise be shortened by the length of the terminal RNA primer. In **simple terminal deletions**, the absent telomere and sub-telomeric region are replaced by a set of enzymes, including telomerase, which "cap" the broken end.

## 9.2. Deletions

A **deletion** involves the loss of a region of the chromosome and is also known as a deficiency. A single break (spontaneous or induced) causes a **terminal deletion**, with the acentric fragment being lost. For example, wild-type

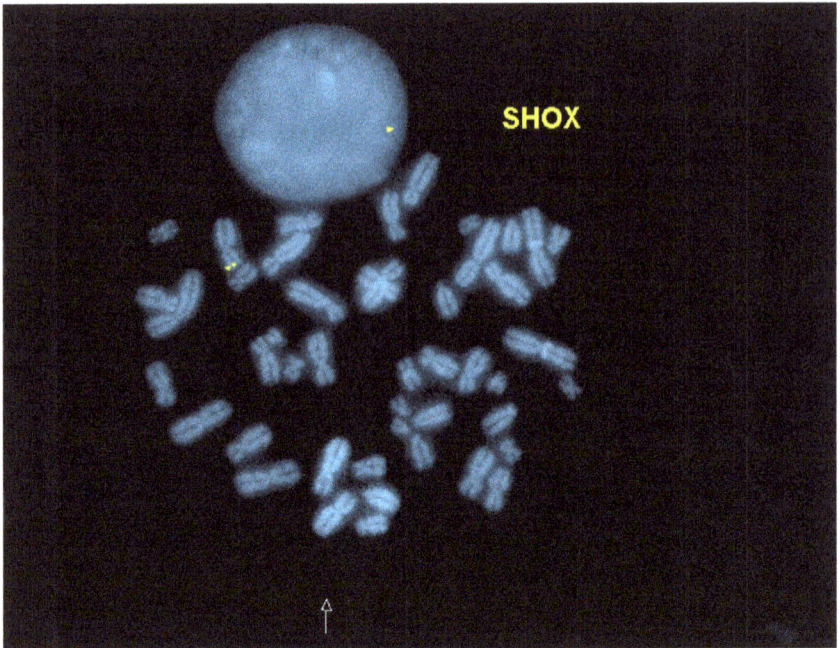

**Plate 9.2.** A SHOX (short stature homeobox) **deletion in the pseudoautosomal region of the human X** in a female, causing short stature. This female has a deletion in one X chromosome (arrowed, bottom), with the SHOX probe showing up the undeleted region in the other X. The interphase cell above also has only one SHOX copy showing. The pseudoautosomal region escapes X-inactivation in females, normally with two active copies of the SHOX gene in males and females giving normal height.

chromosome *ABCDoEFGH* would give centric deletion *CDoEFGH* if there were a break between *B* and *C*, and would give deletion *ABCDo*, with an almost terminal centromere, if the break occurred between the centromere and *E*. Plate 9.2. shows a heterozygous deletion at or near the end of the short arm of the human X chromosome in a female.

To get an **interstitial** (non-terminal) **deletion** requires two breaks and the rejoining of the outer ends. Thus if *ABCDoEFGH* has a break between *E* and *F*, and another between *G* and *H*, one could get the interstitial deletion form, *ABCDoEH*, if the *E* end joins up with the *H* end. As interstitial deletions require two independent breaks and a rejoining, whereas terminal deletions only require one break, interstitial deletions occur more rarely than terminal deletions. The map will change as regions *E* and *H*, previously separated by *F* and *G*, are now adjacent, so that there are fewer crossovers between them. Heritable deletions cannot involve the centromere because acentric fragments are lost.

As deletions lack lengths of chromosome, they usually involve the loss of some essential genes, making them lethal when homozygous. Even in heterozygotes (one chromosome normal, one with a deletion), the deletion may be lethal in the gamete or any other haploid stage if expression of some of the genes lost is needed in the haploid stage. As a heterozygote for a deletion is hemizygous for those loci on the wild-type chromosome which are missing from the deleted chromosome, any recessive lethals or deleterious or beneficial recessives in that region of the wild-type chromosome will be **pseudodominant** (Sec. 8.6) and will show. Except for inbreeding organisms, most diploids carry some deleterious recessives hidden in heterozygotes, so if there are any in the hemizygous part of the wild-type chromosome, there will be a loss of fitness and possible lethality.

In Sec. 8.6, we saw how induced terminal deletions could be used in mapping loci to the visible chromosomes. Deletions are not much used by breeders as they are generally deleterious. Unless they are deleterious in the gametes, they should not directly affect fertility, although they may affect viability. Unless very small, they can often be detected cytologically at meiosis in heterozygotes, because the wild-type undeleted chromosome will have no homologue to pair with in the region of the deletion, causing the wild-type chromosome to form an **unpaired loop** as shown in Fig. 9.1. Such loops can also be seen in *Drosophila* salivary gland chromosomes

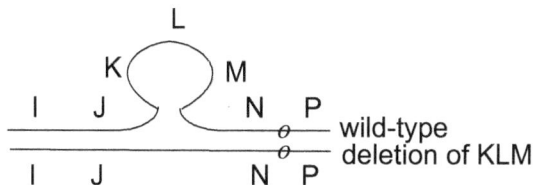

**Fig. 9.1.** The **unpaired loop formed at zygotene** of meiosis in a heterozygote for an interstitial deletion, of *KLM*. The loop is in the wild-type chromosome, *IJKLMNoP*, with *o* representing the centromere. Each strand represents two chromatids if at meiosis, but only one chromosome if this is somatic paring as in *Drosophila* salivary glands. Recessive alleles at loci *KLM* can show **pseudodominance** and be expressed as they are hemizygous.

as they show somatic pairing of homologues, even though they are not in meiosis.

An example of a terminal deletion in humans is the **Cri-du-Chat syndrome**, caused by heterozygosity for a deletion of the end of the short arm of chromosome 5. Affected individuals give a persistent cat-like cry, are severely abnormal, physically and mentally, and usually die young. It is very rare, 1 in 100,000, normally arising anew in the germline of a parent of each sufferer, and would be lethal if homozygous. Plates 12.6 and 12.7 show a terminal deletion of chromosome 1, with the missing part translocated non-reciprocally on to chromosome 12, giving a balanced genotype.

## 9.3. Inversions, paracentric and pericentric; their effects on fertility

An **inversion** involves the turning around of a section of chromosome. It requires one break if terminal or two breaks if interstitial, and the joining of ends. Thus, *ABCDoEFGH* with a break between *C* and *D*, and joining of the unbroken end *A* with broken end *D* after the fragment has turned round, gives *CBADoEFGH*. This is a **paracentric inversion** (para = beyond) as the centromere, *o*, is not within the inversion. If instead *ABCDoEFGH* had two breaks, one between *B* and *C*, and one between *F* and *G*, with turning round of the broken central section and joining of the broken ends, one gets *ABFEoDCGH*. This is a **pericentric inversion**, (peri = around) because it includes the centromere. The map distances between various loci, such as *B* to *F*, or *B* to *C*, change considerably.

Inversions can affect fertility and fitness, and are common in many wild organisms and in evolution. Species often differ from each other by inversions, which can act as partial isolating mechanisms through reducing fertility in heterozygotes. Within a species, inversions may be **fixed** in some populations, with all individuals homozygous for the inversion, and **floating** in others, where the population is polymorphic for an inversion. Wild tulips, flies and grasshoppers have frequent inversions, and they occur in all kinds of organism.

Their effects on **fitness** (not fertility) are variable, sometimes being deleterious, sometimes beneficial, sometimes neutral, and fitness may differ between homozygotes and heterozygotes for the inversion. For example, in *Drosophila pseudoobscura*, an inversion called Chiricahua in the third chromosome had a relative fitness of only 0.4 when homozygous, but 1.0 when heterozygous with wild-type (called Standard), compared with 0.9 for wild-type when homozygous. There was heterozygote advantage for fitness with this inversion, even though it had a reduced fitness when homozygous. There must be cyclic selection during the year, since at Pinon Flats, California, the relative frequencies of Standard, Chiricahua and another third chromosome rearrangement, Arrowhead, changed during the year. Although Standard was the most common type in most months, Chiricahua was the most frequent in June. See references in Ayala and Kiger (1980).

Inversions do not usually affect fertility in homozygotes because in inversion/inversion diploids, both homologues can pair normally with each other, and crossovers have no adverse effects. **Inversion heterozygotes**, however, usually have reduced fertility unless they have adapted by suppressing crossovers in the inverted section. We saw that with heterozygous deletions at meiosis, one chromosome looped out into an unpaired region. In heterozygous inversions, however, both homologous chromosomes are almost fully paired because **both chromosomes loop**, enabling homologous pairing within the inverted region. Segregation is normal if there is no crossover within the inverted region. A crossover within the inverted region, however, produces 50% of inviable gametes, as shown in Fig. 9.2. For simplicity, this shows the two chromosomes as *ABoCD/acobd* (a **pericentric inversion**) and as *oABCD/oacbd* (a **paracentric inversion**), with heterozygosity for inversion of region *bc*, and different centromere positions relative to region *ABCD* in the two kinds of inversion.

Although both types of inversion when heterozygous give only 50% of viable gametes (those without recombination), crossing-over within the inverted regions has different outcomes for heterozygous pericentric and paracentric inversions, as shown in Fig. 9.2. In addition to two chromosomes giving viable gametes, one with the inversion (*acobd*) and one without (*ABoCD*), the heterozygous pericentric inversion gives two chromatids which are simultaneously duplicate for one region and deficient for another (*ABoca* and *DCobd*), hence normally giving inviable gametes. The heterozygous paracentric inversion produces two chromosomes giving viable gametes, one with the inversion (*oacbd*) and one without it (*oABCD*), and one duplicate/deficient acentric fragment such as *DCbd*, which gets lost

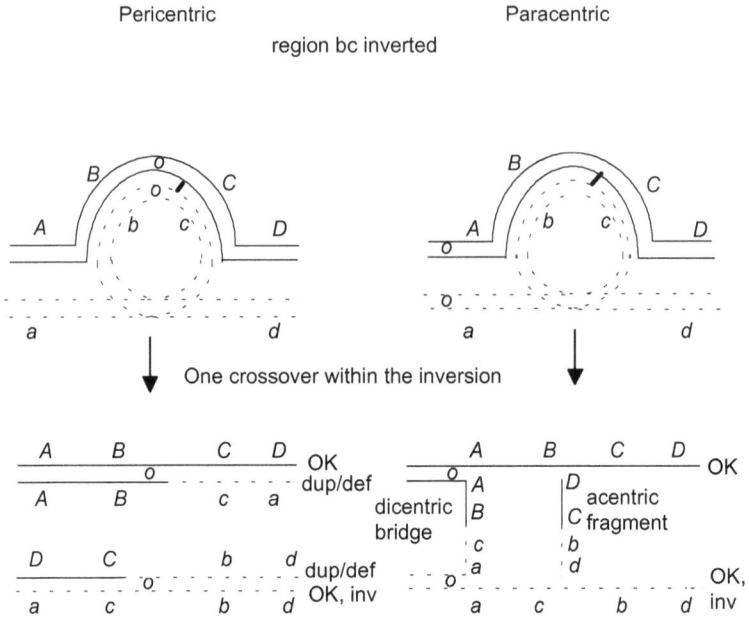

Pericentric                              Paracentric

region bc inverted

One crossover within the inversion

Results:

One strand normal, one inverted, two duplicate and deficient,so only two out of four meiotic products give viable gametes.

One strand normal, one inverted, one dicentric bridge, one acentric fragment, so only two out of four meiotic products give viable gametes.

With both the pericentric and paracentric inversions, after a crossover within the heterozygous inverted region, only the non-recombinant products give viable gametes.

**Fig. 9.2.    The consequences of crossing-over at meiosis within heterozygous inversions,** pericentric and paracentric. Each line represents a chromatid, except for the heavy short line which indicates a reciprocal crossover at pachytene.

as there is no centromere attachment point for segregation in cell division. The fourth product has two centromeres, e.g., *oABcao*. At anaphase segregation of centromeres, one goes to the upper pole and the other to the lower pole, breaking the **dicentric fragment**, giving deficient chromosomes and non-viable gametes.

In plants with heterozygous paracentric inversions, female gamete fertility may be greater than 50%. Three of the four meiotic products degenerate and the egg cell (basal megaspore) may preferentially receive a good chromosome, while bridges and fragments tend to go to the central cells which degenerate.

In these heterozygous inversions, the loss of fertility increases with the frequency of crossing-over within the inversion. There will be strong **selection pressure** in favour of any mutations which reduce crossing-over within that part of the chromosome. If it occurs, that suppression of crossing-over would help to keep together any favourable gene combinations in that region. The fact that inversions are so frequent in evolution suggests that hybrid vigour for heterozygous loci within inversions, and protecting favourable gene combinations from recombination, may be common natural phenomena.

The giant polytene chromosomes in *Drosophila* are excellent for studying chromosome evolution. Different species can be shown to have arisen from inversions, often with a series of overlapping inversions. Thus, a type 1 chromosome may have a sequence of regions *ABCDoEFGH*. An interstitial pericentric inversion from breaks between *B/C* and between *o/E* would give *ABoDCEFGH*. Breaks in that type 2 chromosome between *A/B* and between *G/H* would give a type 3 chromosome, *AGFECDoBH*. Although one can deduce possible evolutionary sequences from the salivary chromosome bands in different species, one cannot directly deduce the order. Thus, the sequence could be type 1 to type 2 to type 3, or 3 to 2 to 1, or 2 to 1 and 3 from different inversions.

Section 8.8 describes how *Bacteroides fragilis*, a dangerous opportunistic parasite of humans, uses **periodic inversions** to evade host immune responses. It varies its surface coat molecules by turning their genes on and off by turning short fragments of DNA through 180°.

## 9.4. Duplications and the origin of new genes

In a **duplication**, one region is present twice on a chromosome, which alters the gene dosage and may have a direct effect on the phenotype because of that. Thus duplicating region 16A of the *Drosophila* X chromosome gives *bar eye*, reducing the number of eye facets from about 700 to about 70 per eye. Duplications are sometimes lethal in gametes, especially male gametes which are more sensitive to gene dosage than are female gametes.

A B C D E F G$_O$

A B C D E F G$_O$ ———→ A B G$_O$

A B C D E F G$_O$

A B C D E F C D E F G$_O$

An illegitimate and unequal crossover between non-homologous regions of two homologues.

One deletion chromosome and one tandem duplication in which the region CDEF occurs twice, adjacently.

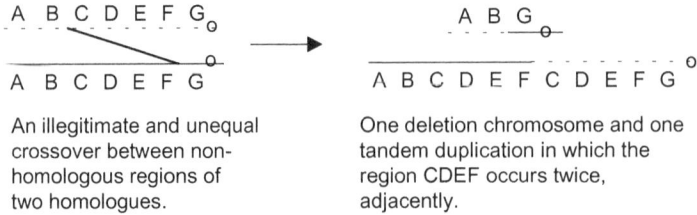

**Fig. 9.3.** **The production of a tandem duplication** and a deletion chromosome by unequal crossing-over between non-homologous chromosome regions.

Duplications can arise by incorporation into a chromosome of a broken fragment from a homologue, needing two or three breaks. For example, chromosome *ABCDoEFGH* might incorporate broken fragment *FG* from a homologue, giving e.g., *ABCFGDoEFGH*, or the fragment might be incorporated into its *FG* region, where it has DNA sequence homology. Duplications which persist do not involve the centromere, as dicentrics usually break at anaphase in cell division.

One common type of duplication is a **tandem duplication** with the two duplicates (e.g., of CDEF) next to each other, in the same order, e.g., *ABCDEFCDEFGo*. This can arise from illegitimate pairing giving rise to unequal crossing-over, as shown in Fig. 9.3. Plate 9.3 shows a duplication in the pseudoautosomal region of the human X chromosome.

Duplications are very important for **long-term evolution**, though less so in the short term, because they provide more DNA, which can evolve **new functions**. For example, if genes *CDEF* are essential in a *CDEF* individual, these four genes cannot evolve new functions without losing the essential old ones. In the duplication ***CDEFCDEF***, one set of the genes can keep the old functions while the second set (which need not be adjacent — it could be the first *C* but the second *D* which change) can evolve new functions. If gene *C* specifies an enzyme, one copy could evolve to specify a slightly different enzyme which might do some further biochemical step not previously possible, and perhaps have co-ordinated controls with the original *C*.

Duplications in some **fungi** such as *Neurospora* and *Ascobolus* get severely attacked immediately before meiosis, with premeiotic deletions or repeated induced point mutations inactivating the duplicated region. In any organisms, duplications could cause further chromosome aberrations from

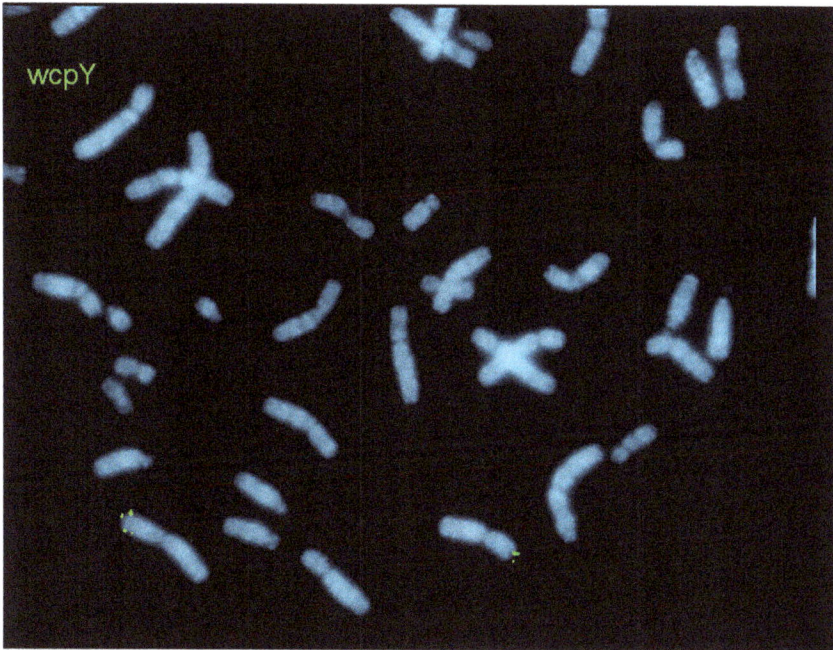

**Plate 9.3.** **A heterozygous duplication in the human X**. Female cell with a duplication in part of the short arm of one X chromosome (bottom left), in the pseudoautosomal region. The probe used is a Y whole-chromosome paint, which has hybridised to the homologous region by which the X pairs with Y at meiosis.

unequal crossing-over, whether they are homozygous or heterozygous, as there are more than two regions of homology which could pair and cross-over. Such illegitimate cross-overs could occur at meiosis or with much lower frequencies at mitosis. They can be between or within chromosomes.

## 9.5. Translocations, single and multiple

A **translocation** occurs when part of one chromosome is inserted in or attached to a non-homologous chromosome. One or two breaks could cause a terminal or interstitial part of one chromosome to break off, e.g., *XYZ* from *PQoRSTUVWXYZ*, and a break in a non-homologous chromosome, *ABCDoEFGH*, could allow it to be inserted with rejoining of the ends, giving e.g., *ABXYZCDoEFGH*, or it could attach to one end of an intact non-homologous chromosome, as in *ABCDoEFGHXYZ*. Those would be

**non-reciprocal translocations**. Plates 1.1, 9.1, 12.6, 12.7 and 12.12 show human non-reciprocal translocations. Many translocations are **reciprocal**, caused by illegitimate crossing-over between non-homologous chromosomes. For example, a crossover could occur between *F* and *G* on *ABC-DoEFGH*, and *W* and *X* on *PQoRSTUVWXYZ*, giving two reciprocally translocated products, *ABCDoEF**XYZ*** and *PQoRSTUVW**GH***. Reciprocal translocations are also called **interchanges** and do not change the number of copies of each locus present. Plates 12.6 and 12.7 show a balanced non-reciprocal human translocation. In humans, **chronic myeloid leukaemia** is produced by a particular translocation between the long arms of chromosomes 9 and 22, producing a Philadephia chromosome with an oncogenic gene combination (Subsec. 12.7.6).

Translocations will obviously change genetic maps, as previously non-syntenic loci are now syntenic and may even be closely linked, as for *W* and *G* in the above example. Even balanced reciprocal translocations which involve no change in which genes are present may affect phenotypes and fitness through **position effects** (see Sec. 9.1). Wild populations polymorphic for translocations are common in snails, grasshoppers, and the plants *Datura* (thorn apple) and *Oenothera* (evening primrose).

Whether or not translocations affect phenotypes, they have strong effects on **fertility** when heterozygous. In homozygotes for translocations, pairing is usually normal but recombination frequencies between pairs of loci may be altered. In heterozygotes for reciprocal translocations, complete homologous pairing at zygotene of meiosis requires that all four chromosomes come together to form a **quadrivalent** as in Fig. 9.4. In the quadrivalent shown, all unbroken lines are paired with homologous unbroken lines, and all broken lines have broken line partners.

Figure 9.4 shows why half the meiotic products of a heterozygous reciprocal translocation are **inviable duplicate/deficient combinations**. One gets partial sterility with distorted pollen grains or sperm. Two crossovers are shown in the figure, in positions having no effect on chromatid segregation. In a diploid with one heterozygous reciprocal translocation, one would get just one quadrivalent at meiosis, plus normal bivalents for the other pairs of chromosomes, so partial sterility plus one quadrivalent at meiosis can be used to identify this condition. More than one reciprocal translocation, if heterozygous, will give more than one quadrivalent, but homozygous translocations just give bivalents (two chromosomes paired).

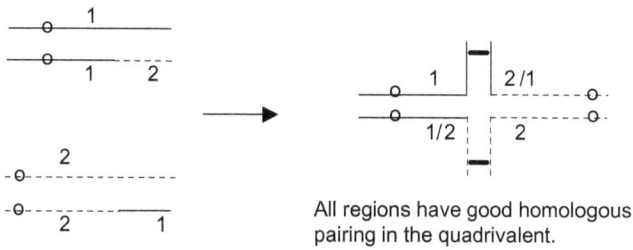

All regions have good homologous pairing in the quadrivalent.

At anaphase, the two centromeres of type1 must go to opposite poles, and the two centromeres of type 2 must go to opposite poles.
If the centromeres of 1 and 2 go to one pole, the centromeres of 1/2 and 2/1 must go to the other pole; both daughter cells have balanced genomes, one untranslocated, 1 and 2, one translocated, 1/2 and 2/1. If the centromeres 1 and 2/1 go to one pole and centromeres 2 and 1/2 go to the other pole, both daughter cells will be duplicate/deficient, giving inviable gametes. For example, a cell with 1 and 2/1 is missing the acentric end of 2 and has two copies in the haploid gamete of the acentric end of 1.

**Fig. 9.4.    Pairing in meiosis in a heterozygous reciprocal translocation** between chromosomes of pair 1 and pair 2, so that there is one normal chromosome 1, one normal chromosome 2, a centric part of 1 plus an acentric part of 2, and a centric part of 2 plus an acentric part of 1.

Extraordinary cases of **multiple translocations** have been identified in a number of plants, especially in the evening primrose genus, *Oenothera*. For example, *Oenothera erythrosepala* cells at mitosis have 14 separate chromosomes, so 2n = 14. In meiosis, one would expect seven bivalents. Instead, one gets one bivalent for chromosomes 1 and 2, and a **ring** of twelve chromosomes because there have been six successive interchanges. Surprisingly, the species is **permanently heterozygous** for the interchanges, with a system of **balanced lethal recessives**.

The 12 chromosomes making up the ring form two separate units, the *velans* and *gaudens* sets of chromosomes. Let us number the chromosome arms, so that the untranslocated set of chromosomes would be 1/2, 3/4, 5/6, 7/8, 9/10, 11/12, 13/14. Some chromosomes have no translocation, but there have been reciprocal (5/6 with 7/8 to give 5/8 and 6/7) and non-reciprocal (e.g., 3/4 with 11/12 to give 4/12) translocations. Meiotic pairing within the ring at zygotene and anaphase I is shown for just four of the chromosomes of the ring in Fig. 9.5. All arms can be fully paired.

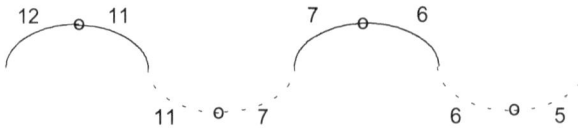

Only four of the12 chromosomes making up the ring are shown.
Unbroken lines: part of the velans set of chromosomes.
Broken lines: part of the gaudens set of chromosomes.
The numbers represent chromosome arms, so that 5/6 and 11/12
are untranslocated chromsomes and 7/11 and 6/7 have translocations.
o represents a centromere.The four chromosomes are shown above
at early anaphase I of meiosis.
At zygotene, they would have been paired as follows:

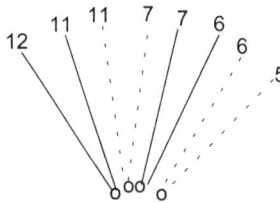

**Fig. 9.5.  Meiotic pairing within part of the ring chromosome complex** in *Oenothera erythrosepala.*

The *velans* complex is made up of chromosomes 3/4, 12/11, 7/6, 5/8, 14/13 and 10/9, and is $l1^-$, $L2^+$, with a recessive lethal allele, $l1^-$, and a dominant non-lethal allele at a different locus, $L2^+$. *Gaudens* is made up of 4/12, 11/7, 6/5, 8/14, 13/10 and 9/3, and is $L1^+$, $l2^-$, carrying the opposite recessive lethal, $l2^-$. These lethals only act in the diploid, not the haploid, and the only viable gamete combination for the zygote is *velans* with *gaudens*, genotype $L1^+$ $l1^-$, $L2^+$, $l2^-$, when each recessive lethal is hidden by the corresponding non-lethal dominant allele. Homozygous *velans* plants die because they are homozygous for lethal $l1^-$; homozygous *gaudens* plants die because they are homozygous for lethal $l2^-$. One would therefore expect only 50% of seeds to be viable, whereas only about 40% are viable. The extra deaths are caused by unbalanced gametes, e.g., from non-terminal crossovers.

For these ring-chromosome systems to survive such high lethality, they must have some major compensatory advantages. The most obvious is that one can have **permanent hybrid vigour** from heterozygote advantage, e.g., with the *velans* set carrying one allele and the *gaudens* set carrying a different allele, for many different loci. The ring system is excellent for hybrid vigour and maintaining already well-adapted lines. It usually prevents favourable gene combinations from being split up by recombination.

The ring can be broken up by certain crossovers, giving new variants. In fact, about 2% of the seedlings from *Oenothera erythrosepala* are so different from their parents that they could be classified as new species or subspecies. Different ways of **breaking the ring** give different forms. Rings can break down into bivalents and bivalents can re-evolve rings. In *Rhoeo discolor*, there are no bivalents at meiosis, just a ring of 12 chromosomes. Some other plants, such as *Campanula percicifolia* and *Paeonia*, can have rings but without a balanced lethal system, so homozygous individuals can occur for each ring complex.

Special multiple translocation stocks are used in some fungi for mapping, especially for assigning loci to chromosomes, e.g., in *Neurospora crassa*.

## Suggested Reading

Ayala, F. J. and J. A. Kiger, *Modern Genetics*. 2nd ed. (1980) Benjamin Cummings, Menlo Park, California.

Bahl, P. N. (ed.), *Genetics, Cytogenetics and Breeding of Crop Plants, Vol. 2. Cereals and Commercial Crops*. (1997) Science Publishers, Inc., Enfield, New Hampshire.

Connor, M. C. and M. Ferguson-Smith, *Essential Medical Genetics*, 5th ed. (1997) Blackwell Science, Ltd., Oxford.

Feuk, L. *et al.*, Discovery of human inversion polymorphisms by comparative analysis of human and chimpanzee DNA sequence assemblies. *PLoS* (2005) 1: 489–498.

Gerson, S. L. and M. B. Keagle (eds.) *The Principles of Clinical Cytogenetics*, 2nd ed. (2005) Humana Press, New Jersey.

Griffiths, A. J. F. *et al.*, *Introduction to Genetic Analysis*, 8th ed. (2005) W. H. Freeman and Co., New York.

Hartl, D. L. and E. W. Jones, *Genetics. Analysis of Genes and Genomes*, 6th ed. (2005) Jones and Bartlett, Sudbury, Maryland.

Jahier, J. *et al.*, *Techniques of Plant Cytogenetics*. (1996) Science Publishers, Inc., Enfield, New Hampshire.

# Chapter 10

## Changes in Chromosome Number: Their Effects and Uses

### 10.1. Background

As we saw in Subsec. 1.3.2, the **ploidy** is the number of copies of each typical chromosome in the nucleus. **Haploids** have one copy of each chromosome, **diploids** have two, **triploids** three, **tetraploids** four, **hexaploids** six, etc., with those having three or more copies being **polyploids**. The number of different types of chromosome (counting the sex chromosomes as one type) in the nucleus is the **basic number of chromosomes**, and a set of one of each type of chromosome makes up a haploid **genome**. The **basic number** has the symbol $X$, unfortunately the same as for a sex chromosome. Symbol $n$ represents the gametic number of chromosomes and $2n$ is the adult number unless the adult is haploid. So $2n$ can be $2X$, $3X$, $4X$, etc. In the rose genus, *Rosa*, different species have $2n$ chromosome numbers of 14, 21, 28, 35, 42 and 56. Clearly, $X = 7$, so these chromosome numbers represent $2n = 2X, 3X, 4X, 5X, 6X$ and $8X$, with all but the diploid being polyploid. Some people use a different system, with $3n$ for the triploid, $4n$ for the tetraploid, etc.

**Euploids** have an exact multiple of the basic number of chromosomes, but **aneuploids** do not have an exact multiple. For example in hyacinths, varieties with 16 chromosomes are euploid diploids, $2n = 2X$, but the variety Rosalie has 17, so it is an aneuploid of the diploid, with $2n = 2X + 1$. Another aneuploid has 23 chromosomes, so $2n = 3X - 1$, an aneuploid of the triploid.

Polyploids make up about 60% of Monocots and 40% of Dicots, and more than one third of all domesticated plants. In animals, there are some polyploid snails and worms, fish, amphibia and reptiles, but polyploidy is

220

very rare in higher animals. In humans, triploids and tetraploids occur at high frequencies in spontaneous abortions (Secs. 12.8 and 12.9).

There is an important distinction between two types of tetraploid. **Autotetraploids** have four of each type of chromosome from a self-doubling in a diploid, so if each type of chromosome is represented by a number, they would be 1111, 2222, 3333, etc. They tend to have reduced fertility as all four of each type of chromosome try to pair at meiosis, often forming quadrivalents in meiosis and having irregular segregations. **Allotetraploids**, however, come from crosses of different species or very different strains, in which previously homologous chromosomes have diverged in evolution. If two diploid species with different genomes, AA in one and BB in the other, cross by chance, they usually form a sterile diploid, A,B; this is sterile because the two sets of chromosomes are no longer homologous and cannot pair at meiosis, giving **univalents** (unpaired chromosomes), not bivalents, in meiosis. Sometimes, in this sterile diploid hybrid, there is a failure of cell division after DNA replication and hence a doubling of the chromosome number. The resulting allotetraploid, genomes AA, BB, is usually fertile because each chromosome of the A genome now has a homologue with which to pair, and so does each member of the B genome; therefore there are **bivalents** (pairs of chromosomes) at meiosis, as well as correct segregation of chromosomes. See Sec. 13.5 for the use of allopolyploids in interspecific hybrids. Polyploids may have different ecological niches and distributions from diploids. If annual diploids form polyploids, these are often perennial. Polyploids are often more vigorous than diploids, and allopolyploids may have a kind of permanent, pure-breeding hybrid vigour, with e.g., genome AA homozygous for one allele of a locus, e.g., *P P*, and genome BB homozygous for an alternative allele, e.g., *p p*. Polyploids tend to be hardier than diploids in extreme conditions, and are often short-day flowering (Subsec. 17.1.1). Woody plants tend not to be polyploid. Polyploids are often evolutionarily a short-term success, but they tend to be stagnant, with a reduced chance of new recessive mutations showing in the phenotype.

In plant breeding, it is often best to do the basic work with diploids or even monoploids (Sec. 10.3), especially if you seek new beneficial mutations. In *Dahlia*, most garden plants are octaploid, 8X, double allotetraploids, but behave like double autotetraploids. According to

Lawrence (1968), not once in 170 years of cultivation has an abrupt new mutation been found. Only dominants with **xenia** (one dominant shows fully in the presence of any number of recessive alleles) would show, and the chance of *A A A A A A A a* ever giving *a a a a a a a a* is remote. Similarly, a mutation from *A A A A A A A A* to *A A A A A A A A′*, with additive action, might go unnoticed. There is a lot of genetic variation in garden dahlias, but it is released primarily through recombination and segregation, not mutation. Their crossovers tend to be localised to the ends of the chromosome arms.

## 10.2. Changes in ploidy

Increases in ploidy usually result from a failure of cell division (mitosis), so that the chromosomes replicate but the cell fails to cleave, and the chromosome number doubles. Thus a haploid yeast cell could fail to divide, giving a diploid vegetative cell, or a diploid rye cell could give a tetraploid cell. These events could affect the next generation if occurring in germline cells. This mechanism produces **even-number multiples** of the basic chromosome number. The plant breeder can treat seeds or growing points (Plate 13.4) with certain chemicals, such as colchicine from the autumn crocus or synthetic colcemid, to double the chromosome number, identifying affected shoots by altered morphology, and then checking their chromosome numbers by microscopy.

**Odd-number multiples** can be produced by fertilisation of a diploid gamete by a haploid gamete, or by fertilisation of a haploid gamete by two haploid gametes. A diploid gamete could come from a tetraploid, or by division failure in a gamete from a diploid. In higher plants, two of the four products of female meiosis usually fuse to form a diploid polar body in the ovule. This diploid nucleus is then fertilised by one of the two gametic nuclei from the pollen, to form a triploid nucleus. In maize and other plants, this divides rapidly to form the **triploid endosperm**, which surrounds the **diploid embryo** and forms a major food reserve for the embryo at germination.

**Unreduced chromosome numbers in gametes** have spontaneously given rise to various cultivars. An unreduced gamete with 14 chromosomes from the diploid raspberry (*Rubus idaeus*, 2n = 2X = 14) fertilised a normal gamete with 28 chromosomes from the octoploid *R. vitifolius* (2n = 8X = 56), giving the hexaploid loganberry, 2n = 6X = 42. The blackberry "John Innes" is tetraploid, 2n = 4X = 28, coming from an unreduced gamete with 14 chromosomes from the diploid *R. rusticans inermis* (2n = 2X = 14), when crossed to the tetraploid, *R. thyrsiger* (2n = 4X = 28) [see Lawrence (1968)].

In some *Hymenoptera* (bees, wasps, ants), fertilised eggs give diploid females, but unfertilised eggs give **haploid (monoploid) males**. In honey bees, 2n = 32 for females, but male bees (drones) only have 16 chromosomes. In the haploid male, meiosis is modified to be more like mitosis, with gametes still having one complete genome and 16 chromosomes, instead of 8. The sex ratio is therefore not 1 female: 1 male, but is strongly biased in favour of females, and is determined by the proportion of eggs which the queen bee allows to be fertilised by the sperm which she stores after mating. The difference between the sterile female diploid worker bees and the fertile diploid female queen is a matter of nutrition, with the feeding of royal jelly to larvae to produce queens. An unfertilised queen could produce a drone and mate with him.

In plants, one occasionally gets unfertilised female gametes developing (**parthenogenesis**), giving haploid progeny, but male gametes do not have enough cytoplasm to form an embryo. This kind of effect has been exploited by potato breeders, with a kind of phantom pollination. **Potatoes** are derived from wild Andigena autotetraploids of *Solanum tuberosum*, but autotetraploids are inconvenient for basic breeding work, especially with recessive alleles. It is possible to go from the tetraploid to the diploid by pollinating the autotetraploids with pollen from selected clones of the diploid *Solanum phureja*. Both the pollen gametic nuclei fuse with the polar bodies, starting endosperm development which stimulates the unfertilised egg to divide and develop parthenogenically, giving the required diploids of *S. tuberosum*. Those diploids can then be crossed with other diploid species to introduce desirable genes, e.g., additional disease resistance. Once the diploid has been suitably improved, it can be returned to the autotetraploid state by treatment with colchicine. Crosses of diploid *S. tuberosum* to compatible strains of *S. phureja*, and making autotetraploids with colchicine from the sterile diploid, could be used to exploit hybrid vigour (see Secs. 13.1 and 13.6). Parthenogenesis in tetraploid *Medicago* (alfalfa) and in potatoes has sometimes given diploid embryos without pollination.

## 10.3. Monoploids and anther culture

**Monoploids**, where the adult is haploid, are sometimes useful to plant breeders because deleterious and beneficial recessives can be detected since dominance does not hide them. Monoploids were first reported in 1924 from natural populations of the thorn apple, *Datura stramonium*, arising from unfertilised eggs. Like most monoploids, they were weak, small, had very low fertility, and soon died out. As there is only one copy of each homologue, meiosis gives only univalents, with irregular segregation of chromosomes, so most gametes have some essential chromosomes missing and are inviable.

Monoploids have been used in breeding maize and tomatoes, particularly for selecting against deleterious recessives. In maize, $2n = 20$, so the monoploid has a complete genome of 10 different chromosomes. Gametes by chance get from 0 to 10 chromosomes, with intermediate numbers being most common. Getting a fertile seed depends on a very rare fertile ($n = 10$) male gamete fertilising a very rare fertile ($n = 10$) female gamete. Crossing monoploids is much less successful than restoring the diploid state by doubling up the chromosome number using colchicine or other agents, as previously discussed. That gives fully homozygous diploids, without the necessity of having to make inbred lines over several generations. They can be used to make F1 hybrids (Sec. 13.1). Fusion of haploid protoplasts from different varieties can be used to make diploid hybrids if there are problems with sexual compatibility.

Using monoploids to make **homozygous diploid lines**, e.g., for use in producing F1 hybrids, can save five to seven generations of inbreeding, which is five to seven years for many herbaceous plants and much longer for fruit trees, forest trees and other plants with a long juvenile phase. In **asparagus**, male plants are more productive than female ones. By obtaining haploids with the Y chromosome, carrying sex-determining gene *M*, and fusing them, YY (*MM*) diploid males can be produced which give all male progeny (XY, *Mm*) when crossed to normal females (XX, *mm*), or which can be propagated vegetatively (Sec. 17.1).

Monoploids can now be made for most plants by **anther culture** (androgenesis). Anthers at a particular stage are put on nutrient agar with certain growth hormones, and are often given a mild heat-shock, so that some microspore cells which have undergone meiosis divide to produce small plantlets. By changing the nutrient agar and hormones, the plantlets can be made to grow roots, and eventually whole rooted plants or explants can be cultured. After the selection of the best haploid phenotypes, colchicine is applied to the growing points, and diploid side-shoots (see Plate 10.1) are selected by eye, then tested cytologically. For a commercial *Brassica* example, see Ockendon (1984). He obtained yields of up to 357 embryos per 100 anthers at the 4 to 5 mm anther length stage, using a thermal shock of 16 hours at 35°. Complete plants were regenerated mainly from hypocotyl explants, and haploids (smaller flowers, < 20% pollen stainability) and diploids (> 70% pollen stainability) were easily distinguished at flowering (Plate 10.2). Although a few plants were triploid or tetraploid, about half were haploid and the other half were diploid. Anther culture of diploid potatoes (see above) can produce haploids. In many species, the success rates from anther culture vary widely between different varieties, and many unwanted albinos are produced.

Anther culture produces a mixture of haploids and diploids. Diploids come from anther wall cells and from pollen in which two haploid nuclei fuse. Haploids are produced from uninucleate microspores, or from pollen when either the generative nucleus (or nuclei) disintegrates and the tube nucleus divides repeatedly, or the tube nucleus degenerates and the generative nucleus (or nuclei) repeatedly divides.

The **standard procedure** for getting haploid plants from anther culture in many species is as follows. The developmental stage of the anther and the physiological status of the mother plant are optimised, sometimes keeping the mother plant at 3 to 10° for 2 to 30 days. Flower buds at the appropriate stage are surface-sterilised, the filaments are dissected away and the anthers mildly ground to release pollen. After filtration through muslin, the pollen can be cultured in liquid medium or on solid medium to get calli. Haploid callus can be induced to form homozygous diploid cells by treatment with 0.5% colchicine for 24 to 48 hours, or haploid plants

**Plate 10.1.**    **Monoploid Brussels sprouts plants produced from anther culture**, showing differences in appearance from each other. The foreground plant has had colchicine applied to the growing point, to induce some diploid side-shoots. Plates 10.1 and 10.2 are from the work of Dr D. Ockendon when at the then National Vegetable Research Station, Wellesbourne, Warwickshire.

obtained from the callus can be treated with colchicine as mentioned for *Brassica* above.

Haploids can sometimes be produced by **chromosome elimination**, as in haploidisation in the fungal parasexual cycle (Sec. 18.4). When two diploid barleys were crossed, *Hordeum vulgare* × *H. bulbosum*, the *H. bulbosum* chromosomes were preferentially eliminated from the diploid

**Plate 10.2.** Flowers on monoploid (left, from anther culture) and diploid Brussels sprouts plants.

hybrid, giving haploid *H. vulgare* plants. Embryo culture is needed because the endosperm fails to develop. This method was used to make the barley variety "Mingo", after doubling up the chromosomes of the selected haploid.

Even in plant monoploids or haploid fungi, not all genes are necessarily expressed. This is true of hypostatic genes in the presence of an epistatic allele at another locus (Sec. 2.2) and applies to inducible genes in the absence of inducers, and to conditional mutants under permissive conditions (Subsec. 3.2.2).

## 10.4. Diploids

**Diploids** are very frequently used for the basic breeding work in plants and in all commercial higher animals. Unless there are chromosome aberrations or sterility mutations, they are usually fertile, with only bivalents at meiosis. If diploids are not inbred, problems in identifying the best genotypes can arise because of dominance.

## 10.5. Triploids

**Triploids** can be produced by crossing tetraploids with diploids, or double fertilisation of gametes by chance, or fusion of a haploid gamete with an unreduced gamete from a diploid. Not all tetraploids have fully fertile gametes. Some fertile gametes will be enough in crosses to diploids to get triploids. Triploids are sexually isolated from their diploid or tetraploid parents, because they are usually largely sterile. With three copies of each chromosome trying to pair at zygotene, there are **trivalents**, sometimes plus bivalents and univalents, with most gametes receiving unbalanced sets of chromosomes and being sterile.

This sexual sterility can be useful if viability, growth and fruiting are satisfactory. In diploid bananas, there are hard black seeds to spit out, but the triploids set fruit very well, with aborted seeds forming small dots in the flesh. The triploid bananas are easily propagated vegetatively (see Subsec. 17.1.5). Similarly, diploid water melons have lots of seeds, and consumers generally prefer triploid (or sometimes aneuploid) varieties, and almost seedless watermelons are usually cultivated now. Some apples are also triploid, such as the Bramley cooking apple.

## 10.6. Tetraploids

The different origins of **autotetraploids** and **allotetraploids** were discussed in Sec. 10.1. Man-made allotetraploids, from doubling up the chromosome number of a diploid hybrid, are sometimes also known as **amphidiploids**. Tetraploids are the commonest natural polyploids, especially autotetraploids, and allotetraploids have been important in evolution. Tetraploids are often hardier and more vigorous than their parental diploids, with larger cells. For example, in the saxifrage, *Saxifraga pensylvanica*, the lower leaf epidermis cell area is 1,600 $\mu$m$^2$ in diploids and 2,740 $\mu$m$^2$ in tetraploids.

Tetraploids sometimes have larger and showier flowers and fruits than diploids, and are useful in horticulture, e.g., tetraploid Easter lilies and *Antirrhinum*s. There may also be useful physiological changes, such as more ascorbic acid (vitamin C) in tetraploid cabbages and tomatoes, while potatoes are normally tetraploid.

If one has a mixture of ploidies and wishes to identify particular types, then initial karyotyping from root tips to find say diploids, triploids and tetraploids is essential. Once they have been characterised, then **other features** which are quicker to determine than chromosome numbers can be used. In sugar beet, the tetraploid beet has a shorter, thicker and darker leaf than the diploid or triploid. If one stains a strip of the lower leaf epidermis with dilute iodine in dilute potassium iodide solution, the tetraploid has more plastids (shown by starch-grain containing bodies) per pair of guard cells than the triploid, which has slightly more than the diploid. In rye-grass(*Lolium perenne*) and turnip (*Brassica campestris*), one can distinguish diploids from tetraploids from the length of the leaf stomata. In members of the Solanaceae such as potatoes (*Solanum*) and tobacco (*Nicotiana*), one can treat pollen grains with sulphuric and acetic acids to get swelling from the germ pores; the diploid plants usually have pollen grains with three germ pores, while tetraploids mainly have four germ pores (some grains have 3, 5 or 6 pores).

The crops best suited for autotetraploidy are those with a low chromosome number, which are harvested for their vegetative parts, which are cross-pollinated, perennial, and are reproducible vegetatively, e.g., ryegrass (*Lolium*), where the tetraploid usually has vegetative reproduction in the grazed sward, but has enough sexual seed for planting out and for breeding experiments.

Although autotetraploids tend to have reduced fertility, breeders can sometimes improve fertility by selection. For example, the forage grass *Dactylis glomerata* is in the wild, a rather variable, partly-fertile autotetraploid, yet breeders at Aberystwyth, Wales, have selected for fertility, producing commercial strains with nearly 100% sexual fertility. In autotetraploids at meiosis, an **adjacent** (four in a ring or square) arrangement of chromosomes usually gives 2:2 segregation of the four chromosomes of a quadrivalent, giving fertile gametes. In contrast, a **linear** arrangement of chromosomes in pairing may give unbalanced gametes from 1:3 segregations to the poles of the cell at anaphase.

Tetraploids of all kinds are **genetically isolated** from their diploid relatives, because even if crosses with them are fertile, they generally produce sexually sterile triploids. Diploid pollen, from tetraploids, cannot usually compete in the style with the faster-growing haploid pollen. Male gametes are much more sensitive to chromosomal unbalance than are female gametes, so unbalanced chromosome combinations are more likely to be transmitted by females rather than males.

If a plant has no known relatives for chromosome number comparisons, how does one tell if it is an autotetraploid or an allotetraploid? Having a chromosome number which is a multiple of four is expected in tetraploids, but it is not diagnostic since many diploids have that. At mitosis, an autotetraploid should have four copies of each visible type of chromosome, such as the nucleolar-organiser chromosome which is often long and has a centromere and a non-staining nucleolus-organising region. Unless the chromosomes of their two parental lines have visibly diverged, allotetraploids should also have four of each type of chromosome. Looking at pollen grains is helpful because autotetraploids usually have many aborted grains, but so do triploids, aneuploids, and plants with pollen-lethal alleles. Looking at meiosis is best as diploids and allotetraploids have bivalents, with the chromosomes in pairs, while triploids have trivalents, with the chromosomes in threes, and autotetraploids have quadrivalents, with sets of four chromosomes. Another method is to look at **segregation ratios** in various crosses.

Autotetraploids have four copies of each chromosome and of each locus, giving five possible genotypes per locus instead of the three for diploids, and will show **tetrasomic inheritance**. If one allele is dominant to any number of the alternative allele, it shows **xenia**; e.g., *A A A A*, *A A A a*, *A A a a* and *A a a a* all have the same phenotype, where *A* shows xenia. The names of the five tetraploid genotypes and the ratios of tetrasomic inheritance are shown in Table 10.1.

The ratios in this table are correct when no crossing-over occurs in the interval between the locus and the centromere, but there are additional complications if all possible crossovers are considered. Monohybrid ratios

**Table 10.1.** Tetrasomic inheritance in autotetraploids.

| Genotype | Name | Gametes if balanced (random combinations) | Test-cross phenotype ratio (when crossed to *a a a a*) | Selfing phenotype ratio |
|---|---|---|---|---|
| *A A A A* | quadriplex | all *A A* | all *A* | all *A* |
| *A A A a* | triplex | 1 *A A*: 1 *A a* | all *A* | all *A* |
| *A A a a* | duplex | 1 *A A*: 4 *A a*: 1 *a a* | 5 *A*: 1 *a* | 35 *A*: 1 *a* |
| *A a a a* | simplex | 1 *A a*: 1 *a a* | 1 *A*: 1 *a* | 3 *A*: 1 *a* |
| *a a a a* | nulliplex | all *a a* | all *a* | all *a* |

such as **5:1 from test-crosses** and **35:1 from selfings** are clearly different from standard diploid ratios.

**Allotetraploids** are normally fertile and include many cultivated plants such as tobacco, plum and macaroni wheats. They are functionally diploid for most characters, unless the two different genome types interact. By incorporating genetic material from two different species, they are wonderful new material for selection and evolution. Even the inbreeding allohexaploids like bread wheat can be highly heterozygous and breed true for that heterozygosity, because different alleles were often present (as homozygotes) in the three species which formed the allohexaploid. See Subsec. 13.5.1 for more on wheat, interspecific hybridisation and allopolyploids.

One can get **segmental allotetraploids** where some segments of the two component genomes are in common, and others differ. At the Royal Botanic Gardens, Kew, *Primula verticillata* (genomes VV, 2n = 18) from the Arabian peninsula spontaneously crossed with *Primula floribunda* (genomes FF, 2n = 18) from the Western Himalayas. The seed gave a hybrid, genomes VF, 2n = 18, which was sexually sterile because the V chromosomes could not pair with the F chromosomes generally. The chromosome number spontaneously doubled to give *Primula kewensis*, 2n = 36, VV,FF. This is only partially fertile because V chromosomes occasionally pair with F chromosomes; there are occasional VVFF quadrivalents giving irregular chromosome segregation. There is clearly some homology left between the V and F genomes, so the new species is a segmental allopolyploid, part-way between an autotetraploid and an allotetraploid.

*Lotus corniculatus*, bird's foot trefoil, forms regular bivalents, with no quadrivalents, so it is strictly allopolyploid at the cytological level. A few loci, however, show tetrasomic inheritance, with a locus for cyano-genesis giving a 35:1 ratio from a duplex selfing, so there must be a few chromosome segments in common to the two genomes of this segmental allopolyploid, but not long enough to cause homologous pairing. Thus, two homologous chromosomes might carry segments *MNOAP*, while two different homologous chromosomes, which do not pair with them at meiosis, carry *CDEAF*, so that there are four copies of the *A* segment or locus between the two genomes, giving tetrasomic inheritance for *A*, but not for *MNOP* or *CDEF*.

## 10.7. Higher polyploids

In some plant groups, polyploidy does not go higher than tetraploidy in nature, but in *Rumex* (docks), it goes up to 20X. Many pteridophytes are highly polyploid.

**Bread wheat**, *Triticum aestivum,* is a **double allohexaploid**, $2n = 6X = 42$, $X = 7$, genomes AA, BB, DD; see Sec. 13.5 for its origin as an interspecific hybrid. Normally, there is only homologous pairing at meiosis, e.g., A1 with A1, A2 with A2, B6 with B6. However, there is a locus, *Ph*, on chromosome B5 which determines whether there is also **homeologous pairing** (pairing of related chromosomes from different genotypes) at meiosis I, in which A6 can pair with B6 or D6, A1 with B1 or D1, chromosomes for which the residual homology is normally too little to cause pairing. With homeologous pairing, one can get hexavalents (six chromosomes attempting to pair) in hexaploid meiosis. This usually causes sterility, but genes such as the one on wheat's B5 could control exchanges between genomes in polyploids. In Sec. 11.2, we shall see how homeologous paring in *Lolium* can cause fertility in a diploid hybrid and sterility in an allopolyploid. The *Ph* gene is sometimes called a diploidising gene because the normal allele makes the polyploid have chromosome pairing in bivalents, like a diploid.

## 10.8. Loss or gain of single chromosomes: aneuploids, monosomics and trisomics

Aneuploids have an inexact multiple of the basic chromosome number, with **monosomics** of diploids having $2X - 1$, **trisomics** having $2X + 1$, double trisomics having $2X + 2$, etc. Aneuploids are often used in horticulture, especially in bulbs such as hyacinths which can be propagated vegetatively. They are also used as intermediates in some chromosome-substitution breeding techniques (Sec. 10.9). Human examples include Turner syndrome, $2n = 45$, with one sex chromosome missing, and Down syndrome, $2n = 47$, with three copies (trisomy) of chromosome 21 (Subsec. 12.8.2 and Plates 12.15, 12.16 and 12.19). Turner syndrome sufferers are normally sterile, as are Down syndrome males, but Down females can be fertile. There cannot be complete chromosome pairing at meiosis in individuals with an odd number of chromosomes.

Monosomics of diploids tend to be weak with reduced fertility, but trisomics of diploids, sometimes, but not always, have increased vigour. In *Datura stramonium*, the thorn apple, there are twelve pairs of chromosomes, $2n = 2X = 24$, and so there are twelve different trisomics, each with a different chromosome in excess. All twelve trisomics can be distinguished from each other and from the diploid from their fruit morphology. In *Datura*, the extra chromosome is rarely transmitted through the pollen, because the n pollen (2.6 mm/h) grows faster than the $n + 1$ pollen (1.9 mm/h from the "Cockleburr" trisomic) down the style.

Compared with diploids, polyploids in plants are usually better balanced against the effects of the loss or gain of a chromosome, so $4X - 1$ is often as vigorous as $4X$, although its fertility is usually reduced. Commercial **sugar cane** varieties are complex polyploid aneuploids from interspecific hybridisation, with chromosome numbers generally in the range of 100–125, where $10X = 100$. They are vegetatively propagated (Subsec. 13.7.7). In **animals**, polyploidy and aneuploidy are usually very harmful, and are not normally used in breeding techniques.

## 10.9. Chromosome manipulations and substitutions

Although artificial allopolyploids have been of limited use in developing agricultural cultivars, they make excellent intermediates in transferring genes from alien species into cultivars. One usually only wants the **desired allele** transferred, but if there is little recombination between the cultivar and the alien chromosome, then the whole chromosome may be retained, or one can try transferring just part of the alien chromosome to a cultivar chromosome by translocation, perhaps following irradiation. Several recurrent backcrosses (Sec. 13.4) of a cultivar × alien species hybrid to the cultivar may be necessary to get rid of unwanted alien genes, while selecting for the desired alien gene.

The main techniques involve **chromosome additions** or **chromosome substitutions**, or a combination of both methods, and may involve manipulation of recombination control of homeologous pairing [see Sec. 10.7 and (Thomas, 1993)]. An example is the production of yellow-rust-resistant bread wheat cultivar "Compair" by Riley and others. Hexaploid "Chinese Spring" wheat, $2n = 6X = 42$, is susceptible to yellow rust, *Puccinia striiformis*, but the wild diploid *Aegilops comosa*, $2n = 2X = 14$, has a gene for resistance to this rust. The two species were crossed, giving a fertile hybrid which was backcrossed for several generations to "Chinese Spring". The backcrossing, with selection for rust resistance each generation, got rid of all *Aegilops* chromosomes, except for the one carrying the resistance. There were therefore all 42 "Chinese Spring" chromosomes and the alien one, M2, which corresponded to homologous group 2 in "Chinese Spring".

The ***Ph* locus** on wheat chromosome 5B suppressed homeologous pairing, so there was no crossing-over between M2 and wheat chromosomes which could introduce the resistance locus into a wheat chromosome, while reducing the presence of unwanted alien genes. The M2 addition line was then crossed to *Aegilops speltoides*, a diploid, $2n = 2X = 14$, in which the *Ph* locus is inhibited, giving a 29 chromosome hybrid which was repeatedly backcrossed to "Chinese Spring". A rust-resistant plant with 42 chromosomes and 21 bivalents at meiosis was isolated, heterozygous for the *Yr8* yellow-rust-resistance gene from *Aegilops comosa* and not expressing undesirable other properties from *A. comosa*. The plant was selfed to obtain plants homozygous for the resistance, thus producing the useful new cultivar "Compair". It did not carry *A. speltoides*'s suppression of *Ph*, which would have made "Compair" have homeologous pairing and thus be mainly sterile.

**Chromosome addition lines** were produced from crossing *Agropyron intermedia* (perennial intermediate wheat grass, a hexaploid with $2n = 42$, but with genomes not closely related to those of *Triticum*) with hexaploid bread wheat. The hexaploid hybrid was weak, with low fertility. It was backcrossed twice to bread wheat, with selection for fertility and for rust resistance. A stable, fertile, rust-resistant octaploid line was obtained with 42 wheat chromosomes and 14 *Agropyron* chromosomes, but it was low yielding. Further backcrosses to bread wheat gave a series of single chromosome addition lines of *Agropyron* chromosomes to wheat, including one with the rust gene. Reducing the alien chromosome contribution improved yield. Addition lines tend to be unstable, often tending to lose the alien chromosomes, especially from monosomic lines such as 2n = 43 = 6X wheat + 1 alien.

Although addition lines are not usually stable enough to use as cultivars, they can be used in making substitution lines, where an alien chromosome is substituted for one from the crop species. Wheat cultivars grown in Germany in the 1930s were shown to be wheat/rye substitution lines, with mainly wheat chromosomes. A *Triticale* (wheat × rye) line was used in a crossing programme, with selection for rye's rust resistance, giving spontaneous substitution lines. Alien chromosomes will only substitute for ones in the same homeologous group.

The following scheme shows how a **stable substitution line** was obtained with 2n = 42, made up of 40 wheat chromosomes and one pair of rye chromosomes in the same homeologous group as the missing wheat chromosome pair (i.e., 6X wheat − 2 wheat + 2 rye). A monosomic line, 2n = 41, gave 20 bivalents and one univalent at meiosis, and gametes with 20 or 21 chromosomes. A disomic addition line had 2n = 44, with 21 wheat bivalents and one rye bivalent, giving gametes with 22 chromosomes, made up of 21 wheat and one rye chromosome. When such a gamete fertilised a gamete with 20 chromosomes from the monosomic line, the plants had 2n = 42, with 40 pairing wheat chromosomes, one unpaired wheat chromosome and one unpaired rye chromosome. This produced gametes with 20 wheat, 21 wheat, 20 wheat + 1 rye, and 21 wheat + 1 rye chromosomes. It was backcrossed to the disomic addition line, and when the latter's 21 wheat + 1 rye chromosome gamete fused with a 20 wheat + 1 rye chromosome gamete, a hybrid was produced with 2n = 43, from 40 paired wheat chromosomes, one unpaired wheat chromosome and a pair of rye chromosomes. When this was selfed, some plants had 40 paired wheat chromosomes, no unpaired wheat chromosomes and one homologous pair of rye chromosomes, giving a **stable substitution line** with good agricultural properties.

By crossing monosomic lines of cultivar A (e.g., $2n = 41 = 6X - 1$) to a euploid cultivar B (e.g., $2n = 42 = 6X$), and backcrossing about seven times to the monosomic line, one can **transfer single chromosomes** from B to A, getting rid of the other B chromosomes to see if any substitution lines have desirable properties.

Often one wants to **transfer a single gene** between species. Addition lines were used to transfer a whole chromosome from *Aegilops umbellulata* to wheat, but only a gene for leaf-rust resistance was wanted. Dry seeds were irradiated to induce a translocation of a part of the *Aegilops* chromosome to a wheat chromosome. A reciprocal translocation achieved this, but gave a duplicate/deficient chromosome. This was tolerated in the hexaploid but would not be tolerated in the diploid. If possible, it is better to get recombination between the alien chromosome and a wheat chromosome, by suppressing gene *Ph* on chromosome 5B, rather than having a translocation.

Autopolyploids can also act as **bridge species** in interspecific gene transfers. Tetraploid rye-grass (*Lolium*) × hexaploid fescues (*Festuca*) can give pentaploid hybrids, from fusion of a diploid gamete with a triploid gamete. These pentaploids can be crossed as males to diploid rye-grass, with backcrossing to diploid rye-grass. Whole fescue chromosomes are eliminated during the backcrossing, but some fescue genes are retained on rye-grass chromosomes after crossovers.

## Suggested Reading

Bahl, P. N. (ed.), *Genetics, Cytogenetics and Breeding of Crop Plants*, *Vol. 2, Cereals and Commercial Crops*. (1997) Science Publishers, Inc., Enfield, New Hampshire.

Jellie, G. J. and D. E. Richardson (eds.) *The Production of New Potato Varieties*: *Technological Advances*. (1987) Cambridge University Press, Cambridge.

Lawrence, W. J. C., *Plant Breeding*. (1968) Edward Arnold, London.

Lupton, F. G. H. (ed.), *Wheat Breeding: Its Scientific Basis*. (1980) Chapman and Hall, London.

Mayo, *The Theory of Plant Breeding*, 2nd ed. (1987) Oxford University Press, Oxford.

Ockendon, D. J., Anther culture in Brussels sprouts (*Brassica oleracea var gemmifera*). I. Embryo yields and plant regeneration. *Ann Appl Biol* (1984) **105**: 285–291.

Poehlman, J. M. and D. A. Sleper, *Breeding Field Crops*, 4th ed. (1995) Iowa State University Press, Iowa.

Thomas, H., Chromosome manipulation and polyploidy, in Hayward, M. D., Bosemark N. O. and Romagosa, I. (eds.) *Plant Breeding. Principles and Prospects*. (1993) Chapman and Hall, London.

# Chapter 11

---

# Supernumerary ("B") Chromosomes

## 11.1. Definition, origins, numbers, size and functions

**Supernumerary ("B") chromosomes** are additional to the normal "A" chromosomes and differ from them in a number of ways, especially in being inessential and variable in number. Their effects on phenotypes and fertility are so variable that it is difficult to generalise about them. For their use in mapping genes on A chromosomes through B-A reciprocal translocations and pseudodominance, see Subsec. 8.6.1. There are reviews by Jones (1995) and Jones and Houben (2003).

### 11.1.1. Origins

B's arise to normal A chromosomes by accidents, especially by breakage and loss of genetic material, particularly at meiosis, when centric fragments may become B chromosomes. Interspecific hybridisation and the formation of allopolyploids can cause extensive genome reorganisation from which B's may be formed.

### 11.1.2. Numbers, occurrence and transmission

B's may be present in some members of a population and absent in others, and may differ between populations too. When they are present, they may be in different numbers and in different individuals. Even in plants and insects regularly having B's, they are not essential. Selection tends to favour a given number of B's in a particular environment, but irregular segregation means that an optimum number cannot be uniformly maintained.

They occur widely in insects such as beetles and grasshoppers, in hundreds of higher plant species, especially cereal grasses, and have been found in some fungi. Wild populations polymorphic for low numbers of B's are found in organisms as diverse as platyhelminths, shrews, fish, rats, insects, gymnosperms, ferns, bryophytes and many grasses. They are common in mammals. The number of B's per cell can even vary within a plant; e.g., in *Sorghum* and *Poa alpina*, B's tend to get lost in root mitosis, but are stable in the stem and reach the germ-line tissues. Karyotyping for B chromosomes from root-tip squashes could therefore be misleading.

Even numbers of B's are more common than odd numbers, and low numbers are more common than high numbers. Diploid rye, $2n = 14$, can have up to 10 B's, while the autotetraploid rye, $2n = 28$, can have up to 12 B's, although increasing numbers of B's usually give increasing loss of fertility. Maize, $2n = 20$, is unusual in that it can have more than 20 B's per cell. B's can be a significant part of the total genetic material: in rye, $2n = 14$, with 8 B's, the B's increased the nuclear DNA by 40%.

B's do not usually pair with A's, not even those they were derived from. They may or may not pair with other B's, and will not do so unless they have regions of homology. Often, they do not segregate regularly even when present in even numbers, and they may be transmitted differently through male and female gametes, in plants and animals. A parent with one B may give 0, 1 or 2 B's in different offspring, because of irregular segregation at meiosis. Most B's have one centromere, but dicentrics also occur. In *Crepis capillaris*, single B chromosomes showed mainly foldback meiotic pairing, giving a symmetrical univalent hairpin loop typical of isochromosomes (chromosomes with two identical arms which are mirror-images of each other). The B's paired extensively within and between themselves, but not with A's.

Jones and Houben (2003) state that B's are sometimes "nuclear parasites with autonomous modes of inheritance, exploiting 'drive' to ensure their survival in populations. Their 'selfishness' brings them into conflict with their host nuclear genome and generates a host-parasite relationship, with anti-B-chromosome genes working to ameliorate the worst of their excesses in depriving their hosts of genetic resources." **"Drive"** occurs when a chromosome gets into gametes more often than expected from Mendelian segregation. For example, in plant female meiosis, B chromosomes can pass preferentially into the nucleus which is destined to form the egg rather than the polar body nucleus, or in pollen mitosis, both B chromatids can fail to separate and both pass preferentially into the generative nucleus. In rye, **directed non-disjunction** in gamete formation is strong and is under the control of the B chromosome. Drive in favour of B's has been shown for about 60% of plant species studied; in other species, it is not known how population equilibrium frequencies are maintained. In maize, alleles *H* (high and preferential transmission of B's) and *h* (low transmission, acting as an anti-B-chromosome gene) on an A chromosome control B transmission.

### 11.1.3. *Size, sequences and function*

Arising by loss of genetic material, B's are on average shorter than A's. They are often, but not always, **heterochromatic**; i.e., their DNA condenses at different times of the cell cycle, compared with normal DNA. Most supposedly unique DNA sequences found on B's were later found, at least in low copy numbers, on the A's or in A's of related species. While B's clearly affect phenotypes, there was no evidence that their genes were transcribed until the report of Leach *et al.* (2005) that one of two ribosomal RNA gene families confined to the B chromosome of the plant *Crepis capillaris* were transcribed in leaves and buds, at a low level. Compared to the numbers of rRNA genes, the number of transcripts was low; perhaps transcription is largely suppressed or transcripts are soon degraded. A giant B chromosome occurs in the cyprinid fish *Alburnus alburnus* and harbours a highly abundant retrotransposon-derived repetitive DNA sequence with strong homology to a retrotransposon in *Drosophila* (Ziegler *et al.*, 2003). In maize, a B chromosome has sequences homologous to the ones in the centromeric region of chromosome 4 of the A set. B's have few identified genes except for clusters of 18S, 5.8S and 25S rRNA genes which can organise extra nucleoli. In a fungal plant pathogen of peas, *Nectria haematococca*, several genes determining pathogenicity have been identified on a 1.6 Mb B chromosome.

## 11.2. Effects on phenotype and fertility

B's were originally thought to be inactive, but C. D. Darlington and colleagues published a paper on "The activity of inactive chromosomes," showing that B's did affect phenotypes. In general, cell volume increases with increasing numbers of B's, but many quantitative characters of economic or ecological interest show strange zigzag patterns when phenotype values are plotted against the number of B's. Odd numbers of B's seem to depress phenotype values, and even numbers tend to increase them, as shown in Fig. 11.1.

Breeders who find unexpected amounts of phenotypic variation between individuals, especially in cereal grasses, should check cytologically for differences in the numbers of B chromosomes between individuals. **Ecological effects** of B chromosomes have been reported. For example, white spruce (*Picea glauca*) has different typical numbers of B's in different geographical locations.

B's usually affect **fertility**, often with high numbers of B's giving sterility, although maize can tolerate 20 B's without losing fertility. In many plants, increasing numbers of B's increasingly delay flowering, so those with high numbers of B's could be partly isolated reproductively from the main population, but with an increased chance of crossing with each other. In *Plantago coronopus*, a plantain, having one B gives male sterility but not female sterility. Pollen with the most B's is sometimes the most successful

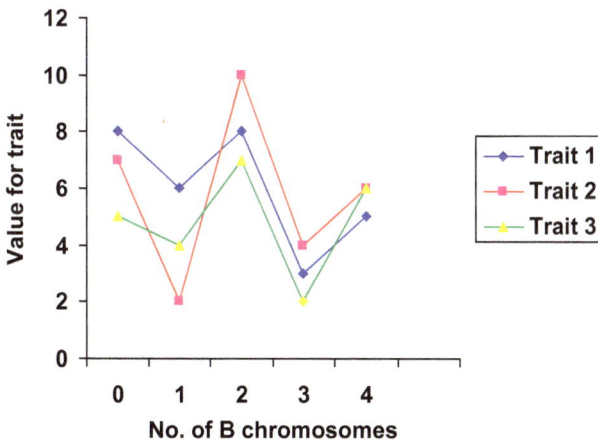

**Fig. 11.1. The effect of different numbers of B chromosomes on plant quantitative characters** in rye (*Secale*). The lines represent different quantitative characters such as straw weight, plant weight and tiller number: the aim is to show typical effects, so it does not matter which line relates to which character. All measures are in arbitrary units.

in fertilisation, as if B's can be advantageous in the gametophyte, even though they are often disadvantageous in the sporophyte.

One important evolutionary effect of B chromosomes, and one which can be of use to plant breeders, is the effect of B's on **chiasma and crossing-over frequencies**. B's sometimes increase chiasma frequencies, as in maize and rye, and sometimes decrease them, as in *Lolium*. If one wanted recombinants in rye, one could obtain them by using plants with B chromosomes, which raise chiasma frequencies, then selecting for recombinants lacking B chromosomes for a more stable product.

In **Lolium**, a widely used meadow rye-grass genus, B's suppress homeologous pairing (pairing between related chromosomes of different genomes), with important effects on fertility. *Lolium temelentum* (genomes TT) and *Lolium perenne* (genomes PP) are both diploids with $2n = 14$. When crossed, they give a hybrid, TP, $2n = 14$. With no B's, there is good homeologous pairing, e.g., T5 with P5, giving a fertile hybrid with seven bivalents at meiosis, with T and P genomes able to exchange genes through crossing-over. If the diploid hybrid has B chromosomes, homeologous pairing is suppressed, giving little pairing of T and P genomes, mainly with univalents at meiosis, so the hybrid is sterile. If one doubles up the hybrid's chromosomes, e.g., with colchicine, the allotetraploid has TT, PP, $2n = 28$. If there are no B's, the allotetraploid behaves like an autotetraploid, with quadrivalents from homeologous pairing of TT with PP chromosomes, and so is largely sterile. If the allotetraploid has B's, then these suppress homeologous pairing, just leaving bivalents from homologous pairing of T with T, P with P, so the allotetraploid is then fertile, behaving like an allotetraploid. B's could, in nature or cultivation, convert largely sterile autotetraploid-behaving hybrids to fertile allotetraploid-behaving hybrids. In planned breeding programmes, one has to take account of the irregular distribution of B chromosomes to the offspring.

## Suggested Reading

Jones, N. and A. Houben, B chromosomes in plants: escapees from the A chromosome genome? *Trends in Plant Science* (2003) 8: 417–423.

Jones, R. N., Tansley Review no. 85: B chromosomes in plants. *New Phytologist* (1995) 131: 411–423.

Jones, R. N. and H. Rees, *B Chromosomes*. (1982) Academic Press, New York.

Leach, C. R. *et al.*, Molecular evidence for transcription of genes on a B chromosome in *Crepis capillaris*. *Genetics* (2005) 171: 269–278.

Ziegler, C. G. *et al.*, The giant B chromosome of the cyprinid fish *Alburnus alburnus* harbours a retrotransposon-derived repetitive DNA sequence. *Chrom Res* (2003) 23–35.

# Chapter 12

# Human and Medical Genetics

## 12.1. Introduction

We are **diploid Eukaryotes**, sharing the principal genetic features of such organisms, e.g., molecular mechanisms, chromosome behaviour and some population genetics. For genetic studies, our life cycle is inconveniently long, family sizes are too small, and desired matings are difficult or illegal to procure. One cannot legally make F1 backcrosses and really close inbreeding is forbidden. Certain types of environmental treatments are forbidden for legal, moral or social reasons.

To balance these **disadvantages**, there are **a few advantages**. In some cases, visible features of families over many generations have been recorded through painted portraits of aristocratic families, e.g., King George III, Queen Charlotte and their 15 children. Humans have also been studied in great detail biochemically, physiologically, anatomically and socially, so many techniques are available and much background information already exists.

Another important advantage is that **extensive statistical records exist**, often made for other purposes, but of interest to human geneticists, e.g., data on human height and weight from army medical records; data from death certificates on causes of death; church or civil records of births, baptisms, marriages and deaths. These records may be inaccurate; e.g., the woman's husband may be recorded as the father of a child when its biological father was different. There are excellent population records from Sweden and

Finland for 300 to 400 years, with data on such things as frequencies of twinning and other multiple births (Eriksson and Fellman, 1996). They include tax and census lists from the 16th century and church archives from the 17th century. From 1749 there are regular annual official population statistics.

The **number of identified human genes** is constantly rising and will rise dramatically as a result of the human genome project. By 1997, more than 5,000 human phenotypes inherited in a Mendelian fashion had been identified, of which about 4,000 were disease-associated. Fewer than 10% are X-linked; more than half are autosomal dominants and about 36% are autosomal recessives, often loss-of-function mutations. The main methods used to work out how human characters are inherited are listed below. For an on-line World Wide Web catalogue of human genes and genetic disorders with over 10,000 entries, contact Online Mendelian Inheritance in Man at http:/www.ncbi.nlm.nih.gov/Omim; it also has mapping details. For human reproductive physiology, see Sec. 17.3 and Sherwood (1995).

Although this chapter is the main one on human genetics, other aspects are dealt with elsewhere, e.g., positive and negative assortative mating in Sec. 6.1, and the population genetics of sickle-cell anaemia in Sec. 15.3. Selection, genetic drift, mutation and recombination are all highly relevant to human medical and population genetics and are considered in other chapters, including human genomics in Secs. 8.8 and 8.9.

In human genetics, some **unusual characters can be locally adaptive**, not harmful abnormalities. For example, Beall *et al.* (2004) studied 905 households living at altitudes of 3,800 to 4,200 m in Tibet and obtained fertility records from 1,749 women. An autosomal locus was segregating, with an allele of a major gene giving 10% more oxygen saturation in the blood compared with normal levels. Women with the high oxygen saturation allele had significantly lower infant mortality, averaging 0.48 infant deaths per woman, compared with 2.53 deaths for women homozygous for the low oxygen saturation allele, with little difference in the number of pregnancies or live births. The women with the high oxygen saturation allele therefore had more surviving offspring, demonstrating a clear case of selection and adaptation to an extreme environment with lower oxygen levels than at lower altitudes.

## 12.2. Finding out how or whether characters are inherited: pedigree studies; twin studies; Hardy–Weinberg analysis; familial incidence

### 12.2.1. *Pedigree studies*

Pedigree studies are used to find out how particular traits are inherited, e.g., whether they are controlled by one locus or more than one, by alleles which are dominant or recessive, and X-linked or autosomal. In **pedigree studies**, one finds someone, the **proband** (sometimes called the propositus if male, proposita if female), showing the character one wishes to study. Then one traces as many relatives of the proband as possible, seeing which members of the kindred have the character and which do not, over as many generations as possible. Because people's memories of relatives may be faulty, one tries to get reliable records of many related individuals from written data or from observations by competent individuals. A set of related individuals is called a **kindred**, which is a larger unit than one family. For the purposes of pedigrees, the term "marriage" includes less formal relationships.

In the **pedigree diagrams**, the earliest generations are nearest the top, with generations having Roman numerals. A horizontal line directly connects two individuals of opposite sex who are "married", with their children, if any, shown coming from a lower horizontal line in age order, the first-born to the left, and they may be numbered for ease of reference, from left to right. Individuals marrying into the kindred have no vertical line above them, and are numbered along with those born into the kindred in that generation. Individuals with the trait have a filled-in symbol and unaffected ones have empty symbols, with a circle for females, a square for males. The proband is indicated by a bold arrow and sometimes with the letter P as well. Deceased individuals may be indicated by a diagonal crossing through of the symbol. Identical (monozygotic) twins have divergent downward lines joined by a horizontal line, while non-identical (dizygotic) twins just have the downward divergent lines connecting them. A diamond with a number inside indicates that number of unaffected offspring, of unspecified sex. Heterozygotes can be denoted by symbols with the left half clear and the right half shaded, or as having a central dot inside the female circle for a heterozygous female for an X-linked character. Other symbols can be used to show individuals examined personally, individuals without

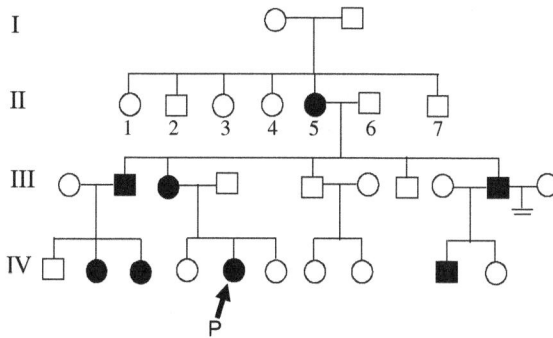

**Fig. 12.1.** **A pedigree for severe foot blistering**. The proband is indicated by an arrow. The symbol downwards between the last two individuals in generation III indicates a childless marriage or relationship. From Haldane and Poole, *J. Heredity*, 1942, 33: 17, but with some individuals omitted.

offspring, consanguineous marriages (a pair of horizontal lines join the male and female symbols) and aborted offspring. A broken line used to be used for children born outside marriage but is now used for offspring with unacknowledged or unexpected parentage.

Some but not all of these symbols are shown in Fig. 12.1, for a kindred with **severe foot blistering**. For example, in generation II, there were five unaffected children and one affected one, who married an unaffected man, and they were the parents of generation III, in which the last-born male had two marriages, to different females. The proband is shown by the arrow and symbol P in generation IV.

When trying to work out the genetics of severe foot blistering, we see that once that trait has appeared in the kindred, all affected individuals have at least one affected parent, suggesting dominance. In families where one parent is affected (and probably heterozygous, not homozygous) and one is unaffected, affected and unaffected children are produced in approximately equal numbers (seven affected, six unaffected), which again fits with dominance, with affected individuals being heterozygous as they have only one affected parent.

The trait can be passed from affected fathers to sons and daughters, and approximately equal numbers of males and females are affected, suggesting that the trait is an autosomal dominant, not X-linked. A dominant X-linked trait could not be passed to a son from his father, as the son inherits the father's Y chromosome, not his X. Additionally, the trait cannot be an X-linked dominant as a father always passes on his one X to his daughters, but an affected father had an unaffected daughter (IV-10).

For such a rare trait, it could not be an autosomal recessive as that would require nearly all the unaffected spouses to be heterozygous carriers, which is highly unlikely. For an autosomal recessive character, both parents of an affected individual must have at least one copy of the recessive allele.

In generation I, both recorded parents were unaffected, yet they had an affected offspring showing this autosomal dominant trait. One explanation is that the gamete from one parent carried a spontaneous mutation from the recessive wild-type allele to the dominant allele. Another explanation is that in generation I, the recorded father is not the child's biological father, with the mother having mated with an unrecorded affected man, producing the affected daughter. It is a fact of human genetics that **adultery** is much more common than **spontaneous mutation**, so the second explanation is more likely than the first, though mutation cannot be ruled out. Other less likely explanations include incomplete penetrance of the dominant allele in one parent, or one parent being a mosaic of normal and mutant tissue.

When looking at pedigrees, one looks to see if the traits are dominant or recessive, and then whether they are autosomal or X-linked. Recessive X-linked characters tend to affect males (hemizygous) much more than females, especially if the recessive alleles are very rare (Sec. 12.6).

Many human kindreds are less extensive than this but can still be conclusive as to the mode of inheritance. Working out pedigrees is exciting detective work but can be frustrating if key individuals are not available through distance, or absence of an address, or through death, or if they are unwilling to be investigated. One can make pedigrees using molecular characters as well as morphological ones, but some individuals may be unwilling to submit samples for analysis.

## 12.2.2. *Twin studies*

Twin studies have been used in man and cattle in deciding the relative contributions of heredity and environment (i.e., heritability) for particular traits. See also Sec. 3.3 for determining heritabilities, including using pairs of monozygotic and dizygotic twins. Twins are either **monozygotic** (MZ, identical), from one egg, one sperm, one zygote, where the embryo divides into two genetically identical embryos, or **dizygotic** (DZ, non-identical, fraternal), from two separate fertilised eggs which are genetically different. MZ twins are almost always of the same sex but DZ twins can be the same or different sexes, just like siblings. DZ twins are no more alike, except in age, than a pair of full siblings born at different times. As MZ twins are of the same sex, one should use DZ twins of the same sex when comparing the two types of twins, or sex differences can be present in addition to other genetic differences. See Sec. 12.15 for human **multiple births**.

One can compare traits in monozygotic (same genotype) and dizygotic (different genotypes) twins reared together (same environment) and reared apart (different environments). With humans, cases of twins reared apart are fairly rare, by chance, but with farm animals one can do controlled experiments. Due to the difficulty in finding enough pairs of twins, especially twins reared apart, there are special "**twin registers**" of pairs of twins willing to cooperate in genetic studies: see Sec. 12.15.

One method used to find relative contributions of heredity and environment for quantitative characters is **average pair differences**; one measures a number of pairs of twins to find out the average difference within pairs. Typical data are shown in Table 12.1. With height, the environment induces small differences within pairs of MZ twins, but genetic effects are more important because the two right-hand data columns, showing pairs with individuals differing in genotype, have much bigger average pair differences than do the MZ twins. MZ twins reared apart are not much more

**Table 12.1.** Average pair differences for three quantitative traits, height, head width, and weight.

| Trait | Monozygotic twins | Monozygotic twins | Alike-sex dizygotic twins | Sibs of alike sex scored at the same age |
|---|---|---|---|---|
| Reared: | Together | Apart | Together | Together |
| **Height (cm)** | 1.7 | 1.8 | 4.4 | 4.5 |
| **Weight (kg)** | 1.9 | 4.5 | 4.5 | 4.7 |
| **Head width (mm)** | 2.8 | 2.9 | 4.2 | No data |

different than are MZ twins reared together, again showing only small amounts of environmental effects. The analysis of head width is very similar to that for height, showing small environmental effects but much larger genetic effects. For weight, however, MZ twins reared apart are much more different than are MZ twins reared together, suggesting a big environmental component of weight. There is a genetic component too, because MZ twins reared together are more alike than dizygotic twins or alike-sex siblings reared together.

For qualitative characters, one uses **concordance scores**. A pair of twins is concordant if they both have the trait, and is discordant if one has it and the other does not. Table 12.2 shows concordance scores from monozygotic and alike-sex dizygotic twins.

If pairs of MZ twins are not always concordant for a trait, that suggests that there is an environmental or chance element to developing that trait. For example, there is a pair of monozygotic twins, one with cleft lip and palate and one with normal lips and palate (picture, Boomsma *et al.*, 2002), showing incomplete penetrance for this multifactorial character. A high concordance score for MZ twins does not necessarily prove a high heritability because they are usually brought up in the same environment, as are pairs of DZ twins. What demonstrates a strong genetic determination for a trait is if MZ twins have a high concordance and DZ twins have a much lower concordance, as for non-insulin-dependent diabetes mellitus and schizophrenia. Where MZ twins are discordant for having a criminal record, the one lacking such a record might not have been criminal, or might just not have been caught (a chance or environmental effect). With stomach cancer, a heritable element was shown by MZ twins having a much higher concordance than DZ twins (27% versus 4%), but the fairly low concordance for MZ twins shows a large chance (environmental) element.

**Club foot** is multifactorial and the higher concordance of MZ twins than of DZ twins shows a clear heritable element, but with only 32% concordance for MZ twins, there must also be a large element of chance (or environmental factors, some of which are unknown) in whether twins with the same polygenes actually develop club foot. The trait has a frequency of 0.1% and a narrow sense heritability of 68%.

The most unexpected figure in Table 12.2 is the 89% concordance for monozygotic twins for **Down syndrome**, when one would expect 100% for a condition caused by having three copies of chromosome 21. Perhaps

**Table 12.2.** Concordance scores for monozygotic and dizygotic twins for various characters.

| Trait | Monozygotic twins Concordance, %* | Dizygotic twins Concordance, %* |
|---|---|---|
| **Age at first walk** | 68 | 31 |
| **Stomach cancer** | 27 | 4 |
| **Scarlet fever** | 64 | 47 |
| **Tuberculosis** | 87 | 26 |
| **Down syndrome** | 89 | 7 |
| **Criminal record** | 68 | 28 |
| **Coffee drinking** | 94 | 79 |
| **Clubfoot** | 32 | 3 |
| **Cleft lip/palate** | 35 | 5 |
| **Hair colour** | 89 | 22 |
| **Eye colour** | 99.6 | 28 |
| **Diabetes mellitus, non-insulin-dependent** | 50–100 | 10 |
| **Diabetes mellitus, insulin-dependent** | 30–40 | 6 |
| **Rheumatoid arthritis** | 30 | 5 |
| **Pyloric stenosis** | 20 | 2 |
| **Handedness (left or right)** | 79 | 77 |
| **Schizophrenia** | 40–80 | 10–13 |
| **Male homosexuality** | 50–100 | 25 |
| **Epilepsy** | 37 | 10 |
| **Manic depression** | 80 | 10 |
| **Alzheimer's disease** | 40 | 10 |
| **Spina bifida** | 6 | 3 |
| **Multiple sclerosis** | 20 | 5 |

*Twins are concordant when they both show the same trait, e.g., both develop stomach cancer, or are both phenotypically similar for a measured character within a small range, e.g., for age at first walk. Data from various sources including Connor and Ferguson-Smith (1997), Strickberger (1985), and Genetic Interest Group, accessed 2005 (www.gig.org.uk).

when a zygotic nucleus with three copies of chromosome 21 divides to give monozygotic twins, one extra copy of 21 sometimes gets lost, giving one normal and one Down's embryo, with one chance non-disjunction of chromosomes correcting an earlier non-disjunction.

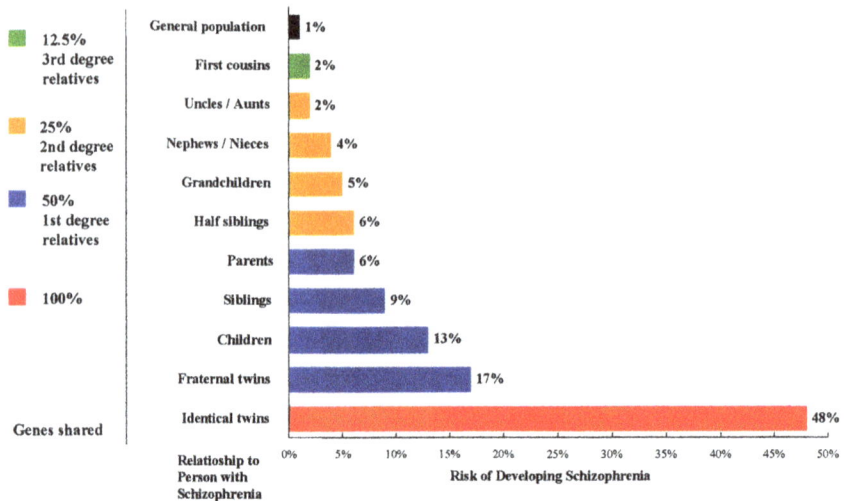

**Plate 12.1.**   **The risk of developing schizophrenia for people of different degrees of relationship to a sufferer**. Courtesy of Irving Gottesman.

The data on **schizophrenia** show a strong heritable element but with some environmental or chance factors also operating, since only 40–80% concordance is shown for MZ twins (Table 12.2 and Plate 12.1). Coffee drinking is probably largely environmental (influenced by family habits rather than family genes), with a slight heritable element shown by MZ twins having a somewhat higher concordance than DZ. Where MZ twins are discordant for a largely inherited character such as eye colour or natural hair colour, one can often attribute the difference to incomplete penetrance. Even where hereditary determinants are shown by concordance scores to operate, the concordance scores do not tell us how the character is inherited, e.g., monogenic or polygenic, dominant or recessive, X-linked or autosomal. Data on concordance may differ sharply in different studies. The concordance for MZ twins for schizophrenia was 80% in one study but 40% in another. Where **MZ twins are not identical** in phenotype, possible explanations include skewed X-inactivation in females, somatic mutations, rearrangements of epigenetic signals in gamete production, and mitochondrial reassortments, as well as effects of disease and accidents, different food intakes, exercise, education and upbringing if separated.

## 12.2.3. *Hardy–Weinberg analysis of populations*

We considered **Hardy–Weinberg** analysis in Sec. 4.2. It can be used to test how human traits are inherited. For example, a population of 420 subjects was studied for the M and N antigens on red blood cells. 137 had the M antigen only, 87 had the N antigen only, and 196 had both antigens. A simple hypothesis for the inheritance of the M and N blood groups is that they are determined by two alleles at one locus, with codominance, 137 being $M\ M$, 87 being $N\ N$, and 196 being $M\ N$. The allele frequency of $M$ is therefore $\frac{(137x2)+(196x1)}{420x2} = 0.56$, so the allele frequency of $N$ is 0.44. One then expects Hardy–Weinberg equilibrium frequencies of the M blood group to be 0.314, the N group to be 0.194 and the MN group to be 0.493. The expected numbers out of 420 would be M group, 132, N group, 81, MN group 207, giving good agreement between observed and expected numbers (a $\chi^2$ test gives $P = 0.28$). The hypothesis of two alleles with codominance fits the data; most other hypotheses do not.

## 12.2.4. *Familial incidence*

A character with a large heritable element will tend to "run in families", with an affected person often having one or more affected relatives. A character being heritable is not quite the same as being **familial**, tending to run in families, because members of a family tend to have more similar environments than unrelated individuals, so familiality could be caused by environmental effects or heritable effects, or both. Speaking a particular language, for example, is familial, but not heritable. For traits with little effect of family environment, one can use familiality as evidence that a character is inherited. For example, people with **vitiligo** (giving unpigmented skin patches; see Plate 12.2 and Lesage, 2002) have one or more affected relatives more often than one would expect from the frequency of the trait in the population, suggesting a heritable element to vitiligo. It affects between 0.4% and 2% of individuals in populations throughout the world, but with different degrees of severity from a few small white patches to complete whiteness all over the face, body and limbs. It is associated with a number of other autoimmune conditions (Alkhateeb *et al.*, 2003). Vitiligo does not follow a simple monogenic pattern of inheritance and various environmental factors

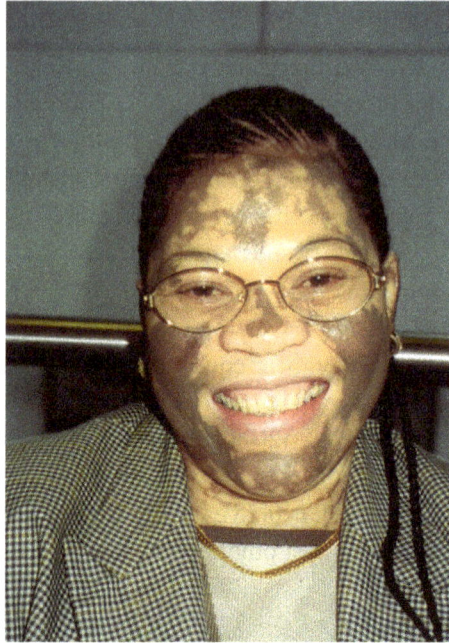

**Plate 12.2.**  **Vitiligo causing white patches** on a dark-skinned lady about four years after she became affected. Her white areas are still spreading. Vitiligo is multifactorial and not infectious. It can cause severe sunburn, embarrassment, low self-esteem and social problems, with people staring and making adverse comments, but is not life-threatening. It greatly reduces a woman's chance of marriage in some communities.

are probably involved, so it is multifactorial; it also has different ages of onset in different people, and different triggering factors.

Another clear case of familiality is **eczema**, although the detailed genetics are unclear. In the work of Wadonda-Kabondo *et al.* (2004) on 8,530 Bristol children, the proportion of children with eczema was 28% if neither parent had it, 40% if one parent had it, and 52% if both parents had it. Although eczema, asthma and hay fever are associated conditions known as atopic disease, the correlation of childhood eczema was stronger with parental eczema than with parental asthma or hay fever. Of the parents, 47% of mothers and 41% of fathers reported having an atopic disease, with hay fever being the most prevalent. Unlike in some studies, these authors found the father's influence on childhood eczema to be as strong as that of the mother. There has been a dramatic rise in the frequency of eczema and asthma over the last 20 years, as shown in these Bristol studies (The Avon longitudinal study of parents and children) over time, presumably with environmental causes, not due to any genetic changes.

To use familiality, one studies the frequencies say of diseases or malformations among genetically related individuals and in the general population. If the trait is heritable, there is a higher frequency of the trait amongst the relatives of affected individuals than in the general population, with the increase in frequency being proportional to the degree of relatedness. This is shown in Table 12.3 for the incidence of cleft lip (with or without cleft palate) in Europe and North America, where the trait has a population incidence of 1 in 1,000 and a narrow sense heritability of 76%. The frequency of cleft lip among relatives of an affected person is much higher than in the general population, especially for first-degree relatives but the decrease in second- and third-degree relatives does not fit expectations of an autosomal dominant (where a decrease of a half is expected for each degree of relation less) or an autosomal recessive. Concordance scores for cleft lip are 40% for monozygotic twins and 4% for dizygotic twins, suggesting an environmental or chance component in addition to a strong hereditary element. For further information on familiality, especially for multifactorial traits, see Gelherter *et al.* (1998).

## 12.3. Single gene characters and disorders, and their treatments

A range of different conditions is described to show different modes of inheritance, different effects, causes, diagnoses and treatments. It is not exhaustive and concentrates on inherited disorders. Some such as polydactyly (extra digits) are expressed early in the embryo so they show at birth, while others may only be expressed much later in life. For example, **Huntington's chorea** (Huntington disease, an unstable length mutation; see Subsec. 7.1.6) has an average age of onset of 37 years, affecting people at any age from the teens to the eighties. It involves progressive nervous and mental deterioration, with chorea (involuntary leg and arm movements). It is eventually fatal. It is an autosomal dominant, affecting about 1 in 20,000 people. The **unstable length mutations** (see Subsec. 7.1.6) such as myotonic dystrophy, fragile-X and Huntington disease are unusual in that amplification of one triplet codon may occur between generations, some-times to different extents in the two sexes, with increasing severity in the disease symptoms. They may be autosomal or X-linked and are usually dominant.

The **frequency of alleles for inherited diseases** often differs sharply between different racial groups. Sometimes the cause for these differences is understood, such as exposure to malaria increasing the frequency of alleles for thalassaemia and sickle-cell anaemia, but the reason for cystic fibrosis being so much commoner in Europeans than in Negroes or Orientals is not clear. **Leber's optical atrophy** (or neuropathy) is unusual in being a mitochondrial mutation and showing maternally inherited blindness.

### 12.3.1. *Autosomal genes*

Section 2.1 covered the general inheritance of such characters, so this sec-tion deals with a range of dominant and recessive traits determined by single autosomal loci, including diagnosis, symptoms and treatment where relevant.

The common form **polydactyly** (isolated polydactyly) is an example of an autosomal dominant allele in otherwise normal individuals (although there are other aetiologies, e.g., trisomy 13, Meckel syndrome, and some autosomal recessive conditions). One in 2,000 Caucasians has extra fin-gers and/or toes. It has variable expressivity, so some sufferers have strong expression, with more than one extra complete toe and/or finger, while oth-ers just have a lump on one side of the hand (Plate 3.1). Sufferers may be homozygous, when both parents should have the condition, or heterozy-gous, which is much more frequent for a rare trait, as expected from the Hardy–Weinberg equilibrium (Sec. 4.2). Homozygotes will have all off-spring affected but heterozygotes, if married to an unaffected person, will have about half their offspring affected. It is slightly disfiguring but not usually serious, and is treated by surgery to remove the extra digits, or by tying them off in the first week of life to cut off their blood supply if they are little more than a small polyp: see Plate 12.3.

Another rare autosomal dominant is **achondroplasia**, giving **dwarfism**, caused by allele *D*. Such dwarves have short stature, relatively large heads, and short limbs, especially in the proximal parts: see Plate 12.4. It is unusual

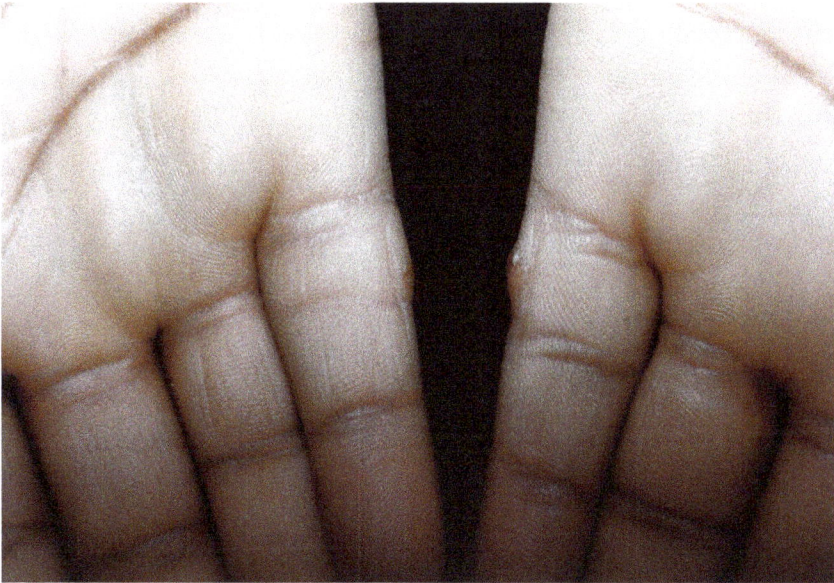

**Plate 12.3.** **Treated polydactyly**. The little fingers bear a small, slightly itchy bump, all that remains of dominant autosomal isolated polydactyly; this was treated on the day of birth by tying off the small extra finger stubs.

for a dominant allele in that about 84% of cases have two normal parents and arise from new mutations in a parent's germ line. Even more surprising is that almost all mutations are in exactly the same nucleotide, in a gene encoding the type 3 fibroblast growth-factor receptor, which mediates the effect of that factor on cartilage growth. Allele $D$ acts as a recessive lethal, although dominant for dwarfism, so that $D\,d$ gives a viable dwarf but $D\,D$ is lethal, giving a very short-limbed individual with a very small chest who invariably dies *in utero*. The frequency of the mutation, and therefore of the trait, is about $6 \times 10^{-5}$ per gamete. Dwarves are fertile, producing a 1:1 ratio of dwarves to normals if mated with non-dwarves, or 2:1 if mated to dwarves, but females have small pelvises and usually need a Caesarean section when giving birth.

**Sickle-cell anaemia** is caused by an autosomal recessive allele, $a$; when homozygous this causes severe and usually lethal anaemia, with the red blood cells going sickle-shaped at low oxygen concentrations and being removed in the spleen. Most sufferers die young but some live 20 years or

**Plate 12.4.  Achondroplastic dwarfism**, showing that the limbs are very short, especially the proximal parts. It is caused by an autosomal dominant allele but which acts as a recessive lethal. Seneb was Chief of the Royal Wardrobe at Gizeh, Egypt, in the 5th Dynasty. He lived from 2563–2423 B. C., and had two normal children by his normal wife. The males are shown darker than the females. Funerary image photographed by the author at the Egyptian Museum, Cairo.

more, with many medical problems. It can be extremely painful when small blood vessels become blocked by distorted red cells. The heterozygote, with sickle-cell trait, is clinically more or less normal in spite of expressing the mutant $\beta^S$ globin gene as well as the wild-type one. Only at very low oxygen concentrations does the blood of heterozygotes show sickle-shaped red blood cells, and people with sickle cell trait sometimes need help from oxygen masks on long-distance flights, as planes have reduced air (and therefore oxygen) pressure. During the second world war, aircrew with African ancestors often suffered blackouts in high-flying unpressurised planes, from the low oxygen, if they were heterozygotes for sickle-cell anaemia.

In sickle-cell anaemia, there is a missense mutation in codon 6 of the $\beta$ globin gene on chromosome 11, changing GAG in the normal allele's mRNA to GUG, and so coding for valine instead of glutamic acid. Most people with the $a$ gene have exactly the same mutation, and its frequency is much too high in some populations to be due to chance mutation. The Amba people of Uganda have about 40% with sickle-cell trait, and its incidence in black Americans is about 9%. It was A. C. Allison who showed that the heterozygotes with sickle-cell trait, $A\ a$, were much less susceptible than normal $A\ A$ individuals to **subtertian malaria**, caused by *Plasmodium falciparum*. In malarial regions, the $a$ allele is maintained by **heterozygote advantage** even though the recessive homozygote suffers severe or lethal anaemia. For the population genetics, see Sec. 15.3. The heterozygotes share their falciparum malaria resistance with heterozygotes for beta thalassaemia (see below) and for glucose-6-phosphate-dehydrogenase deficiency (Subsec. 12.3.2), which is X-linked, unlike the other two traits. Sickle cell trait occurs mainly in people of African malarial region origin but also in those from other malarial areas, such as southern India, Greece and Italy.

There are a number of inherited autosomal recessive diseases associated with **defective phenylalanine or tyrosine metabolism**. In normal individuals, allele $P$ specifies the liver enzyme phenylalanine hydroxylase, converting dietary phenylalanine into tyrosine, which can be converted by tyrosinase to the melanin eye, skin and hair pigments, or by other enzymes to thyroxine growth hormone, or to homogentisic acid. People suffering from **phenylketonuria (PKU)** are $p\ p$ and cannot convert phenylalanine to tyrosine and thence to melanin, so they have lighter pigmentation but are not albinos as the diet provides some tyrosine. The main effect is that phenylalanine builds up in the blood (above 20 mg/dL, compared with 1 mg/dL normal level), with up to a gram a day being excreted in the urine. Most of the amino acid is converted to phenylpyruvic acid, phenylacetic acid (giving a musty smell) and other compounds which damage the developing brain in babies, especially during the first six months, causing severe mental retardation. Two thirds of untreated PKU babies grow up as idiots with IQs of less than 20, plus a range of physical symptoms including a small head. The frequency of the disease is about 1 in 25,000 in Europe.

Fortunately, PKU is easily detected and treated. In developed countries, babies around three days old have a blood sample taken from a heel-prick. The blood is used in a bioassay with phenylalanine-requiring strains of *Bacillus subtilis* which can grow when the blood has abnormally high concentrations of phenylalanine. PKU babies are immediately taken off milk and are given a low phenylalanine diet, with tyrosine supplements, for about 15 years, while the brain is still developing. The treatment is very effective. The heterozygotes, *P p*, can be detected by a test in which phenylalanine is injected into the blood and its rate of removal is monitored, as they have less enzyme than *P P* individuals. PKU women must go on a low phenylalanine diet during pregnancy as the placenta increases phenylalanine levels in foetal blood, so permanent damage to the foetus can occur. The detection of heterozygotes aids genetic counselling (Sec. 12.14).

People who lack the tyrosinase enzyme cannot make the melanin pigments from tyrosine and are **albinos**, *a a*: see Plate 2.1. Albinos have white skin and hair. In strong light, their eyes appear pink from the blood vessels in the retina; the lack of eye pigments usually results in visual problems. There is no treatment but sufferers usually wear tinted glasses and some now wear contact lenses with coloured irises so that they look more normal. Their exposed skin needs sun-block protection from the UV in sunlight. The frequency of albinism is about 1 in 20,000 in Europe, 1 in 3,000 in Nigeria, and 1 in 132 in San Blas Indians in Panama. Like vitiligo (Subsec. 12.2.4 and Plate 12.5), the disease shows up most clearly amongst races with dark skins.

People who lack the enzymes to convert tyrosine to the thyroxin hormone suffer **goitrous cretinism** (incidence, about 1 in 10,000), with stunted growth and mental defects. Treatment with thyroxin tablets is effective if started in the first year of life. Tyrosine is also converted to homogentisic acid, which is converted by the enzyme homogentisic acid oxidase to maleylacetoacetic acid and thence to water and carbon dioxide. People lacking this enzyme have **alkaptonuria** (incidence, about 1 in 250,000). They excrete homogentisic acid in the urine, which goes black in the light or if alkaline. Deposits can cause degenerative arthritis of the spine and large joints, and black or ochre pigments in cartilage and collagenous tissues, including the palate, ears and the sclerae of the eyes. There is no overall treatment but vitamin C may delay the arthritis.

The nature of some of these inherited biochemical diseases was first recognised by the English physician Archibald Garrod, who studied alkaptonuria, pentosuria, cystinuria and albinism. His pioneering book, *Inborn Errors of Metabolism*, was published in 1909 (Henry Frowde, Hodder and Stoughton, London).

**Cystic fibrosis** (#217900, www.ncbi.nlm.nih.gov/Omim) is one of the commonest severe single gene disorders, affecting about 1 in 622 Afrikaners in South Africa, 1 in 2,500 white births in Europe (with 1 in 25 people being heterozygous carriers) but only 1 in 15,000 in African Americans and less than 1 in 30,000 Asian Americans. It is called cystic fibrosis of the pancreas because the part of the pancreas which normally produces digestive enzymes is often replaced by fibrous scar tissue with fluid-filled cysts. A mutation in the cystic fibrosis transmembrane conductance regulator gene results in upset glycoprotein metabolism, giving abnormal secretions. The CFTR protein is located in the plasma membrane and regulates chloride ion transport across the membrane.

The **sweat** is very high in sodium chloride, which can result in salt depletion. This symptom must have been known for hundreds of years, as old Northern Europe folklore has a saying: "Woe to that child which when kissed on the forehead tastes salty. He is bewitched and soon must die". The **lungs** produce thick, viscid mucus which does not drain readily, so sufferers are prone to lung infections such as pneumonia. Recurrent infections can cause replacement of lung epithelium cells by fibrous tissue, decreasing lung capacity and eventually causing death. Because of mucus blocking the ducts, the **pancreas** fails to secrete adequate amounts of three important digestive enzymes: amylase, which converts polysaccharides to disaccharides, trypsin, which helps to digest proteins to amino acids, and lipase, which converts triglycerides to fatty acids and glycerol. With impairment of carbohydrate, protein and fat digestion, malnutrition results. There can be liver problems from biliary tract obstruction and, in newborns, obstruction of the small bowel.

Cystic fibrosis is caused by allele *c* on chromosome 7, and the gene was cloned in 1989. There more than 500 known mutations, of differing severity; a three base-pair deletion ($\Delta$F508) occurs in 70 to 80% of mutations in Northern Europe. There is no cure but the different **symptoms can be treated**, such as giving salt tablets for salt depletion. Antibiotics and vigorous physiotherapy to help drain the lungs are used for the respiratory problems. Diets can be adjusted to ameliorate the digestive problems, and sufferers can be given encapsulated enzymes (usually from dried pig pancreas extract) with buffers, to take with every meal. Untreated children usually die before the age of 10, with an average life expectancy of five years, but with treatment at least 70% now survive to adolescence, with an average survival of 30 years or more. Those who avoid serious lung infections live to 50 or 60, while those infected by the bacterium *Pseudomonas aeruginosa* have a life expectancy of about 30 years, going down to 16 years for those attacked by the related opportunistic bacterial human parasite (but mainly a pathogen of onions), *Burkholderia cepacia*. The females are usually fertile unless mucus blocks the entrance to the uterus, but not the males, who usually lack a functional vas deferens. See Subsecs. 12.12.7, 12.12.9 and 12.17.3 for prenatal and adult diagnosis and gene therapy for cystic fibrosis.

The high frequency of this deleterious gene, with about 1 in 25 white Europeans being heterozygous carriers, has been attributed to **heterozygote advantage** but without much experimental evidence. One theory is that the heterozygote is better able to resist the lethal effects of diarrhoea-causing diseases such as cholera as heterozygotes have fewer chloride-ion channels in the gut than dominant homozygotes, and so excrete less water in diarrhoea, resulting in fewer deaths from dehydration. It has also been suggested that the lower incidence of cystic fibrosis in Africans and Asians comes from the somewhat saltier sweat of the heterozygote than the normal homozygote, with selection in hotter climates against salt-depletion being more severe than in colder climates.

The **thalassaemias** (meaning "sea blood", from the reduced haemoglobin content of the blood, giving anaemia) are diseases caused by a deficiency, but not an absence, of $\alpha$-globin, giving $\alpha$-**thalassaemia**, or of $\beta$-globin, giving $\beta$-**thalassaemia**. $\alpha$-thalassaemia is very common in Southeast Asia, Africa and the Middle East while $\beta$-thalassaemia is very common in people from Cyprus, Greece, Turkey and other malarial (or formerly malarial) regions of the Mediterranean and Southeast Asia. In Cyprus, carriers of $\beta$-thalassaemia recently made up about 16% of the population, with 0.6% of births being of recessive homozygotes, usually dying within the first year of life. This very high frequency of a harmful recessive is accounted for by heterozygote resistance to malaria. Malaria was eliminated from Cyprus after World War II, which should remove the heterozygote advantage and reduce the frequency of the recessive alleles. Genetic counselling of those coming up to marriageable age is helping to reduce the frequency of recessive homozygotes born. The World Health Organisation has estimated that as many as 7% of the world's population are carriers for $\alpha$-thalassaemia or $\beta$-thalassaemia.

Normal adult human haemoglobin (haemoglobin A) is a tetramer of two $\alpha$-haemoglobin and two $\beta$-haemoglobin molecules. In $\alpha$-thalassaemia, there is a large excess of $\beta$ chains, which form homotetramers, giving inclusion bodies of haemoglobin H in the red blood cells, which have much less oxygen-carrying capacity than normal and are small ($50$–$80 \, \mu\text{m}^3$ instead of 90) and reduced in number, giving severe anaemia. In $\beta$-thalassaemia, the excess $\alpha$ chains form homotetramers, but these are insoluble and precipitate, leading to destruction of the red blood cells in bone marrow and spleen. The red blood cells are again small and reduced in number.

$\alpha$-**thalassaemia**. Each chromosome 16 carries two $\alpha$-globin genes in tandem, *HBA1* and *HBA2*, so that a normal individual has four such functional genes, written $\alpha\alpha/\alpha\alpha$. If one of the four genes is mutant, $\alpha\alpha/\alpha$- (where "-" here means nonfunctional, not any allele), one gets the "**silent carrier**" state, which is virtually normal. There are two ways of having two genes inactivated, giving $\alpha$-thalassaemia 1, also called $\alpha$-**thalassaemia trait**, where symptoms are mild, although red blood cell volume is somewhat reduced. In Southeast Asia it is common to find $\alpha\alpha/$- -, with both genes on one chromosome inactivated, but "blacks" usually have one globin gene on each chromosome inactivated, $\alpha$-/$\alpha$-. If three genes are inactivated, $\alpha$-/- -, one gets **haemoglobin H disease** as there are four times more $\beta$ globin chains than $\alpha$ chains. Sufferers have moderate to severe, but not lethal, anaemia when born, often developing an enlarged spleen and bone deformities. The case with all four genes inactivated, - -/- -, gives **lethal Hydrops fetalis**, with stillbirth or neonatal death. When cases of that are detected early enough, blood transfusions have allowed their survival but continued care is needed.

*β*-**thalassaemia**. There is only one *β*-globin gene, *HBB*, on each chromosome 11, where mutations can affect the amount of functioning *β*-globin produced, e.g., from having different levels of transcription, from none to almost wild-type levels. More than 200 different defects are known, with six accounting for 90% of cases and being common in different areas (Thein, 2004; Chan *et al.*, 2004). Mutations include splicing sites, nonsense mutations, other base substitutions and frameshifts. Three levels of severity are recognised. In *β*-**thalassaemia minor**, a heterozygous condition, there are no or very minor symptoms and the genotype has one normal allele and one with no or a reduced function. In *β*-**thalassaemia intermedia**, individuals are anaemic, with symptoms such as an enlarged spleen and bone deformities, but do not usually need blood transfusions. Both *β*-globin genes are mutant, but one mutation is mild, so that a moderate but reduced amount of *β*-globin is produced. In *β*-**thalassaemia major** (Cooley's Anaemia), both alleles are mutant, with no or severely reduced function, usually with death by the age of two. If both are nonfunctional, with no *β*-globin and therefore no haemoglobin A, one gets lethal $\beta^0$-thalassaemia, but if one allele has some function, one gets $\beta^+$-thalassaemia.

People with *β*-thalassaemia major require frequent **blood transfusions** (red blood cells every two to three weeks, about 30 litres of blood a year) or matched bone marrow transplants, or the condition is eventually lethal. It is not usually marked at birth because foetal haemoglobin is still produced. As foetal haemoglobin production tails off during the first year of life, symptoms of severe anaemia slowly develop, together with weakened bones and enlargement of the liver and spleen. If transfusions are not given, death usually occurs within 10 years from anaemia, heart problems, weakness and infection. Progressive transfusions lead to a **build up of iron** which is eventually deposited in the heart, liver and other organs, and can lead to their failure, with death frequently occurring in the teens and twenties. Continuous infusion with iron-chelating agents, especially desferrioxamine (Desferal), requires the wearing of a pump for subcutaneous injection for 50 to 60 hours a week which is inconvenient but prolongs life (www.cooleysanemia.org). Similar problems arise for people with sickle-cell and other severe anaemias. In 2005, Novartis were testing a new once-daily iron chelator, Exjade®, for oral administration, not injection.

## 12.3.2. *X-linked genes*

X-linked recessive deleterious alleles typically affect males — because they are hemizygous — much more often than they affect females, who need two copies of the recessive alleles to show the condition (see Sec. 12.6 for equations). The rarer the recessive allele, the bigger the imbalance between the sexes. The severity of such conditions varies from mild for red-green colour blindness (8% of males), to eventually lethal for Duchenne muscular dystrophy (3 per 10,000 males). See Subsec. 7.1.6 for **fragile-X syndrome**, due to an unstable length mutation.

**Glucose-6-phosphate-dehydrogenase (G-6-P-D)** is an enzyme necessary for the stability of glutathione, which plays a role in cellular respiration. A greatly reduced activity for this enzyme is rare in "whites", but occurs in about 9% of American "blacks" and in some people of Mediterranean origin; heterozygosity for the deficiency confers some malaria resistance as mentioned in Subsec. 12.3.1. Hemizygous males and homozygous recessive females for this X-linked trait are phenotypically normal most of the time, with the condition only showing in enzyme tests. They become ill, with the sudden destruction of many red blood cells giving severe haemolytic anaemia, when they inhale pollen of broad beans (*Vicia faba*) or eat the raw beans (favism), or on taking certain drugs such as sulphanilamide, the antimalarial primaquine or naphthalene (used in moth balls). They recover when the causative agents are removed. The heterozygous females usually have intermediate enzyme levels, but their level may vary from low to normal. There are many different alleles at this locus, unlike with sickle-cell trait. Its incidence has huge geographical variation, from 0 to 6,500 per 10,000 males.

Another X-linked condition is **Duchenne muscular dystrophy** (UK incidence, 3 per 10,000 males), giving progressive muscular wasting. Affected children are usually confined to wheelchairs by the age of 10, with death before 20. Although the functions of the large dystrophin protein are understood and the gene has been cloned, there is no satisfactory treatment at present, although gene therapies are being tested.

**Haemophilia A** is an X-linked recessive condition resulting in spontaneous and wound-induced bleeding in soft tissues and joints, with very poor clotting, so it can be fatal. The gene codes for clotting Factor VIII and is large, about 200 kb long. Factor VIII levels in sufferers vary from 30% of normal (mild symptoms) to less than 1% (severe). One particular inversion (see Lakich *et al.*, 1993) accounts for half of all severe cases and for 20% of all cases. Different missense, nonsense, frame shift and deletion mutations of varying severity account for the other haemophilia A cases. Carrier detection and prenatal diagnosis are possible by blood or DNA analyses. It affects 1 to 2 males per 10,000 and accounts for 85% of haemophiliac patients. Treatment with intravenous Factor VIII is effective. Heterozygous females often have less Factor VIII than normal but are usually symptomless. **Haemophilia B** has similar symptoms to haemophilia A and is also an X-linked recessive, but comes from mutations in a different gene, giving a deficiency of clotting Factor IX. It affects 1 in 30,000 males and is treated by intravenous Factor IX injections. In cats, there are very similar X-linked recessive haemophilia A and B conditions.

**Testicular feminisation** results in individuals who are genetically XY developing as females, with excellent breast development and female external genitalia with a short vagina, but they are sterile, with small internal testes. They often marry as females, unaware of their condition. The single X carries a recessive allele resulting in insensitivity to male hormones. It is difficult to estimate their frequency as they are usually only detected if they report their lack of menstruation and are then karyotyped as XY, 2n = 46. The probable frequency is about 1 in 62,000 XY births.

## 12.4. Polygenic and multifactorial disorders

**Polygenic disorders** or multifactorial disorders arise from the interactions of genes at many loci, some of which may have major effects but many of which have individually small effects. There are often effects of the environment in the aetiology too, so the term "multifactorial" is used. The terms "polygenic" and "multifactorial" are sometimes used interchangeably, although "polygenic" is sometimes used for traits controlled only by polygenes, and "multifactorial" for ones with environmental effects too. There may also be incomplete penetrance and/or variable expressivity (Subsecs. 3.2.4 and 3.2.5), making their genetics difficult to be sure about. An example of a polygenic trait presumed to have little environmental effect is the total number of fingerprint ridges counted over all ten fingers, which is determined early in development and which gives a roughly normal distribution of values.

**Multifactorial disorders** include club foot, cleft lip and cleft palate, coronary artery disease, most congenital heart diseases, hypertension, diabetes mellitus and schizophrenia. They account for about one third of paediatric hospital admissions and about a third of childhood mortality.

Many multifactorial traits give qualitative phenotype variation, where an individual has the trait or does not have it, though there may be different extents of expression amongst those with the trait. For example, babies either have a cleft lip or do not, but those who have it have it to different extents, and some have a cleft palate as well. People at a given time either have or do not have schizophrenia or insulin-dependent diabetes mellitus (type 1 diabetes). For such qualitative traits with multifactorial control, it is thought that there is a more-or-less continuously distributed **genetically determined predisposition** to that trait, with additive effects of polygenes and certain environmental factors, with only individuals exceeding some threshold value developing the symptoms.

As an affected individual usually has a genotype with a high predisposition to the condition, his or her genetic relatives are likely to have a greater predisposition to the condition than are unrelated individuals, with the risk of being predisposed being greatest for the most closely related relatives. Suppose that for a given level of environmentally predisposing factors it requires say 10 cleft-determining polygenes to cause a **cleft lip**, and 20 cleft-determining polygenes to give a cleft palate as well, then the risk of siblings of someone with a cleft lip and a cleft palate having just a cleft lip will be higher than the risk of those siblings having a cleft lip and a cleft palate. The actual risk of a cleft lip in siblings of children with just a cleft lip is 2.5%, but is 6% in siblings of children with a cleft lip and a cleft palate. In contrast, with an autosomal recessive single-gene condition such as sickle cell anaemia or cystic fibrosis, the risk of a sibling of an affected child being affected is one quarter (if both parents are unaffected heterozygotes), and for a dominant autosomal gene like polydactyly, the risk is one half (if one parent is heterozygous and the other is homozygous recessive). For a number of multifactorial diseases, including congenital heart disease and insulin-dependent diabetes mellitus, the risk of a sibling of an affected child being affected (called the **recurrence rate**) is roughly 5% or less. See Gelehrter *et al.* (1998), Chap. 4, for further details, Plate 12.1 and Vieira *et al.* (2005).

**Multifactorial diseases** are much more common than **single-gene diseases**. In the UK, with about 60 million people, multifactorial disease sufferers number about 400,000 for Alzheimer's, 1 million for diabetes, 1.25 million for ischemic heart disease, 1.5 million for cancer, and 5 million for hypertension, compared with 2,500 for Huntington's, 3,000 for Duchenne muscular dystrophy, and 7,000 for cystic fibrosis single-gene diseases. In the USA, with about 297 million people, there are about 1 million with insulin-dependent diabetes and 18 million with non-insulin-dependent diabetes.

**Diabetes** is caused by a deficiency of insulin or a loss of response to insulin, so blood glucose concentrations rise, leading to glucose excretion in the urine, excessive urine production, persistent thirst, poor carbohydrate utilisation and weight loss. Diabetes is familial and can lead to hypoglycaemia and coma, kidney failure, blindness, and other long-term complications. The three main types of diabetes have a combined incidence of about 1 in 50, varying with age. **Insulin-dependent diabetes mellitus** (IDDM, type 1 diabetes) affects 1 in about 400 children with a peak age of onset of 12 years, and also affects adults. It is an autoimmune disease, with T-lymphocytes attacking the pancreas $\beta$-cells which make insulin. It is polygenic, with at least 10 contributing loci, especially an HLA gene on chromosome 6 and the IDDM2 insulin gene on chromosome 11. Treatment consists of several insulin injections each day, plus dietary restrictions. **Non-insulin-dependent diabetes mellitus** (NIDDM, type 2 diabetes, adult-onset diabetes) is caused by either insufficient insulin, or more commonly, a reduced responsiveness to insulin in target cells, from changed receptors. It usually occurs after about age 40, becoming more frequent with increasing age, and is more common in those who had low birth weight (under 2.5 kg). More than 90% of diabetics are type 2 and most can control it by diet and exercise alone. Both type 1 and type 2 are multifactorial: see Table 12.2 for their different concordance scores. A third type of diabetes, **maturity-onset diabetes of youth**, is usually inherited as an autosomal dominant.

## 12.5. X-inactivation and Barr bodies

In diploid mammalian females, one of the two X chromosomes in a cell is inactivated, forming a small, flat structure within the nucleus, a **Barr**

**body**, stainable with dyes for DNA. The inactivation occurs early in the embryo by DNA methylation and the inactivation persists through subsequent cell divisions. In humans, inactivation generally occurs around day three, at about the 8-cell stage. It is a matter of chance which of the two X chromosomes is inactivated in any embryonic cell, so the resulting female is a mosaic of patches of tissue, with one X inactivated in some patches and the other inactivated in other patches. See Chow *et al.* (2005).

This is very obvious in tortoiseshell and calico cats (Plate 12.5), which are females heterozygous for black (*o*) and orange (*O*) alleles at a locus for fur colour; some patches of fur are orange (black-carrying X inactivated)

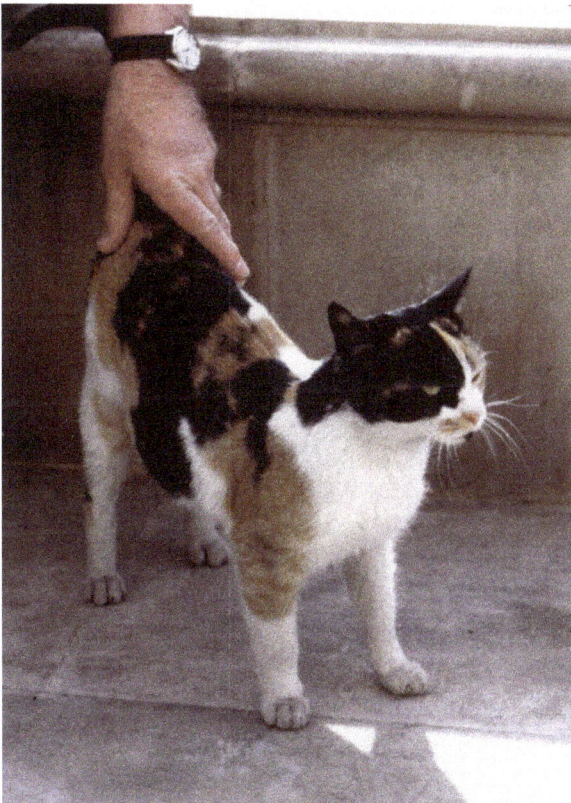

**Plate 12.5. A calico female cat, showing X-inactivation.** She is heterozygous for X-linked alleles *O* (orange) and *o* (black), with different X's inactivated in different body regions. The white areas are controlled by a different system.

and others are black (orange-carrying X inactivated). In calico cats with white spotting, the white spotting causes the orange and black regions to form large patches, with more white causing larger patches. In contrast, in cats without white spotting, the orange and black hairs are intermingled in a brindled appearance, with a few patches of solid colour. Although it is said that calico cats must be female, about 0.6% of calicos are male. One male calico condition is the XX(*Oo*)/XY mosaic which is usually sterile. Most are the equivalent of human Klinefelter syndrome, sterile with XX(*Oo*)/Y. There are also rare X(*O*)Y/X(*o*)Y mosaics, usually fertile orange tabbies with variable sized patches of black tissue, from somatic mutation from the orange coat gene, *O*, to black, *o*, during development. They usually breed as normal orange tabbies (Robinson, 1991).

Human females heterozygous for X-linked **anhidrotic ectodermal dysplasia** (lack of sweat glands) have been shown to be external mosaics, with affected patches sometimes covering the whole face and neck, or one shoulder, or part of one thigh or most of one lower leg, with oval or irregular patches varying from a few centimetres to over 70 cm long. Other cases of detectable "**patchiness**" in females heterozygous for X-linked conditions include patchy retinal pigmentation for retinitis pigmentosa and choroideraemia, and patchy muscle biopsy results for Duchenne muscular dystrophy.

X-inactivation helps to solve the problem of **unequal gene doses** for X-linked loci where the female has two copies of each locus (but only one active copy) while the hemizygous male only has one copy. Thus, for red blood cell **glucose-6-phosphate-dehydrogenase activity** (see Subsec. 12.3.2), the average amounts of this X-coded enzyme are the same in hemizygous males as in females with two copies of the gene. There is only one active X per cell, so an individual with one X has no Barr bodies (normal males and Turner syndrome females — see Sec. 12.8), one with 2 X has one Barr body (normal females and XXY males), an individual with three X has two Barr bodies, etc. The presence or absence of Barr bodies means that normal females and normal males can be sexed even in non-dividing cells just from the number of Barr bodies. This has been used a sex-test for athletes, being less intrusive and embarrassing than a genital inspection. It can also be used for immediate sexing of foetuses in prenatal diagnosis (Sec. 12.12). X-inactivation is not always random if one of the chromosomes carries an aberration.

Although most of one X is inactivated in human females, controlled by the X-inactivation centre in the proximal part of the long (q) arm, **some parts are not inactivated**. These include the pseudoautosomal terminal region of the short (p) arm which is homologous to and at meiosis pairs with part of the Y chromosome. A few other loci are not inactivated, including the Xg blood group and steroid sulphatase loci. Carrel and Willard (2005) found that about 65% of X-linked genes were inactivated in all heterozygous samples; a further 20% were inactivated in some but not all samples, and 15% escaped inactivation in all samples. Surprisingly, most genes that escaped inactivation, including those in the pseudoautosomal region, were not fully expressed from the inactivated X. The proportion of genes escaping inactivation differed dramatically in different regions. Approximately 10% of X-linked genes showed variable patterns of inactivation, expressed to different extents in different inactive Xs. There was a large degree of expression heterogeneity between different females.

The 155 Mb X chromosome has **1,098 functional genes**, only 54 of which have counterparts on the Y, which has fewer than 100 functional genes. The gene density is low on the X. More than 300 diseases have been mapped to the X. The **gene order** on the human X is virtually identical to that of the dog, while rodent sequences show some rearrangements relative to the human sequence. Most genes from the human short arm are found on the chicken chromosome 1, and those from the human long arm on chicken chromosome 4.

Due to vagaries in the extent of expression of X-linked genes in heterozygous females, the terms "dominant" and "recessive" are less clear cut for X-linked genes than for autosomal loci. For example, haemophilia A is normally classed as a recessive condition, but a female heterozygote for this X-linked condition may have a mild bleeding tendency and lower than normal Factor VIII concentrations because the X carrying the wild-type allele has been inactivated in most of her cells, not just in half of them.

## 12.6. Sex differences in disease susceptibility

### 12.6.1. *X-linked diseases*

With X-linked recessive diseases, one expects **more male sufferers** than female sufferers, because males, being hemizygous, only need the one recessive-carrying X, while females need two recessive-carrying X chromosomes to be affected. One needs modified Hardy–Weinberg equilibrium equations to take account of hemizygous males. With a recessive allele $a$, frequency $q$, one gets equilibrium genotype frequencies for females of $A\ A$, $p^2$; $A\ a$, $2\ pq$; $a\ a$, $q^2$, but for males one gets $A\ (Y) = p$, and $a\ (Y) = q$. The ratio of female to male sufferers is therefore $q^2$: $q$, or $q$ female sufferers to one male sufferer. Thus the rarer the recessive allele, the bigger the proportional difference between male and female sufferers. For **red-green colour blindness**, about 8% of males are red-green colour blind, so $q = 0.08$ and $q^2 = 0.0064$, with only 0.64% of female sufferers, with an imbalance of 1 male sufferer to only 0.08 female sufferers, a **12.5-fold** difference. With a much rarer condition, **haemophilia A**, an inability to clot blood in wounds, the frequency of male sufferers is about 1 in 10,000, so $q = 0.0001$, and the frequency of female sufferers expected is 0.00000001, or 0.000001%, or 1 in 100 million. One thus expects about one quarter of a suffering female in the whole of Britain, but about 2,800 suffering males, a **10,000-fold** sex difference. Female haemophiliacs are thus expected to be extremely rare, but in reality have a frequency about 4% of that of male sufferers. This is because homozygotes arise from consanguineous relationships (so Hardy–Weinberg frequencies are not obtained), and because of the vagaries of X-inactivation as mentioned above, so that at the extreme the cell line with one particular active X predominates. Female sufferers may also come from having chromosome rearrangements involving the X chromosome, such an X-autosome translocation. Female haemophiliacs do not die from excessive bleeding at menstruation because menstruation ends by blood vessel contraction, not clotting; they can have children (all sons will be sufferers) but close medical attention at birth is essential to control bleeding.

One can obtain the frequency of the recessive X-linked allele directly from the frequency of male sufferers, which is $a$. If calculating **overall allele frequencies** over both sexes for X-linked loci, $p = \frac{p_{males} + 2p_{females}}{3}$ if males and females are equally frequent. Thus, if we had a non-equilibrium population, 0.2 $A$ (Y), 0.8 $a$ (Y) in males and 0.2 $A\ A$, 0.6 $A\ a$, 0.2 $a\ a$ in an equal number of females, then $p$ for males is 0.2 and $p$ for females is 0.5, so the overall $p = \frac{0.2 + (2x0.5)}{3} = 0.4$, when at equilibrium we would expect 0.4 $A$ (Y) and 0.6 $a$ (Y) males, and 0.16 $A\ A$, 0.48 $A\ a$, and 0.36 $a\ a$

females. Allowances need to be made if there is an unequal number of the two sexes.

## 12.6.2. *Non-X-linked diseases*

There are a large number of diseases or conditions which are not X-linked but which show different incidences in the two sexes. Many are multifactorial traits, or have low heritabilities, with large environmental influences. Diseases **more common in males** than in females, with the preponderances as shown, include: alcoholism, 6-1 (i.e., six times more male sufferers than female ones); amoebic dysentery, 15-1; angina pectoris, 5-1; Asperger syndrome, 9-1; autism, 4-1; cancer of the skin, 3-1; cirrhosis of the liver, 3-1; coronary sclerosis, 25-1; duodenal ulcer, 7-1; gout, 49-1; hernia, 4-1; leukaemia, 2-1; pleurisy, 3-1; sciatica, large; suicide, 13-9. Diseases with **female preponderances** include: anaemia, very high; cancer of the genitalia, 3-1; cancer of the gall bladder, 10-1; gall stones, 4-1; hyperthyroidism, 10-1; bunions, 9-1 in USA; influenza, 2-1; rheumatoid arthritis, 3-1; varicose veins, fairly large; whooping cough, 2-1 (data mainly from Montagu, 1963).

The actual values for these preponderances often vary over time and between populations, social classes and different occupations. **Social factors** are involved as well as biological and purely genetic factors. Exposure during jobs or pastimes to environmental mutagens or infectious agents is involved in some of these differences, such as some largely male manual occupations involving exposure of the skin to oils containing carcinogens. Mothers may often be more exposed to childhood diseases than fathers, through having longer and closer contact with children, which might explain the influenza and whooping cough sex differences. The amounts of stress, exercise, and diet, will affect some of these conditions, but the large excess of females with gall-bladder cancer is difficult to explain. Role-playing by some males as the hard-drinking, go-getting, aggressive stereotype might partly account for the alcoholism difference.

**Autism** affects about 2 in 1,000 individuals, with four times more male sufferers than female ones but with no evidence of X-linkage. It starts before age three, causing poor social interactions and communication, and repetitive behaviour. Lamb *et al.* (2005) found that it had a complex genetic determination, with sex-limited (to males) susceptibility loci on chromosomes 7,

15 and 16. The effects of three other loci depended on whether they had been inherited from the male or female parent.

## 12.7. Causes of human mutation; cancer

Humans suffer mutations from all the causes of spontaneous mutation listed in Chap. 7, plus additional ones such as medical X-rays. The **mutation rate**, $\mu$, is the frequency of mutations per locus per gamete per generation. It is most easily calculated for rare autosomal dominant mutations, for which $\mu = n/2N$, where $n$ is the number of affected patients and N is the total number of births: one uses 2N because mutation at either allele at an autosomal locus could cause the mutant phenotype. The frequency of X-linked recessives can be estimated from the sons of the women in which the mutation occurred.

For most human genes, **typical mutation rates** from wild-type to mutant are between $10^{-6}$ and $10^{-5}$ per locus per gamete per generation, averaging 10–20 per million gametes, although values go up to $10^{-4}$. Higher values are common for larger genes, which present a bigger target. Thus there are high mutation frequencies for neurofibromatosis (which has more than 300 kb of DNA and its NF-1 protein has more than 2,000 amino acids), haemophilia A and for X-linked Duchenne muscular dystrophy, where the gene is more than 2,400 kb long, the longest human gene known. **Mutational hot-spots** within genes include GC dinucleotides, which have 12 times the mutation frequency of other dinucleotides. Some typical spontaneous mutation frequencies are shown in Table 12.4, with some differences between studies.

As shown in Table 12.5, mutation frequencies rise strongly with increasing **age of the male parent**, being 11 times higher in men of 50 than in men of 25. There is little change with the age of the female parent. This is because sperm progenitor cells undergo DNA replication many times in the adult, so those of older men have had many more chances of replication errors than those of younger men; the female reproductive cells finish their mitotic divisions before birth, with many fewer divisions than male reproductive cells.

Most mutagens are also **carcinogens**, so anything causing mutation in humans will probably also cause cancer. According to Morgan (1989), the

**Table 12.3.** **Familial incidence of cleft lip**. Data from Carter (1969).

| Relatives | Percentage of affected relatives | Incidence relative to the general population |
|---|---|---|
| **First degree:** | | |
| Full sibs | 4.1 | ×41 |
| Parents/children | 3.5 | ×35 |
| **Second degree:** | | |
| Aunts and uncles | 0.7 | ×7 |
| Nephews and nieces | 0.8 | ×8 |
| **Third degree:** | | |
| First cousins | 0.3 | ×3 |

**Table 12.4.** **Spontaneous mutation frequencies in human genes**. Largely from Winchester and Mertens (1983) and Cummings (2000).

| Trait | Mutations per million gametes |
|---|---|
| **Autosomal dominants** | |
| Achondroplastic dwarfism | 42 |
| Aniridia (absence of iris in eyes) | 2.6–5 |
| Epiloia (butterfly rash on face) | 6 |
| Marfan syndrome | 5 |
| Multiple polyposis (polyps in colon, often becoming cancerous) | 13 |
| Neurofibromatosis | 50–100 |
| Osteogenesis imperfecta | 10 |
| Polycystic kidney disease | 60–120 |
| Retinoblastoma (tumours on retinas) | 6–23 |
| **Autosomal recessives** | |
| Albinism | 28 |
| Epidermolysis bullosa (skin blisters) | 50 |
| **X-linked recessives** | |
| Haemophilia A | 32 |
| Muscular dystrophy, Duchenne | 43–100 |

**Table 12.5.**   Comparative mutation frequencies in sperm from men of different ages, taking the population average value as 100%. Calculated from a graph in Pritchard and Korf (2003, p. 64).

| Male age, years | 25 | 30 | 35 | 40 | 45 | 50 |
|---|---|---|---|---|---|---|
| Mutation frequency in sperm, relative to the population average taken as 100 | 43 | 63 | 106 | 171 | 291 | 469 |

approximate proportions of **risk of cancer** attributable to different classes of environmental agent were: tobacco, 30%; alcohol, 3%; diet, 35%; food additives, < 1%; sexual behaviour, 7%; occupation, 4%; pollution, 2%; industrial products, <1%; medicines and medical procedures, 1%; geophysical factors, 3%; infection, possibly 10%; unknown, possibly 3%. Morgan (1989) gave the following percentages of average **exposures to radiation** of people in the UK: radon, 32%; gamma rays, 19%; internal, 17%; cosmic rays, 14%; medical, 12%; thoron, 5%; atomic fallout, 0.5%; occupational, 0.4%; nuclear discharge, 0.1%, with 87% being natural and 13% being artificial.

Figures for the United States from the National Radiation Council, Committee on the Biological Effects of Exposure to Low Levels of Ionizing Radiation, 1990, were, in millisieverts, where 1 mSv = 0.1 rem: radon gas, 2.06; cosmic rays, 0.27; natural radioisotopes in the body, 0.39; natural radioisotopes in soil, 0.28, making a total for natural radiation of 3.00 mSv; medical diagnostic X-rays, 0.39; radiopharmaceuticals, 0.14; consumer products including TV, clocks and building materials, 0.10; fallout from weapons tests and nuclear power plants were both less than 0.01, making a total from non-natural sources of 0.63 mSv, and a grand total from all sources of 3.63 mSv. See also Subsec. 12.7.3.

While the exact figures differ between different surveys and in different countries, it is clear that some causes of cancer and mutation are unavoidable, such as those from cosmic rays; others can be controlled to some extent by choice of human behaviour, or by using appropriate safety procedures. Some, such as exposure to radon gas, are determined by local geology, with Cornwall in Britain having unusually high levels of radon gas given off by the rocks.

## 12.7.1. *UV light*

**UV light** is not a source of human germ-line mutations because it cannot penetrate to the sperm or eggs. It does however cause many somatic mutations in exposed skin, often leading to **skin cancer**. There are large **racial differences** in UV-induced skin cancer, controlled largely by skin colour, where the melanins in yellow, brown and black skins give considerable protection and where sun-tanning of white skins is a phenotypically plastic (and usually reversible) response. For example, in America the annual incidence of **malignant melanoma** in the late 1970's was 4.2 per 100,000 for "whites" and only 0.6 per 100,000 for "blacks", with the greatest incidence in both groups in the sunnier states. In the UK, skin cancer is the fastest increasing form of cancer with the number of cases doubling in the past 20 years: each year there are now about 60,000 new cases of non-melanoma skin cancer and 7,000 cases of melanoma skin cancer, causing about 1,600 deaths a year. Increased travel abroad, with more prolonged sunbathing, has been blamed.

Oettlé, head of the National Cancer Association of South Africa, made extensive studies on the different **racial groups** in that country. Skin cancer was 35 times more frequent in "whites" than in "blacks", although the groups were not matched for equal sun exposure. Cancers of the colon and breast were respectively 10 and 5 times more frequent in "whites" than in "blacks", with the former perhaps being diet-related and the latter perhaps being UV-related. Womb cancer was 6 times more frequent in "blacks" than in "whites", possibly related to a greater average number of sexual partners and oncogenic sexually transmitted viruses. Liver cancer was 20 times more frequent in "blacks" than in "whites", possibly from "blacks" eating more food with fungal infections, where fungal aflatoxins are liver carcinogens. There may also be different genetic susceptibilities to various cancers in different races. Subsection 7.3.2 described **xeroderma pigmentosum**, an autosomal recessive disease causing multiple skin cancers because sufferers have impaired excision repair of UV damage to DNA.

## 12.7.2. *Chemicals*

**Diet** influences cancer and germ-line mutations. Tea, coffee and most cola drinks contain caffeine, which is a natural base analogue: as well as being a mutagen (sometimes used in labs for mutating micro-organisms), it can inhibit the enzymes which repair DNA damage. Fortunately for us, it is less mutagenic in mammals than in micro-organisms. Caffeine induces mutations in human tissue culture cells, and there is evidence for germ-line mutations. In one survey, Americans drinking eight or more cups of coffee a day had significantly fewer sons, relative to daughters, compared with the rest of the population. Lethal induced autosomal recessives would not show in the next generation because they would be heterozygous, but lethal X-linked recessives will kill hemizygous sons but not heterozygous daughters, affecting the sex ratio.

There are many other dietary potential mutagens and many possible **chemical mutagens in the environment** which may be picked up through touching, breathing, eating and drinking. Some have been identified, such as benzypyrenes in charcoal-grilled steaks, diesel exhaust particles and tobacco smoke. There was high incidence of lip cancer in smokers who used clay pipes, because of a carcinogen in the clay, and various oils used in industry contained carcinogens. Smoking is responsible for about 15–20% of all deaths in Britain, causing deaths mainly through cancer and effects on the circulatory and respiratory systems.

In India, Taiwan and Sri Lanka, and in some British Asians, **chewing betel** is very common, with the areca nut often accompanied by any of lime, tobacco, betel leaf or betel inflorescence. This palm nut contains arecoline, a stimulant. According to Chang *et al.* (2004), more than 153,000 new cases a year of mouth cancer are recorded in Asia, with mouth cancers being up to half of all malignant tumours in countries where betel chewing is common. Approximately 400 million Asians chew betel. Wu *et al.* (2001) showed in Taiwan that heavy consumers (averaging 20 or more betel quids a day for 20 or more years) had a 9.2-fold greater risk of squamous cell oesophageal cancer than non-chewers, and light users had a 3.6-fold increased risk. Chang *et al.* (2004) looked at the incidence of mouth cancers in male areca chewers in relation to polymorphisms for a $(GT)_n$ microsatellite repeat in the heme oxygenase-1 (HO-1) gene's promoter. The shorter repeat allele ($< 26$ repeats) appeared to be protective against areca nut oral squamous cell carcinoma, compared to the medium repeat allele (26 to 30 repeats), while the long repeat allele ($>30$ repeats) was associated with a higher oral cancer risk. There is therefore a genotype (for *HO*) $\times$ environment (amount of exposure of the mouth cells to areca carcinogens) interaction in the frequency of oral cancer in Asia. Adding tobacco to the betel increases the cancer risk.

Probable **dietary mutagens** include heterocyclic amines (in meat and fish cooked at high temperatures); acrylamide (starchy foods after frying); polycyclic aromatic hydrocarbons (from vegetable oils, smoked and grilled foods, especially red meat); nitrate- and nitrite-derivatives in food (vegetables, fertilizer residues) or added to them (cheese, bacon, preserved meats, salami) can react with amines to give carcinogenic nitrosamines; fungal toxins (e.g., from *Aspergillus flavus*, giving aflatoxins) and [226]radium in bottled mineral water: see Sugimura (2000).

**Environmental mutagenesis** is a very active research field, with many man-made products being tested each year for mutagenicity and carcinogenicity (carcinogens are usually also mutagens, and *vice versa*). One testing system uses the induction of resistance to 8-azaguanine in human tissue culture. In the **Ames test for mutagenicity**, however, quantitative studies are made of the induced reversion to wild-type of various histidine-requiring mutations (in genes *hisG*, *hisC*, *hisD* and *hisO*) of the bacterium *Salmonella typhimurium*: see Subsec. 7.2.2 and Plate 7.2. Both frame shift mutations (revertible by frame shift-causing mutagens) and missense and nonsense base substitution mutations (revertible by base substitution-causing mutagens) are used, together with other mutations which inactivate excision-repair or make the cells more permeable to large molecules. Some strains have plasmids to increase the activity of inaccurate repair mechanisms. As some chemicals become mutagenic only after being metabolised in the liver, the medium incorporates a rat-liver extract and NADP as an energy source. Mammalian oxidation or reduction systems can be used. See Mortelmans and Zeiger (2000).

Many substances have been tested using the Ames test, including industrial chemicals, cosmetics, hair dyes, food additives and pesticides, resulting in bans on certain chemicals and the reformulation of certain cosmetics and hair dyes to reduce their mutagenicity. Tests for direct carcinogenicity, by assessing tumour formation in laboratory animals, are done but are much slower and more expensive than the mutagenicity tests.

## 12.7.3. *Ionising radiations*

These cause mutations, cancer and leukaemia. When ionising radiation strikes water or living cells, highly reactive **free radical ions** are produced and react with DNA and other molecules. Single-strand breaks in DNA are usually repaired efficiently off the other strand but double-strand breaks can cause chromosome aberrations. Free radicals can damage nucleotide bases and cause point mutations in various ways. In human cells in tissue culture, a dose of 0.2 Sieverts gives about one visible chromosome break per cell.

**X-rays** and **gamma rays** are measured in Roentgen (R), where a 1 R dose gives $2.08 \times 10^9$ ion pairs per $cm^3$ of dry air at 0°C and 1 atmosphere pressure. Taking a generation to be 30 years, the natural ionising radiation received per person per generation is about 3 to 5 R, with about 0.8 R from **cosmic rays**, 0.7 R from radioactive elements within our bodies (e.g., potassium isotopes), and 1.5 R from soil and rocks. Radon gas gives about 2,500 cancers a year in Britain. Medical X-rays average about 3 R per generation, with typical doses to the male gonads per irradiation being 0.4 R for a pelvic X-ray if the scrotum is unprotected, but only 0.03 R if protected, 0.001 R for a chest exposure and 0.0001 R for the head, but these figures are decreasing with better application and protection methods. Fallout from atomic weapons testing accounts for only 0.1 R. Other figures were given earlier in Sec. 12.7. A cancer radiotherapy course might involve about 5000 R, with an extra 6000 R for a booster course.

Radiation is also measured in microsieverts (1 Gray = 1 Sievert, and 100 rads = 1 Gray; rads and Roentgens are similar, with 100 R of hard X-rays corresponding to about 93 rads, and 1 rad = 100 ergs of energy per gram of material irradiated), where natural exposure is about 5.5 microsieverts per day. X-ray doses of various treatments, in microsieverts, are about 20 for dental X-rays, 60 for the chest, 100 for joints and limbs, 2,400 for the spine, 1,000 for the lumbar area, 300 for the hip, 5,000 for the stomach and duodenum with a barium meal, 1,000 for the skull, 9,000 for the large bowel, compared with about 100 for a return flight from London to New York in a standard jet, and more for Concorde as it flew at a higher altitude.

In 2005, it was reported that airline pilots were three times more likely than other people to suffer from "nuclear cataracts", those in the centre of the lens. Professor Rafnsson from the University of Iceland said that pilots have more chromosomal abnormalities and skin cancers than others, and he attributed the greater risk of cataracts to **cosmic radiation**, which is 100 times stronger at 36,000 feet (10,973 m) than at sea level.

There is **no "safe level"** for ionising radiations, as damage is approximately proportional to dose over a wide range of doses, with no threshold level. When it was found that X-rays cause temporary male and female sterility, they were actually used on human male gonads as a **contraceptive** but this was quickly abandoned when their harmful effects were found. In about 1950 in Britain, X-ray machines were used in shoe shops to see how well feet fitted inside shoes, with the X-rays directed up through the feet to a green fluorescent screen. They were rapidly abandoned when they were found to harm customers and shop assistants.

In 2005, the radiation protection division of the Board of the Health Protection Agency in the UK reported that the **average annual dose of ionising radiation** to humans had risen from 2.6 milliSieverts in 1999 to 2.7 mSv in 2005, probably resulting an extra 100 UK deaths from cancer a year. New medical diagnostic methods such as computed tomography were responsible. Medical uses of radiation were the largest man-made causes of radiation, about 15% of the total, but the other 85% was natural, with more than 50% from radon gas in buildings.

## 12.7.4. Temperature

Humans have constant-temperature devices such as sweating, but **fevers** can cause hyperthermia, and prolonged exposure to low temperatures can cause hypothermia. In *Drosophila melanogaster*, a rise of 3.3° increases the spontaneous mutation rate by 85%, equivalent to a radiation dose of 40 R. **Wearing trousers** raises the scrotal temperature in man by about 3.3°. If this raises the mutation rate by as much as it does in the fruit fly, wearing trousers could have genetic hazards 10 to 1,000 times those of all sources of radiation, and possibly be a factor in the low active sperm count in man. Central heating and hot baths or showers also raise scrotal temperatures. Men suffering from low fertility have been advised not to wear tight-fitting underwear, and have been recommended to spray their testicles with cold water.

## 12.7.5. *Infection*

Some **viral diseases** such as measles can cause extensive chromosome breakage. That can be mutagenic and cause chromosome aberrations. There are also **oncogenic** (cancer-causing) **viruses**, and various **transposable elements** can interrupt coding sequences, causing mutations. Viruses which can cause cancer include the DNA-based papilloma, hepatitis B, and Epstein-Barr (which is often involved in Burkitt lymphoma) viruses, and the RNA-based retroviruses such as T-cell leukemia virus and Kaposi sarcoma-associated herpes virus.

## 12.7.6. *Cancer genes*

The **development of cancer**, caused by the loss of control of cell proliferation (mitosis) or of cell death (apoptosis), is usually a multi-stage process involving the accumulation of mutations in genes controlling cell division or programmed cell death. Thus, "normal" division-control genes can be **proto-oncogenes**, such as those of the signal transduction cascade whose action helps convey the effect of growth promoters on the cell surface to the DNA in the nucleus to advance the cell cycle. Proto-oncogenes can be changed to **oncogenes** by mutation, with 30% of tumours carrying mutations in *ras* genes; by translocation, as in the origin of the Philadelphia chromosome by a defined reciprocal exchange between chromosomes 9 (giving *abl*) and 22 (giving *BCR*), where the *BCR/abl* fusion protein causes **chronic myeloid leukaemia (CML)**, with increased tyrosine kinase activity; by insertional mutagenesis by oncogenic retroviruses; by amplification, where 10% of tumours have amplified copies of a proto-oncogene making a "double-minute" supernumerary chromosome, or there may be amplified proto-oncogenes within ordinary chromosomes. **Loss of apoptosis factors** can also cause tumours, e.g., with mutation of *Bcl2* as in some lymphomas. Chronic myeloid leukaemia usually gives death within four to six years, has a frequency of about 1 in 100,000 people a year, and accounts for about 15% of adult leukaemias. See Goldman (2004) for imatinib treatment of CML.

Cancers are also caused by mutations in **tumour-suppressor genes**, with deletions and point mutations in *P53* occurring in 70% of tumours. This gene's product is involved in the G1 block in the cell cycle, which allows DNA damage to be repaired; it is also involved in apoptosis. **Retinoblastoma** cancer of the eye occurs in two forms. It can be in one eye only, sporadic, with an average age of onset of 30 months, or in both eyes, familial, average age of onset of 14 months, often with other cancers developing. The wild-type allele, $RBl^{1+}$, is a tumour suppressor blocking mitosis in retinal cells, and it has been proposed that both alleles need inactivating for this cancer to develop. In the familial retinoblastoma, one mutation in this gene can be inherited, so that only the remaining one wild-type allele in any retinal cell needs to be mutated to cause cancer, hence its earlier occurrence, and in both eyes, compared to the sporadic type.

## 12.8. Chromosome number abnormalities and chromosome aberrations

### 12.8.1. *Human chromosome methods*

For studying human chromosomes, one needs only a few drops of blood, say from a fingertip, heel or ear prick. The blood is cultured at 37° with a bean extract containing phytohaemagglutinin, plus nutrients and antibiotics such as penicillin and streptomycin. During three to four days of culture, the nucleated white blood cells (leukocytes, T-lymphocytes) divide by mitosis, then are arrested in metaphase by the addition of colchicine or colcemid. The cells are allowed to swell in hypotonic solution to get good separation of the chromosomes, then are fixed and spread to air-dry on slides. They are finally stained for DNA, e.g., with Giemsa stain.

That procedure gives stained but unbanded chromosomes which can be photographed. The photos can be cut up to make **karyotypes** in which the chromosomes are arranged in order of size and centromere position (see Plate 12.6 for a banded example of a human karyotype). Metacentric chromosomes have the centromere near the middle; submetacentrics have the centromere off-centre and acrocentrics have the centromere near one end. Telocentrics, with the centromere virtually at the end, are not found in normal human karyotypes. With unbanded chromosomes, one could identify about 10 groups of chromosomes based on size and centromere position, e.g., pair 1, the largest metacentrics, pair 2, the largest submetacentrics, etc. Metaphase lengths vary from about 2 $\mu$m (40 Mb of DNA) for chromosome 21 to about 10 $\mu$m (200 Mb) for chromosome 1.

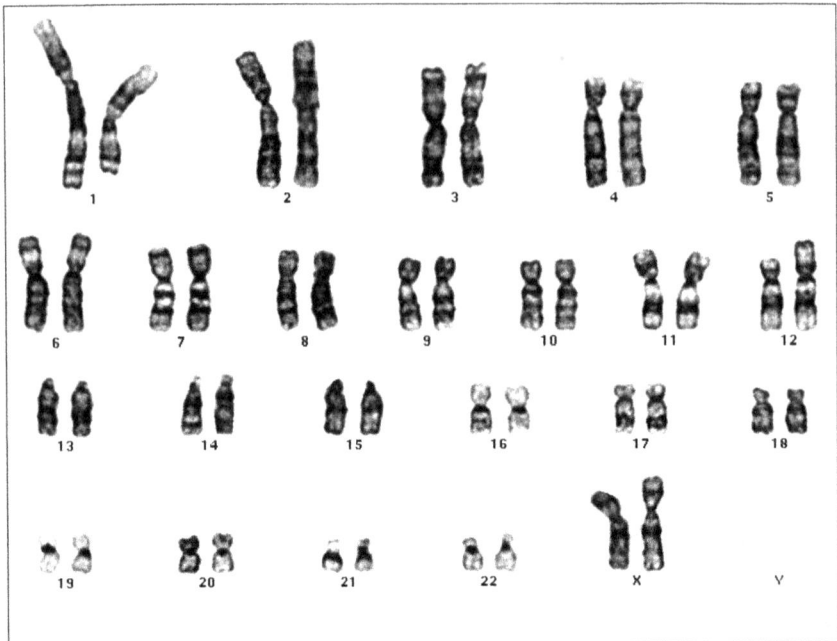

**Plate 12.6.**   **A G-banded human karyotype** of a girl who was slightly dysmorphic and developmentally delayed. 46,XX, t(1;12)(q41;p13.33). The end of the long arm of one chromosome 1 (q41 onwards) has become detached from 1 and attached to the top of the short arm of chromosome 12, at p13.33. See Plate 12.7 for more details. This is **a balanced non-reciprocal translocation**.

Then various methods of chromosome treatment and staining such as **G-banding** were developed so that each chromosome showed a number of distinctive stained bands: see Plates 12.6 and 12.7 for a G-banded karyotype involving a 1:12 non-reciprocal balanced translocation. Giemsa banding gives 300–400 alternating dark and light bands over the whole genome, caused by different condensation, with dark bands being AT rich and containing about 20% of the active genes. All 22 pairs of autosomes plus sex chromosomes X and Y could then be identified separately from length, centromere position and banding pattern at metaphase of mitosis. There are various automatic scanning and image analysis systems which can give computerised karyotypes, even if some chromosomes overlap, but checking

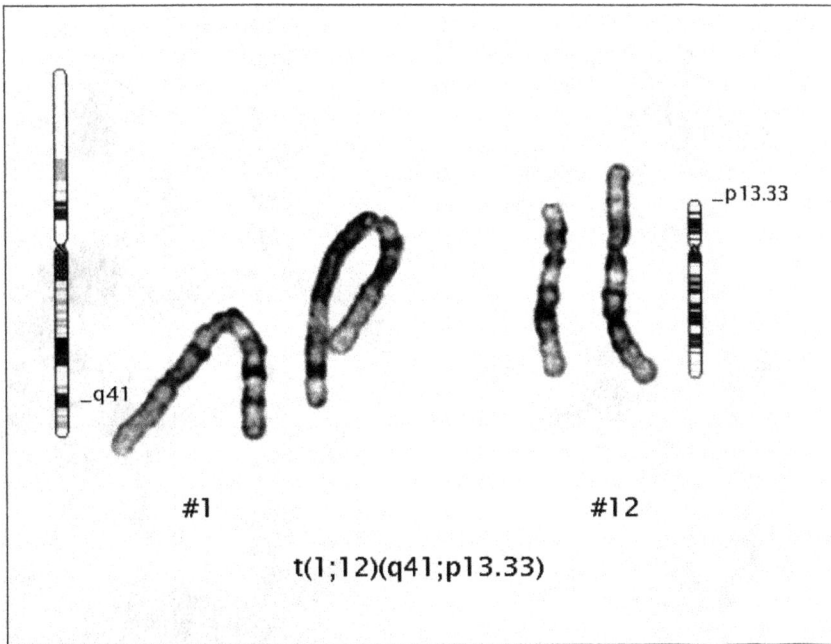

Case: B98-1826   Slide: G-Banding   Cell: 3   Patient:

**Plate 12.7.**   See Plate 12.6 for details. This shows **enlargements of chromosomes 1 and 12, with normal and translocated ones,** and a representation of **the standard Giemsa banding for the normal versions of those two chromosomes.**

of the metaphase by an experienced operator is always advisable, for at least three cells.

The next advance involved **fluorescent *in-situ* hybridisation (FISH).** DNA probes with fluorescent labels were made which could hybridise, under appropriate temperature regimes, to denatured DNA in particular chromosomes or particular regions, of the metaphase spreads; see Plates 9.1 (whole chromosome paint for 5) and 12.8 (sub-telomeric probes for X and 2). Fluorescent light then revealed to which chromosome each probe hybridised, so that a probe for the whole of chromosome 5, with DNA sequences complementary to repeated specific DNA sequences in chromosome 5, would cause all copies of chromosome 5 to fluoresce, in interphase nuclei as well as in metaphase spreads. Using repeated DNA sequences within a chromosome allowed the whole chromosome to be **"painted"**,

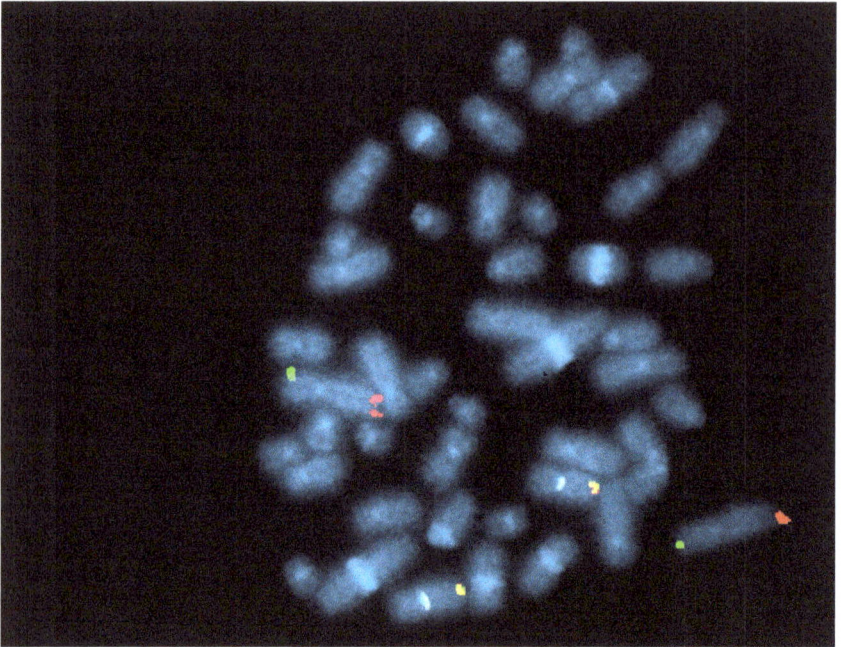

**Plate 12.8.** **Probes for subtelomeric regions**, used on blood lymphocytes in a normal female. Yellow binds to the long arm of the X, red to the long arm of 2, and green to the short arm of 2. Telomeric rearrangements are difficult to see on G-banded preparations as many telomeric regions look similar; subtelomeric rearrangements are seen in about 6% of individuals with idiopathic mental retardation.

which was very useful when looking for translocations or other aberrations. For example, if chromosomes 2 and 3 were "painted" green and pink respectively along their lengths, then a reciprocal translocation near the ends of those two chromosomes would show up as one pink and one green chromosome (untranslocated), one with most of the pink chromosome plus a green tip, and one with most of the green chromosome plus a pink tip, the 2/3 and 3/2 translocated chromosomes.

**Chromosome prints**, as opposed to paints, were developed by Advanced Biotechnologies, based on single copy sequences in particular chromosomes, usually near the centromere, "lighting up" just one region of a specific chromosome in interphase and mitosis, taking about 1 hour instead of about 13 hours preparation for chromosome painting. Plate 12.9 shows a normal XY cell in uncultured white blood cells from a baby with

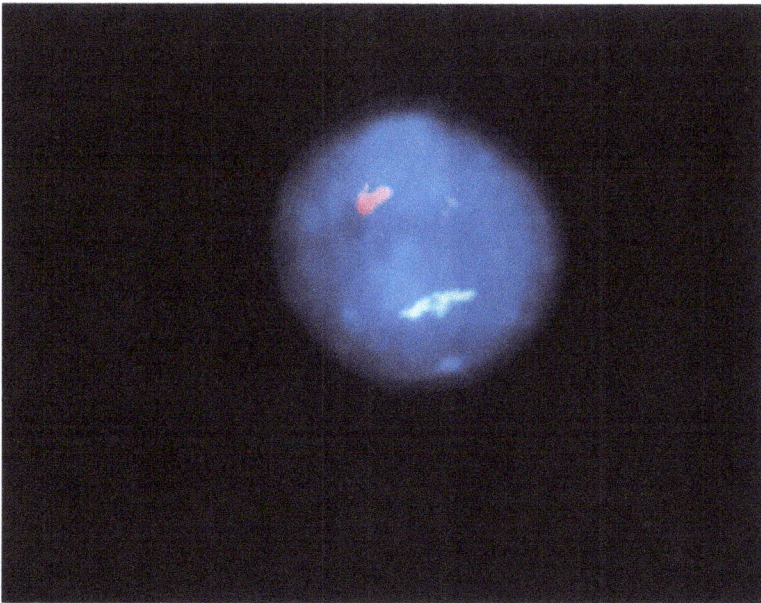

Case: C99-S5   Slide: I   Cell: 8   Patient:

**Plate 12.9.** An interphase cell, an **uncultured white blood cell from a newborn baby with ambiguous genitalia**. FISH with dual-coloured centromeric probes shows the baby to be chromosomally X (red centromere) Y (green-blue centromere), male.

ambiguous genitalia; with these particular dual-colour centromeric FISH probes, the X centromere shows red and the Y centromere shows green.

Some kits are specific for particular abnormalities. Thus the TriGen® Assay is for Down syndrome and X or Y numerical abnormalities, with chromosome 21 fluorescing orange, the X is green and the Y is blue. It can be used directly on cells from amniocentesis without culturing, giving results in 9 to 20 hours as opposed to 7 to 21 days if amniotic fluid cells need culturing and conventional preparation.

Plate 12.10 shows a normal sperm hybridised (in interphase) from a man with a 15/21 translocation. The four probes "light up" three regions of chromosome 15 and one region of 21: (see caption).

Applied Spectral Imaging have developed **Spectral Karyotyping** (**SKY**), using a "probe cocktail" containing painting probes for all 24 human chromosomes, so that each of chromosomes 1 to 22, X and Y shows up a different colour after using special spectral image equipment and software. A

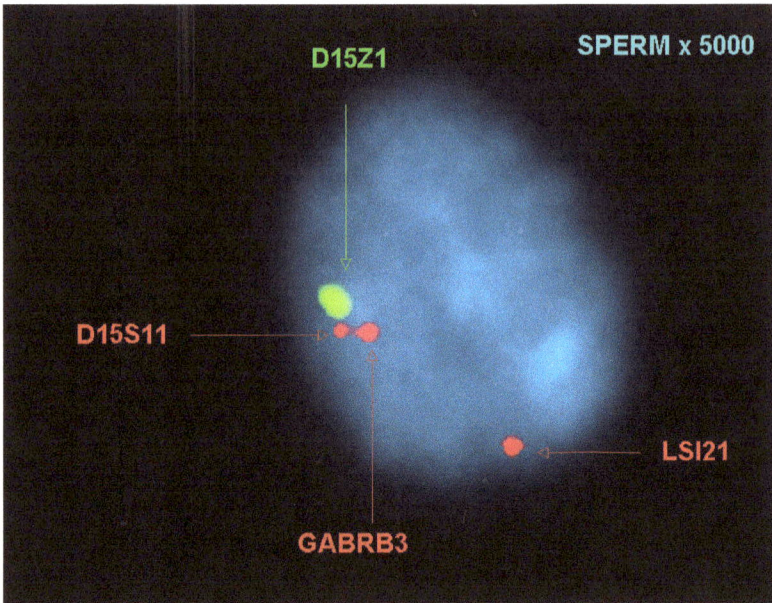

**Plate 12.10.  A normal sperm from a patient with a 15/21 translocation** distal to 15q11.2 and proximal to 21q22; although some sperm were abnormal (Plates 1.2.1 and 1.2.2), this has one copy of 15 (left) and one of 21 (bottom). Probes: LSI21 red at 21q22; D15Z1 green at 15p11, a large repetitive signal; GABRB3 red at 15q11.2; D15S11 red also at 15q11.2 but seen separately from the previous one.

number of different systems are commercially available; Plate 12.11 shows a normal male karyotype. Such systems can give automatic identification of chromosome aberrations, and even diagnosis of complex chromosomal rearrangements in solid tumours or cell lines (Plate 12.12). See Gersen and Keagle (2005) for a recent textbook on human clinical cytogenetics.

**Comparative genomic hybridisation: theory and examples.** This very recent technique (see Pinkel and Albertson, 2005) concerns **DNA copy number**. DNA from normal cells is used as reference DNA and stained with a red fluorochrome, while DNA from the cells to be tested is stained green. These two labelled total DNA extracts are then hybridised to a normal metaphase cell. Regions present in the test DNA in larger quantities than in the normal DNA (e.g., because of an extra copy of a particular chromosome or region) will appear more green than red. Regions present in the test DNA in smaller quantities than in the normal DNA (e.g., because of missing

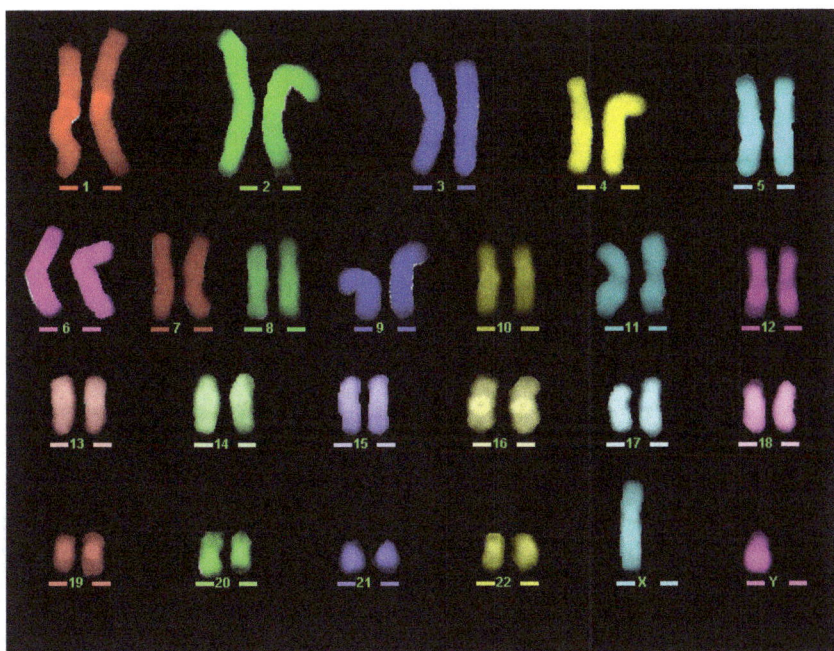

**Plate 12.11.** **Colour patterns in the karyotype of a normal XY male**, using five different fluorochrome labels so that each pair of chromosomes has a unique colour pattern.

chromosomes or deleted regions) appear more red than green, and regions equally present in normal and test DNA show as mixed red/green, as in Plate 12.13, with computer analysis in Plate 12.14 This gives a whole genome screen in a single test.

A number of cells are analysed and a computer program produces a standard banded karyotype with a green bar against chromosomes or regions in the test material in excess of that of the normal reference material, and a red bar against material deficient compared to the normal DNA. It also gives a probability trace in brown for each region, in the black area for normal amounts, in the green region for excess amounts and in the red region for deficient amounts.

## 12.8.2. *Autosomal abnormalities*

**Down syndrome** (Mongolism), see Plates 12.15, 12.16 and 12.19, is due to an extra chromosome 21, 2n = 47. It occurs in about 1 in 600 live

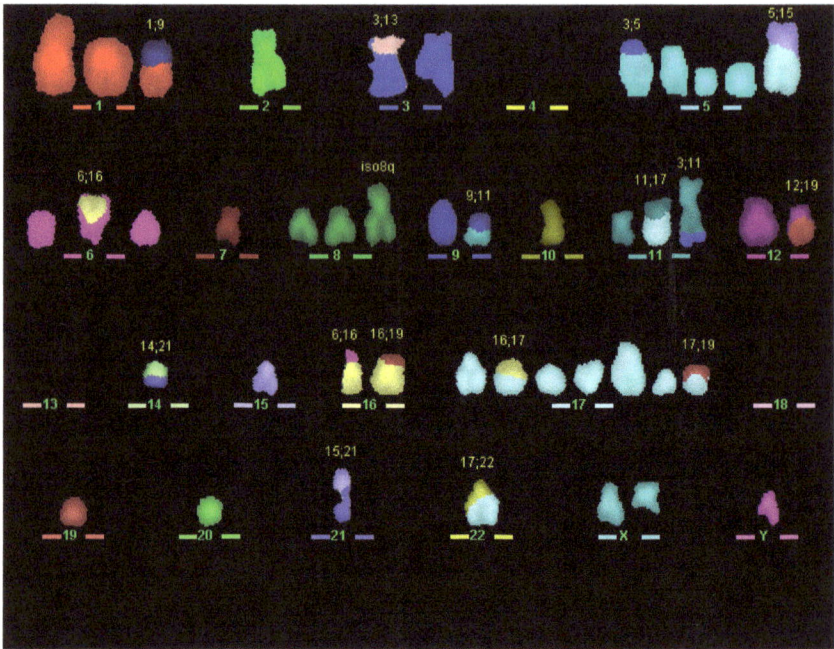

**Plate 12.12.   Complex chromosome rearrangements in a cultured cell line**. Multiplex fluorescent *in situ* hybridisation was used as a single step to identify complex chromosome rearrangements. The numbers above the rearranged chromosomes shows their origins. Such cell lines used in genetics research often evolve in complex ways which are not characterisable by G-banding alone. The pseudo-colours are imposed by the computer software.

births, in both sexes. Intelligence is reduced, with IQs usually in the range 25–75, often 40–50. Affected individuals are of small height, with an extra eye-fold of skin, broad hands with odd palm prints, slow development, muscle weakness, poor short-term memory, and nearly half suffer from heart defects in the valves or septa. The mouth is often open with the large tongue protruding. They have a reduced life span, with about one sixth dying in their first year, and an average span of about 20 years. They have a 15-fold increased risk of leukaemia and are more susceptible to infections. They are universally susceptible to Alzheimer's disease by about the age of 35 years. Some females are fertile but males are usually sterile. In 95% of cases, the extra chromosome is of maternal origin, partly because male gametes are more sensitive to chromosome imbalance but mainly because of a maternal age effect. Even in young mothers the eggs are more than a

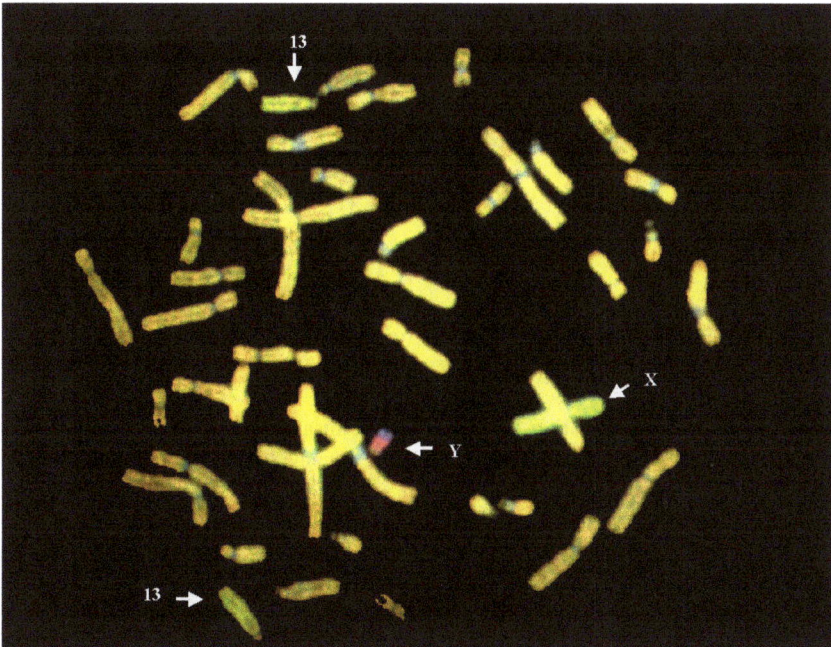

**Plate 12.13.** **Comparative genomic hybridisation**, a cell showing trisomy for chromosome 13, **Patau syndrome**. This shows a metaphase cell from a normal male (XY, two copies of 13) hybridised with labelled DNA (red) from normal male DNA (reference specimen) and DNA (green) from a female (test specimen) with trisomy for chromosome 13. Most chromosomes look red/green; the two copies of 13 look green as the test DNA had three copies of 13; the Y looks red as the female test DNA had no Y. The X shows green as the test DNA was from an XX female while the reference normal DNA was from an XY male. The blue colour shows repetitive sequences of heterochromatin as on the tip of the Y chromosome. The analysis is shown in Plate 12.14. Credit for this plate and 12.14, C. Bedwell.

decade old because oogenesis is largely completed before birth, with the primary oocyte suspended in diplotene until sexual maturity, while sperm only form from puberty.

The condition is easily detected by the symptoms, and confirmed by karyotyping. For genetic counselling, it is important to distinguish between ordinary Down syndrome arising by chance non-disjunction in one parent's meiosis, in which there is no special risk of further siblings being affected,

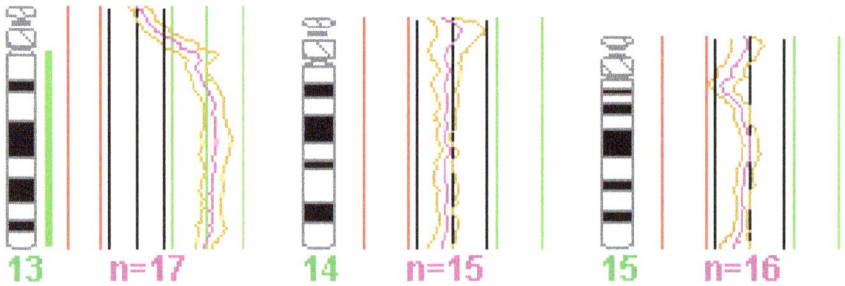

Detail of CGH profile showing extra copy of
chromosome 13 compared with 14 and 15

**Plate 12.14.** **Comparative genomic hybridisation showing trisomy for chromosome 13**: computer analysis. Comparative genome hybridisation profile for chromosomes 13 (three copies), 14 and 15 from the cell shown in Plate 12.13 from a woman with trisomy for 13. The brown line goes into the green area for most of chromosome 13, which has a green bar beside it, while remaining largely within the black boundaries for chromosomes 14 and 15. "n" shows the number of cells scanned for each chromosome.

and the 4% of cases caused by **translocation** of a chromosome 21 onto another autosome, usually 14 but sometimes 13, 15 or 21. A translocation Down's has two normal 21's and one translocated 21, and often one parent (usually the mother) has $2n = 45$, with one normal 21 and one translocated 21, e.g., $14 + 21$, with no external symptoms as all genes are present in the right dose. Such a mother produces four types of gamete in about equal proportions, with the translocated chromosome behaving as if controlled by centromere 14: normal 14 and normal 21, giving a normal child, $2n = 46$; $14 + 21$ and normal 21, giving a translocation Down's child, $2n = 46$; $14 + 21$, giving a translocation carrier, unaffected, $2n = 45$; normal 14, with no 21, which is lethal as it gives monosomy for 21, $2n = 45$.

A couple who have had a translocation Down's child, and where one parent has been found to be a carrier, would be advised that there is a high risk of having a further affected child, but that an affected foetus could be detected by prenatal diagnosis (Sec. 12.12). The risk depends on the chromosome involved and the parent involved. 13/21 and 14/21 female carriers have about a 10% risk of a Down's baby; for male carriers the risk

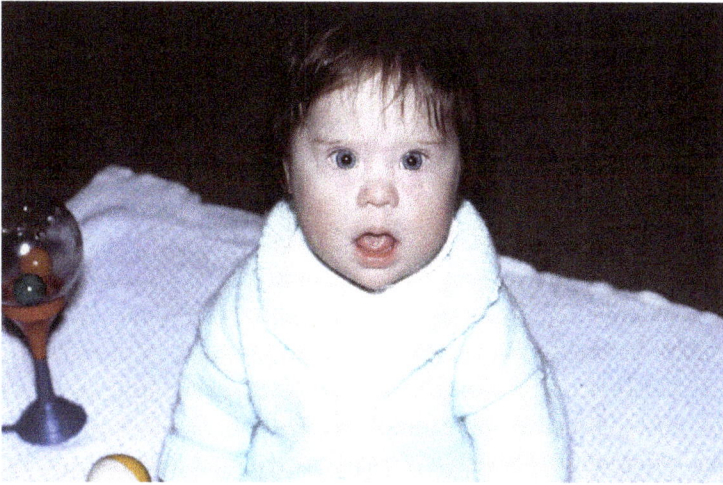

**Plate 12.15.** **A Down syndrome girl,** aged 10 months. Like many Down's children, she had heart defects and needed a major operation, which set back her development still further.

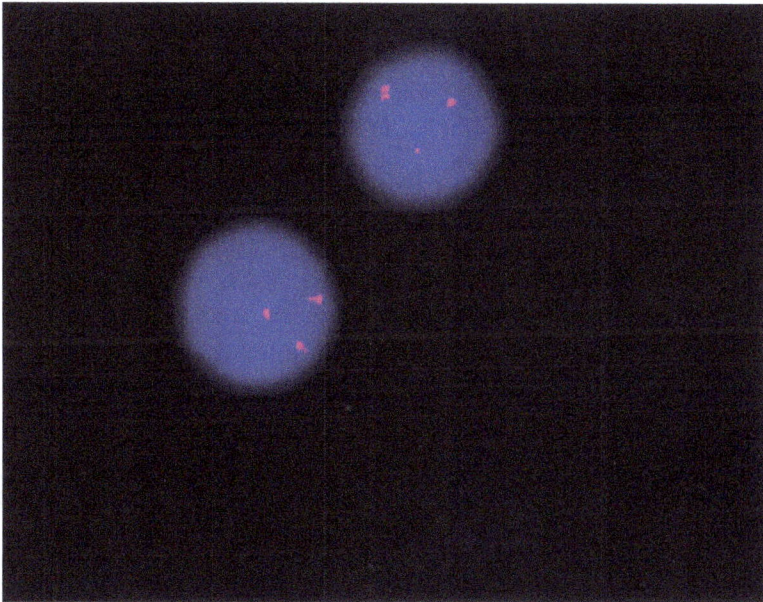

Case: B99–867 Slide: I Cell: 5 Patient:

**Plate 12.16.** An uncultured white blood cell from a **newborn baby with some Down syndrome features.** The "Quintessential" probe for chromosome 21 shows that there are three copies of 21, but it cannot distinguish between a standard Down's and a translocation Down's.

is about 2%. For 21/21 translocations, the risk is 100% for both male and female carriers.

For normal non-disjunction Down syndrome, there is a clear increase in the frequency of affected children with **increasing maternal age**. Typically the risk (the figures differ in different studies) is 1 in about 1,923 for mothers aged 20, 1 in 885 for age 30, 1 in 139 for age 40, and 1 in 32 for age 45. Cuckle *et al.* (1987) give figures ranging from 1 in 1580 for age 15, to 1 in 11 for age 48, and 1 in 6 for age 50. Once one allows for the fact that older husbands tend to have older wives, there is no effect of paternal age on the incidence of Down syndrome. There is no age effect with translocation Down's. The non-disjunction Down's arises mainly (80%) in the first division of meiosis, which may be related to the fact that older women have fewer chiasmata to hold the bivalents together until normal separation at anaphase I. The frequencies of XXX and XXY abnormalities (Subsec. 12.8.3) also rise with maternal age, as do those for most autosomal trisomies, with smaller chromosomes more affected than large ones.

If there is non-disjunction at mitosis in the very early embryo, one can get **mosaics** (Sec. 12.16), individuals with some cells abnormal and some cells normal. If chromosome 21 fails to segregate evenly at the second mitosis in the egg, for example, two cells will be 2n = 46, normal, one will be 2n = 47, with three of 21, and one will be monosomic for 21, 2n = 45, and will die. Depending on what cells are affected and how many, a mosaic's phenotype may vary from normal to somewhat affected.

**Other autosomal abnormalities** occur but are rarer than Down syndrome. **Patau syndrome** has three of chromosome 13 (Plate 12.14), with a frequency of 1 in 5,000 live births, increasing with maternal age. Sufferers have many abnormalities, often including brain development failure and cardiovascular defects, and half die within one month. **Edwards syndrome** has three of chromosome 18, with a frequency of 1 in 3,000 live births, increasing with maternal age, but the incidence at conception is much higher with perhaps 95% aborting spontaneously, especially males. They have multiple malformations and 90% die within one year.

Approximately 1 in 2,000 babies has microscopically visible **deletions or duplications**, usually giving multiple mental and physical symptoms.

Sometimes the mothers carry balanced translocations, but more usually the abnormalities have arisen *de novo*. Heterozygosity for a deletion of about half the short arm of chromosome 5 gives the **Cri du Chat syndrome** (Sec. 9.2). Affected individuals are severely abnormal physically and mentally, and give a typical plaintive, continual, cat-like cry.

**Robertsonian translocations**, also known as centric fusions, are whole-arm exchanges between the five acrocentric chromosomes 13, 14, 15, 21 and 22, and are the commonest human chromosome rearrangements. They arise by breaks or crossovers, usually just above the centromeres, with fusion of the products with centromeres and the loss of the now acentric short arms. They have a high rate of production, about $4 \times 10^{-4}$ per gamete per generation, an order of magnitude higher than the highest mutation rates for human autosomal dominant genetic diseases, and so they contribute significantly to foetal wastage, birth defects, Down syndrome and/or mental retardation. **Rob(13q14q)** (i.e., exchange and loss of the short arms, designated "p" as opposed to long arms, "q", of 13 and 14, in a translocation between one 13 and one 14, to give a single translocated centric product) and **rob(14q21q)** form about 76% and 10% respectively of all Robertsonian translocations. Those two have breakpoints localised to very specific regions of the proximal acrocentric short arms, but others have more diverse breakpoints. Nearly all Robertsonian translocations arise during female meiosis, especially in older mothers: see Page and Shaffer (1997). The short arms of acrocentric chromosomes contain mainly tandemly repeated satellite DNA with no known function, so its loss is usually inconsequential. Band p12 of these chromosomes contains 18S and 28S ribosomal RNA genes, and this homology tends to give associations of these regions, promoting rearrangements. Most Robertsonian translocation chromosomes are dicentric, with one active centromere and one latent one, with a small intercentromeric region. Most **translocation Downs** have the extra 21 in an unbalanced Robertsonian translocation or as an isochromosome — **rea(21q21q)** — with two long arms of 21. Robertsonian translocations are also involved in some blood pathologies and cancers (Welborn, 2004). Human chromosome 2 probably arose from a Robertsonian translocation as its short arm matches bands on the long arm of chromosome 12 in the three apes, while its long arm matches chimpanzee chromosome 13 and gorilla and orangutan chromosome 11.

### 12.8.3. *Sex chromosome abnormalities*

The most common **sex chromosome abnormality** is **Klinefelter syndrome**, about 1 in 1,000 male live births, increasing with maternal age. It results from an XXY, 2n = 47, karyotype, written 47,XXY. The penis is normal but the testes are small, with few sperm. Sufferers tend to be tall, with long limbs. Most individuals have normal intelligence but the average intelligence is reduced because Klinefelter patients are ten times more likely to be mentally retarded than normal XY males. Many sufferers have difficulty with oral communication, even if of normal intelligence. About 40% of sufferers have some breast development. Testosterone replacement therapy can be given from early adolescence to improve sexual characteristics, but XXYs remain sterile unless they are mosaics, with about 15% being 46,XY/47,XXY mosaics.

**Turner syndrome**, 45,X, has a very high frequency of spontaneous abortion, more than 99.0%, with 1 per 5,000 live female births. The intelligence and life span of these females are almost normal but the height is short, averaging 145 cm. The hands and feet are often swollen. The ovaries start to degenerate after 15 weeks gestation, giving failure of sexual development. Approximately 10% menstruate and a very few are fertile, but these are normally mosaics. Sufferers usually have a broad chest, underdeveloped breasts, a broad webbed neck, an increased risk of heart and kidney disease and other problems such as swollen ankles and wrists. Growth hormone treatment from childhood increases adult height by about 4 cm, and sex hormone therapy allows improved sexual development, but not fertility.

Approximately 80% of XOs are of paternal origin, with poor pairing between X and Y allowing non-disjunction to give a nullisomic sperm, O. Although 45,X, from maternal or paternal meiotic non-disjunction, is the commonest type (50%), Turner syndrome can also arise from: having one normal X and one **isochromosome** (a chromosome with two identical arms, containing homologous genes) of the X long arm (and thus having only one copy of the X short arm in their karyotype), (17%); being XX/XO mosaics (20%); being XY/XO mosaics (4%); having one normal X and one ring X (7%); having a deletion in the short arm of one X (2%). The somatic effects are thought to be caused mainly by the lack of a second X-chromosome short arm, while the sexual effects are probably related to the X long arm deficiencies. Two active X chromosomes are required for proper foetal oogonial development, with inactive X's being reactivated in oogonia when meiosis begins in the foetus. Being hemizygous for X-linked loci, Turner syndrome females express X-linked recessive alleles, e.g., for haemophilia and red-green colour blindness, with 8% colour blind compared with 0.67% in XX females.

**47,XYY** individuals are males. The incidence is 1 in 1,000 live male births, with no effect of maternal or paternal age, but is 3 per 1,000 in mentally handicapped adults and 20 per 1,000 in male criminals. The condition arises most often from the production of YY sperm from the father's second meiotic division. Intelligence is 10 to 15 IQ points less than for normal siblings. Sufferers tend to be tall and are often aggressive. They are fertile, with normal sexual development, and sperm usually only carry one sex chromosome.

**47,XXX** individuals are females, with two Barr bodies per cell. The incidence is 1 in 1,000 live female births, with a maternal age effect; about 90% arise from non-disjunction in female meiosis. Many are normal physically and mentally but the average IQ is about 12 points below that of siblings, and learning difficulties are common, especially in language acquisition. Some XXXs are sexually underdeveloped, some have a variety of physical defects and some have delayed emotional maturity. They are often tall, with long legs, and a small head circumference. Approximately 75% are fertile. The lack of clear symptoms means that many go undiagnosed unless they are karyotyped. Women with more than three Xs have more severe symptoms than XXX women.

**46,XX** males occur at an incidence of 1 in 20,000 males, usually with translocation of the sex-determining region (SRY) of the Y to an X chromosome. They are sterile with small testes but are mentally normal. A minority have normal X chromosomes and ovaries but are externally male because of exogenous androgens or defects in adrenal steroid biosynthesis.

**46,XY** females are very rare, with a Y chromosome lacking the SRY region, or with it mutated.

## 12.9. Selection before and after birth

The true frequency of chromosome abnormalities at conception cannot be measured as there is **strong selection** at embryonic and foetal stages against many abnormalities. Probably about half of all conceptions abort, often without the woman realising that she had conceived, but giving a detected miscarriage if occurring later. Of recognised pregnancies, 15–20% end in spontaneous abortion, with about 40 to 60% of these abortuses having chromosome abnormalities if they are lost in the first trimester (three months), and 5% if lost in the second trimester, compared with only 0.6% of the newborn having chromosome abnormalities.

The following figures are from a number of sources, including Jacobs and Hassold (1995), Connor and Ferguson-Smith (1997) and Hartl and Jones (1998), with some differences in the figures between the different sets of data. In studies of thousands of **spontaneous abortuses in the first trimester**, 39% had normal karyotypes and 61% were abnormal. Of the **abnormal types**, about 50% were **trisomics**, with one extra chromosome, especially number 16 (16%, although trisomy 16 never reaches full term). Other common trisomies were for 22 (6%), 21 (5%), 15 and 14 (4% each), and 2 (2%). Trisomy for 1 was never found, so it must be lethal very early in development. There was a small proportion of double trisomics (2n = 48). Monosomics must also generally be lethal, except for Turner syndrome, 45,X, (18%). The high frequency of Turner syndrome abnormalities in these abortuses, compared to only 1 in 5,000 live female births, shows the very strong prenatal natural selection against Turner syndrome. According to Jacobs and Hassold (1995), only 0.3% of Turner's conceptuses survive to birth, and 3% of trisomy 13, 5% of trisomy 18, and 22% of trisomy 21. **Polyploidy** was common, with 20% triploids (2n = 69, with XXX, XXY and XYY) and 6% tetraploids, showing a very large contribution to abnormal zygotes from unreduced gametes or double fertilisations. There were also about 0.06% each of XYY and XXY, and 0.3% of XXX in these first trimester abortuses. **Translocations** were 3%, mostly unbalanced, and mosaics were 1 to 2%. Structural abnormalities such as translocations and deletions were much less frequent than chromosome-number abnormalities.

For **newborn children**, only about 0.6% had abnormal karyotypes, showing strong selection before birth against abnormalities. There were 0.17% sex chromosome abnormalities (47,XYY, 0.1% of males; 47,XXY, 0.1% of males; 47,XXX, 0.1% of females; XO, 0.02% of females). The autosomal abnormalities were 0.14% trisomy for 21 (Down syndrome), 0.03% trisomy 18 (Edwards syndrome) and 0.02% trisomy 13 (Patau syndrome). There were 0.2% balanced translocations, 0.05% unbalanced translocations, and a surprisingly high 1% (excluded from the 0.6% quoted above) heterozygous for small pericentric inversions of chromosome 9, with no phenotypic abnormalities. Heterozygotes for large inversions have a risk of 8% for a carrier mother and 4% for a carrier father of producing offspring with unbalanced chromosomes.

**Abnormalities of chromosome number** are not usually passed on to future generations as they often cause sterility. Structural abnormalities such as translocations can be passed on, especially through the mother, but they often reduce fertility and cause abortions.

**Selection** also occurs, usually after birth (depending on when the deleterious allele is expressed) for X-linked and autosomal genes, such as Duchenne muscular dystrophy and cystic fibrosis, respectively. Some cause premature death and some cause sterility, as in testicular feminisation. Others, such as achrondroplastic dwarfism (Subsec. 12.3.1), allow viability and fertility when heterozygous but are lethal when homozygous, so there will be overall selection against them.

One of the **worries about genetic counselling** is that it might increase the frequency of harmful alleles in human populations by reducing selection. For example, if one reduces the frequency of recessive homozygous sufferers for a deleterious autosomal recessive allele, by counselling heterozygous carriers not to marry other heterozygotes but to marry dominant homozygotes, then that reduces selection against the bad allele, while new bad alleles continue to arise by mutation.

## 12.10. Blood groups, especially ABO and Rhesus

### 12.10.1. *Blood groups and transfusions*

Approximately 400 human blood group antigen systems have been described and the antigens on the surface of red blood cells determine a person's **blood group**. The most important ones are the ABO (chromosome 9) and Rhesus (chromosome 1) groups, which are critical for successful blood transfusions because humans can make antibodies to their antigens (see below). For some other groups such as MN, the antigens are only detected by using antibodies from other organisms, so are not usually a problem in transfusions, but the groups can be used forensically. Groups are named

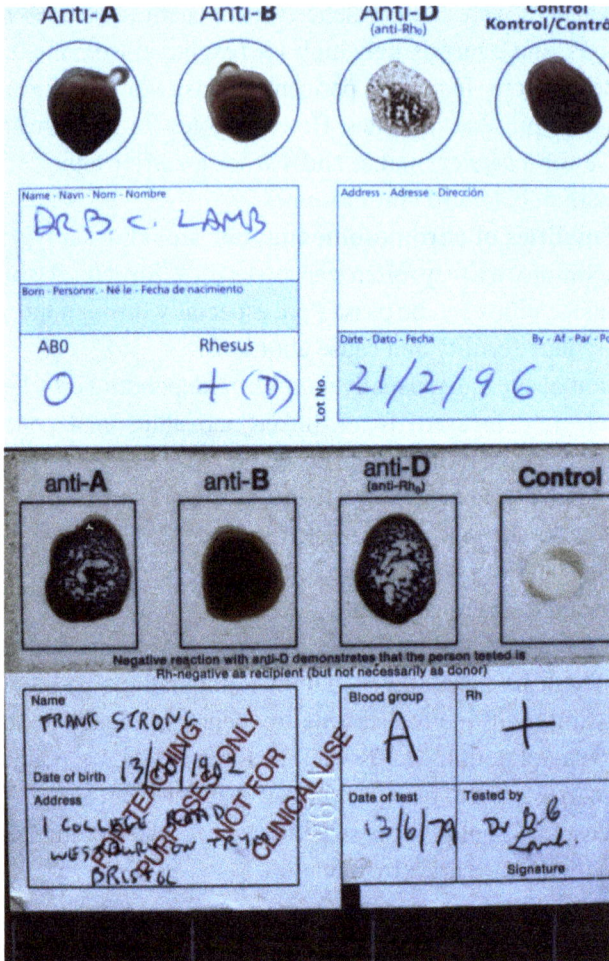

**Plate 12.17.** **Testing ABO and Rhesus blood groups using a card with dried antibod-**
**ies**. The upper card shows an **O Rh positive** group, with only agglutination on the anti-D
(anti-Rh$_o$) panel. The lower card shows strong reactions in the anti-A and anti-D panels, an
**A Rh positive** grouping.

after the antigens. Some other blood groups should be matched for recipients of **repeated transfusions**, who can eventually develop antibodies to
the Duffy, Lewis, MN and S groups. Plates 12.17 and 12.18 show card and
tube tests for the ABO and Rhesus groups, using monoclonal antibodies.
Most blood donations in the UK are typically used within three days.

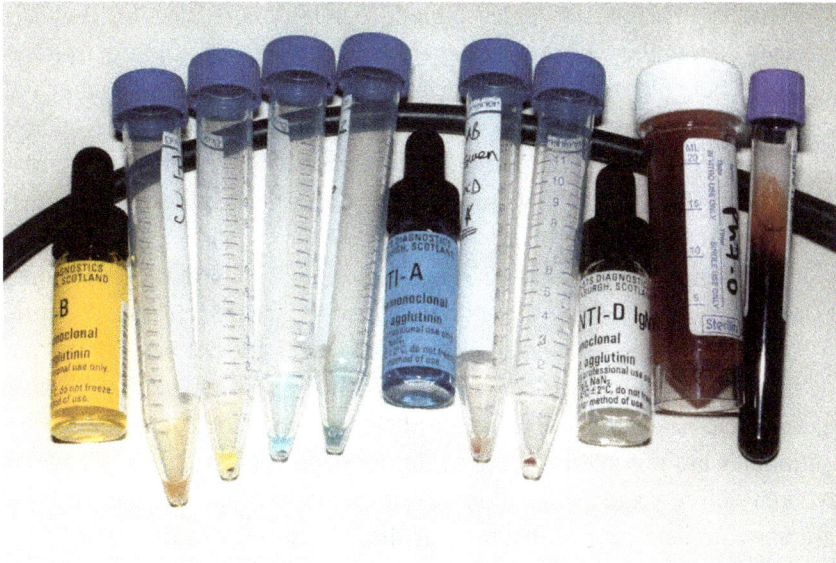

**Plate 12.18.** **Testing ABO and Rhesus blood groups using a centrifuge and mono-clonal antibodies**. Collected blood (the right-most tube) is diluted in phosphate buffer, as in the next tube along, then small samples are mixed separately with the monoclonal antibodies for anti-A (liquid coloured blue), anti-B (yellow) and anti-D (colourless), in different tubes, needing three tubes per blood sample. After mixing and centrifuging the sample, one shakes each tube gently. A positive reaction (agglutination) gives a pellet which stays intact but having no reaction allows the pellet of red blood cells to be resuspended. For each antibody, one tube here has a pellet and one has resuspended cells; e.g., the left-most capped centrifuge tube has no B antigen but the next tube along has B antigen, which has agglutinated with the anti-B antibody to give a firm pellet and clear yellow liquid.

Blood transfusions, even with monitoring for various diseases, always pose a risk of **transmitting diseases** from donor to recipient. McCullough (2003) estimated that the risk in the USA of viral diseases from a single-unit blood transfusion was 1 in 34,000, but with nucleic acid testing the risk was reduced to 1 in 1,900,000 for HIV, 1 in 1,600,000 for hepatitis C and 1 in 210,000 for hepatitis B. There is also a risk of bacterial contamination, giving sepsis in the recipient. Pathogen-inactivated products have been investigated, including plasma (used for some years but later withdrawn), platelets (in clinical trials) and red blood cells (in clinical trials). Blood tests are usually for viral antibodies and may thus miss very recent infections of the donor, or immunologically variant virus strains.

**Risk-reduction strategies** include better donor selection (but that reduces the number of donors), better blood testing, and autologous transfusions where blood collected from a donor is kept for later use on the same donor, say where someone is booked for an operation many months later. In most countries there is a shortage of safe donors. Where donors are paid to donate, there are often many drug-users offering unsafe blood to get money for more drugs.

**Haemoglobin-based oxygen carriers** (HBOCs) have been developed as an alternative to donor blood as they can be disease-free, can be stored for longer than the usual shelf-life of five weeks for red blood cells, and do not rely on donors for their production. They are very useful in medical emergencies such as massive haemorrhaging as red blood cell antigens and antibodies are not involved and so do not need testing for. They increase oxygen-delivery and carbon dioxide removal from tissues via the blood but are not complete blood substitutes as they do not remain in circulation as long as red cells, and lack clotting factors and cellular parts of the immune system. HBOCs are not yet approved for medical use in some countries, but have obvious potential.

## 12.10.2. *The ABO blood group*

People of **blood group O**, 46% of the UK population, are $i^o$ $i^o$ genotype, have no specific ABO antigen on the red blood cells and have anti-A and anti-B antibodies in their serum. **Group A**, 42%, are $i^A$ $i^A$ or $i^A$ $i^o$, with the A antigen on red blood cells and the anti-B serum antibody. **Group B**, 9%, are $i^B$ $i^B$ or $i^B$ $i^o$, with the B antigen and the anti-A antibody. **Group AB**, 3%, are $i^A$ $i^B$, with A and B antigens, but neither antibody. The A and B traits are codominant and O is recessive to A and B. Different countries have different blood group frequencies, with most native South Americans being group O, most Asians being group B, and A being commonest in Norway.

The three alleles determine the activity of a **glycosyl transferase enzyme** which in A and B modifies the H surface antigen, adding $N$-acetylgalactosamine for A, D-galactose for B, and nothing for O, which has a single base deletion (of G at nucleotide position 258) giving a frame shift and an inactive antigen protein. Blood transfusions between members of the same ABO blood group are safe, but if say A blood is put into a B recipient, the anti-A antibody in the recipient's serum will react with the A antigen on the donor's red blood cells, agglutinating the cells, perhaps with fatal consequences. In emergencies, certain other transfusions are usually safe because antibodies in the incoming plasma are partly adsorbed by the tissues of the recipient, as well as being diluted by the recipient's plasma. Group O are sometimes called "**universal donors**" because they lack an active antigen, so their donated red cells are not agglutinated in the recipient, and group AB are called "**universal recipients**", because they lack the two antibodies which could agglutinate donated red blood cells. Thus if O blood was transfused into a group B patient, there is no A antigen for B's anti-A to combine with, hence no agglutination of O's red blood cells, and the donated blood's anti-A and anti-B antibodies would be diluted and largely adsorbed in the recipient, with little effect on the recipient's red blood cells.

**Alcohol-soluble ABO antigens** occur in many tissues besides blood, and people carrying the dominant **secretor allele**, *Se*, also have water-soluble antigens in their secretions such as saliva, urine and spermatic fluid, which has been useful forensically, especially in rape cases before DNA fingerprinting was developed. The secretion only applies to ABO and Lewis-group antigens, and is absent in *se se* individuals. ABO antibodies are **effectively constitutive**, not needing red blood cell antigens to stimulate their production; common bacteria may trigger their formation.

## 12.10.3. *The Rhesus blood group*

This is called **Rhesus** because the antibodies were first found in Rhesus monkeys. The antibody is not constitutive in humans and is produced only after a challenge by Rhesus positive blood. Rhesus positive people have the Rhesus D antigen on their red blood cells and Rhesus negatives do not. Rhesus negatives only have the anti-D Rh antibody if previously exposed to Rhesus positive blood. Rhesus positives have the *D* gene, as *D D* or *D d*, while Rhesus negatives are *d d*, but eight Rhesus alleles can be distinguished using three antisera (anti-D, anti-C, anti-E).

The Rhesus group is very important because children born to Rhesus negative women (*d d*) and fathered by Rhesus positive men may develop **haemolytic disease of the newborn**, giving anaemia and jaundice, with much agglutination and breakdown of the red blood cells. This used to cause the death of most affected children, though survivors usually recovered fully. In the UK each year, more than 80,000 Rhesus negative women have Rhesus positive babies.

In "white" populations as in the UK, about 16% of individuals are Rhesus negative, $d\ d$, so from Hardy–Weinberg $q$ is $\sqrt{0.16} = 0.4$, so $p = 0.6$ for $D$. All babies from $D\ D$ fathers $\times\ d\ d$ mothers will be Rhesus positive, giving $p^2 \times q^2 = 5.8\%$ "at risk" pregnancies, and half the pregnancies in $D\ d$ father $\times\ d\ d$ mother will give Rhesus positive babies, giving $1/2$ $(2pq \times q^2) = 3.9\%$ "at risk" pregnancies, so **9.7% of pregnancies are "at risk"**, yet only about **0.3 to 1%** of pregnancies have suffering babies.

The haemolytic disease does not usually occur in a Rhesus negative woman's first Rhesus positive baby, but can occur in second and subsequent Rhesus positive babies with increasing severity. This is because it usually takes one Rhesus positive baby pregnancy in a Rhesus negative woman to trigger the production of her anti-Rhesus antibodies. During most of the pregnancy, not enough foetal red blood cells leak across the placenta into the mother to trigger anti-Rhesus antibody production, with the main leakage occurring just before or during birth, so the mother usually only makes the antibody just after the birth of her first Rhesus positive baby. **To prevent** the mother from making such antibodies which could harm subsequent Rhesus positive babies, she is now usually given an intramuscular injection of anti-Rhesus antibodies within 72 hours of the birth, to destroy leaked red blood cells from the baby in the mother, before they can trigger her own antibody production.

Rhesus women should only be given Rhesus negative blood in transfusions, to avoid anti-Rhesus antibody production. If a Rhesus positive baby is born with severe haemolytic disease of the newborn, it is usually given transfusions of Rhesus negative blood (without anti-Rhesus antibody) because the mother's anti-Rhesus antibody which has got into the baby could attack Rhesus positive blood if transfused in.

With Rhesus, the anti-Rhesus antibody is **IgG** (immunoglobulin G), which can cross the placenta. The ABO antibodies are **IgM** which cannot cross the placenta, so mother/child incompatibility for ABO does not usually cause any problems. **Mother/baby ABO incompatibility** can help with Rhesus incompatibility if the ABO antibodies destroy the baby's leaked blood in the mother before it can trigger anti-Rhesus production. With a Rhesus positive mother carrying a Rhesus negative baby, if her antigens do get into the baby by leakage across the placenta, there is no danger of the mother being harmed by the baby producing anti-Rhesus antibodies, as babies do not produce the antibody (in response to a challenge) until about six months after birth.

### 12.10.4. *Blood groups in farm animals*

Haemolytic disease of the newborn from mother/baby incompatibility is known in horses and pigs, but not in sheep or cattle. Blood groups are many in cattle, sheep, horses and pigs, and have been used in animal **paternity tests**, e.g., checking whether a female was mated to an excellent sire, or to a cheaper substitute male. There is a cattle blood grouping service in Edinburgh, Scotland, but as in human forensic medicine, DNA analysis is often preferred now to blood group studies.

## 12.11. The major histocompatibility complex

The **major histocompatibility complex** (MHC) includes the **Human Leukocyte Antigen** (HLA) system, which encodes a number of leukocyte antigens. It is the most polymorphic genetic system in man and regulates immune responses. For most loci, no single allele is very common and heterozygosity is extremely frequent. It is a gene cluster containing about 80 genes over 4 Mb of DNA on chromosome 6 (6p21.3). The products from class I (*HLA-A*, *HLA-B* and *HLA-C*) and class II genes (*DP*, *DQ* and *DR*), which are respectively at the telomeric and centromeric ends of the cluster, control the presentation of processed antigens to T-cells, so are very important in immune responses, e.g., to virus infections. Class I products are found on the surface of most nucleated cells, while class II products are found on the surface of B-lymphocytes, activated T-lymphocytes and on antigen-presenting cells including macrophages. The class III genes, in the centre, include genes of unknown and immunological function, including tumour necrosis factors, heat-shock protein 70 and complement proteins.

Multiple alleles are known for most class I and class II genes, and definition of genotypes is normally performed by DNA tests which have largely replaced immunological tests. As the genes are all close, they tend to be inherited together as particular **haplotypes** (the combination of genes on one chromosome is a haplotype). If parent 1 has haplotype A on one copy of chromosome 6 and B on his other copy, and parent 2 has haplotypes C and D, then amongst their children, **the chance of two particular siblings having the same haplotype combination**, e.g., AD, is **one quarter**, because whatever haplotype the first child has, the second child has equal chances of the four possible haplotypes, AC, AD, BC, BD, unless crossovers (which have a frequency of several % within MHC) within the region have produced new haplotypes. This is very important in tissue (e.g., bone marrow) or organ **transplants** as transplants between individuals of the same HLA type are usually accepted, not rejected, by the body's immune system. For example, the half-life of transplanted kidneys is 26 years when fully HLA matched, compared with 12.2 and 10.8 years for grafts from parents and siblings, respectively, without full matching (see Howell and Navarette, 1996).

There is enormous variation between **racial groups** in antigen frequencies for class I and class II genes. For example, *A1*, *A25*, *B37*, *B38* and *B63* are largely confined to Europeans, *A34*, *A36*, *A43*, *B42*, *B53* and *B58* to Negroes, and *B52*, *B54* and *B61* to Orientals. There is extensive linkage disequilibrium, with the frequencies of combinations of alleles often differing from those expected from their individual frequencies (Subsec. 4.2.3). That means that if one HLA allele predisposes people to a particular disease, then other alleles with which that allele is often associated will often have positive correlations (but not necessarily causal) with the disease. The strength of the association of an HLA allele with a particular disease is measured by the **relative risk**, which is the frequency of the disease in those with the allele divided by the frequency of the disease in those without the allele. Thus, HLA allele *B27* occurs in more than 90% of people with ankylosing spondylitis (severe stiffening of the spine), whereas only 8% of the general population has that allele, and the relative risk is nearly 90. The *HLA-B27* allele is used diagnostically for ankylosing spondylitis. The relative risks for various HLA-allele associated conditions are: psoriasis with *DR7*, 43; psoriasis with *B17*, 6; multiple sclerosis with *DQB1*0602*, 36. The notation with * and four or five numbers indicates an HLA DNA marker, while notation such as *DR7* indicates an antigenic marker. Autoimmune diseases such as rheumatoid arthritis (relative risk 3 to 6 with *DR4*) often have HLA allele associations, with self-antigens being treated as foreign, with immune responses against the host cells. See Connor and Ferguson-Smith (1997). The combination in *cis* or *trans* of *DQA1 \*0501* with *DQB1 \*0201*, which both encode the DQ2 antigen, confers a relative risk of about 250 of coeliac disease. See Howell and Holgate (1996) for the relation between HLA genes and allergic disease, including asthma, hay fever and eczema.

## 12.12. Prenatal, neonatal and adult screening

**Prenatal diagnosis** is done to reassure couples if the foetus is normal, and to detect various genetic and chromosomal abnormalities if present, so that the couple can be offered a termination of the pregnancy if they wish. In the UK, prenatal diagnosis is offered in about 8% of pregnancies. Most of the foetuses investigated have no defect, with selective termination being offered in about 7% of cases. Termination is only offered if there are serious defects present, such as Down syndrome. Termination of a baby solely on the grounds of unwanted gender is not allowed in the UK at present: see Subsec. 12.12.6.

### 12.12.1. *Amniocentesis*

In **amniocentesis** at about **16 weeks** pregnancy, about 20 ml of fluid is taken from the amniotic fluid surrounding the foetus through a syringe needle inserted through the woman's abdominal wall, using ultrasound monitoring to avoid damaging the foetus, umbilical cord or placenta. The fluid can be immediately analysed for some biochemical disorders but the main use is of the foetal cells which have been shed and can be spun down. The cells are few and many are dead. They can be immediately sexed from the presence or absence of Barr bodies, or stained with fluorescent dyes to detect the Y chromosome. FISH (fluorescent *in-situ* hybridisation) can be used with specific DNA probes, but for most purposes there are too few living cells, so they need to be multiplied in culture for two to three weeks. Then they can be karotyped, used for DNA analysis or metabolic tests to detect chromosome, gene or enzyme defects in the foetus.

The **risk of miscarriage** or foetal damage is about 0.5% and the procedures are expensive, so amniocentesis is not done routinely in all pregnancies. In Britain it is offered to pregnant women of 35 or over because of their higher risks of Down's and other age-related syndromes, or where there is a known genetic risk in the family. It is particularly useful for women who are translocation carriers, to check whether the foetus is normal, a carrier or a sufferer. The woman (or better, the couple) is usually offered a termination if a bad defect is found, but there is never any compulsion to have one. If the test is only done at 20 weeks and the test results take 3 weeks, a termination would be very close to the legal upper limit of 24 weeks gestation in the UK. However, under the 1990 Human Fertilisation and Embryology Act, termination is permitted up to full term if the defect is serious.

## 12.12.2. *Early amniocentesis*

It was found in 1988 that really skilled people could take 10 ml amniotic fluid at **13–14 weeks** pregnancy. There are more viable cells then than at 16–20 weeks, so karyotyping could be done within 8 days, so any termination could be offered much earlier than with normal amniocentesis. The problems are a slightly higher risk of miscarriage, and taking 10 ml reduces the amniotic fluid by 30%, which can sometimes affect lung development.

## 12.12.3. *Chorionic villus sampling*

A catheter is inserted through the vagina and cervix to take a sample of **chorionic villi**, removing about 20 mg of tissue; any maternal tissue must be dissected away or errors could occur. The chorion is of foetal origin, anchoring the foetus to the uterine wall before full placental development. This sampling can be done at **8–12 weeks** pregnancy, even before the pregnancy is visible externally. The villus cells are dividing rapidly so can be used immediately for karyotyping, DNA analysis and metabolic tests. The advantage is early diagnosis, but there is a risk of about 2% of miscarriage, higher than for amniocentesis. The chorion is sometimes a mosaic of normal and chromosomally abnormal tissue even if the foetus is normal, which can lead to wrong diagnoses.

## 12.12.4. *Ultrasound screening*

**Ultrasound screening** is noninvasive, carries almost no risks, and can detect multiple births and major developmental abnormalities. At **10–12 weeks** gestation, anencephaly (absence of part or all of the brain) can be detected, but **16–18 weeks** is the earliest for detecting many abnormalities of head, body or limb development, including open neural tube defects such as spina bifida (open lower end of the neural tube). Males can be identified by observation of the external genitalia at about **18–20 weeks** gestation.

Ultrasound is particularly useful in detecting multifactorial anatomical traits such as club foot which cannot be done by DNA or biochemical analysis. It is widely used to detect congenital heart disease, kidney and bladder problems, polydactyly, dwarfism, etc. It has led to developments in prenatal surgery, and certainly not all the conditions mentioned here would lead to offers of termination of pregnancy.

## 12.12.5. *Maternal blood sampling*

Pregnant women at **16–18 weeks** gestation are usually screened for their blood **alpha-fetoprotein** (AFP). Foetuses with neural tube defects such as anencephaly or spina bifida tend to leak AFP into the mother's blood. Unfortunately the test is not very accurate as many women have raised AFP levels from other causes, so it is necessary to screen the foetus by ultrasound as well.

There is also the **triple marker** (Bart's) test, done on maternal blood at **16–18 weeks** gestation, measuring three serum components. It is usually only given to women of age 35 or over. It detects 2 in 3 cases of Down syndrome, 9 in 10 cases of anencephaly and 4 in 5 cases of spina bifida. It takes 10 days to get the results and its failure to detect all cases of these conditions can cause great distress.

Foetal cells and **foetal DNA** get into the mother's blood at very low concentrations, with foetal DNA coming mainly from the placenta. Chim *et al.* (2005) studied ways of using **epigenetic markers** to distinguish foetal DNA from the much more abundant maternal DNA so that it can be isolated from maternal blood and analysed to detect foetal genetic abnormalities. They found that the *maspin* gene promoter was hypomethylated in placental tissues and densely methylated in maternal blood cells, the main source of maternal DNA in a mother's blood. That provided the first universal marker for foetal DNA in maternal plasma, permitting measurement of foetal DNA concentrations in pregnancy-associated disorders such as pre-eclampsia and trisomy 13, irrespective of foetal gender and genetic polymorphisms. Foetal DNA has by other methods been used for the prenatal diagnosis of sex-linked diseases, foetal rhesus D blood group type and $\beta$-thalassaemia (references in Chim *et al.*, 2005). While the *maspin* gene methylation case is promising, it does not directly lead to isolation and diagnosis of other foetal genes from circulating foetal DNA.

## 12.12.6. *Pre-implantation genetic diagnosis*

In 1990, Robert Winston and colleagues at Hammersmith Hospital, London, announced the technique of **pre-implantation genetic diagnosis** (PGD), testing cells isolated from pre-implantation embryos for chromosomal analysis and sex determination, and for genetic disorders after PCR to multiply up the DNA for probing for specific conditions. Its first use was to eliminate males when there was a high chance of an X-linked serious recessive disorder. It can only be used on embryos from *in vitro fertilisation* (IVF), with one cell being removed for testing from the eight cells present in three-day-old embryos; at later stages, tight junctions formed between cells make single-cell extraction difficult (Braude *et al.*, 2002). This removal of a cell does not usually harm the embryo. Embryos found to be free of the defect can then be implanted in the mother's uterus. The technique is very successful in families with a known screenable defect and has many applications, but requires skill and IVF.

For example, Mrs Hunter is a carrier of an X-linked recessive condition, haemophilia A (Subsec. 12.3.2). She had an affected father, an unaffected husband and an affected son who had to inject himself with Factor VIII. After two cycles of IVF treatment, four embryos were obtained and tested for the particular haemophilia A DNA defect carried by the mother. Possible embryos were a sufferer male, an unaffected male, a carrier female and an unaffected female. Two embryos were suitable for implantation, one unaffected (male or female) and one carrier female. Both were implanted and a girl was born. It was not known when the story was published (*The Daily Telegraph*, 11/7/2005) whether the daughter was unaffected or a carrier.

Even normally fertile couples trying for children have a pregnancy rate of only 15 to 20% per menstrual cycle; humans produce many abnormal embryos, most of which do not even implant. IVF and PGD can be used to diagnose causes of infertility, e.g., if all embryos have chromosomal abnormalities. If patients have a balanced translocation, embryos with normal chromosomes can be selected. PGD is only usable for single gene and chromosomal defects, not multifactorial ones such as schizophrenia. It can also be used **to select a compatible sibling** for stem cell donation to an existing diseased child (preimplantation tissue typing). PGD is not allowed in the UK for choosing hair and eye colours ("designer babies": Sec. 17.3). In the UK it is not permitted for social choice of sex of baby but that is being reconsidered (2005) for "family balancing", as permitted in Belgium, Jordan and Israel.

Embryos at the eight-cell stage are put in medium free of calcium and magnesium to reduce cell adherence, then are held on a pipette while a hole is made in the zona pellucida with acidic Tyrode solution. A manipulating pipette is used to take out a single blastomere cell. PCR can be used to identify autosomal and X-linked defective alleles and for identifying sex. Each mutation or disease requires a specific PCR test. FISH can be used to visualise chromosomes, especially 13, 18, 21, X and Y as they can give non-lethal aneuploidies. Screened embryos are then selected for implantation, frozen for future use, or discarded. Costs are about £4,000 to £7,000 a cycle for PGD, and there are many technical problems such as mosaicism in an embryo and contamination with other DNA, e.g., from a polar body or sperm.

## 12.12.7. *Foetal DNA screening*

**Foetal DNA**, obtained from chorionic villus sampling or amniocentesis, is amplified in specific regions by PCR (the polymerase chain reaction) for testing with labelled probes for particular disease alleles. Some of the main diseases which can now be tested for in this way are $\alpha$- and $\beta$-thalassaemia, cystic fibrosis, fragile-X syndrome, haemophilia A, Huntington disease and Duchenne muscular dystrophy. For Duchenne muscular dystrophy, the dystrophin gene is amplified and can be hybridised with a specific probe.

A major problem with DNA testing is that probes are normally **allele specific**, detecting only one type of mutation each. There are many possible mutations at the locus for muscular dystrophy, so one probe or even a few probes for the commonest mutations will still fail to detect the other mutations. For cystic fibrosis, where 1 in 25 "whites" is a heterozygous carrier, 70% of mutant genes carry the same mutation, a deletion of a phenylalanine codon at position 508. A probe for this will miss the other 30% of mutations, although the recent DGGE CFTR kit from Ingeny is claimed to detect all mutations. With sickle-cell anaemia, nearly all mutant genes have the same defect in the $\beta$-globin gene, a base-substitution which changes a glutamic acid to valine, and which changes a restriction site, so the mutant gene is easily recognised by DNA analysis.

**Restriction fragment length polymorphisms** (RFLPs) can be used in certain favourable families if the gene for a disease has not been characterised enough for DNA analysis for the mutant allele, and if there is a closely linked RFLP. If one parent is heterozygous both for an autosomal dominant disease and for a very closely linked RFLP marker, the foetuses can be tested for the RFLP marker in samples from amniocentesis.

Suppose that the affected father is *D d* for the dominant disease and *m1/m2* for the RFLP marker, and that the normal mother is *d d, m2/m2*, then one needs to find out whether the father is *D, m1/d, m2* or *D, m2/d, m1*. If the first child born is a sufferer *D d* and *m1/m2*, then the father must have been *D, m1/d, m2*, as the mother provided the *m2* RFLP and *d* in the foetus. If amniocentesis of subsequent foetuses shows any to be *m1/m2*, one can predict that they will be sufferers (*D d*) and termination can be offered, while foetuses which are *m2/m2* should be non-sufferers, inheriting the father's harmless *d* allele. If the first child is normal and is *m2/m2*, then subsequent foetuses with that RFLP combination should be normal, but those which are *m1/m2* have a high risk of having the dominant disease. If the RFLP site is not very closely linked to the disease site, then errors of diagnosis may be made when recombination between the sites occurs.

## 12.12.8. *Neonatal screening (birth to one month of age)*

There is no routine **neonatal** DNA screening. The important routine screening is for **phenylketonuria** (Subsec. 12.3.1) in the first week of life, introduced in 1961. The mother controls the foetus's levels of phenylalanine in the blood before birth, but recessive homozygotes for PKU have blood phenylalanine levels rising to 15 to 30-fold of normal during the first week after birth. This can soon cause brain damage. Around day three, blood is taken from the baby, usually from a heel-prick, and a dried spot is bioassayed for levels of phenylalanine — details were given in Subsec. 12.3.1 — for detection, with a rapid start of dietary treatment for sufferers. PKU has an incidence of about 1 in 10,000. In Denmark it was estimated that **savings on health care** of affected individuals were 28 times those of neonatal screening costs for PKU and hypothyroidism.

Although it would not be economic to screen separately for some very rare diseases, other blood drops from the PKU sample are used to test for galactosaemia (sufferers cannot metabolise galactose, the main sugar in milk; 1 in 70,000 is affected), maple syrup urine disease (1 in 250,000) and homocystinuria (1 in 100,000). All these are treatable to some degree by controlling early diet. Congenital hypothyroidism (1 in 3,500 in UK "whites", 1 in 900 in Asians, 1 in 25,000 USA "blacks") can be detected by testing a dried blood spot for elevated levels of thyroid-stimulating hormone.

## 12.12.9. *Adult or adolescent screening*

This is usually done only where there is a high risk of certain conditions in a particular population. **Genetic counselling before marriage** can be given where such screening is carried out. It is usually uneconomic to screen the general population for inherited diseases, especially just to detect carriers, although "**cascade screening**" (screening relatives of affected individuals) is often done, e.g., for the autosomal dominant gene for familial hypercholesterolaemia in families with a history of premature coronary disease, or mutations in the adenomatous polyposis gene in families with colorectal cancer. In Ashkenazi Jews, **Tay-Sachs disease** usually has a frequency of about 1 sufferer in 3,000, with 1 carrier in 30 compared with 1 carrier in 300 in most other groups. It is due to a lack of hexosaminidase A enzyme, giving death before the age of three, with no effective treatments. Blood enzyme (or DNA) screening can detect carriers and they can be given genetic counselling before marriage, and sufferers can be detected from amniocentesis. Counselling plus prenatal detection and selective abortion reduced by 65% the frequency of sufferers born to Ashkenazi Jews over the period 1970–1980. Also in Ashkenazi Jews, 1 in 400 is a sufferer and 1 in 10 is a carrier for the autosomal recessive **Gaucher disease** which causes glucosylceramide accumulation. Type 1 gives enlarged liver and spleen, skeletal disease and multi-organ failure; type 2 gives neurological complications leading to infant death; type 3 is intermediate. DNA screening can detect homozygotes and heterozygotes carrying the defective gene, so people of Jewish ancestry can be offered screening.

**Cystic fibrosis** genes can be screened for in adults, including carrier detection, but as noted above, one probe only detects one type of mutant gene, so the probe for the most common mutation detects only 70% of cases in Northern Europeans. **Sickle cell anaemia** sufferers and carriers can easily be screened for by electrophoresis of blood haemoglobins, or by a reliable DNA analysis mentioned above.

In Cyprus, malaria resulted in about 16% of the adult population being carriers of $\beta$-**thalassaemia**, with 0.6% of births being sufferers who usually died within one year of birth. With such a high frequency of the deleterious gene, it was worth screening the whole population before they reached marriageable age, so that carriers could be given genetic counselling. Malaria has been eliminated from Cyprus, so the lack of heterozygote advantage should reduce the frequency of the bad gene. Cypriots (and others from malarial areas) who migrated to non-malarial areas such as Britain have a lower chance of producing sufferer babies if marrying a native of that area rather than another Cypriot. A Cypriot/Cypriot marriage has a 1 in 200 chance of an affected child because there is a chance of 2% of such marriages being between two carriers.

With the great advances in molecular genetics, there are **worries** that employers and insurance companies might insist on DNA screening of potential employees or of people wanting insurance, to detect those with genes which might cause increased risks of early death or of illnesses such as heart disease or cancer. In Britain in 1998, the Human Genetics Advisory Commission found only one case of employers using such screening. It was the Ministry of Defence screening potential aircrew members for sickle-cell anaemia, a very sensible precaution (Subsec. 12.3.1). In America in 1998, the American Management Association found nine out a sample of 1,085 employers used genetic testing of staff, and 11 states have passed legislation limiting such genetic testing. For Huntington's chorea, sickle-cell anaemia, muscular dystrophy, thalassaemia, Tay-Sach's disease and some forms of bowel cancer, there are reliable tests for the relevant genes. For dyslexia, some mental illnesses, diabetes, arthritis, heart disease and many cancers there are genes which **predispose** people to having those disorders, but mere possession of the gene does not necessarily mean that disorder will develop.

About 5% of women who develop **breast cancer** have a mutant gene, *BRCA1*, but not all who have that gene get the cancer, and most women who develop breast cancer have no known genetic predisposition to it. In 1997 in Manchester, England, two sisters aged 26 and 28 opted to have both breasts removed in an attempt to avoid breast cancer which ran in their family, with their maternal grandmother dying of it at 26 and their mother dying of it at 31. The sisters were told that they had a 50% chance of having the bad genetic make-up, and that if they had that, there was an 85% chance of getting breast cancer.

There is often a huge **fear factor** associated with genetic testing. Individuals often prefer to have the hope that they do not carry a deleterious gene, rather than be tested, especially for late-developing conditions with no real treatment such as Huntington disease, although testing would aid decisions about whether to have children.

## 12.13. Effects of human inbreeding

**Inbreeding in humans** has the same effect as in plants and animals (Sec. 6.2): it makes deleterious and beneficial recessives homozygous and exposes them to selection. In Japan after World War II, large-scale human genetic studies were made on survivors of the atomic bombs in the Hiroshima and Nagasaki areas, which gave data on the effects of inbreeding as well as on radiation effects. Where parents were first cousins, $F = 1/16$, the frequency of congenital malformations in the children rose from 0.011 to 0.016, a 48% increase compared to children of unrelated parents; stillbirths rose by 25% and the infant death rate rose by 35%. Even the amount of inbreeding involved in first-cousins matings was therefore clearly deleterious.

In data from rural France after World War II, the frequency of deaths before adulthood, including stillbirths, was 0.12 if parents were unrelated and 0.25 if parents were first cousins ($F = 1/16$ for their offspring). The increase in deaths from making one-sixteenth of the heterozygous loci homozygous was thus $0.25 - 0.12 = 0.13$, so making all heterozygous loci homozygous would give $16 \times 0.13 = 1.8$ deaths, so on average all children would be dead almost twice over! As a heterozygote $A\,a$ could go equally to $A\,A$ or $a\,a$, only half its homozygous offspring would die, so from the deaths we register only half the recessive lethal genes, thus the estimated **average number of recessive lethal alleles per person** from these data is 3.6. More modern data give estimates of one to four recessive lethals per person, obviously usually in the heterozygous condition, but the true number must be much higher as those giving death in the embryo or early foetus are probably undetected, especially in figures relating only to stillbirths and to deaths later in development.

**First degree relationships**, with parent-child and full sib matings, are defined as **incestuous**, with half their genes in common. Incest is banned in most communities, but many Asian communities tolerate or encourage first cousin marriages, especially in parts of India, Pakistan and the Middle East. Second degree relationships are between half sibs, aunt-nephew, uncle-niece and double first cousins (where all four grandparents are in common, as when a pair of sibs marries another pair of sibs), with a quarter of their genes in common. Third-degree relationships are between first cousins, half-uncles and nieces, or half-aunts and nephews, with one eighth of their genes in common. The term **consanguineous**, for matings or marriages between close relatives, extends as far as second cousins. To help study the effects of **incestuous matings**, there is a confidential **incest register** in the UK, recording the consequences of mating between very close relatives. Data from the UK and USA on 31 children from father/daughter and brother/sister matings showed that only 13 (42%) were normal and 18 were handicapped, with six dying as young children. All deleterious recessives will be expressed more often in the offspring of consanguineous marriages than if the parents are unrelated, e.g., cystic fibrosis, PKU and galactosaemia.

The **frequency of marriages between relatives** differs greatly between communities, depending on religion, customs, laws and population size. In data from the 1950's, in Baltimore, USA, the frequency of first-cousin marriages was about 0.05%, while in rural India, in Andhra Pradesh, it was 33%, with an average population F of 0.032. That same rural population also had an astonishing 9% of maternal uncle/niece marriages (F of offspring, 1/8), but a much lower incidence of paternal uncle/niece marriages, which is as if it didn't matter if the relationship was only on the female parent's side. In a very small population, or in a small, closed religious group marrying almost entirely within the group, even random mating may involve relatives such as first or second cousins. In Hopi Indians, with much marriage between close relatives, albinism has a frequency of 1 in 121 compared to that in white people, 1 in 20,000. For extensive data on the frequency of consanguineous marriages in different communities, see www.consang.net. Bittles (2002) gave an account of that frequency in India, which differs greatly between religions, castes, social levels and regions, being common in all Muslim communities (all types of first-cousin union), and in South Indian Hindus (especially uncle-niece and first cousin unions between a man and his mother's brother's daughter) but not North Indian Hindus. From about the sixth month of pregnancy to a median age of 10 years, **deaths in first-cousin progeny** exceeded those in non-consanguineous progeny by an average of 44 in 1,000 births. See also Bittles and Egerbladh (2005) for Swedish data.

Mate-choice is restricted in small communities, non-integrating immigrants, small religious groups, and pioneer groups, but in many communities consanguineous marriages are preferred, especially uncle-niece. **Figures for consanguineous marriages** given in www.consang.net. (2005) include 72%, Srikakulum, India; Nyertiti, Sudan, 71%; Nubia, Egypt, 65%; gypsies in Boston, 62%; Riyadh, Saudi Arabia, 31 to 55%. In India, such marriages were very frequent in Muslim and Hindu groups, and in Britain many first cousin marriages are between Muslims. While there are clear genetic disadvantages to such marriages between relatives, **benefits** can include retention of property and goods within the family; the woman's position is strengthened by kinship ties as a deterrent to ill-treatment by the husband or his family; dowry demands may be lessened for relatives; there are fewer problems in finding a mate of a suitable religion, culture, language and social group; better knowledge of the spouse's family background.

Consider **first-cousin marriages**. If an individual is heterozygous for a rare deleterious allele, the chance of a first cousin carrying an identical bad allele is 1/8 (first cousins have 1/8th of their genes in common through recent descent, the theoretical coefficient), and two heterozygotes mating will have an average of one quarter of their children homozygous recessive and suffering the deleterious phenotype. So for a person heterozygous for a deleterious allele and marrying a first cousin, the chance of a child being a sufferer is $1/8 \times 1/4 = 1/32$, about 3.2%. There will be a separate chance of 1/32 of being a sufferer for each recessive deleterious or lethal gene possessed, so with two to three lethal recessives on average per person, and other non-lethal deleterious recessives, it is not surprising that offspring of first-cousin matings suffer increased death rates and other problems, compared with children whose parents are not related. **Offspring of first-cousin marriages** tend on average to have slight reductions in height, girth and aptitudes (IQ down about 4 points), compared to non-inbred children.

The frequencies of sufferers from various genetic disorders who have first cousin parents are: 10 to 20% with albinism, 10% with PKU, 26% with xeroderma pigmentosum, 33% with alkaptonuria and 54% with micro-cephaly. In Iran, 46% of children with hearing loss were offspring of first cousins. The rarer the disorder, the higher the relative frequency of sufferers with closely related parents. The existence of disease-specific lay societies and the bond felt between fellow sufferers from a particular condition can cause **positive assortative mating** (Sec. 6.1) and increased frequencies of sufferers, e.g., for congenital deafness, dwarfism and albinism. There is also positive assortative human mating for perceived facial personality traits (Little *et al.*, 2005).

In one sense, **inbreeding is eugenic** as it exposes harmful recessives to selection, so reducing their frequency. Continued inbreeding in a group should eventually adapt it to inbreeding for this reason. The pharaohs of ancient Egypt practised brother/sister and half-brother/half-sister marriages for many generations and yet were generally successful. As usual in human genetics, one cannot be sure whether the stated royal parents were the actual father and mother.

If **consanguineous marriages** are defined as those between relatives who are second cousins or closer, the frequency of consanguineous marriages is between 20% and 50% in many Muslim societies in north Africa, Asia and parts of the former Soviet Union. In the primarily Hindu southern states of India, consanguineous marriages average 20 to 45%, and when families from these Moslem and Hindu communities migrate to the UK or the USA, they often retain a high degree of consanguineous marriages.

In the UK, the closest legal marriage is usually considered to be between double first cousins (where a pair of sibs marries a pair of sibs, so that both sets of grandparents of the children are in common, not just one set of grandparents in common as for normal first cousins), with $F = 1/8$ in the offspring. However, if **a pair of identical twins marries a pair of identical twins**, the offspring of one such marriage are legally first cousins of the offspring of the other marriage, but genetically they are the equivalent of full siblings, so if they intermarry, their offspring will have $F = 1/4$. Uncle-niece and aunt-nephew marriages are banned in the UK, where the Church of England's Table of Kindred and Affinity is the main basis of the marriage laws. Roman Catholics need special permission for first-cousin marriages. In more than half of the USA's states, uncle-niece, aunt-nephew and first-cousin marriages are banned, as they are in some largely non-Moslem African countries.

In Britain, **first-cousin marriages** are most common in the Moslem community, with the deleterious genetic consequences being partly offset by earlier arranged marriages, earlier first pregnancy and more children per family than in other groups, on average. Marriage within kinships may have economic benefits in keeping property together. In 2005, a Labour MP called for the banning of Asian marriages of cousins in Britain, because the Pakistani community accounted for 30% of births with recessive genetic disorders while only accounting for 3.4% of births. It was estimated that 55% of married Pakistanis in Britain are married to first cousins, rising to more than 75% in Bradford. See Bittles *et al.* (1991), Bittles (2002) and Bittles and Egerbladh (2005), for more on human inbreeding.

## 12.14. Genetic counselling

**Genetic counselling** was defined by Harper (2004) as: "the process by which patients or relatives at risk of a disorder that may be hereditary are advised of the consequences of the disorder, the probability of developing or transmitting it and of the ways in which this may be prevented, avoided or ameliorated". The aim of genetic counselling is to give information to sufferers, carriers or their relatives, or anyone with a high risk of having or passing on a deleterious genetic condition. Under the UK Congenital Disabilities (Civil Liability) Act, 1976, a person whose breach of duty to parents results in the birth of a disabled or abnormal child can be sued, and there is increasing litigation in the USA and UK over genetic disease.

The condition must be **accurately identified**; for example, not all forms of muscular dystrophy are caused by X-linked recessives, and some forms are not heritable. **Treatment** must be given if possible, e.g., by drugs, hormones, special diets, blood transfusions, physiotherapy or surgery. A **prognosis** must be given, informing the sufferer or relatives or guardians what the likely course of the disease will be in future. If the sufferer might be fertile and is capable of understanding the issues, he or she (or relatives or other carers) is given information on the likelihood of having suffering or carrier **children** by different types of partner.

For example, a sufferer from **phenylketonuria** would be advised that all children by a fellow sufferer would be sufferers, that of children by a carrier, about half would be sufferers and half would be carriers, whereas children by a normal dominant homozygote would all be carriers. Information would be given about detecting carriers, about prenatal diagnosis and about the easy detection of sufferers in the first week of life and the effective treatment by diet. Women with PKU are advised to adhere strictly to a low phenylalanine diet during pregnancy because a normal foetus can be adversely affected when growing in an affected mother.

Counsellors need to be aware of phenomena such as: **incomplete penetrance** (Subsec. 3.2.5), through which autosomal dominants such as inherited colon cancer or hereditary pancreatitis can unexpectedly "skip a generation"; **variable expressivity** (Subsec. 3.2.4), e.g., polydactyly; and **variable age of onset**, as in Huntington's disease (incidence, 1 in 20,000 in Caucasians) and adult polycystic kidney disease (incidence, 1 in 1,000). Where penetrance is high for such autosomal dominant conditions, a healthy sib of a sufferer is unlikely to have the bad gene or to pass it on to a child. Where penetrance is low, the chance of carriers of the gene being affected is also low and the risk to their children is low, even if they inherit the defective gene. Even for conditions with full penetrance and constant expressivity, the existence of **gonadal mosaics** (Subsec. 12.16) can complicate the calculations of recurrence risks. An individual might be symptomless because a dominant new mutation was confined to his or her gonads, but the recurrence risk to children could be 50% each if all the gonadal cells carried it. The **recurrence risk** can be defined as the chance that a further child born to two parents who already have an affected child will be affected by that condition. Thus two unaffected parents who have had a child suffering from an autosomal recessive disorder must both be heterozygous carriers, with a risk of 25% that the next child will be affected. If only one parent carried an autosomal dominant and was affected, although the condition has only 60% penetrance, then a child would have a 50% risk of getting the bad gene, but only a 30% risk (60% of 50%) of developing the disorder.

For **autosomal dominants** (Subsec. 2.1.2), a heterozygous sufferer married to an unaffected person has a risk of 50% for a child being affected, but that can be modified by incomplete penetrance, variable expressivity and variable age of onset. For **autosomal recessives** (Subsec. 2.1.2), the recurrence risk from two carriers is 25%. For **X-linked dominants**, affected fathers will pass the condition to daughters only, and heterozygous affected mothers will pass the condition to half their sons and half their daughters. For **X-linked recessives**, affected males married to normal females will have all carrier daughters and normal sons: see Subsec. 2.1.4 for colour-blindness.

It often requires **pedigree information** on relatives to establish the **counselled person's genotype**, or the probability of having a particular genotype, for simple Mendelian conditions. Estimating risks of a couple having further sufferers is much more difficult for multifactorial conditions. For a dominant condition with full penetrance, one can deduce heterozygosity if only one parent is a sufferer, but if both parents are sufferers (which will be very rare for uncommon conditions), their suffering offspring could be heterozygous or homozygous dominant. For an autosomal recessive, one can deduce that both unaffected parents are carriers. If one can deduce genotypes, calculating Mendelian expectations is usually straight forward, but if genotype assessments are just frequency estimates, then further information can change **basic risk** figures to **modified risk** figures, as shown below.

**Myotonic dystrophy** (incidence, 1 in about 13,500 in Caucasians) is an autosomal dominant on chromosome 19 and is the most common adult muscular dystrophy, with very variable clinical symptoms. Definitely affected heterozygous people have a 50% risk of an affected offspring, whatever the number of affected individuals in the family, with a constant risk figure.

For **X-linked recessives** such as Duchenne muscular dystrophy (which results in death in childhood from progressive muscle deterioration), where a female has a carrier mother, that female has a basic risk of 50% of inheriting the defective allele. If she has the defective allele, there is a 50% chance of any one son being affected and of any one daughter being a carrier. Suppose that Jane (the "consultand") is phenotypically unaffected and comes for genetic counselling about her risk of having affected sons. Jane's mother Daphne is phenotypically unaffected, with an unaffected husband, and Daphne has two phenotypically unaffected parents. Daphne has two affected brothers with Duchenne muscular dystrophy, so Daphne's mother Alice must be a heterozygous carrier. The chance of Daphne being a carrier is 50% and Jane has a 25% chance of being a carrier; those are the **basic risks**. In genetic counselling for such X-linked recessive cases, one can modify the basic risk figure in the light of other relatives' phenotypes. Let us consider how that basic risk can be modified by further information in three cases:

*Case 1.*  If Jane has any affected brothers, then her mother Daphne must be a carrier, when Jane's chance of being a carrier is 50%, not 25%.

*Case 2.*  If however Jane has four brothers, all unaffected, then her mother Daphne could still be a carrier but with less than 50% probability, so that Jane's chance of being a carrier is less than 25%.

*Case 3.*  If Jane has an affected son, she must be a carrier and has a 50% risk to further sons.

One does not have to rely on genetic ratios for predictions if **direct tests** are possible, as for conditions where DNA analysis can be done on adults (e.g., for possible heterozygous carriers) or in prenatal diagnosis, or where other tests can be given before birth, e.g., on the mother's blood for alpha-fetoprotein for suspected neural tube defects (Subsec. 12.12.5). Unfortunately, DNA tests are often not reliable if there are many different mutations found at a locus.

Some conditions are difficult to counsel for because they have a number of possible genetic types. A classic case is **osteogenesis imperfecta**, brittle bone disease, incidence about 1 in 7,500, with several subtypes, with severities varying from being lethal before birth to multiple fractures, deafness

and osteoporosis in later life. Many cases are due to new dominant mutations (the recurrence rate in siblings is then extremely low), while others are
autosomal recessive (the recurrence rate for siblings of an affected child is
25%). The different non-lethal forms are difficult to distinguish clinically.

Well-characterised disorders with regular inheritance patterns are the
easiest to counsel for, and sporadic, vaguely-defined conditions are the
hardest. **Sporadic conditions** may be caused by the environment (including infection), by polygenic factors, by rare autosomal or X-linked recessives, by chromosomal changes, or by new rare dominant mutations. With
the unstable length mutation diseases referred to earlier, the severity can
increase between generations in unpredictable ways.

For **autosomal recessive conditions**, the incomplete penetrance and
variable expressivity which complicate counselling for some autosomal
dominant conditions are less frequent. Sufferers from autosomal recessives
from new mutations are expected to be very rare, much rarer than for autosomal dominants. For a well established autosomal recessive when a sufferer
is born to two phenotypically normal parents, one can be almost certain
that both parents are heterozygous carriers and the risk to further children
is 25%. If a rare disorder occurs in one individual only in a family, and is
of unknown genetics, it can be hard to tell whether it is due to a new dominant mutation or to two carriers having produced a homozygous recessive,
especially when family sizes are small.

**Mitochondrial inheritance** occurs for a number of disorders (see
Harper, 2004; Vogel and Motulsky, 1997), of which **Leber's optic atrophy**
is the best known case. It causes rapid vision loss in young adults. The
human mitochondrial genome is circular, more than 16 kb long, encoding
a small (12S) and a large (16S) rRNA, 22 tRNAs, and 13 polypeptides
involved in oxidative phosphorylation. Mitochondria are inherited through
the mother only, through the egg, but not through the sperm which has little
cytoplasm. At fertilisation, the egg cell has about 200,000 mitochondria,
but they do not replicate until the blastocyst stage. Only a few blastocyst cells become embryo cells, and of those that do, very few become
female germline cells. A woman may be uniform (homoplasmic) in her
mitochondria for a particular mitochondrial gene, or may have different alleles in different mitochondrial genomes (heteroplasmic). During
development, different cell lines from a heteroplasmic egg may become

homoplasmic for different mutations, or remain heteroplasmic, and mito-chondrial mutations can render a homoplasmic cell line heteroplasmic. In Leber's optic atrophy, sufferers are usually homoplasmic for the mutation.

This **segregation of mitochondrial types** in different cell lines dur-ing female development makes the genetic counselling for mitochondrial disorders very difficult. A female may be phenotypically unaffected, with her somatic cells containing a small proportion of mutated mitochondria (or even none), but her germ line cells may contain no, a few, or a majority of mutated mitochondria, or a mixture of those types of germ cell. Her children, of either sex, could be unaffected, with all good mitochondria, or unaffected with a small proportion of bad mitochondria (able, if female, to transmit the disease to some offspring), or affected, with all or most bad mitochondria. No males can transmit the disorder, whether they are unaffected non-carriers, unaffected carriers or sufferers. All daughters of affected or carrier females could be affected and could transmit the disor-der, whether affected or not. Their sons are at risk of being affected but not of transmitting the disorder. The proportion of affected sons and daugh-ters is influenced by a number of poorly understood factors. For Leber's optic atrophy, for an affected female, or an unaffected female who has an affected son, the risk to later sons is 50%, while about 80% of daughters are unaffected carriers and 20% are sufferers, so there are more suffering males than females. Those risk figures are not predictable from theory but are found in practice.

Very **strong emotions** can be generated by suggestions of a genetic defect in a family, so great tact and confidentiality are required in coun-selling. A family in which someone is diagnosed as have a defective gene or chromosome sometimes feel that all members' marriage chances may be diminished, perhaps together with their employment prospects, reputation for mental stability, or insurability at normal rates. The affected person can feel guilty about having the condition, and so do the parents who passed on the defective gene(s). What is given by the counsellor is always informa-tion, never commands. In the case of a sufferer from an autosomal recessive disorder, both sides of the family have contributed equally to the defect, so any perceived "blame" for passing on the disorder is shared by both parents, instead of just one parental family being blamed as for an autosomal domi-nant. Relatives of an affected person sometimes refuse to be tested, as they would rather not know whether they carry the bad gene or chromosome. Relatives may have died or be out of touch or living in another country.

**Marriage between two sufferers**, from positive assortative mating (see Sec. 6.1) occurs fairly often for albinism, blindness and severe congenital deafness, where sufferers of one type of defect may be educated together in special schools and/or share social bonds. For counselling such couples, it is desirable to identify which locus is affected in each individual, if possible. With severe oculocutaneous albinism, the autosomal recessive mutations are normally at the same locus, so nearly all children of two sufferers will be homozygous sufferers, unless a very rare crossover occurs within the locus between mutations at different sites within it. **Albinism** exists in three main types: severe oculocutaneous (tyrosinase-negative) albinism (autosomal recessive); mild oculocutaneous (tyrosinase-positive) albinism (autosomal recessive at a different locus), and ocular albinism (X-linked recessive). Since the three types of albinism are determined by different loci, if albinos with different types of albinism marry, their children are phenotypically normal, being heterozygous carriers at both loci, with complementation between the dominant wild-type alleles. If two autosomal albinism loci are designated $c$ and $d$, then two albinos with the same type of albinism could both be $c^+ c^+, d^- d^-$, so all children would be $c^+ c^+$, $d^- d^-$, sufferers. Two sufferers of different types of albinism could be $c^+ c^+, d^- d^-$ and $c^- c^-, d^+ d^+$, with children being $c^+ c^-, d^+ d^-$, unaffected but double carriers.

**Counselling for multifactorial disorders** (Sec. 12.4) can be difficult as several to many loci may be involved, possibly plus environmental factors. For example, type 1 diabetes (insulin-dependent) and type 2 diabetes (non-insulin-dependent) have different susceptibility loci, with different ones for the two types, and with environment (diet and exercise) influencing type 2 diabetes quite a lot, but with much less influence on type 1. Multifactorial disorders are much more common than single gene disorders, as well as being more complex. Where Mendelian inheritance has been shown for some or all of the loci concerned, that is helpful for counselling, but complex interactions between loci, and between genes and environment, make risks hard to estimate.

The concept of **genetic liability** is useful for multifactorial disorders, with the liability being determined by those interactions of genetic and environmental factors. Liabilities for a particular disorder can be envisaged as showing a roughly normal distribution in a population, with a small proportion of individuals at each end of the distribution having very low or very high liabilities, and most having intermediate values. Most sufferers would be in the high liability part of the distribution, but not all in that part would be sufferers as they might not have the right environment for developing the disorder. It takes a combination of genetic and environmental factors to push an individual over the **threshold of liability**, so that they develop the disorder. The heritability of the disorder determines the relative importance of genetic and environmental factors. Harper (2004) gave the following **summary of risk factors**:

- The increased risk is greatest among **close relatives**, decreasing rapidly with the distance of relationship (e.g., see Table 12.3 for cleft lip).
- The risk of recurrence depends on the **frequency of the disorder**, with a rough guide being that the maximum risk to first-degree relatives is approximately the square root of the incidence, so if the incidence is 1 in 10,000, the recurrence risk is 1 in 100 for sibs of sufferers.
- Dominance and recessiveness are often unknown or do not apply (additive action), so the risk to sibs is similar to that for offspring.
- If there is a different incidence in the two sexes, the risk is higher for relatives of a sufferer of the sex which suffers less often from the disorder, as that sex will usually have need to have a higher genetic liability before it shows the condition. For example, **pyloric stenosis** is commoner in males (1 in 200) than in females (1 in 1,000), and the risk for brothers of a male sufferer is 3.8% compared to a risk of 9.2% for brothers of a female sufferer.
- The risk is increased if the disorder is **more severe**, as described for cleft lip, with or without cleft palate, in Sec. 12.4.
- The **more members of a family who are affected**, the greater the risk, as with breast cancer, as that shows a higher genetic liability in that family, unless they have a higher environmental risk. Different populations may differ in risk because of different gene frequencies and/or environmental factors.

Gene therapy is dealt with in Sec. 12.17. Counsellors experience difficulty when genetic advice runs contrary to traditional practices, as with marriage of close relatives amongst Muslims and Hindus (Modell and Darr, 2002).

## 12.15. Twins and other multiple births

While single births are most frequent, twins, triplets, quadruplets, quintuplets and higher multiples all occur naturally, with decreasing frequencies as the number born from the same pregnancy increases. With fertility treatments and implantation of several embryos into a woman after *in vitro* fertilisation, multiple births are then quite common. See Subsec. 12.2.2 for using twins to study how characters are inherited.

Multiple Births Canada gives the following figures for **multiple births** with no fertility treatment. **Twins**, 1 pair in 90 births. **Triplets**, 1 set in 8,100 births. **Quadruplets**, 1 set in 729,000 births. **Quintuplets**, 1 set in 65,610,000 births. A figure for **sextuplets**, from a different source, is 1 set in five billion births, with only six authentic records before fertility treatments began. The number of multiple births in Canada is rising, with 16% of multiple births coming from **fertility treatments**, including 60% of triplets, 90% of quadruplets and 99% of quintuplets. Between 1974 and 1990, multiple births in Canada increased by 35% in successful pregnancies, with the incidence of triplets and quadruplets going up by 250%.

In 1895, Hellin suggested that twins occurred about once in 89 births, triplets about once in $(89)^2$ births, which is one in 7,921, quadruplets about once in $(89)^3$ births, which is one in 704,969, quintuplets about once in $(89)^4$, which is one in 62,742,241, and sextuplets would then be expected at a frequency of one in 5.6 billion. These figures agree remarkably well with the Canadian data above.

The overall **frequency of twins births** differs in different populations. A Nigerian Yoruba population had 4.5% of all births being twins, while this frequency was as low or lower than 0.8% in several South American populations and among Chinese and Japanese. In the USA in 1964, before *in vitro* fertility treatments became common, 1 in 106 (0.94%) "white" births were of twins, compared with 1 in 73 (1.37%) "black" births. In Norway, in a valley near Trondheim Fjord, different family lines of several isolated agricultural communities had very different frequencies of twinning. They varied from no twins in 800 births in one pedigree (0%) to 107 in 2,840 births (3.8%), with an overall frequency of 1.2%. According to Eriksson and Fellman (1996), Eskimos, Ainus and American Indians frequently used to kill one or both twins and now have very low frequencies of twinning, while amongst Yorubas, twins were highly regarded and the twinning frequency is very high. Boklage (1990) suggested that more than 12% of all natural conceptions start as multiple pregnancies, with only 2% surviving to term as twins and the rest giving single births or no live births.

All figures for frequencies of multiple births are for cases where at least one baby was born alive. As multiple births involve **higher mortalities** than single births, the frequency of multiple births at conception is probably much higher than those recorded at birth, with many spontaneous abortions of all the babies in a pregnancy. Recent American research with ultrasound scans of women in very early pregnancy suggests that identifiable twin embryos are present in about **15% of conceptions**. In most cases, however, one twin is lost very early in pregnancy without this being noticed, seeming later to the mother like a normal single birth.

Twins can be **identical**, from the division of a single fertilised egg, so that they carry identical genes (Plate 2.1), or **non-identical**, from two separate fertilised eggs, with different sperm, and with only half their genes in common. Identical twins are of the same sex but non-identical twins can be same sex or opposite sex, and apart from their time of birth, non-identical twins are no more alike than a brother or sister, or two brothers or two sisters. Identical twins occur at similar frequencies all over the world, at about 3.5 per 1,000 births. An exception is the village of Umri, in Uttar Pradesh, India, with 34 sets of twins, mostly identical, in a population of 900, a frequency of identical twins more that 10 times the normal frequency. **Non-identical twins have a more variable frequency**, from a low rate of 2 to 7 per 1,000 births in many Asian communities, to 45 per 1,000 in Nigeria. In studies of interracial crosses in Hawaii, the frequency of non-identical twins was highly correlated with the **mother's race** but was not correlated with the father's race, as expected if the frequency of multiple egg production is the key factor in such twinning.

From these statistics, one can calculate for different populations the **chances that a pair of twins will be identical, not non-identical**. The figure is 1 in 2.3 for Chinese and Japanese, 1 in 2.7 for "White" Americans, 1 in 3.9 for "Black" Americans, and 1 in 11 in Nigerians such as Yoruba.

Multiple births from fertility treatments are mainly of non-identical sets, as the fertility treatment increased egg-production or involves implantation of several separate fertilised eggs.

Apart from the use of ovarian stimulants in fertility treatments, a major factor in the frequency of non-identical twins is the **mother's age**, with the frequency of this type of twinning rising rapidly as women increasingly delay pregnancy, say for career reasons. For adolescent mothers, the chances of a multiple birth are 6 per 1,000 births, rising with the mother's age to a peak of 16 per 1,000 for ages 35 to 39, going down to 13 per 1,000 at 40 to 44, then down to 8 per 1,000 in women of 45 and over.

**Spontaneous identical twin formation** from a single fertilised human egg can occur at different stages of its development, up to two weeks after fertilisation. If the two daughter cells separate after the first division of the fertilised egg, one can get identical twin embryos implanting in the womb far apart or close together. If separation into two halves occurs at a later stage, there may be a single implantation, but with two embryos developing. Two thirds of identical twins share a placenta, and one third have their own placentas, as do all non-identical twins. Almost all twins, of both types, develop in their own separate amniotic sacs within the womb.

Although such events are extremely rare, fertilised eggs have occasionally divided into three or more, giving for example three healthy red-headed **identical triplet boys**. The **Dionne quintuplets** were born in Canada in 1934 and were the first recorded case in modern times of all five members of a set of quintuplets surviving birth and adolescence. By blood tests and other tests, they were all shown to be identical genetically, from one fertilised egg. Most cases of quintuplets which have been analysed appear to be combinations of identical and non-identical babies. Triplets can come from one, two or three separate eggs, requiring one or more egg splitting if from two or one eggs. **X-inactivation** usually occurs before separation of identical twins, so that identical female twins who are heterozygous for X-linked characters such as haemophilia, Duchenne muscular dystrophy or colour blindness may differ in those characters. For example, two of the five Dionne identical quintuplets were colour blind and three were not.

**Non-identical twin formation** comes from the mother maturing two or more eggs in one menstrual cycle, and two of them being fertilised and implanting in the womb. Although **non-identical twins** always have the same mother if from natural births, they may rarely have **different fathers** if the mother mated with different men in a short space of time. There was a classic case when Nazi Germany had taken over Austria, with discrimination against Jews, when it was advantageous to be shown not to have Jewish origins. A non-Jewish mother of non-identical twins had a Jewish husband but claimed that her 25-year-old twins were conceived during an affair with a non-Jewish man, making the twins non-Jewish. All five people were available and were tested for blood groups ABO and MN (where heterozygotes are MN). The husband was B, M; the mother was O, M; the ex-lover was A, MN; the twin brother was B, M, and the twin sister was A, MN. That is consistent with the husband being the father of the twin brother and the ex-lover being the father of the girl twin, providing her with the A and N alleles which the husband and mother lacked. In a later case of non-identical twins with different fathers, one father was white and the other was black, giving rise to very different twins.

If twins are of opposite sex, they must be non-identical. If they are of same sex and differ in many characteristics, then they are probably non-identical. DNA testing is the most accurate way of finding out, although blood groups and finger prints have been used. Approximately 95% of the variation in the **total number of fingerprint ridges** is genetic, with about 5% being environmental or chance, so because of that 5%, identical twins do not have fully identical finger prints but are much more alike than for non-identical twins.

In spite of their name and although they have identical genes, **identical twins are often not physically identical**, from chance environmental factors in development, before or after birth. Within the womb, one twin may get a better supply of nutrients from the placenta or be better placed within the womb for space. One may suffer birth trauma when the other does not. After birth, twins may get different diseases, or if separated, be subjected to different environments. **Identical twins may therefore differ** in size and shape at birth, and afterwards. There was a case of two Asian boy identical twins where at the age of about 14, the top of the shorter one's head only reached the bottom of his twin's nose. There was a case of two white boys, where one identical twin had a cleft lip and palate and the other had neither. Identical twins sometimes show "**mirror imaging**", where one may be left-handed and the other right-handed, and one may show clockwise hair whorls on the back of the head when the other shows anticlockwise whorls.

Fraga *et al.* (2005) studied the global and locus-specific differences in **DNA methylation and histone acetylation** in a large cohort of identical twins. They found that twins are **epigenetically indistinguishable** during their early years, but older monozygotic twins showed remarkable differences in their overall content and genomic distribution of 5-methyl cytosine DNA and histone acetylation, affecting their profiles of gene expression. The within-pair profiles for twins aged three were nearly identical while those for 50-year-olds were very different. The twin pairs who had spent less of their lifetime together, or who had different lifestyles, also showed the greatest epigenetic differences, suggesting effects of environment on epigenetic changes, as well as random epigenetic drift occurring with increasing age.

**Epigenetic effects**, from DNA methylation, have been increasingly evoked to explain many phenomena, such as a pregnant woman's dietary deficits increasing her offspring's risk of diabetes, stroke and heart disease later. Certain nutritional supplements fed to pregnant mice altered coat colour in their offspring from methylation of the agouti gene. Behaviour is also claimed to be affected by epigentic effects.

**Siamese twins (conjoined twins)** are very rare, about 1 in 50,000 births, and most pairs die young. They arise from incomplete splitting of one embryo, so are identical. They are named after two famous conjoined twins, Chang and Eng, born in 1811 in Siam (Thailand), who later made a living in the USA exhibiting themselves in curiosity shows before switching to tobacco farming. They married a pair of sisters, and being joined did not inhibit their love lives, as Chang had 10 children and Eng had 12. Eng was unaffected when Chang had a stroke at the age of 61, but when Chang died of bronchitis two years later, Eng died two hours later. They were joined only by a tissue bridge 10 cm wide from the lower end of the breast bone (sternum) to the navel. It contained liver tissue connecting the two livers, and separation surgery in those days would probably have been unsuccessful, although it should succeed today.

**Surgery** to separate conjoined twins is normally carried out and success depends on what parts are joined and what organs are shared. For example, in 2001, staff at the Birmingham City Hospital made the world's third, and Britain's first, successful separation of twins joined at the spine. Girls **Eman and Sanchia Mowatt** weighed a combined 4.5 kg when born by Caesarian section at 36 weeks' gestation. They had emergency surgery to separate a joined piece of gut, then were allowed to grow stronger for three months, doubling in weight before the main operation to separate them. A silicone balloon was placed under the skin 10 weeks before the operation and was filled with enough liquid to stretch the skin so that it would cover the separation wounds on both twins. In the 16-hour operation, the surgeons separated their two spinal canals and their lower spinal cords which were joined in the lower back. The operation was a success. The mother noted that these identical twins already had very different personalities, with one more dominant, vocal and demanding than the other.

Multiple births obviously put **much more strain** on the mother and on the babies than do single births, with some effects on the father too. As one would expect from the fact that a mother has only limited resources for feeding babies in the womb, the more babies per pregnancy, the **lower the average birth weight** of each child and the higher the risk of defects and the greater the need for neonatal intensive care. In Canada, multiple births represent 2% of all live births, but 16% of the low birth-weight infants. Almost half of all twins are low birth weight and/or premature, while that figure rises to more than 90% for triplets, quads and quins. Children in multiple births are five times more likely than those in single births to have birth defects and/or disabilities. Whilst many families with multiple births manage perfectly well, there are extra difficulties, practical, physical, emotional and financial.

Twins, triplets, etc., have to **compete with each other** for food resources and space within the womb, and for food, care and attention after birth. With any kind of twin, there are higher incidences of congenital malformations and lower average birth weights than in children from single births. The IQ (intelligence quotient) of twins, from any social class, is on average 5 to 6 points lower than for their non-twin brothers and sisters. Identical twins, on average, suffer even more than non-identical twins from a higher death rate around the time of birth and from a higher frequency of congenital malformations, possibly relating to identical twins more often being very close to each other in the womb, with more crowding than for non-identical twins. Most twins are healthy, however, and often catch up to more normal weights as they grow.

Approximately 50,000 children a year are born world-wide from *in vitro* **fertilisation (IVF)**. A study of Swedish children by Strömberg *et al.* (2002) mentioned that 2% of babies in that country were born after IVF, with 45% of those by intracytoplasmic sperm injection, with routine implantation of two embryos. Their study compared 5,680 IVF children with 11,360 children born after natural fertilisation, and 2,060 twins born after IVF compared with 4,120 twins from normal fertilisation. The IVF babies were three times more likely to suffer **brain damage**, especially cerebral palsy, and four times more likely to have **slow development**, than babies from normal fertilisation. Twins born after IVF were not significantly different in these respects from twins from normal fertilisation. The researchers concluded that these differences in fitness were not due to the IVF process but were due to the high number of twins from IVF, with resulting lower birthweight and lower gestational age. They recommended only implanting one embryo at a time, not two. As 30% of all the IVF children were born before 37 weeks and 7% were born before 32 weeks of gestation, compared with 11% and 2.6% for IVF single births, an implantation of only one embryo would reduce prematurity by about 60%, although it would lead to more failures to produce any births.

In another Swedish study, Bergh *et al.* (1999) reported that in most countries, **assisted births** such as IVF resulted in **20–30% multiple births**, with 30–40% of all IVF babies being born as a result of multiple pregnancies, compared with 2–3% in the general population. In Sweden, multiple births occurred in 27% of IVF pregnancies, compared to 1% of normal pregnancies. More IVF babies were **born early** (less than 37 weeks) than were normal babies, 30% and 6%, respectively, and more had **low birthweights** (less than 5 $1/2$ lbs or 2,500 g) than normal babies, 27% and 4.6% respectively. There were also proportionally more neural tube and oesophageal defects in the IVF babies. The frequencies for delivery by **Caesarian section** were 33% for single births, 53% for twins, and 88% for higher order multiple births.

With **IVF in Britain**, in 2002–2003, 9% of cases had one embryo implanted, 76% had two embryos and 15% had three, resulting in 76% single births, 24% twins and 0.5% triplets. In 2005, the Human Fertilisation and Embryology Authority (HFEA) decided to examine whether Britain should follow many other nations in only allowing one implanted embryo, to **minimise the risks associated with multiple births**. In January 2004, the HFEA limited clinics to two embryos at a time to women under 40 and a maximum of three to those over 40.

Our knowledge of human genetics has benefited enormously **from the use of twins in medical research**. As previously mentioned, by comparing the differences between identical and non-identical (of the same sex, to avoid sex differences biasing the results) twins reared together and reared apart, one can study the extent to which variation for particular conditions results from genetic variation and from environmental variation. For this purpose, many countries keep special **twin registers** of identical and non-identical twins who are willing to participate in such studies, either generally or for particular areas of study. For example, there are twin registers in Italy (120,000 twins), Korea (154,783 twins), Britain (at the Twin Research and Genetic Epidemiology Unit, St Thomas' Hospital, London), Sri Lanka and Australia (Boomsma *et al.*, 2002). The wide range and power of twin studies are shown by these extracts from the Information Leaflet from the St Thomas' Unit. Their twin research has shown that: osteoporosis is 75% genetic; disc degeneration causing back pain is 60% genetic; familial diabetes is related to stomach fat; smoking alters skin thickness; genes control body shape; timing of the menopause is largely genetic; musical ability is 80% genetic; "wacky" humour is due to upbringing; short-sightedness is 85% genetic; blood clotting properties are mainly genetic; moles and freckles are strongly genetic.

## 12.16. Mosaics, chimeras and hermaphrodites

Mosaics have already been mentioned several times in this chapter. **Mosaics** are individuals coming from a single zygote but consisting of **two or more genetically different cell lines** that diverged in development by mutation, new chromosome aberration or chromosomal non-disjunction. This can produce a range of phenotypes, depending on the genetic change and how early it arose in development. For example, 1% of Down syndrome cases are mosaics (Plate 12.19), with varying severity of the disease depending on the proportion of trisomic and normal tissue. Mosaics sometimes have distinctive patterns of differently pigmented skin, Blaschko's lines, which may only be visible under UV light.

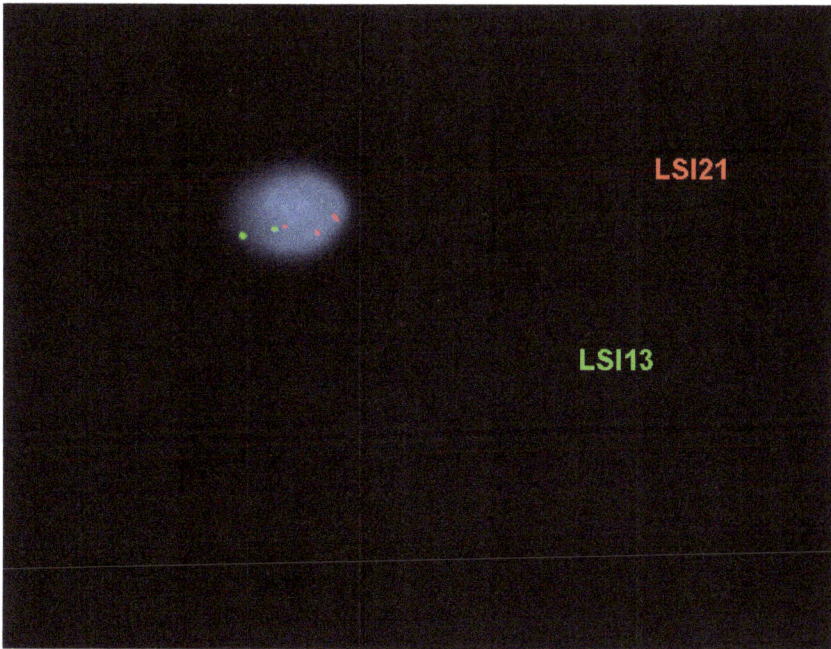

**Plate 12.19.** **A female mosaic with Down syndrome in buccal swabs but not in blood cells.** This female has a classical Down syndrome phenotype but is a mosaic. Blood cells are normal (two copies of 21) but in buccal swabs as shown here, 70% of cells have trisomy 21. The probes are 21q22 probe LSI21 (red) and 13q12 probe LSI13 (green). This demonstrates the use of FISH on interphase cells which cannot divide in culture and which cannot have their chromosomes examined normally. Clinical applications include the detection of mosaicism and examining sperm.

**Chimeras** also have more than one cell line, but they originate from at least **two different zygotes**. They are rare and their true frequency is unknown as they are usually only discovered by accident. For example, an individual may have more than one type of blood group or ambiguous genitalia. Individuals may have signs of hermaphroditism or be microchimeras, with say some blood cells of different ancestry. Transplants, including bone marrow transplants, are a source of chimeras: see Plate 12.20.

**Natural "whole" chimeras**, involving many tissues, form from the fusion of two independent fertilised ova (of the same sex or of different sexes), or from fertilisation of a mature ovum and its first or second polar body. If the first polar body is fertilised by a sperm, the two zygotes will

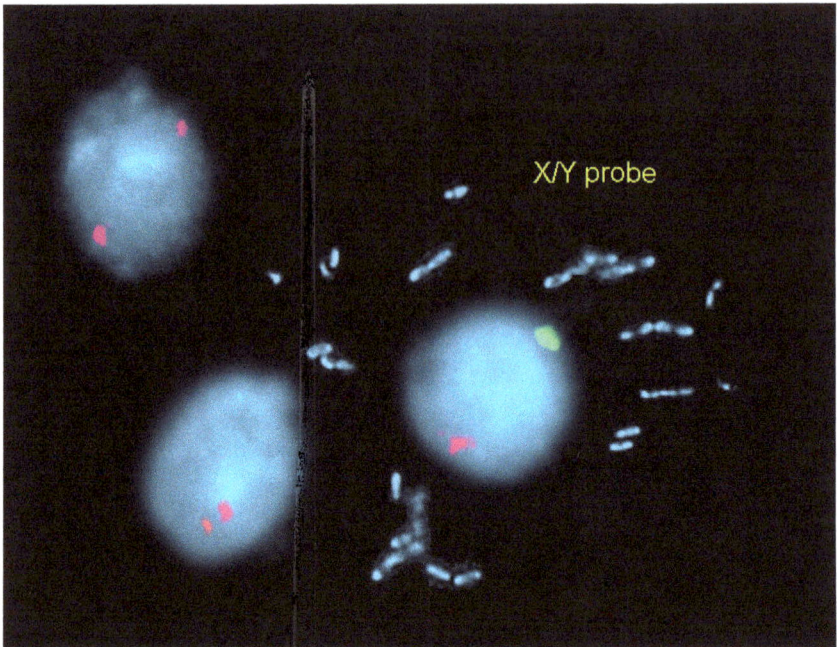

**Plate 12.20.**  A **human female chimera with male cells from a bone marrow graft**. Buccal swab with cells probed with red DXZ1 for X centromere and green DYZ1 for region q12 on the Y. The small rods are mouth bacteria. This female had a bone marrow transplant from a male donor. Following a graft-versus-host reaction, nucleated white blood cells were found in the buccal sample. This shows two of her XX buccal cells and an XY donor white blood cell.

have different maternal centromeres; if the second polar body is fertilised by a sperm, the two zygotes will have identical maternal centromeres. Such "tetragametic" chimeras have mixed tissues of two different genotypes and are most often detected if the fused twins are of different sex. Different eye colours and skin colours are possible. A fertile male blood donor of normal appearance was found to have XY bone marrow and reproductive system but his skin and other tissues were partly XY and partly XX.

In rare **germline chimeras**, the main body tissues, including blood cells, may be of one genotype and the gonads of another. Mayr *et al.* (1979) studied a mother of blood group AB who had two children of group O, which is unexpected. Her family were all typed for various blood group, enzyme and HLA markers. The woman's blood and skin were incompatible with her being the genetic mother of any of her four children, while her husband and her parents had appropriate genotypes. The mother was deduced to be a dispermic chimera; her gonads had a different HLA type from her skin, of a type compatible with her being the genetic mother of the four children, as was her gonadal blood group genotype.

The increase in *in vitro* fertilisation, with more than one embryo implanted at a time, has led to more multiple births as mentioned in Sec. 12.15; such twins would normally be dizygotic. The 30- to 35-fold increase in dizygotic twins must increase the frequency of chimeras. The fusion of two monozygotic zygotes back to one would be undetectable and would not give a chimera.

**Microchimeras** are more common, where cells from a different individual occur at a low level in the body, say just the blood. Microchimerism can occur when cells are transferred through blood between **mother and foetus**, **non-identical twins** in the uterus, or through **blood transfusions**. Foetal cells can persist in the mother's blood for many years, and may cause autoimmune-type diseases such as scleroderma (Nelson *et al.*, 1998). The son's blood would be XY and could be of different blood group and HLA type to the mother. Foetal CD34 and CD38 cells can differentiate into immune competent cells; they can be detected in the mother at week 34 in more than 90% of pregnancies. **Twins** often share a placental blood supply and blood stem cells may pass into the other twin's bone marrow, seeding some of the twin's blood cells. About 8% of dizygotic twins have **chimeric blood** with different blood groups, e.g., their own A cells but O cells from their twin (Van Dijk *et al.*, 1996); acquired tolerance of the twin's haematopoietic cells permits long-lasting cell propagation. In one case, a young male twin had 86% of his own blood cells and 14% of his twin sister's blood cells, changing after 15 years to 63% of his own and 37% of his sister's, while she only had 1% of his blood cells at each test.

Many single births come from multiple conceptions where an unknown twin is lost early in pregnancy, leaving the single-born individual a **blood chimera** with traces of the unborn twin's blood circulating inside it. That is the explanation of a woman whose routine blood test showed some male blood cells mixed with her female ones.

**True hermaphroditism**, with an individual having both testicular and ovarian tissue, is very rare, often giving ambiguous genitalia. There may be one ovary and one testis or both types of tissue in one gonad, an **ovotestis**. Hermaphrodites could be chimeras or mosaics, including translocations of Y onto X. An XX/XY chimeric boy had breast development, a rudimentary uterus with a right fallopian tube and ovary, with a functional left testicle. His father contributed different marker alleles to each of the different cell lines while only one of the mother's marker alleles was present in both cell lines, so the boy was a dispermic chimera with fertilisation of the mother's egg and second polar body. Very rare cases have included hermaphrodites with a **diploid/triploid karyotype**, such as 46XX with 69XXY, with many abnormalities.

**Human/animal chimeras** have been made artificially; half way through gestation, the immune system is not developed enough to recognise foreign cells, which can be incorporated into the developing body. Zanjani *et al.* (1992) made sheep/human chimeras by injecting liver cells from therapeutically aborted 12–15 week old human foetuses into 48 to 54 day-old sheep foetuses. From 42 such transplants, 22 animals were born alive and five were sheep/human chimeras. This was part of a study to see if human stem cells, say from bone marrow, could be used in animals such as sheep or pigs, to get "humanised" organs such as liver for transplants, although disease transmission could be a problem.

## 12.17. Gene therapy

### 12.17.1. *Introduction*

**Gene therapy** is the treatment of disease by the transfer of therapeutic DNA or RNA into the patient. See Chap. 14 for general aspects of genetic engineering. In theory, one can use wild-type genes to correct mutant genes causing human diseases, if one can get them to the right tissues and expressed at the right time. If the mutation is dominant, it would be very difficult to get added wild-type recessive genes expressed unless the dominant alleles could be removed or silenced, which is generally impractical once a defective baby has been born. With recessive mutations, added dominant wild-type genes should be expressed. Only somatic tissue changes are permitted, not germ-line therapy.

The introduced DNA is not usually found at the homologous locus, occurring instead in widely diverse places in the genome. It does not replace the defective gene, but it can mask its effect if the transgene is dominant. Normally only some of the treated cells receive the transgene, and even fewer actually express it.

The **criteria for effective human gene therapy** were laid down in 1984 by one of the pioneers, French Anderson:

- A cloned healthy gene must be available.
- Transfer of this gene to diseased recipient cells must be possible, e.g., by retrovirus vectors or electroporation.
- The target tissue must be accessible, e.g., you can remove bone marrow cells (for blood-disease treatments) for manipulation, then transfuse them back in blood into the body, or you can give viral aerosols into the lungs to treat cystic fibrosis.
- Treatment must not harm the patient. For example, if treating $\beta$-thalassaemia with $\beta$-globin genes in bone marrow cells, one would not want white blood cells expressing the globin genes intended for red blood cells. Random integration of retroviruses into host DNA usually gives insertional mutagenesis, disrupting a gene; this would not matter in most cells, but would if it caused tumour formation.
- Treatment must lead eventually to a large improvement in the patient's health. If only a few cells were transformed, or if there was poor expression of the therapeutic gene, the procedure would not be worthwhile.

A major aim in gene therapy is to have a **tissue-specific vector** with easy administration, especially by intravenous infusion.

## 12.17.2. *Methods for somatic gene therapy*

**Micro-injection** of somatic cells is usually impractical because far too many cells would need treating and many are inaccessible, e.g., lung cells. **Electroporation** is the use of short pulses of high voltage to open up membrane pores to DNA for short periods. It can be used on cells outside the body, such as bone marrow cells, but is difficult to use *in situ*.

**Retroviruses**, with RNA genetic material and **reverse transcriptase** to form DNA for integrating into host cells, can be effective vectors for therapeutic genes with a high rate of random integration into host chromosomes. They are the usual choice, once oncogenic sequences have been deleted.

One does not want the retroviruses replicating in the host, so replication-defective strains (lacking *gag*, *pol*, and *evn* but possessing the packaging region) are grown up with the aid of "helper" strains (with *gag*, *pol*, and *evn* but lacking the RNA signal for packaging its genome in protein coats) in packaging cell cultures to provide the engineered viral particles for infection of the host. Unfortunately retroviruses can only carry inserts up to about 8 kb long, too small for some genes, and they might recombine with endogenous retrovirus strains already in the host to give infective replicative forms, causing disease. A serious problem for treatment of fully differentiated cells is that integration of the therapeutic gene from retroviruses normally only occurs when the host cell replicates. Integration within a host gene will usually knock out that gene and with some types of gene may occasionally cause tumour formation.

Other possible virus vectors include **adenovirus** which does not need host cell division for integration and has a low risk of insertional mutagenesis, but gene expression is transient as the viral DNA is an unintegrated episome, and hosts may have or develop immunity to such viruses. Adeno-associated virus, herpes, vaccinia and influenza viruses have also been investigated as possible vectors, especially Moloney leukaemia virus (MLV). It may be possible to modify virus capsids to include tissue-specific ligands to make the vectors more tissue-specific.

**Lentiviral vectors** differ from most other retroviruses in that they can transfect a high proportion of non-dividing cells and so are promising for terminally differentiated tissues. In a trial by the VIR×SYS firm, they have been used to transfect autologous T cells with an antisense code directed against HIV envelope RNA.

Retroviruses (including lentiviruses) and adeno-associated viruses integrate into the genome of the target, so the transgene can be retained and passed on during cell division. MLV typically integrates near the start of a gene, and as the transgene will usually contain a promoter, it could turn on the gene at which it integrates. HIV, however, preferentially integrates within a gene which it would then disrupt and turn off. Adeno-associated viruses can integrate but unlike retroviruses can also exist as replicating episomes. Adenoviruses and poxvirus, also usable as vectors, exist and replicate only as episomes and may be more immunogenic than the integrating viruses. Episome vectors can transfect dividing and non-dividing cells, frequently with a high copy number, but transmission to daughter cells is not reliable.

Of non-viral vectors, **direct injection of DNA** into muscle can be effective, with good expression of the therapeutic gene after transformation of the recipient cells. DNA can be coated with cationic lipids to make **liposomes** which can fuse with cell membranes, releasing the DNA, e.g., after the particles have been blown into the lung. Various membrane-disruption methods are used to get DNA into cells, including electrical, chemical or pressure shocks. Complexing DNA with particular surface-receptor ligands may permit targeting of DNA to specific cell types. DNA delivered directly rather than through viral vectors rarely integrates and thus is only expressed transiently. Some schemes now use transposons to aid the integration of naked DNA. For example, a plasmid containing the transgene is flanked by transposable elements and is cotransduced with a plasmid coding for transposase such as "Sleeping Beauty" which excises the gene and randomly transfers it into a chromosome at frequencies at least 100 times that of integration of naked DNA.

Experiments are also taking place using **Human Artificial Chromosomes** as vectors, with human telomeric DNA, genomic DNA and arrays of $\alpha$-satellite DNA (which has centromeric functions). They can be transformed into the target cells by standard transfection methods. The Chromos' ACE system artificial chromosome has about 70 acceptor sites for transgenes, with use in gene therapy predicted by 2006–2008.

A very recent development is the use of **organically modified silica nanoparticles (ORMOSIL)** as a nonviral victor for *in vivo* gene delivery and expression, initially in the mouse as a model for humans. The aim is to use this for **human gene therapy for cerebral and cerebrovascular disorders**, including stroke, brain injury, neurodegenerative diseases, and developmental and neoplastic brain disorders. Viral vectors can reach the brain, but adenoviruses can cause excessive immune responses and retroviruses can cause insertional mutagenesis. Vector viruses could also recombine with endogenous viruses. Bharali *et al.* (2005) devised and prepared stable aqueous suspensions of highly monodispersed nanoparticles which were surface-functionalised with amino groups for binding DNA. The particles were complexed with plasmid DNA encoding green fluorescent protein and injected into mouse brains *in vivo*. Transfection was achieved with an efficiency as great or greater than with viral vectors, and with no toxicity. Another experiment with transfection using a plasmid expressing the nucleus-targeting fibroblast receptor type 1 showed that the nuclear receptor could control the proliferation of stem/progenitor cells in the brain. They reactivated adult stem cells on the floor of the brain ventricles, so that healthy neurons could be induced from them. The authors state that these ORMOSIL nanoparticles are promising for therapeutic manipulation of the neural stem/progenitor cells and well as for targeted brain gene therapy.

Another very recent discovery is that perhaps 30% of human protein-coding genes are regulated by **microRNAs**. Each microRNA can target several genes and one gene may be regulated by several different microRNAs (Twyman, 2005). They probably work by binding to the ends of mRNAs and stopping translation, or by causing mRNA destruction. Some have been implicated in unregulated growth of cancer cells. The potential use of microRNAs to regulate genes for therapeutic purposes is the subject of much research.

### 12.17.3. *Progress*

The first human gene therapy attempt was made in 1990 by French Anderson and Michael Blaise in the USA on a four-year-old girl with **adenosine deaminase deficiency** (ADA). This very rare autosomal recessive disease gives severe combined immunodeficiency disease (SCID), making sufferers very susceptible to infectious diseases, usually causing death in early childhood. Such children used to have to live in a germ-free protective chamber, but now they can live more normally if given weekly injections of polymer-coated enzyme (PEG-ADA) which slowly releases the enzyme into the blood.

This treatment was kept up during the gene therapy. Large samples of T-lymphocytes were taken from the girl's blood and were multiplied in cell culture. They were then transfected with mouse leukaemia virus carrying a normal human ADA gene. The transfected T-lymphocytes were then returned to the body in eight blood transfusions over 11 months. There followed six months without transfusions, then transfusions every four months. The inserted gene was expressed well and the girl recovered well from ADA deficiency and went to a normal school.

A team at Great Ormond Street Hospital, London, treated a boy, Musaf, who was born in 2000 and developed ADA; he did not respond to the PEG-ADA treatment and no matching bone marrow donor could be found. In 2003, a sample was taken of his bone marrow; stem cells from it were treated with engineered mouse virus with a normal ADA gene. He was given mild chemotherapy to suppress his defective bone marrow cells and then was given the treated stem cells. This was successful and good ADA enzyme was made. In 2005, the boy was healthy, attending pre-school and living without many restrictions, having previously been a "bubble boy" living in a protective chamber.

**X-linked severe combined immunodeficiency (SCID-X1)** differs from ADA-deficiency in being caused by mutations in the common cytokine-receptor $\gamma$ chain, affecting the development of T lymphocytes and natural-killer cells, and affecting B-lymphocyte function. Gaspar *et al.* (2004) treated four children in London by somatic gene therapy with a gibbon-ape-leukaemia-virus pseudotyped gammaretroviral vector. Autologous CD34-positive haemopoietic bone-marrow stem cells were transduced *in vitro* with the complete coding region a wild-type human $\gamma$c gene cloned into the vector. All patients showed large improvements in clinical and immunological symptoms, with no serious side effects, and prophylactic medication was no longer needed by two patients. This phase I/II clinical trial was therefore successful, making a total in December 2004 of 18 successful SCID gene therapies, 14 for SCID-X1 and four for ADA deficiency, a very small world total.

Gene therapy for **cystic fibrosis** (Subsec. 12.3.1) is under trial, using an aerosol with treated adenovirus or liposomes into the lungs. Unfortunately, the treated cells are continually shed, requiring repeated treatments, perhaps every three weeks. Although the treatment agents reach the lung epithelium, they do not usually reach the submucosal cells in which the cystic fibrosis transmembrane conductance regulator protein gene is most strongly expressed. Treatment does not transform all the lung cells, so some continue to secrete viscid mucus, and immunity to the treatment agent, especially viral coat proteins, may build up during successive treatments. Although liposomes are easily absorbed by endocytosis of the cell membrane, only about one DNA particle in 1,000 reaches the nucleus. Furthermore, treating the lungs does not help the sweat gland or pancreas symptoms. Therapy for cystic fibrosis has so far not been nearly as successful as with ADA deficiency. Lung **transplants** offer an alternative, or liver transplants if the liver is badly affected.

**Trials** are under way for gene therapy for other diseases, including rheumatoid arthritis in knuckles, but there are doubts as to whether gene therapy for some diseases will ever be cost-effective. Of the 1,020 clinical trials in progress by January 2005, 66% were for **cancer treatments**. One with promising initial results has been for **malignant melanoma**, a skin cancer. Skin tumours were directly injected with plasmid DNA engineered to express the HLA class I gene B7, with a mismatch to the patient's own HLA type. This is to stimulate the patient's immune system to attack the melanoma tumour cells.

The first government-approved gene therapy for cancer was **Gendicine**, approved in China in 2003 for use from 2004 in treating head and neck squamous cell carcinoma (Peng, 2005). It uses a recombinant adenovirus encoding a human p53 tumour-suppressor gene (rAd-p53), and has been trialled since 1998, using direct injection into the tumour.

For some diseases, exact **regulation of the amount of product** is not crucial, e.g., Factor VIII for treating haemophilia A (Subsec. 12.3.2), but if one were treating $\beta$-thalassaemia by introducing the normal $\beta$-globin gene into bone marrow cells, one would want accurate transcriptional control of the level of expression. An excess or a deficit of $\beta$-chains compared to $\alpha$-chains could cause precipitation of one type of chain and result in anaemia. Gene therapy might work well for more diseases in future, but progress has so far been slow and at a high financial cost.

There have been **setbacks**. In 1999, Jesse Gelsinger died of multiple organ failure four days after receiving adenovirus-based therapy for a rare liver disorder. In 2002, a child developed leukemia after retroviral therapy for X-linked severe combined immunodeficiency, with further leukemia cases later in the same trials. In June 2004, promising trials of gene therapy for haemophilia B were halted temporarily because of adverse immunological reactions, with patients having mild inflammatory responses to the adeno-associated virus vector and to the Factor IX peptides produced. Such problems led to the temporary halting of a number of gene therapy trials, and to much increased regulation and monitoring. Increased regulatory burdens are very understandable for safety but have proved costly and time-consuming, which has deterred a number of smaller companies and academic institutions from continuing gene therapy work. In the USA, patients now have to be monitored for 15 years for insertional mutagenesis-caused leukaemias.

Gene therapy trials are underway for **eye diseases** such as **retinitis pigmentosa** where photoreceptor cells in the eye slowly die, with a vision loss of about 5% a year. Immortalised retinal pigment epithelial cells engineered to secrete ciliary neurotrophic factor (CNTF) are placed in small capsules inside the eye, allowing nutrients to diffuse in and CNTF to diffuse out, but protecting the cells from the immune system.

**Alternatives to gene therapy** such as transplants from matched donors, so-called "designer babies" conceived by *in vitro* fertilisation, with pre-implantation selection for tissue matches, are covered in Sec. 17.3, on human reproduction. The very successful new technique of cord blood transplants is described in Sec. 12.18.

## 12.18. Stem cell therapy, including the use of cord blood

**Stem cells are totipotent**, able to differentiate into all kinds of cell, or **pluripotent**, able to differentiate into many kinds of cell, as well as producing more stem cells. They therefore have potential for regeneration, for treating many kinds of disease, injury and even ageing, often with few or no side effects.

**Human embryonic stem cells** are obtained from the inner cell mass of a blastocyst-stage embryo five or six days after *in vitro* fertilisation, but research on them for possible therapeutic uses is controversial and in the early stages (Okie, 2005). **Foetal stem cells** are obtained from aborted foetuses aged three to six months; their use is also controversial. **Adult stem cells** can be obtained by biopsy of bone marrow or obtained from peripheral blood after cytokine injections to get them from bone marrow. Their use is usually autologous, with the donor as the later recipient, to treat blood cancers, blood diseases or to restore blood cells after high doses of chemotherapy. Such treatment has been widely available since the 1960s but is expensive to provide. Haematopoietic stem cells in **bone marrow transplants** have been used to treat blood diseases such as aplastic anaemia, leukaemia and lymphoma, by repopulating a sufferer's bone marrow with donor cells, but bone marrow transplants require carefully matched donors and recipients. It is often difficult to find suitable HLA-compatible donors, and taking and using the donation can be complicated.

**Umbilical cord blood stem cells** have been used since 1988 to treat blood diseases and blood cancers, and are now rapidly finding wider uses. **Umbilical cord blood** is very easy, cheap and painless to collect non-invasively. Immediately after the birth, it is drained from the placenta and umbilical cord into a sterile collection bag. It is processed to remove excess red blood cells and plasma, keeping the nucleated cells including **stem cells**. It can be tested for infectious diseases, genetic diseases, HLA type, blood groups such as ABO and rhesus, then is given dimethyl sulfoxide as a cryoprotectant, before being frozen for storage under liquid nitrogen. When required, it can be thawed and given to a compatible sufferer from a genetic disease or leukaemia. Parents often give permission for its collection if the possible use for later treating that baby is pointed out.

There are many centres with large public and private collections of cord blood stem cells, including in the USA, the National Marrow Donor Program, Cord Blood Transplantation Study Banks and the New York Blood Center. The Cord Blood Registry in California stores a quarter of a million units. In 2004, there were an estimated 2,000 cord-blood transplantations worldwide.

Another advantage over bone marrow is that cord blood is less restrictive in its HLA-compatibility requirements, but a disadvantage is that the 40 to 100 ml collected has many fewer haemopoietic progenitor cells than a bone marrow donation, and one unit may be insufficient to treat an adult (Steinbrook, 2004). As with bone marrow donations, the recipient of a successful cord blood transplant is a **chimera**, containing live cells from two quite different individuals.

The first cord blood transplant was on a six-year-old American boy with severe Fanconi's anaemia. He received cryopreserved umbilical cord blood from an unaffected HLA-identical younger sister in 1988 in France. Today he is healthy, with no disease symptoms, so the transplant gave him stem cells which successfully spread and grew in his bone marrow, making good red blood cells. Adult and children can be treated, with the method becoming increasingly used (Laughlin *et al.*, 2004).

Most treatments have been for **siblings of the donors**, who have a good chance (25%) of HLA compatibility (Sec. 12.11), but they can be used between **suitable unrelated people**. For example, in Britain George Young had the rare IPEX syndrome affecting his immune system, so he spent his first 18 months inside a 2.5 × 2.5 metre sterile "bubble", with his mother unable even to kiss him for fear of infecting him. His only hope of survival was a bone marrow transplant but no relatives matched. A search of international data bases produced one good match, from cord and placenta blood from a baby born in New York in 2002. Following the successful cord blood transplant in Newcastle General Hospital in 2004, the boy was able to go home, a lively child who might grow strong enough to go to nursery school.

For treatments, if white blood cells are not removed, HLA matching is needed, and there is about a 10% risk of graft-versus-host disease. More modern practice is to have full removal of white blood cells, when HLA matching is not required. Purified cord blood stem cells have shown widespread value in treating many heart, lung, skin, stroke, autoimmune, eye and viral conditions, diabetes, neurological disorders, blood diseases and blood cancers. The cells can differentiate into neurons and heart cells, for example. The American Association of Blood Banks listed many **conditions suitable for treatment** with whole cord blood instead of bone marrow transplants. They include malignant diseases such as acute lymphocytic leukaemia, chronic myelogenous leukaemia and neuroblastoma, and nonmalignant diseases such as Fanconi anaemia, thalassaemia, sickle cell anaemia, severe combined immunodeficiency and Hurler syndrome.

For example, **infantile Krabbe's disease** gives progressive neurologic degeneration and death in early childhood. It is an autosomal recessive disorder from a deficiency in the lysosomal enzyme galactocerebrosidase, giving failure of myelination in central and peripheral nervous systems. Bone marrow transplants can be used but it is often hard to find a matching unrelated adult donor in time to treat this rapidly progressive disorder. Escolar *et al.* (2005) treated asymptomatic affected newborns (whose family history permitted early diagnosis before symptoms developed) with umbilical cord blood from unrelated donors after myeloablative chemotherapy. There was partial HLA-matching. The treated babies showed good survival and cognitive function, but with some motor impairment, while untreated babies generally died within two years. Treatment once symptoms developed was much less successful.

Smith and Webbon (2005) reported that stem cells from the sternum could be used in **race horses** to speed repair of tendon injuries. Mesenchymal stem cells from affected horses are grown in culture to provide **autologous transplantation inocula** of more than $4 \times 10^6$ cells for implantation into the central core tendon lesion. The method might suit human tendon injuries, especially in sportsmen and sportswomen.

## Suggested Reading

Alkhateeb, A. *et al.*, Epidemiology of vitiligo and associated autoimmune diseases in Caucasian probands and their families. *Pigment Cell Res* (2003) 16: 208–214.

Baraitser, M. and R. Winter, *A Colour Atlas of Clinical Genetics*. (1988) Wolfe Medical Publications, London.

Beall, C. M. *et al.*, Higher offspring survival among Tibetan women with high oxygen saturation genotypes residing at 4,000 m. *Proc Nat Acad Sci USA* (2004) 101: 14300–14304.

Bergh, T. A. *et al.*, Deliveries and children born after *in vitro* fertilisation in Sweden 1982–95: a retrospective cohort study. *The Lancet* (1999) 354: 1579–1585.

Bharali, D. J. *et al.*, Organically modified silica nanoparticles: a nonviral vector for *in vivo* gene delivery and expression in the brain. *Proc Nat Acad Sci USA* (2005) 102: 11539–11544.

Bittles, A. H. *et al.*, Reproductive behavior and health in consanguineous marriages. *Science* (1991) 252: 789–794.

Bittles, A. H., Endogamy, consanguinity and community genetics. *J Genet* (2002) 81: 91–98.

Bittles, A. H. and I. Egerbladh, The influence of past endogamy and consanguinity on genetic disorders in Northern Sweden. *Annals Hum Genet* (2005) 69: 549–558.

Boklage, C. E., Survival probability of human conceptions from fertilization to term. *Int J Fertility* (1990) 35: 75–94.

Boomsma, D., A. Busjahn and L. Peltonen, Classical twin studies and beyond. *Nat Genet Rev* (2002) 3: 872–882.

Braude, P. *et al.*, Preimplantation genetic diagnosis. *Nat Rev Genet* (2002) 3: 941–955.

Carrel, L. and H. F. Willard, X-inactivation profile reveals extensive variability in X-linked gene expression in females. *Nature* (2005) 434: 400–404.

Carter, C. O., Genetics of common disorders. *Brit Med Bull* (1969) 25: 52–57.

Chan, K. *et al.*, A thalassaemia array for South East Asia. *Brit J Haematol* (2004) 124: 232–239.

Chang, K.-W. *et al.*, Polymorphism in *heme oxygenase-1* (*HO-1*) promoter is related to the risk of oral squamous cell carcinoma occurring on male areca chewers. *Brit J Cancer* (2004) 91: 1551–1555.

Chim, S. C. *et al.*, Detection of the placental epigenetic signature of the *maspin* gene in maternal plasma. *Proc Nat Acad Sci USA* (2005) 102: 14753–14758.

Chow, J. C. *et al.*, Silencing of the mammalian X chromosome. *Annu Rev Genom Hum Genet* (2005) 6: 69–92.

Connor, J. M. and M. A. Ferguson-Smith, *Essential Medical Genetics*, 5th ed. (1997) Blackwell Scientific, Oxford.

Cooper, D. N. and M. Krawczak, *Human Gene Mutation*. (1993) Bios, Oxford.

Cuckle, H. S. *et al.*, Estimating a woman's risk of having a pregnancy associated with Down's syndrome using her age and serum alpha-fetoprotein level. *Brit J Obstet Gynaecol* (1987) 94: 387–402.

Culver, K. J. W., *Gene Therapy — A Handbook for Physicians*. (1994) Mary Anne Liebert Inc., New York.

Cummings, M. R., *Human Heredity. Principles and Issues*, 5th ed. (2000) Brooks/Cole, Pacific Grove.

Edwards, J. H., M. F. Lyon and E. M. Southern, *The Prevention and Avoidance of Genetic Disease*. (1988) The Royal Society, London.

Eriksson, A. W. and J. O. Fellman, What genealogical sources tell us about twinning and higher multiple births. Genealogica and Heraldica: report of the 20th International Congress of Genealogical and Heraldic Sciences, Uppsala, August 1992, (1996), pp. 44–56, The Swedish National Committee for Genealogy and Heraldry, Stockholm.

Escolar, M. L. *et al.*, Transplantation of umbilical-cord blood in babies with infantile Krabbe's disease. *New Engl J Med* (2005) 352: 2069–2081.

Fraga, M. F. *et al.*, Epigenetic differences arise during the lifetime of monozygotic twins. *Proc Nat Acad Sci USA* (2005) 102: 10604–10609.

Fraser-Roberts, J. A. and M. E. Pembrey, *Introduction to Medical Genetics*, 8th ed. (1985) Oxford University Press, Oxford.

Gaspar, H. B. *et al.*, Gene therapy of X-linked severe combined immunodeficiency by use of a pseudotyped gammaretroviral vector. *The Lancet* (2004) 364: 2181–2187.

Gelherter, T. D., F. S. Collins and D. Ginsburg, *Principles of Medical Genetics*, 2nd ed. (1998) Williams and Williams, Baltimore.

Gerson, S. L. and M. B. Keagle, *Principles of Clinical Cytogenetics*, 2nd ed. (2005) Humana Press, New Jersey.

Goldman, J. M., Chronic myeloid leukemia — still a few questions. *Exp Hematol* (2004) 32: 2–10.

Harper, P. S., *Practical Genetic Counselling*, 6th ed. (2004) Arnold, London.

Hartl, D. L. and E. W. Jones, *Genetics Principles and Analysis*, 4th ed. (1998) Jones and Bartlett, Maryland.

Hartl, D. L. and E. W. Jones, *Genetics Analysis of Genes and Genomes*, 6th ed. (2005) Jones and Bartlett, Maryland.

Howell, W. M. and S. T. Holgate, Human leukocyte antigen genes and allergic disease, in Hall, I. P. (ed.) *Genetics of Asthma and Atopy, Monographs in Allergy*, Vol. 33, (1996), pp. 53–70, Karger, Basle.

Howell, W. M. and C. Navarette, The HLA system: an update and relevance to patient-donor matching strategies in clinical transplantation. *Vox Sanguinis* (1996) 71: 6–12.

Jacobs, P. A. and T. J. Hassold, The origin of numerical chromosome abnormalities. *Adv Genet* (1995) 33: 101–133.

Jones, K. L., *Smith's Recognizable Patterns of Human Malformation*, 6th ed. (2005) Elsevier Saunders, Oxford.

Lakich, D. *et al.*, Inversions disrupting the factor VIII gene are a common cause of severe haemophilia A. *Nat Genet* (1993) 5: 236–241.

Lamb, J. A. *et al.*, (International Molecular Genetic Study of Autism Consortium), Analysis of IMGSAC autism susceptibility loci: evidence for sex limited and parent of origin specific effects. *J Med Genet* (2005) 42: 132–137.

Laughlin, M. J. *et al.*, Outcomes after transplantation of cord blood or bone marrow from unrelated donors in adults with leukemia. *New Engl J Med* (2004) 351: 2265–2275.

Lesage, M., *Vitiligo. Understanding the Loss of Skin Colour*, 2nd ed. (2002) The Vitiligo Society, London.

Little, A. C., D. M. Burt and D. I. Perret, Assortative mating for perceived facial personality traits. *Personality and Individual Differences* (2005) 40: 973–984.

Mayer, W. R., V. Pausch and W. Schnedl, Human chimera detectable only by investigation of her progeny. *Nature* (1979) 277: 210–211.

McConkey, E. H., *Human Genetics, the Molecular Revolution*. (1993) Jones and Bartlett, Boston.

McCullough, J., Progress towards a pathogen-free blood supply. *Clin Infect Dis* (2003) 37: 88–95.

McKusick, V. A., *Mendelian Inheritance in Man*, 12th ed. (1998) Johns Hopkins University Press, Baltimore.

Modell, B. and A. Darr, Genetic counselling and customary consanguineous marriage. *Nat Rev Genet* (2002) 3: 225–229.

Montagu, A., *Human Heredity*, 2nd rev. ed. (1963) Signet, New York.

Morgan, D. R., A guide to living with risk. *Biologist* (1989) 36: 117–124.

Mortelmans, K. and E. Zeiger, The Ames *Salmonella*/microsome mutagenicity assay. *Mutat Res* (2000) 455: 29–60.

Nelson, J. L. *et al.*, Microchimerism and HLA-compatible relationships of pregnancy in women with scleroderma. *Lancet* (1998) 351: 559–562.

Okie, S., Stem-cell research — signposts and roadblocks. *New Engl J Med* (2005) 353: 1–5.

Page, S. L. and L. G. Shaffer, Nonhomologous Robertsonian translocations form predominantly during female meiosis. *Nat Genet* (1997) 15: 231–232.

Peng, Z., Current status of Gendicine in China: recombinant human Ad-p53 agent for treatment of cancers. *Hum Gene Ther* (2005) 16: 1016–1027.

Pinkel, D. and D. G. Albertson, Comparative genomic hybridization. *Annu Rev Genom Hum Genet* (2005) 6: 331–354.

Pritchard, D. J. and B. R. Korf, *Medical Genetics at a Glance*. (2003) Blackwell Science, Oxford.

Roberts, J. P., Gene therapy's fall and rise (again). *The Scientist* (2004) 18: 22–29.

Robinson, R., *Genetics for Cat Breeders*, 3rd ed. (1991) Pergamon Press, Oxford.

Scriver, C. R. *et al.*, *The Metabolic and Molecular Basis of Inherited Disease*, 7th ed. (1995) McGraw-Hill, New York.

Sedivy, J. M. and A. L. Joyner, *Gene Targeting*. (1992) W. H. Freeman and Co., New York.

Sherwood, L., *Fundamentals of Physiology. A Human Perspective*, 2nd ed. (1995) West Publishing Company, St. Paul/Minneapolis.

Skirton, H., C. Patch and J. Williams, *Applied Genetics in Healthcare*. (2005) Taylor & Francis, London.

Smith, R. K. W. and P. M. Webbon, Harnessing the stem cell for the treatment of tendon injuries: heralding a new dawn? *Brit J Sports Med* (2005) 39: 582–584.

Steinbrook, R., The cord-blood-bank controversies. *New Engl J Med* (2004) 351: 2255–2257.

Stern, C., *Principles of Human Genetics*, 3rd ed. (1973) W. H. Freeman and Co., New York.

Strachan, T. and A. P. Read, *Human Molecular Genetics*, 3rd ed. (2003) Garland Science, Oxford.

Strickberger, M. W., *Genetics*, 3rd ed. (1985) Collier Macmillan, London.

Strömberg, B. *et al.*, Neurological sequelae in children born after *in vitro* fertilisation. *The Lancet* (2002) 359: 461–465.

Sugimura, T., Nutrition and dietary carcinogens. *Carcinogenesis* (2000) 21: 387–395.

Thein, S. L., Genetic insights into the clinical diversity of Beta Thalassaemia. *Brit J Haematol* (2004) 124: 264–274.

Twyman, R., Small RNA: big news. *Wellcome Sci* (2005) 1: 20–21.

Van Dijk, B. A. *et al.*, Blood group chimerism in human multiple births is not rare. *Am J Med Genet* (1996) 61: 264–268.

Vieira, A. R. *et al.*, Medical sequencing of candidate genes for nonsyndromic cleft lip and palate. *PLoS* (2005) 1: e64, 1–9.

Vogel, F. and A. G. Motulsky, *Human Genetics, Problems and Approaches*, 3rd ed. (1997) Springer-Verlag, Berlin.

Wadonda-Kabondo, N. *et al.*, Association of parental eczema, hayfever, and asthma with atopic dermatitis in infancy: birth cohort study. *Arch Dis Childhood* (2004) 89: 917–921.

Weatherall, D. J. *et al.*, *Oxford Textbook of Medicine.* (1995) Oxford Medical Publications, Oxford.

Welborn, J., Acquired Robertsonian translocations are not rare events in acute leukemia and lymphoma, *Cancer Genet Cytogenet* (2004) 151: 14–35.

Winchester, A. M. and T. R. Mertens, *Human Genetics*, 4th ed. (1983) C. E. Merrill, Columbus.

Wu, M.-T. *et al.*, Risk of betel chewing for oesophageal cancer in Taiwan. *Brit J Cancer* (2001) 85: 658–660.

Young, I. D., *Introduction to Risk Calculations in Genetic Counselling.* (2000) Oxford University Press, Oxford.

Zanjani, E. D. *et al.*, Engraftment and long-term expression of human fetal hemopoietic stem cells in sheep following transplantation in utero. *J Clin Investig* (1992) 89: 1178–1188.

# Chapter 13

# Plant and Animal Breeding Methods and Examples

This chapter is about various **methods used in plant and animal breeding**. For reproductive physiology and crossing methods, see Chap. 17. For types and uses of selection, see Chap. 5, with Sec. 5.13 for meat characters. For genotype, phenotype and breeding values, see Sec. 3.4. **Breeding objectives** depend on the role of a breed in an animal or plant production system, for example, whether an animal breed is a general purpose one, one for making F1 hybrids, a maternal line or a terminal sire line. In maternal lines, there would be emphasis on reproductive performance; in terminal sire lines, selection would be for growth and carcase characteristics; in lines for F1 hybrids, complementarity and hybrid vigour are crucial.

## 13.1. Using hybrid vigour

### 13.1.1. *Definition of hybrid vigour (heterosis)*

**Hybrid vigour** (heterosis) occurs when the hybrid offspring are better than either parental line for a particular trait or traits. The parents considered are usually inbred lines but could be different species. The classic case is the mule ($2n = 63$) which is hardier than either parent, a mare (female horse, $2n = 64$) and a he-ass (male donkey, $2n = 62$), although it is sexually sterile as not all chromosomes can pair at meiosis. Nearly all maize grown in the USA is hybrid maize, from crossing selected inbred lines, because the F1 hybrids have faster growth and higher yields than pure lines.

**F1 hybrids** are used extensively in maize, kale, sprouts, sheep, cattle, pigs, tomatoes, and many garden flowers and vegetables. The inbreeding

**Plate 13.1.   Hybrid vigour in eucalyptus timber.** The two parental lines are on the lower stands at the sides. Wood from the two reciprocal cross F1 hybrids, on the higher middle stands, was harvested at the same age as that from the parental lines.

of plants which normally have self-incompatibility, such as many *Brassica* species, may require special techniques such as hand-pollination at the bud stage before the self-incompatibility develops. Hand-pollination is expensive to do. Plate 13.1 shows commercially useful hybrid vigour in eucalyptus timber.

## 13.1.2. *Explanations of hybrid vigour*

There are two explanations for hybrid vigour, both of which are correct some of the time; they are not mutually exclusive. On the **dominance hypothesis**, the **masking of harmful recessive alleles** by dominant alleles improves yield. For example, inbred line 1 might have genotype *A A, b b, C C, D D* and inbred line 2 might have *A A, B B, C C, d d*, where *b* and *d* are harmful recessives. The F1 hybrid would then be *A A, B b, C C, D d*, in which harmful recessives *b* and *d* are both hidden by the corresponding dominant alleles, giving the F1 hybrid a better yield than either inbred parent. On selfing the F1 hybrid, the average yield would be reduced because the harmful recessives would, in some of the offspring, be homozygous and expressed.

On the **overdominance (heterozygote advantage) hypothesis**, the hybrid is fitter than either parent because the heterozygote is phenotypically different from and superior to both homozygotes; i.e., $B\ b$ is fitter than either $B\ B$ or $b\ b$. This could happen if the two alleles specified different products, with one dose of two different products being better through complementation than are two doses of a single product. It could also happen if the recessive allele is inactive, and one dose of the dominant allele product is better than two doses. On selfing the F1 hybrid, the average yield would again fall, as on the dominance hypothesis, but because heterozygosity was being lost.

While the practice of using F1 hybrids works whichever hypothesis is most often applicable, the two hypotheses have different long-term implications for breeding aims and breeding economics. On the dominance hypothesis, one could breed an ideal high-yielding pure-breeding type by breeding out harmful recessives, to get say $A\ A,\ B\ B,\ C\ C,\ D\ D$. On the overdominance hypothesis, that would be impossible, as the ideal type would be highly heterozygous and could not be pure breeding, e.g., $A\ A,\ B\ b,\ C\ C,\ D\ d$. The hybrid would have to be made afresh each generation, unless it could be propagated vegetatively.

If plant or animal breeding firms produced a pure-breeding ideal type, they could sell it initially to farmers, but the farmers would never again have to buy that line from the breeders, as they could propagate their own stock for future generations. If the ideal type was highly heterozygous, the farmers would have to buy new stock each generation from the breeding firms which grew the parental inbred lines. One Dutch spinach breeder was describing the advantages of his F1 hybrid spinach seed, and his first-listed "benefit" was that "The hybrid excludes the growing of further generations independent of the original breeder." The breeding firms therefore prefer the overdominance hypothesis to be right, while the farmers would prefer the dominance hypothesis. The experimental evidence generally suggests that the dominance hypothesis is more often right than the overdominance one. For plant breeders' rights, see Sec. 13.9 and Subsec. 19.2.6.

### 13.1.3. *Typical F1 hybrid breeding programmes*

The breeders obtain a number of lines of the chosen crop, from a range of diverse habitats and geographical areas. They propagate and inbreed these lines, selecting for desirable characters, qualitative and quantitative, and against undesirable ones. The inbreeding exposes desirable and undesirable recessive alleles, which might have been hidden in heterozygotes, to selection. Over a number of generations this produces a series of more or less homozygous pure-breeding lines with desirable characters but they are usually fairly low yielding because of inbreeding depression (Sec. 6.2).

Different pairs of inbred lines are then crossed to produce a range of F1 hybrids. These will be genetically uniform within one cross and will be highly heterozygous if the parental inbred lines were quite different genetically. Many will show hybrid vigour, with increased yields compared to the inbred lines. Crossing inbred line 1 with inbred line 2 produces F1 hybrid 1,2; if this has a high yield, it can be tested for commercial use. Such a simple **two-way cross** has the disadvantage that the F1 hybrid seed is itself produced on a low-yielding inbred line, so the seed or stock will be expensive, in addition to all the developmental costs. One solution is to use **three-** or **four-way** crosses, so that the final F1 hybrid seed (it is still called F1 hybrid seed even if it is really F2) is borne on high-yielding hybrids. For example, lines 1 and 2 can be crossed to produce hybrid 1,2, which is used as female in crosses to line 3, producing three-way cross hybrid 1,2,3. Alternatively, 1 can be crossed with 2, and 3 can be crossed with 4, then 1,2 is crossed with 3,4 to give a four-way cross F1 hybrid 1,2,3,4. In both cases, the final seeds for sale, 1,2,3 or 1,2,3,4, are borne on high yielding F1 hybrids, reducing seed costs, but there will be some loss of uniformity compared with a two-way cross, as segregation can occur in the hybrid parent (1,2 and/or 3,4) of the three-way or four-way cross hybrid seed.

Even six-way crosses have been used in some crops such as kale because of complications due to self-incompatibility. An F1 hybrid **coconut tree**, intermediate in height (about 9 m) between its normal (25 m) and dwarf parents (2 to 3 m), is shown in Plate 13.2. Unlike typical F1 hybrids, it is not produced by crossing inbred lines, but its advantages include easier, cheaper, harvesting of coconuts from its lower height, and bearing nuts at least five years earlier (from year 3 or 4 until about year 23) than do tall palms (from about years 12 to 55), as well as some hybrid vigour. The yield is similar to that of tall palms. A recent variety, Nicobar, is only 2 m tall, with a very early yield and a life of about 8 years, yielding about 40 nuts a year compared with 100 from a tall coconut; it has a high yield of copra (flesh).

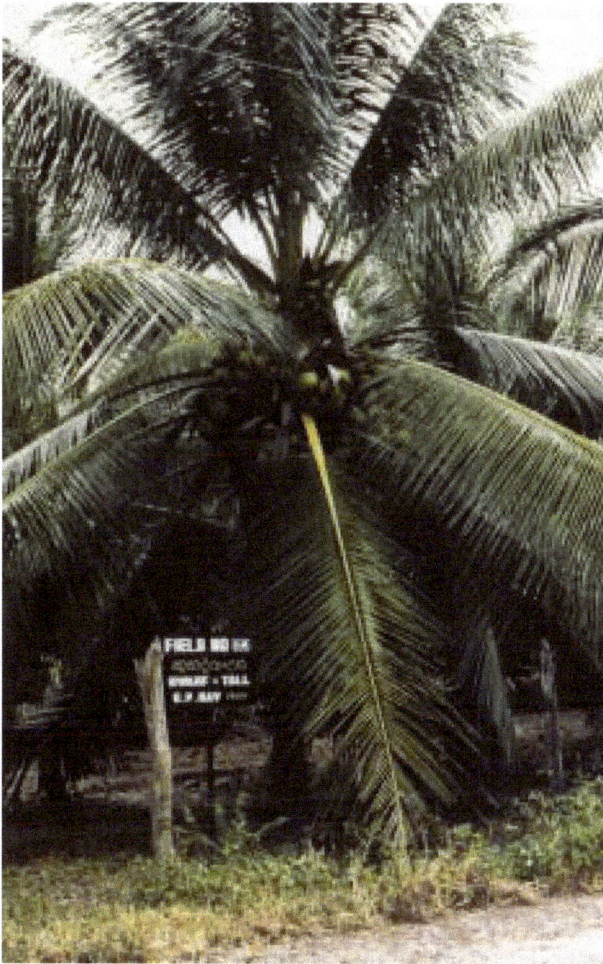

**Plate 13.2.** **An F1 hybrid coconut plant**, from crossing tall x dwarf trees, at the Coconut Research Institute, Lunuwila, Sri Lanka.

In **potatoes**, attempts with F1 hybrids were made by producing inbred lines and intercrossing them, but there was a progressive loss of pollen fertility which caused the programme to be abandoned (Dunnett, 2000). As potatoes are vegetatively propagated once the sexual crossing has been done, F1 hybrids could have been propagated vegetatively with no loss of heterozygosity or hybrid vigour, with great commercial advantages.

In farm animals, one cannot self-fertilise an individual, so mating of close relatives, such as sib-mating, parent-offspring or cousin-mating, is practised to get inbred lines for making F1 hybrids. **Hybrid vigour in farm animals** is often achieved by crossing selected lines of different breeds, rather than making inbred lines within breeds. For example, the Pig Improvement Company UK claim that: "Compared to the traditional Large White sire, the PIC Hampshire provides 100% diverse genes to complement the Large White, Landrace and Duroc dam-lines. This positively influences growth and efficiency, but most particularly robustness. … This hybrid vigour creates hardy, streetwise finishers."

### 13.1.4. *Hybrid maize production*

*Zea mays* is a natural outbreeder, producing separate terminal male inflorescences near the top of the plant, and female inflorescences in the lower leaf axils, with the stamens ripening before the stigmas. **Maize** plants of diverse genetic backgrounds are obtained from different areas and are selfed for number of generations, selecting for desired characters such as disease resistance and grain colour and composition. The breeder tries some selection for yield, as inbreeding this natural outbreeder gives inbreeding depression. Deleterious and beneficial recessives are made homozygous and are then expressed, permitting selection. This initial selection within and between inbred lines gives definite improvements, but the big increase in yield occurs at the hybrid stage.

Not all crosses of inbred lines give increased yields, so the breeders test their best selected inbred lines for **general combining ability**, usually by a "top cross" to a common tester pollen stock, itself an inbred line. The inbreds giving the best general combining ability are then crossed to each other in different combinations to find which pairings give the best **specific combining ability**. Two-, three- and four-way crosses are tried. The best combinations are then multiplied up and tested over a number of years in a range of agricultural situations and areas, to find the best commercial F1 hybrids to sell for particular areas.

The results of a typical maize F1 programme are shown in Table 13.3. It is obvious that the big increase in yield comes at the F1 hybrid stage, with 2-, 3- and 4-way crosses all increasing yield by a factor of about 2.7 compared

**Table 13.3.** **Results of a typical maize F1 hybrid programme.**

| Cross type | Lines crossed | Yield in bushels per acre (1 bushel per acre is about 90 litres per hectare) | | | | |
|---|---|---|---|---|---|---|
| | | Average yield of parents, $F=1$ | Average yield of F1 hybrid, $F=0$ | Offspring yield after random pollination of F1 hybrid | % decline in yield after random pollination | F after random pollination |
| **2-way** | $1 \times 2$ | 24 | 63 | 44 | 48 | 0.500 |
| **3-way** | $1,2 \times 3$ | 24 | 64 | 49 | 37 | 0.375 |
| **4-way** | $1,2 \times 3,4$ | 25 | 64 | 54 | 26 | 0.250 |

with the fully inbred parental lines. If the seeds from random pollination of the F1 hybrids are sown, the yield declines, as shown in Table 13.3, because heterozygosity is lost (overdominance hypothesis) and/or because deleterious recessives can become homozygous (dominance hypothesis). The decline in yield is different for 2-, 3- and 4-way crosses, with the percentage decline roughly matching the inbreeding coefficient after the random pollination.

This **decline in yield** after using the original F1 seeds is why fresh hybrid seeds are needed each generation. One drawback of using F1 hybrids in some less developed parts of the world is that farmers often will not buy fresh seeds each year, just using seeds from their F1 hybrids, with declining yields. This is especially true if farmers are initially supplied with free or cheap F1 hybrid seeds under an aid programme but are later expected to pay commercial rates.

Hybrid maize doubled yields in the USA between 1925 and 1955 and accounts for nearly all maize planted there now, but there were no commercial **hybrid wheats** until 1975, and most wheats are still pure lines, not F1 hybrids. Wheat and maize are both wind pollinated but pollination is less efficient in wheat. Hand-pollination is uneconomic for cereals, so inbred lines to be hybridised must be planted in close proximity, with some system for avoiding self-fertilisation. In wheat, one has to plant alternate rows of the inbreds acting as male and female, so only half the rows bear seeds. In maize, one only needs one row of males to two rows of females. The system used in wheat to make males and females only gives 75% pollination, and it is difficult to ensure that males and females are fertile simultaneously. For such reasons, F1 hybrid seed in wheat is about four times as expensive as pure line seed, which is usually too much compared with the increased yield of the F1 hybrids.

The system devised to get separate males and females in maize is very ingenious and even with one row of males to two rows of females, it gives almost 100% pollination. It involves cytoplasmicly-determined male sterility, where cytoplasmic factors $S$ (male sterile) and $F$ (male fertile, normal) show **maternal inheritance**, being transmitted through the ovules but not the pollen. There are two nuclear alleles, $R$, a dominant restorer of male fertility even with $S$ cytoplasm, and $r$, non-restorer. In a 4-way cross, using $1,2 \times 3,4$, one could have inbred line 1 with $S, r\,r$, line 2 with $F, r\,r$, line 3 with $S, r\,r$, and line 4 with $F, R\,R$.

In an area away from other maize crops, line 1 would be interplanted for crossing with line 2. Due to $S$ with $r\,r$, line 1 would be male sterile and with $r\,r$, $F$, line 2 would be male fertile. Seeds collected on line 1 cobs must therefore be hybrid, 1,2, with genotype $S, r\,r$, and seeds collected on line 2 must have been selfed. Similarly, line 3 would be interplanted with line 4 for crossing. Due to $S, r\,r$, line 3 would be male sterile, while line 4, $F, R\,R$, would be male fertile. Seeds collected on line 3 must be hybrid, 3,4, while seeds from line 4 would be selfed. The hybrids 1,2 ($S, r\,r$, male sterile) and 3,4 ($S, R\,r$, male fertile because of the restorer gene) are interplanted for crossing, again away from any other maize plants. Seeds collected from male-sterile 1,2 are the commercial F1 hybrid seed, 1,2,3,4, and seeds from 3,4 are selfed. As 3,4 is heterozygous, $R\,r$, the commercial seed consists of about equal numbers of male-fertile $S, R\,r$ and male-sterile $S, r\,r$ types, but both are female-fertile and having half the plants male-fertile is enough to pollinate both genotypes.

Similar cytoplasmic male-sterile plus dominant nuclear restorer systems have been used in **other organisms**, such as onions, sugar beet, sorghum, rye and sunflowers. In biennial crops harvested for their vegetative parts, as in onions and sugar beet, there is no need for the F1 seeds to give sexually fertile plants. As one would expect, male-sterility is usually selected against, and it is very difficult to find good combinations of nuclear restorers and cytoplasmic male sterility genes in the wild or in crop organisms. Once a suitable system has been found in a crop, it tends to be used widely, which may have unfortunate consequences from pests or pathogens. In **farm animals**, self-fertilisation does not occur, so no precautions are needed against selfing: one just has one sex of one line mated to the opposite sex of the other inbred line.

In the early 1970s, most maize stocks had the same cytoplasm, from a Texas strain, and the **Southern Corn Blight** fungus (*Helminthosporium*) developed the ability to attack plants with that cytoplasm, costing about 20% of maize production in 1970 in the USA. As a result, many growers abandoned the cytoplasmic male sterility system for ensuring crossing between inbred lines. Instead, the lines to be used as male-sterile parents had their tops cut off mechanically before the anthers matured, as all the male flowers are in the terminal tassels; hybrid seeds were collected from the detasselled lines. In sugar beet F1 hybrid production, Owen cytoplasm is used world-wide, but fortunately there has been no disease problem like the one associated with maize Texas cytoplasm. For alternative systems of F1 production in rice, see Subsec. 13.1.6. See Sec. 14.7 and Bisht *et al.* (2004) for a genetically engineered nuclear system of male sterility and fertility restoration in oilseed rape, *Brassica napus.*

## 13.1.5. *Hybrid sprout production*

**F1 hybrid Brussels sprouts** are widely grown but so are pure lines. To do selfing requires hand pollination at the bud stage to overcome self-incompatibility, which makes the inbred line seed very expensive.

Although these data are old, from 1974 at Asmer Seeds, Ormskirk, Lancashire, they show the relative costs of different seed very clearly. Non-F1 commercial seeds cost £11 a kilo, needing 1.1 kg per ha. The seeds gave very variable plants and only the best were transplanted into the fields from the seed beds, so seed costs were about £12 per ha, plus transplanting costs. F1 hybrid seeds cost £176 a kg, but plants were uniform and good, so only 0.14 kg were needed per ha, giving seed costs of about £25 a ha, with no transplanting costs as they could be planted directly in the fields. Unlike with onions, seed costs are not a major part of the sprout production costs and the increased yield of the F1 hybrid compared with non-F1 lines makes it often worthwhile to buy F1 hybrid seed. Due to the expense of hand bud-pollination, the parental inbred lines' seeds for making the F1 hybrids cost about £550 a kilo. To avoid bastardisation of hybrid seed production from wild brassicas, two boys were trained to go out with binoculars to spot and destroy any such plants within a mile (1.6 km) of the seed fields. See Sec. 10.3 and Plates 10.1 and 10.2 for the production of homozygous diploid sprout lines from monoploids from anther culture.

## 13.1.6. *Hybrid rice production*

As rice is the staple food for about three billion people, its genetics are important globally. An **interspecific hybrid**, **NERICA** ("New Rice for Africa") combines the hardiness of the African species *Oryza glaberrima* with the productivity of the Asian species *O. sativa*. It required embryo rescue methods (Subsec. 13.5.1) and shows good hybrid vigour.

Most rice is **pure-line**, with many varieties available. **F1 hybrid rice** was first produced in China in 1974 when a male sterility gene from wild rice was transferred to get a **cytoplasmic male sterility** (CMS) line to use as female, together with a maintainer line and a restorer line. This is a three-line system. There are also two-line systems involving an **environmentally-sensitive genetic male sterile** (ESGMS) **line** to use as female for the F1 hybrid seed. There is no need for a maintainer or a restorer line. The ESGMS line is multiplied by allowing selfing in a fertility-inducing environment, then for use as an F1 hybrid parent, it is grown in a sterility-inducing environment. 15 million ha of rice, about half China's total, is now F1 hybrid rice, with national rice yields rising from 3.5 to 6.2 tonnes/ha (www.rice2004.org).

## 13.1.7. *Hybrid animals*

In **fowls**, first crosses are made between different breeds or strains. The F1 are mated to a third breed or strain (three-way cross) or to another first cross (four-way cross) to produce vigorous second-cross birds for egg or meat production. Alternatively, inbred lines are made within breeds for egg production, and various F1 hybrids are tried between inbred lines. The most productive crosses (those which "nick" best) are used commercially.

In **cattle**, the most obvious effects of hybrid vigour are an increase in female fertility and milk yield, and in the growth rate of calves. The use of F1 cattle for meat is partly due to the fact that the improved milk supply gives faster growth and earlier maturity of the offspring. In Britain many beef herds consist of cows from interbreed crosses mated back to bulls of one parental breed or to bulls of a third breed. In British beef breeds, the F1 hybrids are better than the two parental breeds' average by about 5% for growth and viability, and about 8% for cow fertility. Hybrid vigour is two to three times greater than that when British cattle are crossed with the Zebu *(Bos indicus,* Brahmin cattle ; see Subsec. 13.5.1). In American results with Hereford cows × Shorthorn bulls, with the F1 cows mated to Aberdeen-Angus bulls, the hybrid calves weighed 212 kg at weaning compared with 176 kg for the pure breeds, with a final weight of 469 kg versus 414 kg for the pure breeds. See Subsec. 13.7.9 for hybrid sheep examples.

## 13.2. Selection methods for inbreeders

### 13.2.1. *Single line selection*

**Inbreeding**, by self-fertilisation or crossing of close relatives, eventually gives **pure-breeding lines** homozygous for all loci. Not all individuals in habitual inbreeders will be homozygous for all loci, because there might have been some outcrossing or mutation. A diploid individual heterozygous at only 10 loci could give $2^{10} = 1024$ different homozygous inbred lines.

If one has a population of inbreeders with some genetic variation, one can take a large number of superior selected individuals from that population, say of plants, and raise progenies from selfing them, each plant's progeny constituting a line. Further selection by the breeder between many lines is practised each generation, for qualitative and quantitative characters, killing off unsuitable individuals and only planting seeds from the best lines. Once a line has become homozygous at all loci, nothing can improve it further except new mutations or outcrossing to other lines.

The best lines after a number of generations of single line selection are used in replicated field trials for possible commercial use. This kind of single line selection has been used to improve the garden bean, *Phaseolus vulgaris*, and other inbreeders.

## 13.2.2. *Using a mixture of lines, with agricultural mass selection*

Another method is to plant a mixture of different lines of an inbreeder in the same field, and allow them to self-fertilise each generation. The breeder destroys any inferior-looking plants before flowering, and harvests the bulk seed, not recording individual plant or line yields, unlike in the previous method. The seeds are sown for the next generation, with some destruction of visibly poor plants, and seeds are harvested. This continues for several generations and involves **agricultural mass selection**. This works because the highest yielders will be proportionately better represented in the next generation than lower yielders, because those having the most seeds, and seeds with the highest viability, will give rise to the most plants in the next generation.

The selection is therefore a combination of the breeder's deliberate actions and the agricultural environment selecting for plants best adapted to that particular set of field conditions and climate. Unlike the previous method, there is no labour-intensive recording of line yields. The final selected plants may often be a mixture of the best lines, and variation is beneficial from a disease-resistance aspect. Such a line mixture may have a disadvantage in that the breeder may not be able to register breeder's rights (a patent, see Sec. 13.9 and Subsec. 19.2.6) on a mixture, depending on local registration rules. It would have a mixture of homozygous genotypes from selection between different inbreeding lines.

## 13.2.3. *Bulk population breeding*

This is similar to the previous method except that one starts by hand-cross-pollinating **two different compatible inbreeding lines**, to be sure of having genetic variation on which to select. The F1 are allowed to self for several generations, with the practice of agricultural mass selection on the segregating offspring, plus the breeder "roguing out" undesirable types such as ones which are small, weedy, deformed or disease-susceptible. This method is very good for **adapting proven varieties** to new localities and environments, and has been widely used in wheat, other cereals, and beans. The initial cross between different lines could occur naturally in inbreeders which have occasional outcrossing, such as wheat.

## 13.2.4. *Pedigree breeding*

In **pedigree breeding** for inbreeders, one takes account of line averages as well as individual yields. One crosses two varieties which complement each other in useful ways, selfs the F1, then in the F2 one selects single plants with desirable traits, including good yield. One continues with selfing through several more generations, but recording individual yields. In the F3, 4, 5 and 6, one selects the **best plants** from the parents in the previous generation which gave the **best average yields**. For example, in the F4, one line might have an average yield of 20 units, with individual yields ranging from 14 to 27. Another line might average 25 units, ranging from 16 to 40, and a third line might average 30, with individuals ranging from 18 to 40. The best plant to select on the pedigree evidence would be plants yielding 40 units from the line averaging 30 units as they are superior individuals from a superior family.

By the F6, most lines will be homozygous at nearly all loci, and significant further progress is unlikely. There will be little further response to selection between lines, so the best lines are bulked up for commercial testing. The best lines can also be crossed to some other good line, and another cycle of pedigree selection can be started. See Sec. 5.6 for general aspects of pedigree breeding.

## 13.3. Selection methods in outbreeders or random-maters

The techniques described here can be used with organisms having some degree of outcrossing even if they are mainly selfing, such as wheat, as well as with outbreeders or random-maters. Rather confusingly, the term **"outbreeders"** is sometimes used just to mean non-inbreeders, such as random-maters.

In outbreeders, there will often be deleterious and beneficial recessives hidden from selection in heterozygotes. To get these to show for selection in plants or animals, one often uses some enforced inbreeding by selfing or mating of close relatives, followed by selection as recessives become exposed in the phenotype. During this inbreeding, there will often be inbreeding depression, so one selects for vigour as far as possible to counter it. Having selected for particular desirable characters, one then restores some degree of heterozygosity at other loci by crossing different selection lines. This is what is done in F1 hybrid production, but it can also be used to produce superior lines which are not remade as F1 hybrids each generation.

An important long-term method, which is especially suitable for annual grain crops with at least some outcrossing, is the **composite cross** with **agricultural mass selection**. To make a composite cross, one plants together several to many different strains of the crop and allows random pollination, even if some individuals self-pollinate. One harvests the seeds and sows them next season, then continues harvesting and sowing for several generations. The random pollination makes an enormous number of different crosses between types and between their descendants. There is no labour of hand-pollination, nor of recording yields.

The agricultural mass selection works because individuals with the highest yields and most vigorous seedlings and plants are proportionately better represented in the seed and in the next generation's plants than are less good types. This easy method can be used to adapt existing strains to new environments or to new or prevailing agricultural practices. For example, if you want earlier ripening stocks, you harvest the seed early each season, when the earliest maturing types will be the only ones with viable grain.

The composite cross plus agricultural mass selection has even been successful with wheat and barley, which usually self but have a little outcrossing. The more the outcrossing, the faster the method will work, so a regular outcrosser like maize responds faster than wheat. Maize yields went up 22% in four generations in one trial and 33% in three generations in another. With barley, 25 years of mass selection in California gave 19 important new commercial varieties, and the whole trials paid for themselves from the sale of surplus seed each year. A composite cross initially produces very variable individuals but mass selection by agricultural practices imposes stabilising selection for critical characters.

**Breeding new varieties** often consists of crossing two existing varieties and looking for good progeny in the F1, F2 or later generations. When crossing two genetically different **pure-breeding lines**, all first generation plants or animals will be identical to each other except for chance mutations or environmental effects. Segregation and recombination at meiosis in the F1 will then cause a wide range of types to select from in the F2 and later generations. If the parental lines are **not pure-breeding**, the F1 will be

**Plate 13.3.** **Lord Lambourne dessert apples**, from a James Grieve × Worcester Pearmain cross by Laxton Brothers in 1907. Photographed at the Brogdale Horticultural Trust, Faversham, Kent.

genetically heterogeneous and selection can start in the F1. **Apple trees are not pure-breeding and cannot be selfed because of self-incompatibility mechanisms (Subsec. 17.1.2). Crosses between any two compatible apple varieties can therefore give a whole range of new genotypes in the F1 from pips which will be highly heterozygous, and successful new varieties can be propagated vegetatively by bud-grafting onto suitable rootstocks (Subsec. 17.1.5). Thus variety "Lord Lamborne" (Plate 13.3) was selected as a high quality dessert apple from an F1 seedling raised in 1907 by Laxton Brothers from a "James Grieve" × "Worcester Pearmain" cross.

Modern selection methods often involve testing for **molecular markers** as well as for conventional characters such as yield and flavour. Plate 8.1 shows a tray of apple seedlings awaiting early selection from molecular markers before growing on. Where there are known molecular markers (e.g., DNA, enzyme isozymes or antigenic characters) associated with known commercial characters, say a DNA RFLP marker associated with fruit quality, then selection for that marker **at the seedling stage** reduces the number of plants to be grown for further selection at the fruiting stage. For

**marker-assisted selection**, one needs heterozygosity for the agricultural character and the DNA marker, and close linkage between them.

## 13.4. Recurrent backcrossing for gene transfer

**Recurrent backcrossing** is repeatedly crossing offspring back to one parental type, so that all genes except the desired ones from one parent are replaced by those of the recurrent parent. It can be used in plants, animals and microbes. It is used to transfer single qualitatively-acting alleles, especially ones for disease resistance. In an inbreeder, it may require hand cross-pollination. In sheep, it was used to transfer a gene with additive action, $Fec^B$, for a higher lambing rate, from the Booroola Merino into less fertile breeds, backcrossing to those less fertile breeds to maintain their other breed characteristics.

Suppose one has a good line for yield but which is susceptible to a particular disease ($r\,r$), and that there is a line available which is poor for yield but resistant to this disease, with **dominant resistance** ($R\,R$). One crosses the two lines. All the F1 should be $R\,r$, resistant to the disease, and one tests this. The F1 are then backcrossed to the $r\,r$ recurrent parent, the one with good yield, so the next generation should be $1\,R\,r : 1\,r\,r$. The individuals are tested for disease resistance, and resistant ones are backcrossed to the recurrent parent. This is continued for about six generations, or as long it takes to get the progeny yield up to that of the recurrent parent.

During this process, the genes of the poor line are replaced by ones from the recurrent parent, with about half the remaining poor-parent genes being replaced each generation, especially ones unlinked to $R$. The final product is only $R\,r$, so is selfed or crossed with others of that type to get some $R\,R$ genotypes, which can be distinguished from $R\,r$, by test-crossing to $r\,r$. This has been used to transfer salt-resistance from a seashore wild tomato into commercial lines for growing in partly saline soils, selecting each generation for salt-tolerance, fruit colour and flavour.

If the character to be transferred is **recessive**, say $a$, one needs a more complicated crossing scheme, with selfing to allow identification of the recessive homozygote. To transfer beneficial recessive allele $a$ into $A\,A$ plants, one crosses donor $a\,a$ to recurrent parent $A\,A$, then backcrosses the resulting $A\,a$ to $A\,A$. The next generation will consist of $1/2\,A\,A$ and $1/2\,A\,a$, with the same phenotypes; the plants are selfed. One eighth of the progeny will be identifiable phenotypically as *a a*. Those individuals are backcrossed to the recurrent parent, *A A*, and the cycle of crosses is repeated. In plants, this cycle can be shortened; the selfing does not contribute to the getting rid of unwanted donor parent genes. In the generation with $\frac{1}{2}$ *A A* and $\frac{1}{2}$ *A a*, each plant can have some flowers selfed and some backcrossed to the recurrent parent. Seeds from the backcross are only used from plants whose selfing showed them to be *A a*, not *A A*.

Recurrent backcrossing is also used in **mutation breeding**, to reduce the number of unwanted harmful mutations, using a wild-type as the recurrent parent. One mutates the wild-type to get the M1 generation (Chap. 7). This can be selfed or intercrossed to get the M2 generation in which recessive and dominant mutations should be recognisable. Favourable mutations are selected for, and the wanted M2 types are then subjected to recurrent backcrossing to the wild-type to get rid of the many unwanted mutations which will also have been induced by the mutagen. For recessive mutations, generations of selfing need to be alternated with the recurrent backcrossing to identify homozygous recessives.

## 13.5. Interspecific and intergeneric hybrids

### 13.5.1. *Interspecific hybrids*

**Interspecific hybridisation** has occurred naturally many times (*Primula kewensis*; see Chap. 10.6) and can be practised by the breeder. Many cultivars, especially of garden plants, are man-made interspecific hybrids, including many rhododendrons, roses and dahlias. See Sec. 10.2 for examples with raspberries, blackberries and loganberries.

In animals, such hybrids may occur in the wild or zoos, e.g., sheep (2n = 54) × goat (2n = 60) gives a shoat, male lion × tigress gives a liger, male tiger x lioness gives a tigron, and horse × zebra gives a zebroid. Plate 13.4 shows a **zeedonk**, produced by a female zebra and a male donkey. Although the offspring of such crosses are usually healthy, the chromosomes are not usually sufficiently homologous to pair well at meiosis, so the offspring are generally infertile. In the case of ligers and tigrons, the male is usually completely sterile but females may be partly fertile. The zedonk has a mane like a zebra and striping on the legs, but the main body is plain like

**Plate 13.4.   The zeedonk**, from a female zebra × a male donkey. It has the mane and striped legs of the zebra and the body of the donkey. It is sterile. Picture courtesy of Groombridge Place Gardens, Kent.

a donkey's. As we saw in Sec. 13.1, a female horse (2n = 64) × a male ass (donkey, 2n = 62) gives the hardy and useful mule (2n = 63), but it is sterile and has to be remade each generation by interspecific crosses.

In the **cats** (including genera *Felis*, *Acinonyx* and *Panthera*), the chromosome number is usually 2n = 38, e.g., for the African wild cat, bob cat, cheetah, domestic cat, fishing cat, jaguar, leopard, lion, lynx, puma, serval and tiger, so that hybrid sterility is not due to different chromosome numbers. A few cats have 2n = 36, as in the ocelot; two small chromosomes fused to give one larger one per genome.

In the **camelids**, all of which have 2n = 74, there are the Old World Camels (*Camelus dromedarius*, *C. bactrianus*), the New World domesticated species llama, *Lama glama* (domesticated from the guanaco by about 4000 B.C.), and alpaca, *L. pacos*, and the New World wild species, guanaco, *Lama guanicoe* and vicuña, *L. vicugna*. Although the New and Old World species probably diverged 10 to 20 million years ago, a team in Dubai in 1998 produced the world's first viable camel-guanaco hybrid. They used sperm from dromedary camels to inseminate a guanaco female. The male calf, Rama, has the coat and face of a guanaco and the long tail and coat of a camel. He was still too young for fertility testing when this was reported (Laboratory News, May 1999, pp. 2). All four New World camelids are inter-fertile with fertile offspring (www.llamaweb.com/Guanacos/Info.html, accessed 9/11/2004). In plants, even intergeneric hybrids (Subsec. 13.5.2) have been used in some conifers, orchids and cereal grasses, e.g., *Triticale* from wheat (*Triticum*) × rye (*Secale*).

There are a few animal examples of a **successful interspecific cross** resulting in good fertility and very useful commercial herds. **Bonsmara cattle** in Southern Africa were started in 1937 by Bonsmara who crossed *Bos taurus* (Hereford/Shorthorn) with *Bos indicus* (Indian Brahman and native cattle), with Bonsmara cattle being genetically about five-eighths *B. taurus* and three-eighths *B. indicus*. They combine the good yield of the European cattle with the hardiness of the African ones. **Argentinean cattle** are famous for beef and are similar interspecific hybrids. Genetically they are usually about 85% British (Aberdeen-Angus, Shorthorn, Hereford) and 15% native Indian/Brahmin cattle, with the latter included for resistance to heat and pests, especially insects.

Interspecific hybridisation occasionally occurs in the wild for plants where it can lead to the production of successful new types, whether diploid or allopolyploid, but the breeder has access to all kinds of species, from different regions of the world, which would not normally meet or cross. Breeders have many exciting and exotic possibilities as they can cross related species, even from different hemispheres.

With plants, interspecific hybrids may be **sexually sterile**, not setting seed even when they produce wonderful flowers; except in seed crops, sterility may not matter if the plants can be propagated vegetatively, e.g., by cuttings, offsets, grafts, etc.; see Subsec. 17.1.5. Grape vines (Plates 17.4 and 17.5), apples and roses, for example, are usually propagated by grafts onto genetically different selected rootstocks, not grown from seed. The sterility of the initial hybrid can sometimes be overcome by making allote-traploids, e.g., by treatment with colchicine, so that each chromosome has a homologue with which to pair at meiosis. Although colchicine-induced allotetraploids are excellent intermediates in transferring genes between species (see Sec. 10.9 and the tomato examples below), not many are in wide

Diploid ancestor, 2n = 14 ⟶

Triticum urartu
2n = 14, genome A A

Aegilops speltoides
2n = 14, B B

Triticum monococcum
einkorn wheats,
c. 12,000 years
ago

Sterile hybrid, univalents at meiosis
2n = 14, A B

Spontaneous doubling

Fertile allotetraploid, bivalents at meiosis
2n = 28, A A, B B

Triticum turgidum
durum pasta wheats
c. 8,000 years ago

Aegilops tauschii
2n = 14, D D

Sterile triploid, 2n = 21, A B D

Spontaneous doubling

Fertile double allohexaploid
2n = 42, A A, B B, D D

Triticum aestivum
bread wheats, c. 7,500 years ago

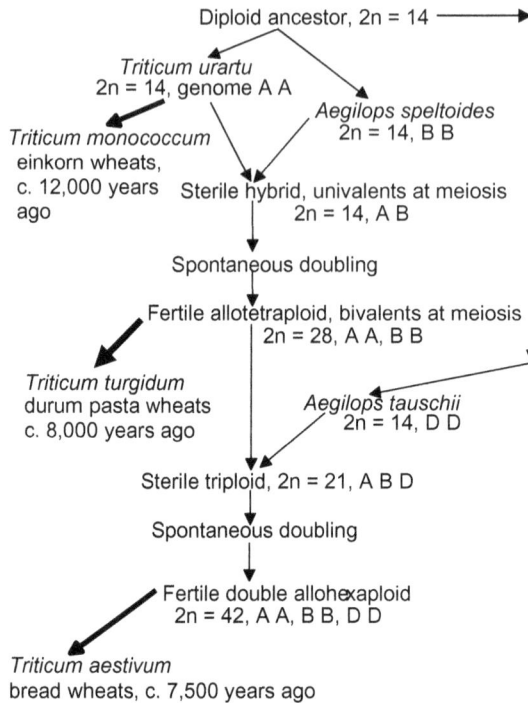

**Fig. 13.1. The origins of the modern wheats.** Normal arrows indicate natural events and heavy arrows indicate selection by man. Different sources name different species but the principle is the same.

use. Those which are include tetraploid rye-grass (*Lolium multiflorum* × *L. perenne*) and Triticale (wheat × rye; see below).

The best example of natural successful interspecific hybridisation utilised by man is **wheat**. Its history was worked out by Riley and others from cytogenetic studies of possible ancestors and relatives. It is illustrated in Fig. 13.1. Two different sterile hybrids, giving unpaired chromosomes in meiosis, had a spontaneous doubling of the chromosome number, giving fertile allopolyploids. From a diploid, an allotetraploid and a double-allohexaploid, man selected, with further breeding and agricultural selection, the modern einkorn, durum and bread wheats, respectively. By knowing the ancestry of the wheats and their genomes, wheat breeders can easily plan further breeding, e.g., by crossing to wild relatives with useful disease resistance.

The wild diploid **teosinte** (*Fuchlaena mexicana,* synonym *Zea mays* ssp. *mexicana*, and there is also ssp. *parviglumis*), 2n = 20, is the presumed ancestor of **maize**. It is fertile when crossed to the modern diploid maize, *Zea mays*, 2n = 20, so useful genes can be bred into maize when required. In crosses between the two species, Doebley and Stec (1993) used molecular marker loci to map quantitative trait loci. They found that a relatively small number of loci with large effects were involved in the evolution of the key differences between teosinte and maize.

An example of man-made successful interspecific hybridisation is the **strawberry**. An old octaploid, 2n = 56 = 8X, *Fragaria virginiana*, from Eastern USA, was of poor quality. Another octaploid, *F. chiloensis*, 2n = 56 = 8X, from Chile, had large fruits of poor quality and a dioecious habit (separate male and female plants) reduced yield. In the early 19th century, English breeders crossed the two octaploids, selecting among the progeny for large, good quality fruit and hermaphrodite flowers. They succeeded. See Lawrence (1968).

An interspecific hybrid of **lavender** occurred in commercial fields by natural bee cross-pollination between spiked lavender, *Lavandula latifolia*, which smells strongly of camphor, and the true lavender (fine-leaved lavender), *L. angustifolia*, whose subtle scent is much prized by perfume producers. The hybrid, lavendin, is the most widely grown form in France, making large areas of Provence purple with its flowers in late summer. It has a yield of essential oil of 5 to 6%, compared with only 1% from the true lavender, thus showing hybrid vigour.

As mentioned in Sec. 10.9, making allopolyploids is an excellent intermediate step in **transferring genes** from wild species into cultivated ones before repeated backcrossing to get rid of unwanted other genes from the wild relatives. This has been done extensively to transfer genes into the cultivated **tomato**, *Lycopersicon esculentum*. They include genes for fungal resistance from *L. hirsutum*, *L. pimpinellifolium* and *L. peruvianum*, genes for virus resistance from *L. chilense* and *L. peruvianum*, insect resistance from *L. hirsutum*, quality characters from *L. chmielewskii*, and adaptation to adverse environments from *L. cheesmanii*. In the case of nematode resistance, it took 12 years to separate that character from poor fruit characters which were introduced in the same cross to the wild relative: see Esquinas-Alcazar (1981).

One problem with interspecific and intergeneric hybrids is **sterility**; if this arises from a lack of pairing in meiosis, one can try making allopolyploids, as twice occurred naturally in wheat: see Fig. 13.1. Other causes of sterility can often be treated. One can try **reciprocal crosses**: with species A and species B, one crosses A as the female (ovule) parent × B as the male (pollen) parent, and B as female × A as male, because sometimes only one of the two reciprocal crosses is fertile, with unilateral incompatibility. One can try **hormone treatments**. For example, pear tree (*Pyrus communis*) as female × apple tree (*Malus pumila*) as male is usually sterile but can be made fertile by applying 40 parts per million of the hormone naphthoxy-acetic acid to the ovary and style. This increases growth of the apple pollen and reduces early abscission of the flowers and fruit. If failure of endosperm development, or of late embryo development, occurs, one can dissect out the early embryo and grow it on suitable nutrient agar, as in *Lilium lankon-gense* hybrids. This **embryo rescue** technique has been used in making hybrids between distantly related rice varieties, followed by anther culture to stabilise breeding lines via monoploids for easy selection. In *Lathyrus* (sweet pea) and *Pyrus* (pear), stylar incompatibility within these genera has sometimes been overcome by amputating the style and pollinating the style base directly.

In conifers, many species have the same chromosome number, 24, so that chromosome number differences are often not a problem in natural or man-made hybrids between species or genera. The main forestry **larch** now planted in Britain is the Dunkeld larch, *Larix × eurolepis*, 2n = 24, which excels its parents in vigour. It arose naturally in about 1900 at Dunkeld House, Perthshire, Scotland, from cross-pollination between the European larch, *Larix decidua*, 2n = 24, and the widely planted Japanese larch, *Larix kaempferi*, 2n = 24, which was introduced into Britain in 1861.

## 13.5.2. *Intergeneric hybrids*

Natural and man-made **intergeneric crosses** have been used successfully in conifers, orchids and cereal grasses. The **Leyland Cypress**, *x Cupressocyparis leylandii* (Plate 13.5), arose by natural cross-pollination between its two parents, *Chamaecyparis nootkatensis* and *Cupressus macrocarpa*. In 1888 at Leighton Hall, Welshpool, Wales, six seedlings

**Plate 13.5.** **An intergeneric hybrid, the Leyland Cypress,** *x Cupressocyparis leylandii*, which arose in 1888 by natural cross-pollination between *Chamaecyparis nootkatensis* and *Cupressus macrocarpa*.

from *C. nootkatensis* were visibly different from others and were deduced to have *C. macrocarpa* as their pollen parent. From these seedlings came the commonest cultivar, cv Haggeston Grey. In 1911 the cross happened naturally again, but with seedlings from *C. macrocarpa* with *C. nootkatensis* as pollen parent. This reciprocal cross gave the cultivars Leighton Green and Naylor's Blue. Later crosses gave plants with yellow foliage such as cv Castwellan and Robinson's Gold. "Leylandii" is one of the fastest growing conifers, often growing a metre a year and reaching 30 m; it is an excellent shelter-belt tree and hedge but grows too large for most small gardens, hence its notoriety in disputes between neighbours.

*Triticale* is a man-made allopolypoid from crossing bread wheat (*Triticum*, 2n = 6X = 42) with rye (*Secale*, 2n = 2X = 14), and doubling up the chromosome number from 28 to 56. It combines the high yield of wheat with the hardiness of rye, and is often planted in areas not good enough for wheat.

Many **orchids** grown for showy flowers are hybrids from crossing two or even several different genera. Such intergeneric hybrids include *Vuylstekeara* and *Odontocidium* from South America, which are available in a wide range of colours. Intergeneric hybrid *Oncidium* orchids are widely grown in Asia for the cut-flower market.

## 13.6. Making polyploids

It is useful to be able to **make polyploids** to see whether they are better than diploids or to overcome sterility problems in interspecific hybrids. Whether one is making autotetraploids from a diploid, or allotetraploids from interspecific hybrids, any of the following can be tried.

One can treat seeds, germlings or buds (anything with a growing point, especially one giving rise to both somatic and germ-line tissue) with **chemicals** which allow chromosome replication but sometimes inhibit cell cleavage, thus doubling the chromosome number. Colchicine, from the autumn crocus, or synthetic colcemid are commonly used. Colchicine was used to make commercial polyploids in beets and various grasses. Other chemicals which have been used include chloral hydrate, sulphanilamide and ethyl mercury chloride. See also Sec. 10.2. **Heat shock** has been used to make tetraploid maize and cold-shock was used to get tetraploid fruit flies (*Drosophila*). Even mechanical methods can work, e.g., decapitating young tomato plants, where some regrowths may be polyploid if the damage interfered with a cell's cleavage.

One can identify polyploid shoots from chromosome analysis of dividing cells, but often a polyploid shoot can be identified from its looking different in growth habit, thickness or colour, or by analysis under a micro-scope for cell area, number of plastids per guard cell, stomatal pore length, etc., where values have been established for diploids (Sec. 10.6).

## 13.7. Examples of plant and animal breeding programmes

Although people sometimes get the idea that breeding programmes must involve high technology, most of today's commercial plants and animals come from straightforward methods based on the principles considered in much of this book. For a recent genetic engineering plant case history, see the end of Sec. 14.7, for the transfer of an aluminium-tolerance gene from wheat to barley, enabling that sensitive species to grow on acid soils. See Chap. 14 for genetic engineering in plants, animals and micro-organisms, Sec. 17.2 for animal reproductive physiology, and Subsec. 17.2.7 for animal cloning. In 2005, there were about 1,000 cloned cattle in the world, especially clones of older, extensively progeny-tested elite bulls.

### 13.7.1. *Semi-dwarf wheats and rice — the "Green Revolution"*

In Mexico in 1945, wheat yields averaged 750 kg/ha. This had risen to 2,790 kg/ha by 1967, mainly through the use of **semi-dwarf wheats**. After the Second World War, a short-stemmed wheat, **Norin 10**, was introduced from Japan to the USA. It was only one third of the height of normal wheat but was high yielding because high levels of nitrogenous fertiliser could be used without the plants lodging (being knocked down by wind and rain). It was taken to Mexico and subjected to **cyclic selection** (Sec. 5.4) to improve environmental tolerance. Seeds were planted just above sea level in October, with new seed being harvested in March. This was taken to a second site at an altitude of about 2,440 metres above sea level, planted in April, and harvested in September. Those seeds were returned to the first site for planting in October, and this cyclic selection with two generations a year was continued for five years, which is 10 generations. This **two-centre agricultural mass selection** gave daylight-independent flowering, and the cyclic selection gave environmental tolerance, giving good yields when the seeds were tried in diverse places, though the plants often needed further breeding to adapt them against locally important diseases.

From 1966, dwarf or semi-dwarf wheats were rapidly introduced in many areas of the world, including Africa and Asia, for example doubling yields in Pakistan. Apart from the higher yield in each generation, the extra benefit of reduced sowing-to-harvest time allowed farmers in warmer areas to grow two cereal crops a year, e.g., wheat and sorghum or wheat and maize. This "Green Revolution" was a big agricultural advance but required extra inputs of nitrogenous fertilisers and pesticides.

In the mid 1960s, the International Rice Research Institute released the first high-yielding **semi-dwarf rice** varieties, leading with further breeding to higher yields, with yields in 1991 being roughly double those of the early 1960s. Peta (a tall Indonesian variety with high seed vigour, seed dormancy, and resistance to several insect pests and microbial diseases) was crossed with Dee-geo-woo-gen (a semi-dwarf — due to spontaneous recessive mutation $sd_1$ — Taiwanese cultivar with high yield and heavy tillering). The tall F1 were selfed and semi-dwarf F2 plants were selected for resistance to rice blast (the fungus *Pyricularia oryzae*; see Chrispeels and Sadava, 2003), with selection in further generations to get **variety IR8**. This had high yield if given enough nitrogenous fertiliser, was short, sturdy and lodging-resistant, with upright leaves and good light interception, insensitivity to photoperiod, a short maturation time ($\sim$125 days) allowing multiple cropping, moderate seed dormancy and some disease resistances. It was still sensitive to some diseases and grain quality was not ideal. It was widely distributed across south and south-east Asia in 1967, needing breeding with local varieties for consumer acceptability of grain type and further disease resistances.

Further yield increases came with better water control, more fertiliser application and better pest and pathogen control. When farmers were able to grow rice all the year round, there was a big increase in many Asian countries in Grassy Stunt Virus (GSV) which is spread by the brown planthopper (*Nilapavata lugens*). In the early 1970s it destroyed about 116,000 ha of rice. Resistance to GSV was found in wild rice, *Oryza nivara*, and was bred into cultivated rice species by recurrent backcrossing and selection; resistance to the brown planthopper was also incorporated into the improved strains. For genetic engineering in rice and Golden Rice, see Paine (2005) and

Sec. 14.7. For comparative mapping in wheat and rice using EST markers, see Yu *et al.* (2004). For modern wheat cytogenetics, see Gupta *et al.* (2005).

## 13.7.2. *Broad beans*

In the early 1970s, the procedure for **broad beans** for canning and freezing was to mow the plants with ripe pods, then to leave them for two to three days for evaporation to reduce the bulk of the plants before the vining machines came to separate the beans from the waste to be dumped. The breeding aim was to produce smaller plants so that only a single combined mowing/vining operation was required.

A commercial variety, "Triple white" (white flowers, stipules, etc.) was crossed to a dwarf bean plant (15 cm high). The breeder selected for intermediate height and the production of several pods per node, as the smaller plants had fewer nodes and "Triple white" only had one pod per node. Selection was also made for more beans per pod. The new variety only needed a single mowing/vining operation.

## 13.7.3. *Semi-leafless combining pea*

**Cultivar Countess** is a semi-leafless white **marrowfat protein pea**, specially bred for combine harvesting to meet the expanding demand for protein for compounding in animal feedstuffs, although it could be used for canning for human consumption. The genes for being semi-leafless convert leaflets into tendrils, which twine together and help support the plants. There is less leaf trash to separate out at harvesting and the leaf area prone to disease is reduced by more than 50%, allowing light to penetrate the open canopy, and better air circulation. It is fast drying and easy to harvest because it stands up well, and the interlocking tendrils help to prevent lodging in bad weather. Countess is well adapted to a wide range of soil types and is particularly drought-resistant on light soils.

It is high yielding, with yields 109% of control variety Protegra in the 1987 NIAB (National Institute of Agricultural Botany) Recommended List for General Use, and in 27 independent trials from 1982 to 1986 had a yield 116% relative to cultivars Protegra, Maro and Birte. It is resistant to Race 1 of pea wilt (*Fusarium oxysporum*) and to the most common pathotype 4 of downy mildew (*Peronospora viciae*). Countess was bred by Booker Seeds, Essex, England.

### 13.7.4. *Coca plants for cocaine*

Plant breeding can also be used for illegal purposes, as in the case of **cocaine production from coca** (*Erythroxylon coca*) in South America. It was reported in 2004 (*The Daily Telegraph*, 27/8/2004) that Colombian drug cartels have developed a new strain of coca plant yielding up to four times more cocaine than normal strains. It was estimated that drug traffickers spent £60 million in research and development, crossbreeding Peruvian strains with the best Colombian varieties, and using genetic engineering. The new strain grows to a height of three metres instead of 1.5 metres, has a higher yield and a more pure yield than normal strains. Drug traffickers can produce a kilo of cocaine for less than £1,500, then sell it for £14,000 in Miami, £34,000 in London and £50,000 in Tokyo, so the breeding costs were quickly recovered.

### 13.7.5. *Potato breeding and disease-resistance*

Wild **potatoes**, as found in South America, often have bitter, poisonous tubers and the berries of even cultivated potatoes are poisonous. The first selections by natives of the Andean Highlands must have occurred centuries or millennia before the first Europeans arrived there, choosing plants with more palatable, non-poisonous tubers for cultivation. Potatoes do not breed true but good material can be propagated vegetatively from the tuber eyes. "Seed potatoes" are tubers for vegetative reproduction and not are produced by sexual fertilisation.

European potatoes, *Solanum tuberosum*, are Andigena autotetraploids, and Sec. 10.2 describes how the diploid *Solanum phureja* can be used to get diploids from those autotetraploids. The diploids can then be intercrossed to transfer genes into *Solanum tuberosum*, which can have colchicine used to get back autotetraploids for commercial use. However, Dunnett (2000) stated that breeding at the diploid level around the world for 30 years had produced nothing of practical value. His book provided most of the information in this section.

**Photoperiod** is important in tuber formation in potatoes, as it is in flower production in many plants. In the Andes from which our potatoes originally came, from the Alto Plano of Peru and Bolivia, day lengths are much more nearly equal all year round than in Europe, as those Andean regions are closer to the Equator than is northern Europe. Andean varieties planted in Northern Europe do not even start forming tubers until the days shorten in early autumn, giving masses of foliage but little tuber crop. There must have been selection in Europe for earliness of tuber production, for starting tuber production during days with long hours of daylight.

The **number of years needed for a potato breeding programme** was described by Dunnett (2000). He wrote that there has to be a gap of up to five years between generations, to allow time for selection of the desirable characters and make the desired sexual crosses, growing new stock from sexual seed and taking it through to tuber production. He estimated that it took at least ten years to breed and release a new variety, then several years of multiplication before any income is obtained, with extremely few new varieties ever making it though all the necessary trials to get on national approved or recommended lists. Dunnett stated that he could raise and test 5,000 seedlings a year, while larger organisations could manage up to a million. He allows one year for crossing, one year for raising seedlings, then three years for his selections before giving his best lines to commercial seed potato growers to try out in a range of conditions, with the best lines being trialled in 20 countries, before deciding which new varieties to submit for independent testing for National List Trials. A typical potato breeding programme could easily take 20 years before general commercial sales of a new variety, with perhaps only two or three generations of crossing and selection.

In the UK, if a new variety is better than existing ones in some respects and has no bad flaw, it can be placed on the **National List** (Sec. 13.9) for sale of certified seed potatoes. Outstandingly successful varieties can be placed on the **UK Recommended List**, as decided by the National Institute of Agricultural Botany, with about 24 varieties of potato on that list.

For **sexual crosses**, it is essential to have flowers. Pollen parents are allowed to grow very tall (~3 m) to promote fertility. A pioneering Scottish potato farmer and breeder in Dundee, William Paterson (1810–1870), noticed that all potato varieties in his fields and garden were degenerating, getting weaker and less productive. In 1853, he tried improving potatoes from fruits with sexual seeds, but most British stocks were almost sterile. He imported various strains from England, South Africa, Australia, USA and India. By planting the best tubers near a stream for a damp atmosphere, he induced most of them to flower and to set sexual seed. He obtained many new, better varieties, including the best one of its day, Victoria. Some of his selections are ancestors of excellent modern varieties.

**Disease-resistance or tolerance** is important. In vegetatively propagated species like potatoes, a build-up of viruses often leads to performance loss in the long term. Potato viruses are not passed through the sexual seed, so in breeding programmes any plants raised from seeds from the berries are virus-free. Virus-free stocks can also be obtained from micropropagation, and Britain has a seed-potato virus testing scheme for certification. The King Edward variety was introduced in 1903, suffered a decline from virus build-up, but was restored to vigour after meristem culture.

**Pests and diseases** to which potatoes are prone include blight (*Phytopthora infestans*), virus Y, leafroll virus, scab, wart, blackleg, and eelworms (yellow-cysted *Globodera rostochiensis* and pale-cysted *G. pallida*, which together do about £50 million damage a year in Britain, infesting more than 60% of potato-growing land). The H1 resistance gene gives complete resistance against *G. rostochiensis* but is ineffective against *G. pallida* which is increasing and is difficult to control with chemical nematicides. Some virus-resistance genes were obtained from wild potatoes. Blight resistance was found in a wild Mexican species, *Solanum demissum*, but hybrids between that and the commercial *S. tuberosum* were unstable, with reduced fertility. The problem was overcome by using *S. phureja* as a bridging species, with the triple hybrid *S. demissum* × *S. phureja* × *S. tuberosum* being stable. It was backcrossed to *S. tuberosum*, with successful selection for blight-resistance and other desirable characters.

Dunnett (2000) began his breeding programme by crossing leading variety Desirée (a Dutch red-tubered UK maincrop type, but lacking resistance to eelworm and scab) with Maris Piper (susceptible to scab but resistant to yellow-cysted eelworm). The hybrids were crossed to another leading variety, Pentland Crown, for its resistance to scab, virus Y and leafroll virus. Resistance to the pale-cysted eelworm was then crossed in from *Solanum vernei*.

One of the problems of breeding for disease-resistance is that **major gene resistance** is often soon broken by new mutants of the fungus or virus. British variety Pentland Dell had three different major genes for blight-resistance, but a new blight race developed which could overcome those resistance genes. Lower levels of resistance which are **multigenic** are usually more stable against pathogen change, as with partial resistance to blight and to pale-cysted eelworm. Disease susceptible varieties can be very successful if grown in the right environment, e.g., King Edward, even though it is susceptible to potato blight. A new very blight-resistant variety was released for sale in 2004: Sarpo Mira, bred by the Sárvári Research Trust at the University of Wales, Bangor (www.thompson-morgan.com). It is an early maincrop for most types of soil, with high yields, tasty, floury flesh, and weed-suppressing foliage. It is claimed also to be resistant to slugs, wireworms and viruses, strongly recommended for organic cultivation.

**Quality estimation** is obviously important in any crop, including yield, disease-resistance and consumer acceptability. Dunnett (2000) quotes a Potato Marketing Board Consumer Survey and Report comparing his new varieties Nadine and Stemster with established varieties Maris Piper, Wilja, and Desirée. There were about 13 **criteria of consumer acceptability**, all estimated twice, once before storage and once after storage. They included skin colour, tuber shape, flesh colour, appearance, freedom from disintegration, freedom from discolouration, boiling, chipping, baking, roasting, texture, flavour, all rated on a scale of 0 to 10. The results enabled any two varieties to be directly compared on each criterion, and an overall rating was calculated for each variety.

**Coloured potatoes** are becoming more popular and are now easily available for growing (e.g., www.thompson-morgan.com; www.marshalls-seeds.co.uk). They include Shetland Black (black outside), Salad Blue (a vivid ultramarine) and Highland Red Burgundy (a bright beetroot red). Carroll's **Heritage Potatoes** (www.heritage-potatoes.co.uk) specialise in producing old variety potatoes of different colours, shapes, textures and tastes (Plate 13.6). Some examples (with the year of introduction after the name) are: Fortyfold 1836, small, round with purple and white streaked skin, nutty flavour; Pink Fir Apple 1850, long, narrow and very knobbly, skin is part pink, part white, with waxy flesh; Ratte 1872, long, white skin, yellow waxy flesh; Red King Edward 1916, oval, red skin with white flushes,

**Plate 13.6. Old coloured potato varieties which are still available**. Courtesy of Carroll's Heritage Potatoes, Northumberland.

floury texture; Salad Blue early 1900s, oval, blue skin, blue-purple floury flesh; Shetland Black 1923, oval to long, black skin; yellow flesh with a markedly blue vascular ring; Highland Burgundy 1936, oval and long, dull russet layer over a bright burgundy skin, red flesh with a ring of white flesh under the skin.

For the many people on the Atkins diet, the **low-carbohydrate Adora potato** is being promoted as it has carbohydrates reduced by one third and calories reduced by one quarter compared with most standard breeds. It was developed in the mid 1980s by HZPC in Holland and reaches maturity in 80 days compared with 140 days for most breeds, allowing less time for starch development.

## 13.7.6. *Lime-tolerant Rhododendrons*

**Rhododendrons**, in the Ericaceae, normally have to be grown in acid soils as they are lime-intolerant. The German Research Station for Garden Plants at Ahrensburg carried out selection from more than 1.8 million seedlings over 20 years to get a lime-tolerant rootstock which can be grown in almost any type of soil, even heavy alkaline clays. The beauty of developing a special rootstock is that many kinds of ornamental Rhododendron can be grafted onto it, instead of each type having to be bred for lime-tolerance. The rootstock also gives very robust growth. By 2004, over 250 varieties of lime-tolerant Rhododendron were being marketed under the INKARHO$^{®}$ brand with a wide range of flower colours and sizes.

## 13.7.7. *Sugar cane*

**Sugar cane** is thought to have started as a wild grass, *Saccharum robustum*, growing in New Guinea and used for thatch and fencing. According to O'Connell (2004), local people chewed the canes and selected sweeter ones, producing *Saccharum officinarum* which was taken to India about 8,000 years ago and hence to other warm regions. In the 1920s, existing stocks were badly affected by red rot fungus, *Colletotrichum falactum*. Dr Brandes and others went to places like Papua New Guinea and collected 130 varieties, but they showed little improvement. In the 1960s there was a programme to broaden the genetic base, with collections in New Guinea, Indonesia and India of wild species *Saccharum spontaneum*. Many crosses were made between different isolates, and between selections from those crosses and commercial *Saccharum officinarum* varieties which were generally octaploid, 2n = 8x = 80. The four cultivated species all intercross readily. Vigour, sugar production and disease resistance were selected, giving yields in Barbados of up to 23% sugar, much higher than before. Most modern commercial sugar cane varieties are **aneuploids of polyploids** from these interspecific hybrids, with chromosome numbers ranging from 100–125, which is 2n = 10X or more.

In **sequencing and gene function analysis**, copy DNA from mRNA has been sequenced to find ESTs (Expressed Sequence Tags; see Chap. 8), with about 300,000 ESTs corresponding to 50,000 genes. The **Sugar Cane EST Genome Project**, mainly based in Brazil and France, aims to uncover what genes control what functions, in order to improve breeding programmes. Whether much higher yields are physiologically possible remains to be established, with conventional cross-breeding and selection having already been so successful. By 2003, more than 800 RFLP and 900 AFLP markers had been located onto about 100 cosegregation groups, forming 10 linkage groups, and genomic colinearity had been established with maize and sorghum (www.cirad.fr), with which sugar cane will cross.

**Sugar cane propagation** is vegetative, from stem cuttings of immature (8–12 months-old) canes. These "setts" are best taken from the upper third of the cane, with younger buds and being less likely to dry out. They are planted at a 45-degree angle or horizontally in furrows, then are lightly covered in soil. The perennial plants are usually cropped for three to six years before replacement. The cane is cut close to the ground as the lower stem has the most sugar, and new canes emerge from the stems left behind. Flowering in this wind-pollinated outbreeder is not wanted in plantations as it ends growth and sugar production. It can be avoided by extending the day length by exposing the canes to electric light for short periods at night in the appropriate season. Inbreeding is very deleterious. Sugar cane yields the greatest number of calories per hectare of any plant, with up to 10 tons of sucrose per ha in Barbados. The juice has up to 10–23% sucrose. Other products and derivatives include molasses, wax, ethanol, rum and bagasse fibre; that is the main fuel in sugar factories and usable in paper, cardboard and other boards.

### 13.7.8. *Hops in Britain — dwarf "hedgerow" hops*

Varieties Goldings (from the 1700s) and Fuggle (1785) were the basis of the British **hop industry** for beer, with about 29,000 ha in the early 1900s. There are now only about 2,000 ha, with Fuggle badly hit by *Verticillium* wilt fungus. Hops have been traditionally grown up 5.5 metre poles, with harvesting by cutting the hop vine at the base to bring the whole plant down for stripping out the hop bracts.

Breeders at East Malling Research Station in Kent have bred "**hedgerow hops**", with rows of 2.5 m high permanent hedges which can be easily harvested without being cut down and need minimal pesticide application. Dwarf plants were selected and crossed by hand-pollination to the well-flavoured traditional tall Whitbread Goldings variety (1911). This produced First Gold in 1992, resistant to hop aphid, then produced Boadicea. Both these hedgerow varieties are now in commercial use for beer, with others being trialled. A revival of the British hop-growing industry has started with these products of traditional cross-breeding and selection.

## 13.7.9. *Sheep, including cross-breds and Border Leicesters*

Sheep are often **cross-bred** for hybrid vigour and to make the best use of particular breeds or hybrids in particular environments (Croston and Pollett, 1994). For example in Britain, the **Blackface** breed (about 110 lambs born per 100 ewes mated) is very hardy and does well in poor hilly country, such as much of Scotland's uplands. Surplus ewes are taken from the hills to slightly better land for mating with rams of breeds such as the Border Leicester which are larger and more fertile, with genes for a higher milk yield and therefore rearing capacity. The **Scottish Greyface** males produced are reared for meat and the ewes are taken to good pasture land in the north of England. The ewes (more than 150 lambs per 100 ewes mated) are crossed to rams of specialised meat breeds such as the Suffolk or Hampshire Down, to produce fat lambs for slaughter.

In 1991, a **Border Leicester Group Breeding Scheme** was set up, with an "**Elite Flock**" in Caernarfon, Wales, with 48 ewes selected from flocks all over Britain. They were in-lamb to top sires, and had two crops of lambs taken from them to continue the elite flock. Selection is made for prolificacy (number of lambs), mothering and milking qualities, plus conformation and growth rate. Border Leicester rams give high survival rates of crossbred lambs, which inherit the thick coat of wool. They give rapid growth rate, uniformity of carcase, easy lambing even when used on small breeds such as Welsh Mountain ewes, high prolificacy, abundant milk, good mothering, and a large yield of prime meat. Over one million breeding ewes in the UK are sired by Border Leicester rams. Border Leicesters were developed from Leicester Longwool and Cheviot sheep in the England/Scotland borders in the early 19th century, and were recognised as a breed in 1869, with a herd-book from 1898.

The **Southdown** sheep (Plate 13.7) can be used as a pure bred, with very fine fleece, excellent meat flavour and fast growth. It can also be used as a sire on large ewes of other breeds to give a smaller birth weight lamb with fast growth and high quality. For an account of common sheep breeds, see Maijala (1997).

**Plate 13.7.**   **Southdown lambs** at the South of England Show. The many awards shown by the cards increase the animals' value.

   **Pedigree breeding** is very important in pure-bred sheep, including **Suffolks**, which are excellent for weight gain and are used for meat and short wool. An example of a **prepotent "super-ram"** is Pankymoor Prelude, a Suffolk ram bought as a six-month-old lamb in 1993 for a then record price of £23,100. By 1999 he had fathered 189 top-flight rams and 350 pedigree daughters, worth more than £1 million, and still had years of reproductive life left. In 1995, one of his sons sold for a record £71,400 while ordinary Suffolk rams only fetch £250–£500.

   In Mediterranean areas such as Greece, which has 10 million ewes, sheep are often kept for **milk** for cheese-making, with the owners' incomes coming about 2/3 to $^3/_4$ from milk and 1/3 to $^1/_4$ from meat. Sheep milk fat and milk protein have fairly high heritabilities, 0.5 to 0.6, with a high correlation between them, $+0.7$, but with a slight negative genetic correlation. As one might expect from competing physiological demands, milk yield is negatively correlated with % milk protein ($-0.4$) and % milk fat ($-0.3$): see Barillet (1997).

   In **wool breeds**, the critical characters are fleece weight per year, the fineness of the wool fibres, and the number of lambs weaned per year. The

phenotypic merit of a breeding ewe could be expressed in dollars per year
from an index such as (clean fleece weight per year, kg, $\times \$7.2$) + ($-$ wool
fibre diameter, $\mu$m, $\times \$1.2$) + (number of lambs weaned per *year* $\times \$7.2$),
where cleaned fleece fetches $7.2 a kg and coarser wool is penalised at $1.2
an extra $\mu$m diameter (figures from Kinghorn, 1997). Colourless fleeces
are usually preferred to coloured ones. For wool fibre characters and for
tables of heritabilities, breeding aims and hybrid vigour in farm animals,
see Dalton (1980). For sheep and cattle breeding, see Simm (1998). For
transgenic animals, see Bishop (1999). See Sec. 8.8, genomics, for the
National Scrapie Plan which is rapidly changing sheep genotypes for scrapie
resistance in Britain.

## 13.7.10. *Cattle, including Ayrshires (dairy) and Aberdeen-Angus (beef)*

See Sec. 3.4 for breeding values, Sec. 5.14 for meat selection and Sec. 17.2
for reproductive physiology in cattle. Details of many livestock breeds can
be found on www.ruralindex.net/livestock.html. In Britain in about 1950,
farm labour was abundant and cheap. Most dairy cattle herds were small,
with fewer than 20 cows, and 70 was considered a very large herd. The
proportion of herds milked mechanically rose from 5% in 1936 to 10%
in 1939, to 50% by 1949, and to 70% by 1956. Today there are far fewer
herds, but larger, with scarce labour and much more mechanisation. In one
**high-intensity system** seen by the author near Toronto, Canada, cows stay
indoors in stalls and are fed individually by a computerised system which
adjusts the amount and composition of each feed in accordance with that
cow's recent milking and reproductive record.

Breeding has become more scientific, but bureaucratic interference has
become more acute. Under EU rules, Britain cannot produce all the milk
it needs, with strict quotas on production, with knock-on effects on dairy
products such as cheese. In Britain in the 1950s, most milk was sold to
a national body, the Milk Marketing Board, which then paid a pool price
for all milk regardless of its nutritional quality. That gave breeders and
farmers no incentive to improve milk protein or fat content, except for
farm-bottled milk, where the rich milk of **Guernsey cattle** commanded
a price premium. AI (Artificial Insemination, see Subsec. 17.2.5) became

more widely used, with Guernsey inseminations from the Whiligh Cattle Breeding Centre rising from 1,000 in 1946 to 78,000 in 1964. In the South East, a Dairy Progeny Testing Scheme was set up in 1964.

**Cow size** is important as it affects growth, weight gain, milk production and feeding and housing costs. The maintenance costs are higher for larger cows and can become critical under low food availability, as in Alberta, Canada, in the long cold winters. A Charolais or Simmental cross needs 10 to 30% more energy for maintenance than the smaller Angus × Hereford cross. Smaller cows are preferred where temperature or moisture availability restrict pasture growth and feed availability, but where nutrition is not limiting, cows above average size are the most productive. A calf's weaning weight and preweaning average daily weight gain are strongly determined (50 to 60%) by the mother's milk production, which affects beef and dairy cattle.

**Ayrshire cattle** (Plate 13.8) started in South West Scotland several hundred years ago and now number more than two million around the world, with breed societies in Australia, Brazil, France, Finland, Kenya,

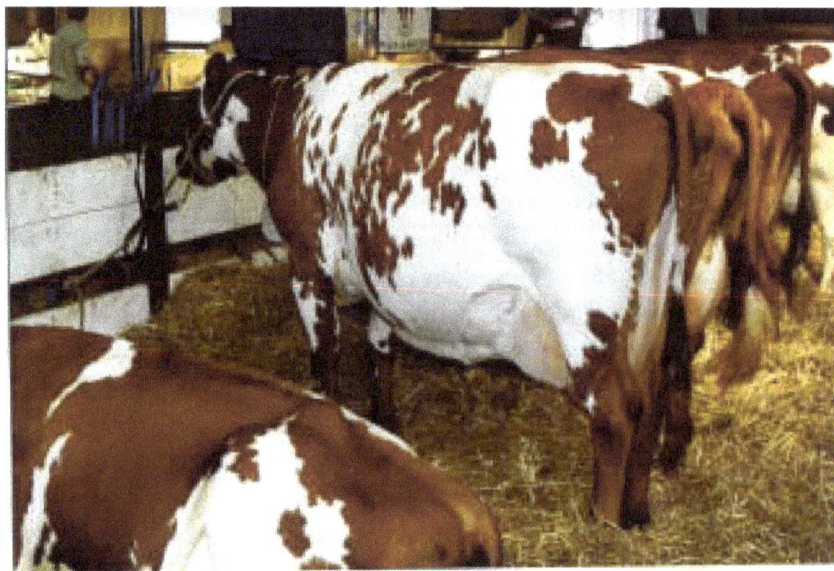

**Plate 13.8.** **Ayrshire cows** from the Pylon herd, at the South of England Show, Ardingly, Sussex. Subsec. 13.7.10 includes a description of show points for this dairy breed.

UK, USA, former USSR countries and several other countries, and a World Federation of Ayrshire Breed Societies based in Ayr, Scotland. The World Federation organises conferences, promotes the exchange among countries of semen, livestock and embryos, and promotes the breed internationally to bodies such as governments. The exchanges widen the gene pool available for selection.

Approximately 450 young Ayrshire bulls are **progeny-tested** annually throughout member countries, and the INTERBULL Centre in Sweden takes into account the different environments in which they are tested. Its INTERBULL rankings allow comparison of bulls in different countries without a need for conversion formulae. The Finnish Ayrshire breeders have an Embryo Transfer Breeding Programme, "ASMO", with a target of 900 embryos a year to market in Finland and 300 for export. Only the top 1% of cows are used as bull dams. 70 heifers a year are evaluated for production and functional traits, with the best 20 selected as embryo donors. With the OPU method, collecting oocytes directly from the cow's ovaries, about 60 embryos can be obtained per donor per year. See also Subsec. 17.2.6.

On the marketing side, "Ayrshires"$^{®©}$ milk is available in the UK as a breed-branded milk, with a six-fold sales increase since the launch in 1996. In South Africa, milk from 2,000 Ayrshires (one third of the country's total) supplies a major retailer, Woolworths. Ayrshires are dairy cattle and the breed's national milk records over a lactation of 305 days include: America, 22,067 kg; Canada, 15,082 kg at 4.1% butterfat and 3.2% protein; UK, 12,441 kg at 4.5% butterfat and 3.2% protein. In Britain, the South East Ayrshire Breeders Group introduced progeny testing of young bulls in 1970. One of their best bulls, Dunlops Shanghai, was sold in 1979 for £6,500 to the Milk Marketing Board's AI service, so his semen was very widely distributed. One of his daughters, Ingram's Gladys Jane, produced well over 100 tonnes of milk over 11 lactations. It was only in 1989 that the South East Ayrshire Club changed the rules of its Herd Competition to allow the protein content of milk to be included as a show criterion.

According to an undated leaflet, *Your Concise Guide to Successful Showing*, published by The Ayrshire Cattle Society of Great Britain and Northern Ireland, Ayr, available in 1999, the following are some **aims of Ayrshire breeding and showing**.

*"A head in proportion to the rest of the cow, displaying femininity with a bright, alert appearance and broad strong muzzle. The udder, depending on the age of the cow, of medium size but always carried above the hock. The rear attachment must be high and wide, displaying a well-defined central ligament with the fore attachment firm and strong, blending smoothly into the body; teats should be about 6 cm long, placed in the centre of each quarter. The texture of the udder should be soft and silky with prominent veining and a tortuous milk vein carried well forward from the udder. The shape of the cow should provide generous heart room for strength and durability, ample body capacity through length and depth for a high forage intake and weighing 550–600 kg, with a level topline giving strength through the loin. She should be clean and sharp, with an indication of her production potential; a minimum height at the shoulder of 54" (137 cm) is desired. When viewed from the side, the legs should exhibit a small degree of set but when viewed from the rear they should be straight. The bone should be flat, the hock clean, with short strong pasterns. The cow can be any shade of red or brown including mahogany and white, although either colour may predominate. Each colour should be distinctly defined."*

In contrast to Ayrshires, Limousin (Plate 13.9), Aberdeen-Angus, Charolais (Plate 2.5), Saler and Simmental cattle are **beef breeds**. **Simmental** bulls can be finished on cereals (weight gain about 1.5 kg a day) for slaughter at 10 to 12 months, or on a concentrate/high quality silage ration to finish at 12 to 15 months. Males can be left intact (not castrated) and slaughtered at about 600 kg, with very good feed conversion rates. Simmental suckler cows (i.e., ones for feeding calves, not for milk for humans) typically calve in spring, with weaning in autumn. The bulls have good hindquarters and are ideal for the restaurant trade which wants lean meat.

The **Aberdeen-Angus beef breed** was developed in the early part of the 19th century from the polled and predominantly black cattle of North East Scotland. The Herd Book dates from 1862 and the Aberdeen-Angus Cattle Society from 1879. By line breeding and selection for type, the early pioneers established this famous beef breed in Angus, Aberdeenshire, Speyside and the Laigh of Moray before it was spread to many parts of Britain. It was established in the USA by the end of the 19th century where by 1901 they were registering more pedigree cattle than in Britain and

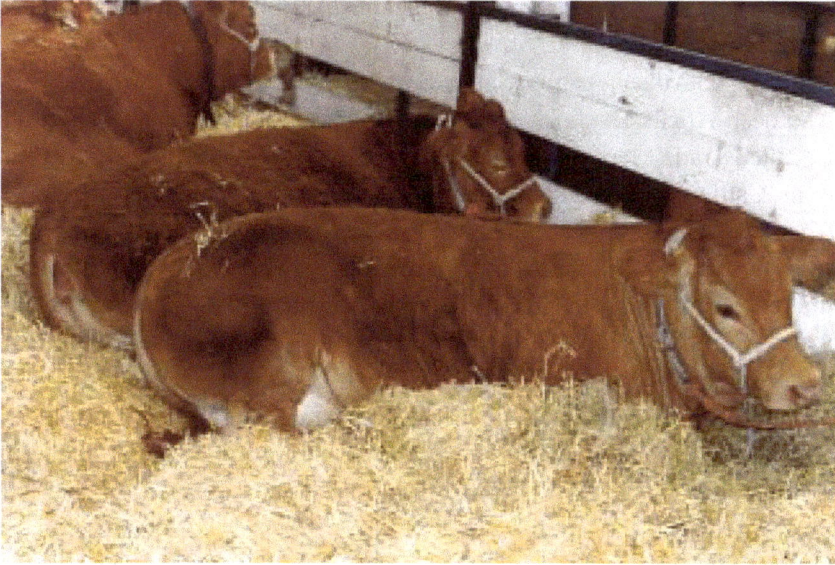

**Plate 13.9.** **Limousin cows**, a beef breed, at the South of England Show. The cow with the crinkly coat was first in her class and the foreground smooth-coated cow was third.

now register 40 times more. The breed spread to other English-speaking countries such as Canada, Australia, New Zealand and South Africa, and to South America, especially Argentina. Today, all those countries have greater populations of pure Aberdeen-Angus than the British Isles; Argentinean beef has a very high reputation.

During the first half of the 20th century, Britain was the centre for breeding Aberdeen-Angus and exports were very successful. Unfortunately, over this period the size of the cattle was reduced, to the disadvantage of commercial producers. The 1960s saw the introduction to the UK of large-muscled Continental cattle and the sale of beef through supermarkets where quality was neglected in the interest of cheapness. Export markets also disappeared. In the 1960s larger animals were selected in Britain, plus some use of New Zealand and Australian imports. For veterinary health reasons, imports for the USA were impossible but from 1972 onwards Britain imported bulls and cows from Canada, many of whom were offspring of leading American sires. Semen and embryos were later imported directly from the USA. The breed today is thus a blend of top North American bloodlines of Scottish pedigree and the British lines. It has calving ease and the good temperament, fleshing and marbling that makes it the most popular temperate beef breed in the world.

The **Aberdeen-Angus** Cattle Society administers its own Certified Aberdeen-Angus Beef Scheme, which licenses the use of the Society's Certified Trade Mark under the Department of Trade and Industry regulations. Cattle sold through the scheme must be sired by a pedigree-registered Aberdeen-Angus bull and must meet special standards in terms of quality, farm assurance and authenticity. Premium prices are paid for cattle meeting the required quality standards and can be as high 20–25 p extra per kilo. A premium of even 15 p/kg is worth £45 on a 300 kg carcase of a comparable weight and grade and can be the difference between profit and loss. Aberdeen-Angus cattle mature early under natural conditions to achieve a good balance of fat and lean with a small proportion of bone. Leanness is often cited as the main characteristic to be sought in selecting beef, but beef which is best for eating should be well "marbled" with fine threads of fat interwoven through the lean. This ensures tenderness when cooked and brings out flavour and succulence. Beef which is too lean becomes hard and leathery.

**Limousin** cattle (Plate 13.9) are early maturing, with medium-weight carcases ideal for butchers and supermarkets, while light-boned frames mean more meat from the same live weight. They can achieve weight gains of 1.5–2.0 kg a day with high food conversion efficiency. Limousin females are excellent suckler cows, with large pelvic openings and light-framed calves, giving easy calving, and they are fertile and milky. They are hardy, so can be used in upland or lowland conditions. Limousin bulls pass on their conformation when crossed to other breeds, their light-boned frames allowing easy calving, so they make good terminal beef sires. For Limousins, the critical carcase characters are: killing-out percentage about 55%, lean-to-bone ratio about 4.2, salable meat about 73% of the carcase, and higher priced cuts about 45% of the carcase (figures from the Meat and Livestock Commission). In August 2004, Limousin bull Graham's Unbeatable obtained the top price at an important auction in Carlisle, fetching 35,000 guineas (£36,750), with the buyer planning to put his semen in the Cogent firm's collection for export.

**Saler** cattle are one of the oldest and most genetically pure breeds of cattle in the world, from the southern half of the Massif Central in the

Auvergne, France, and their mountain origins make them suitable for poor conditions and rough ground. The carcase kill-out is greater than 60%, with high-quality marbled meat. They have the largest pelvic area of the major beef breeds and a short gestation period, with easy calving and less need for human intervention. They have milky, fertile cows showing early maturity (with the first calf at two years of age) and capable of calving progeny from more muscled terminal sire breeds. They have a long working life, are hardy, disease-resistant, and can use rough forage.

Other **beef breeds in Britain** include:

- **Polled Herefords**. Polled (hornless) cattle were selected from Herefords to give this excellent beef breed with a white face and red-brown coat (Plate 2.5).
- **Belgian Blue**. These are very muscular bulky cattle with above average daily weight gain, with high beef yields; they are often used for cross-breeding, using AI as the bulls are very heavy and could harm smaller breed cows at mating.
- **Scottish Highland**. These are native to the rough terrain of northern Scotland, very hardy and able to produce beef on poor land. The beef is lean, well marbled and flavoursome, with little outside waste fat because long hair, not fat, insulates them against the cold.
- **Charolais**. The cattle are a distinctive creamy white to wheaten colour (Plate 2.5), and revolutionised the British beef industry after their introduction from the Continent in the late 1950s to become (according to the British Charolais Cattle Society; www.charolais.co.uk) the leading terminal beef sire, with bulls noted for muscling, with excellent loins, good hindquarters and deep second thigh, while females are less heavily muscled with well developed udders. Growth rate and killing-out percentage are high, with intramuscular fat aiding meat eating quality. Selection based on BLUP and EBVs (Sec. 3.5) has been for production traits and ease of calving.

**Other milk breeds**: in 1999, a bull calf — Mr Frosty — was produced using the last sample of 37-year-old frozen semen collected in 1961 from a champion **British Friesian** bull, Horwood Janrol. The latter was born in 1950 before widespread antibiotic use on cattle, and long before BSE. The breeder, Mr Booker of Avon Breeder Services, Bath, is reviving the British Friesian breed to produce cattle for export which can thrive on a variety of forage and are BSE-free. British Friesians went out of fashion as imported black and white Holsteins were used to breed **Holstein-Friesians**, which produce most of Britain's milk. Mr Booker said that Holsteins were high-performance animals but needed a lot of expensive bought-in rations and had high vets' bills, unlike British Friesians.

The selection criteria are very different for beef cattle and for dairy cattle. **Breeding for bullfighting** is different again. In Spain, the *ganaderos* (breeders of fighting bulls, *Bos taurus ibericus*) select for characters such as ferocity, stamina, eyesight, horn configuration, promptness in responding to visual stimuli and fixity in following a target (the bullfighter's cape). Each ranch has developed its own line of bulls. By Spanish law and by Catholic Church law since 1565, a bull cannot be used more than once in the fighting ring, which makes testing difficult: a bull which performs brilliantly in the ring and is killed there can obviously not be used for breeding after that ultimate test. Family selection is practised, with an emphasis on the mother's characteristics. The bulls never see a matador or cape before arriving to fight in the ring when they reach maturity. They are very intelligent and would soon learn to attack the matador, not the cape.

## 13.7.11. *Peruvian guinea pigs*

In 1970, animal breeders at La Molina University in Lima, Peru, started a **guinea pig breeding programme** as guinea pigs are a dietary staple in Peru with about 65 million eaten there each year. Using conventional crossbreeding methods, the new "Raza Peru" variety has been produced; by 2004 it had proved very popular in Peru, with exports to overseas Peruvian communities and to Japan and America. The traditional Peruvian guinea pig weighs about 1.5 lb/0.68 kg, but the new variety weighs up to 2.5 lb/1.13 kg, with more protein and less cholesterol than beef, pork, lamb or chicken (see Subsec. 19.2.4 for "health foods").

As always, customer choice is important, and even Peruvian restaurants in London are doubtful about the acceptability of eating guinea pigs in Britain, where about 100,000 guinea pigs are kept as pets. Just as British consumers usually reject the idea of eating horsemeat although it is widely eaten in France, they would probably refuse to eat a species regarded here as a children's pet.

## 13.7.12. *Chinese crocodiles*

Although most of these examples are of successful plant or animal breeding programmes, it is also useful to consider a less successful programme and the reasons for its problems. Crocopark Guangzhou in south east China is one of the world's largest **crocodile farms**, with about 65,000 animals raised for crocodile meat, and high-value leather for handbags, shoes, purses, belts, etc. The farm uses stock imported from Thailand, but in the cooler climate in that part of China, at times going down to 10° at night, the crocodiles often suffer from impotence, obesity, runny noses and expensive appetites, according to Keith Bradsher (article, *New York Times*, selected for the *Daily Telegraph's New York Times* section, 28/10/2004, pp. 1).

In the cold, the male crocodiles eat more in late autumn than they would in Thailand, becoming very fat and showing little interest in the females in the crucial spring mating season. This reduces the number of young born and the increased feeding costs are made worse by the animals rejecting the abundant local ducks and fish, refusing to eat anything but chicken breasts. Diseases are also common, requiring expenditure on antibiotics and on people having to go into the crocodile pools to inject sick animals. As the farm is not profitable from the sale of crocodile products, it has now been opened to the paying public, where children can pay to feed the crocodiles using chicken breasts on a rope from a bamboo stick.

## 13.7.13. *Pig breeding to meet the buyer's specifications; meat quality factors*

Production, processing and genetic factors all affect meat quality and quantity. In **pork**, important characters are colour, pH, water-holding capacity (which depends on levels of glycogen and protein in muscle), tenderness and intramuscular fat, all affecting attractiveness and palatability as well as losses during processing and storage. The decline in eating quality of modern pork is associated with the reduction in **intramuscular fat**, which has reduced juiciness. Dry cured ham needs more intramuscular fat than cooked ham. See Sec. 5.14 and references in Chap. 5.

This case history comes from Coates (2005), describing how PIC (The Pig Improvement Company UK) helped a US customer with 15,000 sows in herds of 600–2,400 sows **to meet a buyer's contract requirements**. The buyer wanted carcass weight increased by 4–5 kg, 3 mm less backfat and 5 mm more eye muscle depth. The eye muscle is the meat in the centre of a pork chop; it is the main muscle running from the shoulder to the hind leg. A measurement taken in the centre is a good indication of the lean to fat ratio of the whole carcase (Pam Brunning, personal communication).

Changes were made to **feed formulations** to accelerate lean growth to produce leaner, heavier pigs. A **genetic survey** showed that a wide variety of terminal sires was being used, with a mixture of natural service and AI with on-farm semen processing. PIC moved to using only a central boar stud for all terminal sires, using their PIC337 boars with centralised AI facilities. The mean EBV (**estimated breeding value**) was calculated for boars for average daily feed intake, days to reach 110 kg, backfat thickness between the 11th and last rib 6.5 cm off the midline, loin depth, loin muscle pH, and 'Minolta', a measure of meat colour. A selection index was drawn up with relative ratings of loin depth 80%, backfat depth 17%, days to 110 kg 2.5% and feed intake 0.5%. The index included several molecular markers such as PT1 for lower backfat. It was applied to all the stud boars and their replacements, with improvements each generation. Selection was also applied to the sow herd, replacing older sows and those of the wrong breed composition. Carcass weight rose by 2–3 kg in four months, and a further 2–3 kg over the next six months. Over two years the backfat decreased by about 3 mm and loin muscle depth increased by about 5 mm, achieving buyer's specification by a combination of improved nutrition and selection.

**Daily weight gain** in pigs depends on many factors, genetic and non-genetic. They including age, appetite, food energy content, food amino acid content (especially lysine), disease and environment.

Important **factors in consumer meat choice** include juiciness, tenderness, taste, appearance, colour, size, price, brand name and food safety (Klont, 2005). For repeat purchases, taste, tenderness and juiciness are important. For the initial purchase, the price, colour and general appearance of the meat are most important, followed by cut size, packaging, amount of juice released and meat-cabinet lighting. The main underlying meat factors affecting consumer purchase are ultimate pH (pH 24 hours after slaughter), colour, water-holding capacity and intramuscular fat. Pale, soft, exudative meat often results from a rapid breakdown of muscle glycogen to lactic acid, giving a rapid fall of pH in pigs, especially those with the halothane (stress) gene. A low ultimate pH gives meat proteins with reduced water-holding and a lighter colour, while a higher pH gives a darker colour and less drip loss. Drip loss in a pack is disliked by consumers. Ultimate pH affects eating quality features of juiciness, tenderness and taste. A higher ultimate pH gives more water retention during storage and hence juicier meat.

Longer periods of feed withdrawal before slaughter result in less muscle glycogen and therefore a higher ultimate pH. Berkshire pigs have a high ultimate pH. Genes are known to influence ultimate pH, including the *RN* gene, for which there has been selection in many commercial lines. **Rare breeds** such as Gloucester Old Spot (Plate 15.1), Tamworth and Berkshire have exceptional meat quality but with poor production efficiency and low overall lean meat. However, using appropriate selection indexes, DNA markers and good nutrition, meat quality can be improved within fast-growing leaner commercial lines.

The PIC firm, a Sygen company, lists 41 patented **exclusive gene markers**, broadly categorised as meat quality (MT) or performance trait (PT). For example, MQ54 is a marker for meat quality, so progeny of stud boars carrying the good allele have juicier, more tender meat. PT28 gives faster growth, lower backfat and greater muscle depth, resulting in lower production costs and higher carcase prices.

## 13.8. Breeding for shows; breeds and varieties

The value of an animal is greatly enhanced if it has done well in competitive shows with recognised expert judges: see Plates 2.5, 13.7, 13.9 and 13.10. There are many shows of varying degrees of importance. In England, a win at the Royal Agricultural Show would increase an animal's value considerably, as would a win at the South of England Show for a bull up to two years of age in the Sainsbury's Super Beef Bull competition or for a cow in the Unigate Super Dairy Cow competition. A lot of wins for one breed in multi-breed classes would boost the reputation of that breed. In the case of plants, a potato show entry might be judged on skin colour, flesh colour, skin finish (from dull, blemished and scabby to clean and bright), tuber shape, etc. See Sec. 5.6 for breed records and pedigree selection.

The **ideal type** is defined in the **show rules** or the **breed handbook**. For example, cat types are specified in Britain by the Governing Council of the Cat Fancy, and dog types by Cruft's. For pets, the criteria are aesthetic but for farm animals such as Hereford cattle, the required characters are a mixture of commercially relevant ones such as size and body proportions, plus ones of largely aesthetic value such as tail length and body colouration. Each breed society lays down its ideal type against which individuals can be judged on "**breed points**" as well as on performance where that is included. For example, a cow may be judged on her appearance and on her milk record (litres per lactation, percentage protein and butter fat), and reproductive record. See Subsec. 13.7.10 for Ayrshire cattle show points.

One problem in **cattle** is that the cash value of a bull may reflect his performance in winning awards at shows, whereas his capacity as a sire in breeding may not be closely related to that ability. His breeding value may be much less than his phenotype value. There may be conflict between achieving a good level for some historic aesthetic character and for some modern commercial character. Thus for show purposes a bull might need to be very large and deep chested, whereas a lighter body size might be optimum for modern commercial types. Show rules are not always updated to follow commercial developments.

For a **dog** with no working function, purely a pet, all characters are "fancy", as with a Chihuahua but some dogs are or have been working animals. The following is a typical show description, from a Cruft's dog show catalogue, of the **Saint Bernard dog**. This used to be a working dog for rescuing people from deep snow in the Alps. Some of the description relates to functional characters for that work, such as size and muscles, but others such as facial expression and thigh feathering are largely irrelevant.

*"The expression should be benevolent and kindly. The head is very massive, the circumference of the skull being more than double the length. From the stop to the tip of the nose it is moderately short, and the muzzle is square. The lips are deep; the eyes rather small and deep set, of darker colour and not too close together. Neck long, muscular and slightly arched, with much dewlap. Chest wide and deep; loins slightly arched, wide and*

*very muscular. Forelegs perfectly straight, with huge bone, and hind-legs should not be cow-hocked. Feet large and compact. The minimum height of a dog should be 31 in. [79 cm], but the taller ones are preferred. In the rough variety, the coat should be dense and flat, rather fuller round the neck, and some feathering on the thighs. The colours may be red, orange, various shades of brindle, or white with patches on the body. The muzzle, the blaze on the face, collar round neck, chest, forelegs, feet, and end of tail — white."*

In show animals such as cats, dogs and rabbits, and in farm animals, an **established breed** has a particular general conformation (body structure and proportions, musculature, stance, head shape, tail length and thickness, ear length and position, such as erect or drooping), coat type (e.g., rough or smooth, long or short) and coat colour and markings, and may also have different behaviours, such as noisy calling by Siamese cats. There may be **varieties** within breeds, such as different coloured Burmese (dark seal brown, medium chocolate brown, blue, lilac, golden red, cream) and Siamese **cats**. Blue cats include the British blue, the Russian blue and the Korat. Some breeds are hybrids, such as the Burmilla from crossing a lilac Burmese to a Chinchilla, followed by selection for a fairly uniform breed. **Rabbits** with large floppy lop ears may be of several types such as the English lop, French lop, dwarf lop or cashmere lop. Some breeds may have a large variation permitted in coat type, with Friesian cattle varying from all black coat to widely varied amounts of white spotting.

While most breeders can manipulate the required qualitative characters for a show breed, the champion animals or plants are often at one **extreme for polygenes** or modifiers for important quantitative characters. In cats, for example, any longhairs are homozygous for the recessive $l$ $l$ alleles, with mongrel longhairs usually having coarse long hair with moderate hair density. A good show Longhair (sometimes known as **Persian**) will be $l$ $l$ but will have been bred to have many polygene alleles giving even longer, more dense hair which is really soft to the touch. Maine Coon cats are also longhairs but with a very different body conformation and behaviour from Persians, and of different geographic origin. At the other extreme, Sphinx cats when adult are almost entirely hairless. Ginger, marmalade and red cats all have the sex-linked dominant allele $O$, which changes black melanin to yellowish phaeomelanin. The show Red Tabby has a superb intense rich red tabby coat, from selection for many modifying rufous polygenes. In orange-eyed white cats, a very deep orange eye is very desirable for shows, and appears to be a polygenic extreme of the yellow to orange type, with most mongrel cats only having yellow irises.

Some breeds originated from a **single spontaneous qualitative-acting mutation**: see Subsec. 7.2.1 for rex and Manx cats. Some cat breeds cannot or should not breed true. The **Manx cat** is *M m*, where *M* gives dominant tailless, the main feature of the breed, but is also a recessive prenatal lethal, so two Manx cats on mating give an average of one normally-tailed non-Manx cat, two Manx cats and one prenatal death of the homozygote, *M M*. The adult **Scottish fold cat** has the tip of the ears bent forwards, caused by a dominant allele, *Fd*. The heterozygote is usually healthy, with the ear fold, but the homozygote *Fd Fd* has a cartilage overgrowth affecting the tail and legs, causing much discomfort in walking. It is therefore best to avoid producing *Fd Fd* homozygotes. One can cross *Fd fd* heterozygotes to non-fold *fd fd* animals of Scottish fold stock, giving half the kittens as fold and half as non-fold.

The differences between **show performance and commercial considerations** can be very important. For example, a white-skinned potato variety might do very well in shows, but many white varieties show undesirable greening after two days under the strong lights in supermarkets, whereas potatoes with coloured skins are slower to go green under lights.

## 13.9. Breeding programmes from crosses to selection, to local and national trials, possible commercial release, and approved lists

The information here comes largely from the Home-Grown Cereals Authority and the British Society of Plant Breeders Ltd. Certification procedures differ between countries, especially outside the EU. From making initial crosses to the release of a proven new variety typically takes 10 to 15 years even for an annual plant such as wheat, and up to 18 years for potatoes. **Seed potatoes** are often grown in Scotland where the lower incidence of aphids reduces aphid-transmitted potato viruses.

Breeding will be illustrated with a **winter wheat** example. The initial crosses are planned to get: high yields for cost-effectiveness; quality characteristics for specific markets, e.g., bread-making, biscuit-making, food markets (e.g., domestic flour) or exports; disease-resistance to minimise fungicide inputs; predictable uniform maturity; straw strength for lodging resistance and ease of harvest. In early trials, the wheat breeder looks for improved pest and disease resistance, greater standing power and straw stiffness, early maturity date, drought and frost resistance. Genetic fingerprinting for molecular markers may be used to predict some aspects of agronomic performance (**marker-assisted selection**; Sec. 8.7). In the early years, breeders will sow millions of lines and select them down to a few hundred. The best lines have to be shown to be stable, then after the breeder's own trials, are entered in national trials over two years. Specific milling and baking tests are carried out in Britain by the Campden & Chorleywood Food and Research Association. The National Association of British and Irish Millers, and British Cereal Exports, help to devise suitable testing methods. Trials by the main testing bodies and a range of regional collaborators provide **national and regional recommendations** for particular varieties of a crop. If successful, the new varieties join the national recommended listsof varieties and can be sold commercially.

To make the F1 generation, two or more parental lines with good characteristics are crossed. Hundreds of different matings may be made, and even if only two parents are used, if one or both are not totally homozygous, segregation and recombination will produce many different genotypes in the F1. In the F1, perhaps 1,100 crosses might be made, with perhaps 2,000,000 individual F2 plants grown. In the F2 the best plants are selected, say 50,000. The seed from each selected plant is grown in a small row, giving 50,000 F3 "ear rows". The best plants are again selected, giving say 3,000 F4 families. The selection process is repeated, with pedigree selection as well as individual plant selection. In later stages, yields of lines are accurately measured. Samples of each line are grown in special plots where they are exposed to particular diseases; they may be sprayed frequently with water when testing for resistance to *Septoria* fungus.

Up to the F5, trials are usually on one site, but by the F6 and F7, trials are often on three sites, perhaps with different soils or mesoclimates. By F8, there may be trials of the most successful lines in eight locations. During the generations F6 to F10, stabilising selection is used to "**purify**" each new line so that all plants are alike within a line. Each wheat line is **multiplied** during testing, with each kilogram of seed yielding about 30 to 60 kg seed. The breeder multiplies up only the best lines. During National Trials, the breeder will probably grow about 0.15 ha of a line, building up to 3 ha and then 80 ha for commercial sale of the seed if the line is approved.

By about the F9, lines are ready to be entered into **official trials** undertaken in Britain by the National Institute of Agricultural Botany for the Department for Environment, Food and Rural Affairs and its Northern Irish and Scottish counterparts. Lines are assessed over several years for merit, as "Value for Cultivation and Use", which for wheat includes yield, straw stiffness (lodging can halve crop yields) and resistance to the main cereal diseases. Seeds and plants are also checked to see whether the line is "Distinct, Uniform and Stable", i.e., that it is a distinct new variety giving a uniform crop.

The **National List trials** were set up under the Plant Varieties and Seeds Act 1964 and the European Communities Act 1974. Approximately 30 wheat varieties tested in the first year might be selected down to 15 in year two, to six in year three and only three might make the final recommended list. Successful performance in the National Trials means that a new variety is added to the UK National List and to the EU's Common Catalogue, allowing them to be marketed anywhere in the EU. These lists help the farmer compare the new varieties with control varieties (see Subsec. 13.7.3 for a pea example). It takes two years of National List trials before a variety can be marketed, and at least one more year before a variety can be added to the NIAB/HGCA Recommended Lists. Most of the costs of the trials are borne by the breeders, so with producing a new variety taking over 10 years and incurring very large costs, breeding is a financially risky business. The breeder's return comes from a small royalty charged on seed. For cereals, it is about £2.50 an acre, or just over £6 a hectare (Subsec. 19.2.6).

UK wheat **yields**, in tonnes/ha, have risen from 4.05 in 1963/67 to 7.20 in 1990/94, with much of the increase being due to plant breeders, and over the same period barley yields have gone from 3.67 to 5.40, and oat yields from 3.10 to 5.16. There has been much better disease resistance, better resistance to adverse weather, better lodging resistance, less premature grain shedding before harvest, and earlier ripening allowing harvesting before autumn rains. Varieties are aimed at much more specific markets, e.g., malting barley for brewing is quite different from animal feed barley. These advances have made the UK the world's sixth largest wheat exporter, instead of being a net importer as previously.

Big advances are being made with other crops. **Oilseed rape** was once grown for limited industrial markets, with 40–50% erucic acid, but since it has been bred for the food oils market, with much less erucic acid, the acreage has increased enormously (from about 48,000 ha in 1977 to over 400,000 ha in 1996 in the UK), making vast patches of yellow across the landscape. Rape oil is now the third most important oil crop in the world. Rape is unusual in including several species, *Brassica napus*, *B. rapa*, *B. carinata*, *B. juncea* and *B. nigra*. It is now being bred for biofuels and for specialised industrial lubricants, by changing its oil composition. As rape is partly (30–50%) cross-pollinated and partly selfed, it needs bagging for protection from stray pollen during breeding. Cross-pollination threatens **varietal purity**, making the use of farm-saved seed risky.

**Linseed** was grown mainly for quick-drying oils for the paint industry but is now being bred for use in animal and human foodstuffs. More cold-tolerant sunflowers and lupins have been bred for growing in Britain. When farmers are planning what to grow, the availability of **subsidies** (Subsec. 19.2.1) is important, with newer crops often being encouraged to avoid overproduction — and hence lower prices — of traditional cereals and dairy products. In the **EU Set-Aside scheme** (see Ansell and Tranter, 1992), farmers can even be paid to grow nothing, to avoid surpluses. Some of the absurdities of the EU system were shown in 1999 (*The Daily Telegraph*, 25/5/99, p 5). A farmer in Wales grew 29 ha of linseed fibre flax, receiving a 100% EU subsidy of £15,000 to produce 100 tons of fibre flax. In previous years he had harvested the crop but was ordered to destroy it as the 350-mile round trip to the processing plant in East Anglia made it uneconomic. In 1999 he was fined £1,500 for ploughing the crop back in to save £1,300 harvesting costs.

## 13.10. Selection in feral animals; feral and farmed animals

While most animal production is from **farmed herds or flocks**, with animal breeders often involved in their improvement, some is from wild (**feral**) or semi-wild herds which have varying degrees of management. Goats and sheep, say in Mediterranean countries, or cows in Africa, may be allowed to wander over very extensive rough grazing, but there is some selective culling to maintain or improve stock, as well as harvesting and the control of numbers to suit the available food. With birds such as wild geese or wild duck, there is hunting but no artificial selection or directed breeding, although there is natural selection and the possibility of selective responses to hunting. In Britain, **game birds** such as pheasants and partridge occur and breed naturally in the wild, but most such birds for organised shoots are now specially bred and reared before being released into the local environment. On big estates, most birds shot there have been raised there but can fly to other places.

**Deer, reindeer, elk and moose** provide good examples of animals which may be largely feral, although they can also be farmed. Deer were traditionally wild animals that were hunted in forests or on large estates, with some degree of management or regulation to prevent over-hunting and to gain income, often with hunters paying the owners for a licence. Commercial deer farming began in Europe and New Zealand in the early 1970s. By 1993, New Zealand produced 22,000 tonnes of farmed venison (carcass weight) compared with only 1,000 tonnes of feral venison, with even more farmed venison now. Most of the information in this section comes from www.maf.govt.nz/mafnet/rural-nz, accessed 22/10/2004.

World **farmed deer** numbers are estimated at 6.5 million, with Russia (mainly reindeer) having 60%, New Zealand (mainly red deer) 14%, and Scandinavia 10%. Reindeer are the most common types of farmed deer, with the herds foraging over large areas of tundra in Russia, Scandinavia and Canada. They are usually counted as being farmed rather than feral as the herds have owners, with Lapps and others often following their herds over long distances. They are only occasionally driven into pens, for selective culling, harvesting antler velvet or weaning the young. In contrast, many of the farmed red deer in New Zealand are permanently penned in large fields and raised on managed grass. The main commercial breeds of deer are reindeer (63%), red deer (14%), sika (8%), wapiti/elk (6%), fallow deer (6%), and a few others such as muntjak. See Subsec. 19.2.4 for deer antler velvet production and Subsec. 1.3.2 for deer chromosome number variation.

The **managed feral herds** of Europe are normally culled by licensed hunters in autumn and early winter. Only about 20% of the meat is sold commercially, 80% being consumed by the hunters, their families and friends, or sold locally. About 45,000 tonnes of venison are harvested annually in this way in Europe, especially in Germany, Sweden, Austria, Britain and France.

**Culling** is essential to control numbers as deer grazing and browsing can prevent growth of new trees and damage wild and commercial forests and crops. Deer in recreational areas such as Richmond Park near London, which has about 350 fallow deer and 300 red deer, have to be culled to avoid damaging the environment through overgrazing. Culling in the park is selective, targeting old, sick or deviant animals. It is remarkable among the fallow deer there how much variation there is in coat colour from almost white to very dark brown/black, with very variable amounts of white spotting. One can observe **age-related sexual selection** each October, especially in the red deer, where the older stags have bigger antlers and generally gather much bigger harems of females than do young males.

## 13.11. DNA fingerprinting

The term "**genetic fingerprinting**" is now used loosely to cover minisatellite fingerprinting, microsatellite typing, restriction fragment length polymorphisms (RFLPs) and random amplified polymorphic DNA (RAPD). The most frequent type of variation encountered within and between breeds is for alleles of major genes and polygenes. One kind of variation which is of little consequence for phenotypes and performance, but which can be very useful for research, is in repeated DNA sequences in "**minisatellites**". These are segments of DNA, typically a few thousand base pairs long, consisting of tandem repeats of sequences perhaps 30 bases long. There is frequent variation in the number of repeats, perhaps from incorrect alignments in replication or crossing-over, as well as variation within the sequence of a repeated unit. By using the Polymerase Chain Reaction to amplify the appropriate region many times, the sequences can be studied from very small amounts of DNA, e.g., from mouth cells, hair roots or sperm. They have been used extensively by Professor Jeffreys of Leicester University for "**DNA fingerprinting**" of humans, with forensic applications. The method can be used on humans, animals or plants to trace ancestries and relationships between individuals or breeds. The high rate of change in the number of repeats means that most individuals are different from one another for a particular sequence, so there is an enormous amount of genetic variation for minisatellite DNA, but with little phenotypic or ecological consequence. Variation for such DNA is largely lost through drift, not selection. Unusually, most minisatellite mutations are generated in the male line, at spermatogenesis. For one very unstable minisatellite sequence, about one sperm in seven carries a new mutation. **Microsatellites**, being smaller, are better for PCR methods than are minisatellites.

DNA fingerprinting using microsatellite markers is increasingly used in livestock for **pedigree analysis**. In syndicate mating, where in sheep or cattle a group of sires mates in a paddock with a group of dams with no records of individual matings, parentage can be established. Horse breed registries in most countries rely now on such genetic testing to protect studbook integrity (Tozaki *et al.*, 2001). Cattle and horse breed societies are starting to implement routine DNA fingerprinting to confirm the parentage of newborn animals before registration. These methods are less used in species such as sheep where there are many individuals of low individual value. Molecular markers are useful in animal breeding (Beuzen *et al.*, 2000).

## Suggested Reading

Abbott, A. J. and R. K. Atkin (eds.), *Improving Vegetatively Propagated Crops*. (1987) Academic Press, London.

Ansell, D. J. and R. B. Tranter, *Set-aside: In Theory and in Practice*. (1992) University of Reading, Reading.

Bahl, P. N. (ed.), *Genetics, Cytogenetics and Breeding of Crop Plants*, *Vol. 2. Cereals and Commercial Crops*. (1997) Science Publishers Inc., Enfield, New Hampshire.

Barillet, F., Genetics of milk production, in Piper, L. and A. Ruvinsky (eds.) *The Genetics of Sheep*. (1997), pp. 539–563, CAB International, Wallingford.

Beuzen, N. D., M. J. Stear and K. C. Chang, Molecular markers and their use in animal breeding. *The Vet J* (2000) 160: 45–52.

Bishop, J., *Transgenic Mammals*. (1999) Longman, Harlow.

Bisht, N. C. *et al.*, A two gene-two promoter system for enhanced expression of a restorer gene (*barstar*) and development of improved fertility restorer lines for hybrid seed production in crop plants. *Mol Breeding* (2004) 14: 129–144.

BSPB, *Cereals. Breeding a Brighter Future*. The British Society of Plant Breeders Ltd., Cambridge.

Cameron, N. D., *Selection Indices and Prediction of Genetic Merit in Animal Breeding*. (1997) CAB International, Wallingford.

Chrispeels, M. J. and D. E. Sadava, *Plants, Genes, and Crop Biotechnology*, 2nd ed. (2003) Jones & Bartlett, London.

Coates, A., Customised genetics — a case history. *Pig Genomics* (2005) 3: 1–2.

Croston, D. and G. Pollott, *Planned Sheep Production*, 2nd ed. (1994) Blackwell Scientific Publications, Oxford.

Dalton, D. C., *An Introduction to Practical Animal Breeding*. (1980) Granada, London.

Doebley, J. and A. Stec, Inheritance of the morphological differences between maize and teosinte: comparison of results for two F2 populations. *Genetics* (1993) 134: 559–570.

Dunnett, J., *A Scottish Potato Breeder's Harvest*. (2000) North of Scotland Newspapers, Caithness.

Esquinas-Alcazar, J. T., *Genetic Resources of Tomatoes and Wild Relatives*. (1981) IBPGR-FAO, Rome.

Gupta, P. K. and R. K. Varshney (eds.) *Cereal Genomics*. (2004) Springer, Dordrecht.

Gupta, P. K. *et al.*, Wheat cytogenetics in the genomics era and its relevance to breeding. *Cytogenet Genome Res* (2005) 109: 315–327.

Hammond, J., J. C. Bowman and T. J. Robinson, *Hammond's Farm Animals*, 5th ed. (1983) Arnold, London.

Hartman, H. T. *et al.*, *Plant Propagation: Principles and Practice*, 6th ed. (1997) Prentice Hall International (UK) Ltd, London.

HGCA, *Varieties. Breeding and Testing for Improvement*. Home-Grown Cereals Authority, London.

Kinghorn, B. P., Genetic improvement of sheep, in Piper, L. and A. Ruvinsky (eds.), *The Genetics of Sheep*. (1997), pp. 565–591, CAB International, Wallingford.

Klont, R., Influence of ultimate pH on meat quality and consumer purchasing decisions. *Pig Genom* (2005) 3: 5–6.

Lawrence, W. J. C., *Plant Breeding*. (1968) Edward Arnold, London.

Lupton, F. G. H. (ed.) *Wheat Breeding: Its Scientific Basis*. (1980) Chapman and Hall, London.

Maijala, K., Genetic aspects of domestication, common breeds and their origin, in Piper, L. and A. Ruvinsky (eds.) *The Genetics of Sheep*. (1997), pp. 13–49, CAB International, Wallingford.

Mason, I. L., *A World Dictionary of Livestock Breeds, Types and Varieties*, 4th ed. (1996) CAB International, Wallingford.

Mayo, *The Theory of Plant Breeding*, 2nd ed. (1987) Oxford University Press, Oxford.

Moreno-Gonzàlez, J. and J. I. Cubero, Selection strategies and choice of breeding methods, in Hayward, M. D., N. O. Bosemark and I. Romagosa (eds.) *Plant Breeding. Principles and Prospects*. (1993), pp. 281–313, Chapman and Hall, London.

O'Connell, S., *Sugar: The Grass that Changed the World*. (2004) Virgin.

Paine, J. A., Improving the nutritional value of Golden Rice through increased pro-vitamin A content. *Nat Biotechnol* (2005) 23: 482–487.

Piper, L. and A. Ruvinsky (eds.) *The Genetics of Sheep*. (1997) CAB International, Wallingford.

Poehlman, J. M. and D. A. Sleper, *Breeding Field Crops*, 4th ed. (1995) Iowa State University Press, Iowa.

Robinson, R., *Genetics for Cat Breeders*, 3rd ed. (1991) Pergamon Press, Oxford.

Simm, G., *Genetic Improvement of Cattle and Sheep*. (1998) Farming Press, Ipswich.

Tozaki, T. *et al.*, Population study and validation of paternity testing for thoroughbred horses by 15 microsatellite loci. *J Vet Med Sci* (2001) 63: 1191–1197.

Van Vleck, L. D., E. J. Pollak and E. A. B. Oltenacu, *Genetics for the Animal Sciences*. (1987) W. H. Freeman and Co., New York.

Yu, J. *et al.*, EST derived SSR markers for comparative mapping in wheat and rice. *Mol Genet Genomes* (2004) 271: 742–751.

# Chapter 14

# Genetic Engineering in Plants, Animals and Micro-Organisms

## 14.1. Introduction

**"Genetic engineering"** is now used for *in vitro* genetic manipulations involving the artificial joining of two different DNA molecules to generate genomes with particular desired properties. It aims to replace or supplement traditional methods of plant, animal and microbial improvement from random mutation and recombination, with directed modification of the genome. In essence, two DNA molecules are cut at specific base sequences by restriction endonuclease enzymes, are rejoined in a desired way, and are reintroduced into a living cell which can sustain and multiply the cloned DNA; selective techniques are used to identify cells which carry this DNA. For expression in another organism, a **transgene** (a gene modified *in vitro* for reintroduction to an organism) will be constructed to have a highly active promoter (preferably tissue-specific in higher Eukaryotes), a protein-coding sequence, a signal sequence to get the protein product moved to the right part of the cell, and a terminator sequence. More details are given in Subsecs. 18.6.3 and 18.6.4 on baker's yeast and industrial enzymes. **Human gene therapy** involves genetic engineering and is covered in Sec. 12.17, with the use of stem cells and cord blood therapy in Sec. 12.18.

For a multifactorial or polygenic human disorder such as club foot or heart disease, influenced by many genes of small effect and by environment and chance, genetic engineering offers little in the way of diagnosis or treatment. It is not the universal "cure-all" as sometimes represented in the media. The current use of genetic engineering in farm animals for those animals' own traits was summed up by Maijala (1997): "However, there appeared to be many problems and, since the conventional selection methods with modern statistical techniques give good rates of progress, molecular-genetic techniques have not yet led to new breeds in sheep".

Whether new varieties are produced by genetic engineering or classical breeding, it is essential that they be **rigorously tested** in a range of environments over a number of seasons to prove their commercial reliability and to determine under what conditions they are better than existing cultivars (Sec. 13.9). Even a very deliberate molecular change in one gene may have unpredictable and unfavourable effects on other aspects of an organism's performance (see Sec. 14.10, Bergelson *et al.*, 1998). For more examples of genetic engineering than are given below, refer to the suggested reading list.

There are now some very sophisticated systems available to **control the activity of individual genes** in higher Eukaryotes, quantitatively and reversibly, with regulation factors of up to $10^5$. For example, Clontech Laboratories UK Ltd have introduced gene activators which are switched on by the presence of tetracycline, and transcriptional silencers where genes can be turned off by tetracycline, to different extents in different transformants.

## 14.2. Restriction endonucleases and ligases

**Endonucleases** cut chains of nucleotides within the chain, while **exonucleases** remove bases from the chain ends. **Ligases** can join up two ends of DNA, making an intact chain by joining a terminal $5'$ phosphate to a terminal $3'$ hydroxyl group.

**Restriction enzymes** were given that name when they were found to restrict the range of strains of *Escherichia coli* which were attacked by a particular bacteriophage such as *lambda* ($\lambda$). $\lambda$ grown on *E. coli* strain C could grow well on that strain but had a very low frequency of infection on *E. coli* strain K, while $\lambda$ grown on strain K had a high frequency of infection on both strains C and K. In 1970, Meselson and Yuan showed that the effect was due to strain K having a restriction endonuclease which could break up unmethylated DNA, and to strain K having a **DNA methylase**, while strain C lacked both enzymes. $\lambda$ grown on C had unmethylated DNA and was able to attack strain C, but when it injected its DNA into strain K, DNA was usually destroyed by the endonuclease. If $\lambda$ DNA managed to escape this fate and was replicated in strain K, then the $\lambda$ produced had methylated DNA.

The methylation protects the DNA from the endonuclease, so λ grown on K could infect K and C easily. When an organism produces an endonuclease, it usually has the corresponding DNA methylase which protects its own DNA from being attacked by its own endonuclease. If strain K had the endonuclease without the methylase (methyltransferase), it would cut up its own chromosomal DNA.

Restriction enzymes each recognise a specific DNA base sequence, usually of four to six bases, while the associated DNA methyltransferase enzymes recognise the same sequence and methylate a key base within that sequence, thus preventing the DNA from being cleaved by its own endonuclease. There are two main types of restriction endonuclease. **Type I** recognises a specific sequence in the DNA, but cleaves the DNA rather randomly after moving along the chain from the recognition sequence. They include the *E. coli* K restriction enzyme. **Type II** cuts at specific sequences and so it always produces the same fragments from the same DNA, with the same exposed ends. They are found in all types of organism.

Type II enzymes usually recognise rotationally-symmetrical sequences in the DNA and either make **blunt** (flush) **end breaks** by cleaving the two chains at the same point, or make **staggered breaks** by cleaving the two chains at different points, leaving "sticky ends" of unpaired bases, with complementary sequences at the two ends. Thus, a blunt-end break enzyme from the bacterium *Arthrobacter luteus* recognises the sequence:

5′ AGCT 3′

3′ TCGA 5′, which has a central axis of rotational symmetry between the C's and G's, with the sequence going away from this in the top strand being the same as the sequence going away from it on the opposite side in the other strand. The enzyme cuts each strand between the C and the G, leaving blunt ends,

— AG 3′ on one fragment and 5′ CT — on the other fragment.
— TC 5′                             3′ AG —

The most useful enzymes for genetic engineering are those making staggered breaks. *Eco*RI from *E. coli* recognises the sequence

5′ G*AATTC 3′

3′ CTTAA*G 5′, cutting the upper chain after the G and the lower chain before the G, at the positions marked by asterisks, leaving the left-hand fragment ending in

— G 3′ and the right-hand fragment ending in 5′ AATTC —
— CTTAA 5′                                                    3′ G —

These are **sticky ends** in that the left- and right-hand ends now have complementary single-stranded base sequences and could pair up, with DNA ligase rejoining each of the two chains. While DNA ligase from phage T4 can join blunt ends, unlike most ligases, there is no specificity as to which blunt ends join which other ones; with sticky ends from staggered breaks, the complementary base pairing between the sticky ends ensures **specificity** in which ends are joined.

The main reason that genetic engineering is possible is because the same **type II staggered-break restriction enzyme** will make the same sticky ends in any type of DNA containing its recognition sequence. If one wants to insert a particular fragment of human DNA into a bacteriophage genome, one has to find a restriction enzyme which will cut the human DNA twice, once near each end of the desired fragment and once in the bacteriophage genome. There are thousands of restriction enzymes that could be tried, each with a unique recognition sequence of bases. If *Eco*RI is the enzyme chosen, it will make the same sticky ends in the human DNA as in the bacteriophage DNA, so the two types of fragment can be brought together and allowed to base-pair at the sticky ends, before DNA ligase is used to join the fragments covalently. Not all molecules will be of the desired type, as the phage DNA could circularise by pairing of its own sticky ends with no insert, or several molecules could join, or the insert could be joined in the wrong orientation. The techniques for overcoming these problems are beyond the scope of this book.

## 14.3. Vectors

The **vector** is the carrier molecule into which the desired DNA is inserted. Ideally, it should only have one copy of the recognition site for a particular restriction enzyme, but it can have just one recognition site for each of a number of different restriction enzymes, so that the user has a choice of enzymes. Common vectors include *E. coli* plasmids, phages such as λ and M13, and derived plasmids such as cosmids, which are made using, for example, the cohesive ends of λ, with other DNA sequences.

One has to be able to get the vector plus insert back into a living cell. Cell walls may need to be enzymatically removed from plants and filamentous fungi before vectors can penetrate into the cells. With the **biolistic** method, DNA-covered gold or tungsten particles are shot by helium pressure from a gene gun into tissues or fungal hyphae without needing protoplasts. With plasmid or cosmid vectors, one can use the DNA to transform *E. coli*, but as that bacterium has no active DNA uptake system, one has to use **passive DNA uptake** after treating the culture with calcium chloride to make the walls and membranes permeable to the DNA, or one can use **electroporation** (Sec. 14.6), but the bigger the molecule, the less efficient the transformation. With phage DNA, one can use **transformation** or one can add empty phage coats to package the DNA into **infective phage** particles.

Transformation efficiency is usually low: for $\lambda$, it is about $10^{-4}$ to $10^{-6}$; for plasmid ColE1, it is $10^{-3}$, but goes down to $10^{-5}$ or $10^{-6}$ with a large insert. With such low efficiencies, one needs a selective system to detect and isolate bacteria which have taken up and expressed the vector DNA plus insert. Ideally, one has an insert with an antibiotic resistance gene, e.g., to tetracycline, and plates the bacteria on medium with the antibiotic, so the only survivors are the desired bacteria plus inserts.

A vector should also have an origin of replication so that it can be replicated inside the bacterium or the other organism used. "**Shuttle vectors**" have a bacterial origin of replication and often a yeast origin of replication so that they can replicate in Prokaryotes and in Eukaryotes.

Plasmids are used for cloning small fragments of DNA, e.g., five to ten kilobases. $\lambda$ phage particles are of a defined size and will only package DNA which is within the range of 80–105% of a normal $\lambda$ genome length, which is 50 kb. That would only allow an insert length of 5% of that, but by deleting 20% of the $\lambda$ genome which is not essential for infection, $\sim$25% of the $\lambda$ genome length can be used for inserts, i.e., eight genes. 20 kb is the largest insert usable in $\lambda$, but cosmid vectors, which are plasmids, can take up to about 45 kb inserts. Deletions of phage P1 can accept up to 85 kb of insert.

An example of a commercially available cloning vector is the **pBluescript plasmid** (Stratagene Cloning Systems, La Jolla, California), which is circular, with 2,961 base pairs. It has two origins of replication, one from plasmid ColE1 (a high copy-number plasmid, so that one can get $\sim$300 copies per cell of the vector), and one from single-stranded DNA phage f1. When bacterial cells containing a recombinant plasmid are additionally infected with f1 helper phage, the f1 origin enables packaging of a single strand of the inserted fragment into phage particles, which is useful for *in vitro* mutagenesis. The vector carries resistance to the antibiotic ampicillin for selection of the transformed cells which have taken up the vector, and a **multiple cloning site** within part of the *E. coli lacZ* gene. The multiple cloning site (nucleotides 657 to 759) contains one copy of each of the recognition sites for ~23 different restriction endonucleases. Having this within part of the *E. coli lacZ* gene is very convenient for identifying cells which have an insert into the multiple cloning site, as that makes them Lac$^-$ compared with cells without the insert, which are Lac$^+$; with the $\beta$-galactoside X-gal, Lac$^+$ colonies are deep blue, while Lac$^-$ colonies are white.

A very recent development is to use **organically modified silica nanoparticles** as a nonviral victor for *in vivo* gene delivery and expression, initially in the mouse as a model for humans. The aim is to use this for gene therapy for cerebral and cerebrovascular disorders, including stroke, brain injury and neurodegenerative diseases. See Subsec. 12.17.2.

## 14.4.  Getting a particular piece of DNA into a vector, and recognising a clone containing it

If one cuts DNA with a particular restriction enzyme, the number of fragments will depend on the frequency of that enzyme's recognition sequence in that DNA. If one cuts a small molecule like $\lambda$ DNA (about 50 kb), there might be as few as 10 fragments, whereas cutting *E. coli* DNA might produce thousands of fragments, and cutting the much larger human DNA might produce a million different fragments. Random cloning of the cut DNA into a vector could therefore produce anything from a few different "vector + insert" combinations to millions of different such combinations, each with a different insert.

The simplest cloning selection can be done if the wanted DNA insert directly confers a selectable phenotype to the organism receiving the "vector + insert". One could select for the insert being a *histidine A* gene by cloning from an organism which was *hisA*$^+$, putting it into *hisA*$^-$ bacteria, and plating on minimal medium, so that only *hisA*$^+$ bacteria could grow. Antibiotic resistance or phage resistance genes could be selected on medium with the antibiotic or the phage.

If direct selection is not possible, it may be possible to purify the DNA before adding it to the vector. The DNA after treatment with the restriction enzyme may be subjected to electrophoresis, which can partition the fragments according to the size and electrical charge. It would be useful if the size of the desired restriction fragment were known, so that only fragments of that length were added to the vector. If the DNA sequence is known at the ends of the desired fragment, and if the DNA length to be used is less than ∼50 kb, the **polymerase chain reaction** (PCR) (PCR) can be used to amplify the DNA of interest many thousands of times. Oligonucleotide DNA primers approximately 20 nucleotides long are artificially synthesised, one complementary to the left hand end sequence and one to the right hand end sequence of the target DNA. With thermal cycling to denature the target DNA into single strands (∼95°), to anneal the primer oligonucleotides (∼55°) to the target DNA, and to allow for elongation of each primer (∼70°) by DNA synthesis off the complementary single target strand, only the target DNA is multiplied exponentially over a series of cycles, while the rest is not. If a sufficient number of PCR cycles is used, with $n$ cycles theoretically giving $2^n$ copies of the target DNA, most vectors should pick up the target DNA, even though other fragments will also be present. The taq polymerase used in PCR is very heat-stable, coming from the heat-resistant bacterium *Thermus aquaticus*, and so survives this thermal cycling for many rounds of replication.

Most Eukaryote genes are rather long compared with Prokaryote genes, because the Eukaryote coding sequences in the **exons** of the gene are usually interrupted by non-coding **introns**. This often makes the whole gene too long to amplify by PCR and may involve several copies of a restriction enzyme's recognition sequence within the one gene, so that it is difficult to isolate the whole gene. When working with Eukaryote genes, especially if one eventually wants to have them expressed in Prokaryotes, which cannot excise the introns and would therefore not produce the right proteins, it is often best to isolate already processed messenger RNA, with the introns removed in the Eukaryote. This processed mRNA can then be transcribed into **complementary DNA (cDNA)** by **reverse transcriptase enzyme**.

Isolating fully processed mRNAs for a particular gene is best done by using a tissue in which that gene is most strongly expressed, such as the oviduct of mature hens for chicken ovalbumin mRNA, coding for the main egg-white protein.

It is usual to use a vector to obtain a large set of different clones, referred to as a **library**, and then to identify which particular clones carry the gene that one is interested in. This may require having labelled DNA or RNA probes for that gene, which can be applied to DNA from each clone, and will only hybridise with those having complementary sequences. If the protein produced by the gene is known, clones having and expressing that gene can be detected by using labelled antibodies to that protein. Other useful techniques include map-based cloning for genes originally known only from their phenotypes, e.g., cystic fibrosis, and transposon-tagging for that purpose in plants.

## 14.5.  Site-directed mutagenesis

As spontaneous and induced mutations are random, it is usually very difficult to obtain a specific mutation at a specific place by those methods. If a gene has been cloned and its base sequence is known, then there are methods for direct DNA manipulation to obtain a **particular mutation**. Suppose it is desired to change a particular **GGA** DNA codon (CCU in mRNA, proline) to a **GAA** DNA codon (CUU, leucine), then one synthesises an oligonucleotide containing DNA complementary to that region of the target gene, with CTT for the desired complementary codon.

The target gene can be cloned into a single-stranded DNA plasmid which has two antibiotic resistance markers, e.g., ampicillin and tetracycline resistance, where the ampicillin-resistance gene has a single base substitution mutation, making the double-stranded form confer the phenotype Amp$^S$, Tet$^R$ on its bacterial host. The single-stranded plasmid DNA is annealed with two synthesised oligonucleotides (20 to 50 bases), one carrying the desired mutant sequence complementary to that part of the target gene, and the other carries a sequence complementary to the part of the *amp* gene, but with the *amp$^R$* sequence, not *amp$^S$*. When the two oligomers have annealed to the plasmid, they act as primers for synthesising the second strand of the DNA by DNA polymerase plus ligase. This gives a plasmid with two base-pair mismatches, one in each oligomer region, with CTT opposite GGA in the target gene, giving mispair T/G.

The plasmid with the two mismatches is transformed into *E. coli* cells which carry *mutS*, and therefore lack a mismatch repair system, and $amp^S$, $tet^S$. At replication, the double-stranded molecule segregates, one daughter molecule having the unmutated target gene sequence and the $amp^S$ gene. The other, derived from the synthetic oligomers, carries the mutated form of the target gene (GAA paired correctly with CTT) and the $amp^R$ gene. When the bacteria are plated on medium with ampicillin and tetracycline, that selects for cells carrying DNA derived from the mutated strand, as only they will be $amp^R$ and $tet^R$. Most will have the desired target gene mutation. There are many other ways of carrying out *in vitro* mutagenesis. The term **reverse genetics** is used for studies in which a known DNA change is made to see its effect on the phenotype, instead of finding a mutant phenotype, and then finding what DNA change was responsible for it.

## 14.6. Gene targeting; cosuppression; RNA interference in humans, animals and plants

**Gene targeting** is the modification of a gene in a chromosome by the use of **homologous recombination** with DNA of an introduced vector. It has been particularly used in mouse **embryonic stem (ES) cell lines** established from the inner cell mass of blastocysts from preimplantation mouse embryos. If treated correctly to prevent differentiation, these ES cells can resume normal development and contribute to all cell lineages, including germ-line cells of the resulting **chimeric** (having different cells with different genotypes; Sec. 12.16) mice, and so they can be expressed in the next generation. In practice, insertion of the vector only occurs occasionally in the desired gene, and at other times, elsewhere in the genome.

The ES cells are given the targeting vector by **micro-injection** or by the easier but less efficient **electroporation**, which is transformation aided by short electrical pulses. Depending on the vector design, homologous recombination can yield an insertion, a replacement or a deletion in the target locus. As only one ES cell in $10^2$ to $10^5$ may undergo the desired targeting event, selective methods are needed to detect the right cells. Often, a **positive-negative** selection is used. There is positive selection for cells which have incorporated the vector anywhere in the genome, often using neomycin ($neo^R$) resistance to antibiotic G418. There is a negative selection against cells which have randomly integrated the vector into regions lacking homology to the vector, often using the hypoxanthene phosphoribosyl transferase gene *(Hprt)*. In the presence of the base analogue 6-thioguanine, only the $Hprt^-$ cells survive, those in which the desired homologous recombination occurred.

When the desired mutation in the targeted ES cells (e.g., from a brown mouse) has been selected, the cells are grown on and injected into the blastocoel cavity of a pre-implantation mouse embryo (e.g., from a black mouse), and the blastocyst is transferred to the uterus of a foster mother (any colour). If the injected cells survive, the mice born will be chimeric, composed of cells from the donor ES cells and of ES cells from the host embryo. The mice may be identified as chimeric if the coat shows brown and black regions. If the mutated donor cells have helped to form the germ line, then mating the chimeric mouse to a wild-type mouse can yield some mice in which all cells are heterozygous for the mutation, and further breeding can give homozygotes, giving expression even of recessive changes in the target gene.

Suppose it was desired to inactivate gene 1 in an $Hprt^-$ cell by insertion of a neomycin resistance gene within gene 1. The vector could be made with part A of gene 1, then the $neo^R$ gene, part B of gene 1, and an $Hprt^+$ gene. If there is a double crossover of the vector with the chromosomal intact wild-type gene 1, with one crossover within part A and one crossover within part B, then the chromosomal gene will be interrupted by the $neo^R$ gene, but the chromosomal gene will not incorporate $Hprt^+$ as that is beyond the double crossover region, in a region without homology to the target area. Cells which have had this event can be selected on the neomycin analogue, G418. Cells in which the whole of the vector was randomly inserted, without the double crossover in regions of homology, will carry the $Hprt^+$ marker from the intact vector, and they will be killed when the cells are grown with 6-thioguanine.

This kind of gene targeting, where genes can be disrupted (knocked out) by insertions or deletions, or given replacement DNA sequences, is very useful in studying how genes work and what they do. Systems for gene targeting are better developed for the mouse than for most other organisms.

The fact that potential **germ-line** cells can be treated is important for breeding purposes in plants and animals.

One technique with wide applications in plants is the use of **anti-sense RNA**. Normally, the promoter ensures that only the correct "sense" strand of DNA is transcribed, but by moving the promoter in relation to the gene, the other strand, "anti-sense", can be transcribed, making RNA complementary to the "sense" mRNA, and which can therefore bind to and inactivate the "sense" mRNA. Anti-sense mRNA was engineered into **tomatoes** to reduce the production of the enzyme polygalacturonase which is involved in fruit softening during ripening. The transgenic tomatoes have a longer shelf-life before softening, and bruise less easily than normal tomatoes. Extended shelf-life has also been obtained in tomatoes by introducing a gene for a bacterial enzyme which reduces ethylene production; ethylene is a major ripening agent in fruits (Subsec. 17.1.4).

**Homology-dependent gene silencing (cosuppression)** is now often used for the same purpose as anti-sense mRNA, and some cases of "anti-sense" suppression may actually be gene silencing by RNA interference (see later in this section). People seeking to enhance the expression of an endogenous gene often introduce additional copies of that gene with more active promoters. Such transgenes are not always expressed in the expected way and sometimes actually **reduce the activity of the introduced gene** and of the homologous endogenous gene, hence the term "homology-dependent gene silencing". The gene silencing can affect repeats in transgenes, or transgenes with sequence homology to endogenous genes. In plants, both transcriptional and post-transcriptional silencing have been observed (see Grant, 1999 and Waterhouse *et al.* 1998). In petunia, *Petunia hybrida*, the red and purple flower pigments are due to anthocyans, with the enzyme chalcone synthase being limiting in pigment production. When additional copies of the *Chs* gene were engineered in to get deeper colours, the flowers had a much lighter colour and 50-fold lower *Chs* mRNA, because of **cosuppression by transcriptional silencing**.

**RNA interference (RNAi)** is a recently discovered method of gene-action control with great **therapeutic and disease-control potential** in humans, animals and plants, as well as occurring naturally in development control. Small segments of double-stranded RNAs can suppress genes with sequence homology by degrading targeted RNA transcripts in a sequence-specific manner (Shankar *et al.*, 2005). It has been used for gene-silencing in human cell lines and for targeted knockouts to study gene function. RNAi can be induced in cells by transfecting in small synthetic double-stranded RNAs or by using viral vectors to express small stem-loop RNAs (short hairpin RNAs). Small double-stranded RNAs of 21 to 23 nucleotides long transfected into mammalian cell lines have efficiently silenced target genes, without triggering interferon production. Experiments with rodents show a great potential for **treating human diseases** of the liver, CNS, eye, kidney, lung and various types of tumour and viral disease, including DNA (herpes, papilloma) and RNA viruses (HIV, influenza, polio, foot and mouth), and Alzheimer and Huntington diseases (see Shankar *et al.*, 2005). Another way of silencing genes is via **virus-induced gene silencing** (VIGS), which involves a recombinant virus engineered with a piece of host DNA to get the viral gene and the homologous host gene simultaneously silenced by plant responses to viral infection (see Tang and Galili, 2004).

RNA interference has been proposed as a method for improving the **nutritional value of plants** (Tang and Galili, 2004), e.g., by decreasing the levels of catabolic enzymes. RNAi vectors are designed to generate long dsRNA species with the same sequence as the target genes, or to express hairpin RNAs. For example, it is often desired to increase **lysine levels** in seeds, but lysine synthesis is regulated by feedback inhibition of the initial enzyme, DHPS. Mutating its gene to make the enzyme insensitive to this inhibition increases lysine in all plant organs, including the seeds, but high lysine levels in vegetative tissues often cause abnormal growth and flower development. Targeted expression of transgenic DHPS in seeds of several crops, by using seed-specific promoters, has overcome this problem with high lysine levels only in the seeds. RNAi used to reduce lysine catabolism specifically during seed development improves seed germination.

RNAi has also been used to suppress the **caffeine synthesis** gene to lower the caffeine content of coffee. It has also been used to make **high lysine maize** by reducing the expression of the gene for the lysine-poor 22-kD zein storage protein; the resulting maize is better than one carrying the high-lysine *opaque* 2 mutation (Subsec. 15.4.1) which reduces seed quality and yield. With tissue- or organ-specific RNAi vectors, there is minimal interference with normal plant metabolism.

## 14.7. Genetic engineering in plants

For a review of genetically modified crops, see Halford (2003). Transgenic crop plants have been produced by using the *Agrobacterium* **Ti plasmid**, by direct uptake of DNA by protoplasts, by high-velocity DNA-coated particles, and by micro-injection of DNA into cells. Plant genetic engineering is largely based on the tumour-inducing (**Ti**) plasmid from the bacterium *Agrobacterium tumefaciens*, a soil bacterium causing crown gall disease of dicotyledonous higher plants, where tumours (galls) develop on the stem near the soil level. The plasmids are circular, double-stranded and large, about 200 kb. They have an origin of replication, a 25 kb region called T-DNA, a *virulence* region for T-DNA transfer into the plant genome (more or less at random, without a need for homology) and a region for nopaline (or octopine) utilisation. The transferred part, the T-DNA, codes for the production of tumours (by uncontrolled plant cell division) and for the synthesis of opines such as nopaline by the plant under the control of the plasmid. These opines are used by the bacterium, helped by the opine-utilising genes on the plasmid.

Natural Ti plasmids have few suitable restriction sites, are too large, and the tumour production is unwanted. For plant genetic engineering, parts of the plasmid are deleted, including the tumour-inducing region and the real engineering is done with an **intermediate vector** which is inserted into the deleted Ti plasmid, forming a **cointegrate plasmid**. A typical intermediate vector has the insert of interest in a cloning region with a variety of restriction sites available, a selectable bacterial gene such as spectinomycin resistance ($spc^R$), a bacterial kanamycin resistance gene ($kan^R$) engineered for expression in plants, plus a region of the Ti plasmid enabling integration of the intermediate vector into Ti by homologous recombination. Spectinomycin is used to select *Agrobacterium* cells which have Ti containing the intermediate vector, which cannot replicate on its own in the bacterium. Cointegrate plasmids are being superseded by binary vector systems.

These bacteria are then used to infect plant cells, usually in punched-out leaf disks. The T-DNA of the Ti plasmid is transferred from the bacterium to exposed plant cells, integrating into the plant chromosomes at random. Transfected plant cells are selected on medium with kanomycin and callus clumps are produced. These can be induced to form roots and shoots for growing up in soil as **transgenic** plants, i.e., plants with inserted external DNA. The DNA insert then behaves as a normal plant chromosomal gene in mitosis and meiosis, and usable in conventional breeding.

Proteins can be **targeted** to organelles such as chloroplasts by the addition of particular signal sequences to the DNA insert. This is important for many instances of herbicide resistance, where the target enzymes are located in the chloroplasts. **Transgenic chloroplasts**, from targeting transgenes to the chloroplast's double-stranded circular genome, have many advantages over transgenic nuclei because there are many genomes per chloroplast, so transformation of chloroplasts can introduce thousands of copies of foreign genes per cell, with high levels of protein production. There is improved containment as the transgenes in chloroplasts usually only go through the maternal line, so they are not spread in pollen. Transformation vectors use two targeting sequences that flank the foreign genes and insert them through homologous recombination at a precisely predetermined place in the chloroplast genome, giving uniform transgene expression (Ruf *et al.*, 2001).

The inserted genes may be of use themselves in the modified plant, or they may be used to study the action and regions of expression of normal plant genes, by enabling their action to be visibly "reported". The upstream promoter regions of the genes of interest are fused to the **reporter genes** and are inserted into a plant via the Ti plasmid, so that whenever and wherever the plant's gene expression mechanisms are turned on, the reporter gene becomes active and its action can be visualised. One system uses the bacterial $\beta$-glucuronidase gene (*Gus*), which when active turns compound X-Gluc blue, or the bacterial $\beta$-galactosidase gene (*lacZ*), whose action turns X-Gal blue. Alternatively, there is the firefly luciferase enzyme which produced visible light from luciferin plus ATP.

The *Agrobacterium* Ti system seems to need plants with a wound response. Although Ti is sometimes ineffective in Monocots, it works well in Monocots with a good wound response such as asparagus, but not on cereals, which have insufficient wound response, and not on meristems.

Another way to get desired DNA into plants or other organisms is biolistics, using a "**gene gun**" which fires DNA-coated gold or tungsten particles ($\sim$0.6–1.6 $\mu$m diameter) at high velocity into cells. The transforming DNA may be incorporated into a cell's DNA by recombination events, especially if there are regions of homology between the shot DNA and the plant. See Klein *et al.* (1987). This led to the first recovery of transgenic maize plants, from embryonic cell cultures. In the gun, helium gas is used to make an aerosol of gold particles in the DNA-laden water, then the aerosol is accelerated rapidly by gas pressure (100–2,200 psi) in a Pitot's tube.

Various plant protoplast systems are transformable with DNA but there are often problems in regenerating whole plants. Genes with no selectable phenotype in culture can sometimes be used in **co-transformation**, when closely linked to selectable markers.

The first commercial genetically engineered plants were used in 1985. **Glyphosate** is the active ingredient in many widely-used weed-killers of agricultural importance, such as "Roundup". Glyphosate is N-phosphonomethyl-glycine, which interferes with aromatic amino acid biosynthesis. It is harmless to animals and is quickly broken down by soil micro-organisms. Resistance to glyphosate was induced by mutation and selected in the bacterium *Salmonella*; it was cloned into *E. coli* and re-cloned into the T-DNA of the *Agrobacterium* Ti plasmid. It was then introduced into crop plants such as soya beans, maize, tobacco and cotton, so that fields can be sprayed with glyphosate, killing weeds but not the crops. Roundup-resistant soya beans (*Glycine max*, a diploid with 2n = 40) have been very successful in the USA, and have reduced herbicide usage and the number of sprayings necessary. The fact that it is very largely self-pollinated reduces the risks of it cross-pollinating weeds and other plants, with cross-pollination of less than 0.02% at distances more than 4.5 metres from the pollen source. Transgenic tomatoes, carrying a tobacco mosaic virus coat-protein gene, were highly resistant to TMV virus.

Another application has been to utilise Ti plasmids to put into crop plants, particularly, *Bacillus thuringensis* **Bt toxin** genes, which when combined with a suitable plant promoter, produce toxins which are specific to particular **insect pests** of plants. After three years of testing in Europe and America, Ciba Seeds maize strains resistant to the European corn borer (*Ostrina nubialis*), because of appropriate Bt toxins, were approved for use in 1996 in some countries, including the USA where the corn borer caused losses of nearly $1 billion a year. Some countries ban such crops because of consumers' safety concerns or on ecological grounds (Sec. 14.10).

Another development has been the production of genetically engineered systems of nuclear, not cytoplasmic, **male sterility and nuclear fertility restoration** for making F1 hybrids in brassicas such as *Brassica napus*, the widely-grown oilseed rape. Male sterility is caused by an extracellular RNA nuclease, **barnase**, from the bacterium *Bacillus amyloliquefaciens*. The bacterial structural gene was fused with the regulatory sequence of gene *TA29*, which has tissue-specific expression in the tapetal cells of the anther. The combination was transformed into brassicas via the Ti plasmid, resulting in the destruction of tapetal RNA, killing the tapetal cells, such that no pollen matures. Another *Bacillus amyloliquefaciens* protein, **barstar**, is used in the restoration of fertility, with its gene also fused to *TA29* for tapetum-specific expression. Barstar protein combines with barnase enzyme to form an inert complex. Barstar on its own is harmless and restores male fertility in the presence of barnase. When a male-sterile inbred line carrying *TA29-barnase* is crossed with a male-fertile plant, *TA29-barstar*, all seed harvested on the male-sterile plant must be crossed seed, F1 hybrid. The F1 plants will be male fertile. See Mariani *et al.* (1990 and 1992). The barnase method has also been applied to cabbage, chicory, cotton, maize and oilseed rape.

Attempts have been made to genetically engineer plant tissue cultures to produce **high-value compounds** such as the anti-cancer agents, vinblastine and paclitaxel (Taxol®, normally obtained from yew bark), the anti-viral castanospermine and the anti-malarial artemesinin. There have been problems with genetic instability, low yields and the high costs of tissue culture.

According to the British Society of Plant Breeders, the **main aims** today for UK crops involve the following modifications, including the use of genetic engineering: **maize** — insect resistance, herbicide tolerance; **oilseed rape** — modified oil, herbicide tolerance; **sugar beet** — modified sugar content, herbicide tolerance; **wheat** — modified starch, disease resistance; **potato** — modified starch, disease resistance, insect resistance; **tomato** — slower ripening/longer shelf life; **apple** — disease resistance, slower ripening; **field vegetables**, pest resistance; **soft fruit**, slower ripening.

Another development on which a number of firms are working is "**terminator technology**", so that cereal seed grown by farmers can be sold for food, but will not germinate. That would force farmers to buy fresh seed from the producers each year. Several groups are working on **"plant vaccines"**, to produce engineered plants giving cheap vaccines against cholera, HIV, rabies, hepatitis B, Norwalk virus and travellers' diarrhoea, *Exterotoxigenic E. coli*. Potatoes, tomatoes and bananas are being tried and could deliver vaccine directly to the intestinal surfaces. In Japan, an allergenic gene was made giving a protein inducing allergy to cedar pollen. The aim was to make people eating the "allergenic rice" resistant to hay fever. It is being trialled for possible release in 2007, if Japanese government approval can be obtained.

For more information on transgenic plants, see Galun and Breiman (1997), with an appendix on intellectual property rights and the commercialisation of transgenic plants, and Owen and Pen (1996) for plant systems for producing industrial and pharmaceutical proteins.

An example of a potentially very useful piece of plant genetic engineering in agricultural species is the transfer of high-level **aluminium tolerance** to barley from wheat, using the *ALMT1* gene (Delhaize *et al.*, 2004). Of the Earth's arable land, ~40% consists of acid soils with plant growth often limited by toxicity from aluminium solubilised by acidity to the toxic $Al^{3+}$ cation, which causes poor root growth and poor uptake of nutrients and water. Liming such soils takes many years to correct acidity in subsoil layers, and many crop and pasture plants lack Al-tolerance. Wheat (*Triticum aestivum*) contains the *ALMT1* gene, coding a malate transporter associated with malate efflux and Al-tolerance, acting as a single major gene for tolerance. Barley (*Hordeum vulgare*) is among the most Al-sensitive cereal crops, and is economically important in many countries.

By a series of cloning steps, the wheat gene with a ubiquitin promoter and a terminator were put into binary vector pWBVec8 and hence into *Agrobacterium tumifaciens*, which was used to transform barley cultivar Golden Promise. From 25 primary transformants, the three highest-expressing lines were analysed. Expression was comparable to that in wheat, giving robust root growth in hydroponic culture at aluminium concentrations which greatly inhibited the roots of non-transformed barley controls. The Al-tolerance was also expressed in soils. With the ubiquitin promotor, the transgene *ALMT1* was expressed constitutively in meristematic and mature regions of the barley roots. While transgenes for insect resistance or pathogen resistance can be overcome by mutations in the pest organism, genes combating abiotic stresses such as aluminium toxicity should not break down like that. This methodology should enable a number of normally sensitive agricultural species, such as barley, to be grown on a wide range of acid soils, but the GM product has to be acceptable to consumers.

**"Golden Rice"** was introduced in Asia in 2001 with increased vitamin A, to combat blindness, depressed immune systems and other vitamin-A deficiency problems said to cause a million deaths a year in developing countries. Two genes from daffodil and one from a bacterium were put into rice, increasing the level of beta-carotene which the body converts to vitamin A. The original Golden Rice was not very effective but Syngenta Seeds Golden Rice 2, replacing a daffodil phytoene synthase gene with one from maize, is now claimed to accumulate more than 20 times the beta-carotene of the original (Editorial, 2005). Other recent transgenic rice strains have improved texture after cooking, or herbicide resistance.

Recent **trials in China** have suggested that **GM rice** could increase yields by 6 to 9%, and reduce pesticide poisoning which affects $\sim$50,000 farmers a year. Strain Xianyou 63 was engineered to be resistant to rice stem borer (which affects 70% of the rice growing areas) and leaf roller by insertion of a BT gene, with a 9% increase in yield over typical strains of rice. Youming 86 was insect resistant by introduction of a resistance gene from cowpea, with yield up 9%. Pesticide usage was only 20% of normal, with some farmers using none.

Genetic engineering to obtain plants resistant to viruses has sometimes employed "**pathogen-derived resistance**", using inserted viral nucleic acid sequences whose expression interferes with the viral life cycle. Coat protein genes have been the most successful, leading to some commercial lines (Schillberg *et al.*, 2001).

## 14.8. Genetic engineering in animals

**Transgenic manipulations** in animals (see Bishop, 1999) are done in order to introduce selected isolated genetic material from species or varieties which could not cross with the target animal. The gene's promoter is usually altered to remove the normal one, and replace it with a specially designed one to switch the gene on at a desired time in development and in the correct tissues. Germ-line transgenic animals can be produced by embryonic stem-cell blastocyst injection technology, as described in Sec. 14.6, or by micro-injection of nuclei in one-celled embryos.

In the **micro-injection method**, the cloned DNA is inserted directly into fertilised eggs which are then implanted in the recipient's uterus. It is difficult to control into which chromosome the DNA will be inserted and whether the whole engineered sequence will be incorporated; there is a rather variable expression in different offspring. The success rate is usually less than 5% of the treated eggs giving fully-expressing offspring, although it can reach 20% in mice. See Subsec. 17.2.7 for **cloning of animals, and its problems**.

In 1999, Transgenic Technology Services of Imperial College London offered a micro-injection service in which they prepared the DNA and did the micro-injection of mouse oocytes and the implantation into foster mice. They could use DNA from plasmids, phage P1 artificial chromosomes, bacterial artificial chromosomes, and yeast artificial chromosomes. They claimed 8% of offspring were transgenic with plasmid DNA and 5% with yeast artificial chromosomes, at a cost of £300 per session, each session giving about 13 mice born.

In **sheep**, many of the injected embryos die, and only ~1% become transgenic lambs (Wilmut *et al.*, 1997). The site of integration of the trans-gene appears to be random, usually within genes which are disrupted, giving lethal mutations in more than 5% of the cases, with the lethality often only showing once the gene becomes homozygous. **Levels of expression** in mice and sheep are very variable, depending on where the transgene is inserted. The recombinant proteins from human genes in sheep are not identical to the ones produced in humans because of differences in **post-translational processing**. In sheep, commercial applications of genetic engineering have been for pharmaceutical proteins, and not for agricultural purposes.

Somatic or germ-line transformations are possible with plasmids, and also with retrovirus vectors inserting DNA, made by **reverse transcriptase** from their RNA, into animal host chromosomes. If present, tumour-causing genes must be deleted from the retroviruses, and that deletion makes room for desired inserts.

For making proteins such as hormones or drugs for use in humans, one advantage of engineering genes into mammals rather than into bacteria is that mammals can carry out various post-translational modifications of proteins which prokaryotes cannot do, so that the final proteins are much more like normal human proteins than are the ones grown e.g., in *E. coli.*

Table 14.1 shows various products being made or developed in **transgenic farm animals**, including the value of that product per animal per year. Production of products in blood is less useful than in milk, as animals need to be bled or slaughtered to obtain the product. The production of tissue plasminogen activator, for treating blood clots as in coronary thrombosis, is advantageous in the milk of sheep or goats, because the yields in ∼ five litres of milk a day are as much as in a 1,000 litre bioreactor using mammalian cell cultures. The genes are placed under the control of a $\beta$-lactoglobulin promoter which is only active in mammary tissue, and

**Table 14.1.**   Products produced in farm animals after genetic engineering.

| Product | Animal | Value/animal/ year (in 1995), US$ | Use of product |
|---|---|---|---|
| **Haemoglobin** | pig | 3,000 | Blood substitute in transfusions |
| **Alpha-1-anti-trypsin** | sheep | 15,000 | Treatment of inherited emphysema |
| **Lactoferrin** | cow | 20,000 | Infant formula feed additive |
| **Factor VIII** | sheep | 20,000 | Treatment of haemophilia A |
| **Factor IX** | sheep | 37,000 | Treatment of haemophilia B |
| **Tissue plasminogen activator** | goat | 75,000 | Treatment of blood clots |
| **Cystic fibrosis transmembrane conductance regulator** | sheep | 75,000 | Treatment of cystic fibrosis |
| **Human protein C** | pig | 1,000,000 | Anticoagulant to treat blood clots |

females carrying the desired genes (after micro-injection of fertilised eggs) are identified by PCR. Expression in milk is often at high levels and the required proteins are easily purified. Cell cultures were previously used to make human $\alpha$1-antitrypsin, but production in milk of sheep gives large amounts of glycolsylated bioactive protein at low cost (see Wright, 1991). Approximately 200 g are needed per patient per year, to treat people with genetic disorders causing their livers to under-produce this protein. These disorders are common in Caucasian males, giving emphysema which is often lethal if untreated. In 1998, PPL Therapeutics announced that human blood clotting factor IX expression levels of 300 mg/l had been obtained in milk from transgenic sheep. The current annual value of factor IX production is ~£100 million.

Unfortunately, some products from engineered human genes have caused problems when used on humans. The use of insulin to treat diabetes has been known since 1921, with pig and cow insulin being used for injections. Beef insulin is longer acting with a smoother action than pork insulin. In the early 1980s, insulin was obtained from a bacterium into which it had been cloned a human insulin gene. In 1991, it was reported that 700 diabetics in Britain wanted to sue for damages when they suffered serious side-effects from human-gene insulin. They and their families claimed that they suffered a range of side-effects including paralysis, permanent memory damage and even death, as a result of being transferred from natural animal insulin to the genetically-engineered "human" insulin that became available in the mid 1980s.

This problem had not yet been solved even till 2006. One in four diabetics on human insulin experienced severe problems. A psychiatrist said that human insulin made her feel like a zombie, unable to concentrate, and with the slightest exercise severely reducing her blood sugar level. After two years of suffering, when she was considering early retirement because of those problems, she managed to switch from using human insulin to pig insulin, and she recovered completely (see www.iddtinternational.org).

Farm animals may also be modified by genetic engineering to affect their own normal products, not just therapeutic proteins for humans. This includes casein for cheese-making and milk with low levels of lactose for people with lactose intolerance. Sheep can have improved wool production from bacterial genes for improved dietary cysteine utilisation.

The gene for **human growth hormone** has been fused with regulatory sequences from the mouse metallothionein-I gene and put into pigs and sheep where it is expressed in cell types which usually produce metallothionein-I, including the liver. The hormone increased the growth of pigs and reduced their fat content, but overproduction of the hormone gave problems from premature aging, sterility and arthritis in pigs, and diabetes in sheep. Hormone balances are so crucial and so difficult to control for engineered products, that the use of such inserted genes to improve the production of farm animals has largely been abandoned. Some success has been achieved with the **sockeye salmon growth-hormone gene** in the coho salmon, giving faster growth and maturity.

Attempts are also being made to incorporate human genes into **pigs**, to lower the rejection rate of pig organs when used in **transplants** into man, as pig organs are of a suitable size and physiology, and there is a serious shortage of human donors. Transplant rejection is not the only problem, given that pig diseases might be transmitted during transplanting. **Transgenic chickens** have been produced which express a gene for a protein from avian leukosis virus, giving protection from that virus.

**Genetically modified fish** are being developed to produced human therapeutic proteins. In 2004 (Avasthi, 2004), it was announced that a freshwater fish from African lakes, the **tilapia**, had been modified to produce in its liver, and hence, in its blood, human blood clotting Factor VII to treat a rare form of haemophilia. It is currently under test for safety. Professor Maclean of Southampton University works with AquaGene of Florida and said that they had a list of 20 other human therapeutic proteins that could be produced via fish to treat lung disease, liver problems and even tumours. Factor VII can be purified from human blood, but would carry a risk of transmitting diseases. The alternative product, NovoSeven, is grown in GM hamster cell culture, but can cost $10,000 for a single injection. Tilapia is a fast-growing, fast-breeding fish that is widely farmed for food. Maclean added a genetic switch from tilapia to the inserted human gene for Factor VII, so that it is switched on in the fish liver and secreted in the blood. Human blood has about 500 nanograms of Factor VII per ml, and it hoped to make 10 times this amount in tilapia blood.

An advantage of fish is that fish diseases are not transmissible to man, and fish are much cheaper to work with than cattle, sheep or mammalian cell cultures. Other groups are looking at producing engineered proteins from plants, chicken eggs, silkworm larvae and various farm animals, including cattle.

## 14.9. Genetic engineering in micro-organisms

Genetic engineering has been very successful in bacteria, for the production of bacterial drugs and some human proteins, with a huge variety of vectors and methods available. For fungi, see Anke (1997), several papers in Arora and Khachatourians (2003), and references in Chap. 18, especially Subsec. 18.6.3, on improving yeast, and Subsec. 18.6.4 on improving industrial enzyme production. The fungus baker's yeast (*Saccharomyces cerevisiae*) is very suitable as there is a convenient 2 $\mu$m length plasmid, 6.3 kb, which is transmitted through mitosis and meiosis. It can be transformed into the cell. Bacterial plasmids can also be transformed into yeast cells. As bacterial plasmids do not have a suitable origin of replication to multiply autonomously in yeasts, they are used after integration by crossing-over into a yeast chromosome. The yeast **2-micron plasmid** can replicate autonomously in the nucleus, so it does not need to integrate. It can be given a bacterial origin of replication too, making it a shuttle vector which can grow in bacteria and yeasts. This plasmid can also be given a yeast centromere, ensuring that both daughter cells inherit it after cell division. In yeast, "**ars**" vectors containing a replication origin, or autonomously replicating sequence (ars), are also used. Various industrial yeasts have been engineered to change the flavours in beer or the $CO_2$ production in baker's yeast.

**Yeast artificial chromosomes** (YACs) can be constructed from a linearised plasmid, a centromere, yeast replication origins and yeast telomere sequences at both ends. Heterozygosity for a locus on two homologous YACs can show proper Mendelian 2:2 segregation ratios at meiosis. YACs have been widely used as cloning vectors for large segments of Eukaryote DNA, such as genes or regions up to 1,000 kb long. The longest human gene is that for Duchenne muscular dystrophy (3 sufferers in 10,000 males), which is 2.4 Mb long and takes about 16 h to be transcribed, but exons only make up 0.6% (about 14 kb) of this gene.

Genetic engineering is often the method of choice for altering highly adapted microbial genomes, if information is available on what changes might be beneficial. If it is desired to knock out a particular yeast gene for brewing or baking, or to change the base sequence, then site-directed mutagenesis will change only that gene, whereas random mutation with radiation or chemicals could affect many genes and has a low chance of making the desired change to the target gene. Many characters such as yield, however, are polygenic, with many of the genes unidentified, when genetic engineering is often of little use.

Of all known natural antibiotics, more than 60% come from Streptomycetes. DNA manipulation has been extensively used in them to improve titres of antibiotics, and to generate new antibiotics.

Many **vaccines** against viral diseases consist of inactivated virulent strains or live attenuated strains. Attenuated strains must be non-revertible to wild-type when replicating in the recipient, to avoid chance infections as has happened rarely with live polio vaccine. Engineered large non-reverting deletions are often used. A recombinant vaccinia-based vaccine against rabies is now being employed, distributed on chicken heads, to vaccinate wild foxes in Europe against rabies.

## 14.10.  Some dangers of genetic engineering; the amount of genetically engineered crops grown

There are many other engineered plants, animals and micro-organisms under production or testing, including the ones for the production of vaccines, pharmaceuticals and high-value biochemicals. There is, however, much **public opposition** to genetic engineering and the use of genetically engineered products such as foods. The case was stated admirably by Attfield and Bell (2003):

*"Certainly the use of r[recombinant]DNA methods would need to result in strain properties that provide clear benefits to consumers as well as producers, for to date, regulatory authorities have tended to give approval for GMOs while the public have not. A major problem lies with our scientific community's inability to overcome the irrational and emotional arguments associated with 'Frankenfood' imagery. The anti-GM lobby is far better prepared in terms of political lobbying and public relations, and even though many of its arguments are based on alarmist myths, the perception created is difficult to break down by using unemotional scientific facts".*

Polls in the UK in 1999 showed that only 1% of the public was keen to see development of the biotechnology industry, and 96% did not want GM foods. In England in 1999, 87,000 packets of organic maize tortilla chips were destroyed as they were found to contain genetic contamination from cross-pollination of the organic maize by genetically modified maize carrying a promoter from cauliflower mosaic virus. Maize pollen from genetically modified (GM) crops ("Bt-corn") has been shown in the USA and Switzerland to kill caterpillars of the monarch butterfly and the aphid-eating beneficial lacewings. This led to Professor Beringer, a government advisor and chairman of the Advisory Committee on Releases to the Environment, to call for the withdrawal of licences for maize with Bt toxin, if the research was substantiated.

The position in Britain in 2006 is confusing, with some voluntary embargoes on commercial plantings of GM crops and some statutory ones, while licensed trials take place. At present, there are no EU-approved GM crops grown in Britain. Some supermarket chains and national retailers have banned from their shelves all products from genetically modified organisms (GMOs). The Local Government Association advised members to ban GM foods from schools and hospitals, because of health concerns. In May 1999, the British Medical Association's report, *The Impact of Genetic Modification on Agriculture, Food and Health*, stated that: "As we cannot yet know whether there are any serious risks to the environment or human health, the precautionary principle should apply." The BMA called for an indefinite moratorium on the planting of GM crops. Government advisors from English Nature expressed worries about the effects on wildlife of growing herbicide-resistant crops, and asked the biotechnology industry to produce crops which could not cross-contaminate wild plants. Putting herbicide-resistance genes into chloroplast genomes, instead of the nuclear genome, would help to restrict contamination by cross-pollination as chloroplast genes in most flowering plants are maternally transmitted, and not through pollen.

In the European Union, EC Directive 90/220 (as amended) controls the deliberate release into the environment and the marketing of GMOs. According to a MAFF Joint Food Safety and Standards Group document in February 1999, *Genetic Modification of Crops and Food*, no GM crops had then been approved in the UK for commercial cultivation. It stated that herbicide-resistant GMOs were the only type of GMO likely to receive market approval in the UK. Evaluations were to be carried out using herbicide-tolerant spring and winter oilseed rape, maize and sugar beet. The document stated that the GM foods and ingredients currently on sale in the UK included tomato paste, maize and soya as ingredients in a variety of foods, and chymosin (an enzyme from a GM micro-organism) used in cheese manufacture as a "vegetarian" alternative to rennet from calves' stomachs.

The first GMO approved in the USA by the Food and Drug Administration, in 1992, was the Flavr Savr$^{®}$ tomato, with a gene for softening (specifying polygalacturonase) switched off by antisense genes, but it was not a commercial success for other reasons. In the USA, there is not much opposition to GM crops. Approximately 81% of soya grown there is GM, as is much of the cotton and ∼15% of maize; they are resistant to Monsanto's "Roundup" herbicide which can be used on seedlings instead of at pre-planting time, so that farmers need not spray so often: yields are higher and costs are up to 30% lower. In soya, the gene for herbicide-resistance acts as a single Mendelian dominant gene. Monsanto is a major firm involved in GM crop production as well as in herbicide production. It is producing herbicide-resistant ("Roundup Ready") potatoes, wheat and rice, and has produced pest-resistant potatoes (which kill Colorado beetles but not other insects) and cotton. Sprays of "Roundup" drifting onto land used for non-resistant crops have caused losses in those crops.

According to BIO TECH International (September 2004): "*... given the myopic nature of the GMO debate in Europe, it is frequently forgotten that elsewhere in the world GMOs have been a remarkable success story. It is only eight years since the first GM crop was introduced, but since then there are no fewer than 7 million farmers in 18 countries who are regularly planting almost 70 million hectares of GM crops. This is probably the most rapid adoption of crop technology ever*".

By 2005, 81 million hectares, about 5% of the Earth's cultivated crop land, was planted with GM plants, especially maize, soya and cotton. More than 8 million farmers in 17 countries planted GM crops in 2004, with 90% of them being poor farmers in developing countries. In Britain and the EU, however, research funds for agriculture have been cut, causing Syntgenta and several other large agricultural biotechnology and plant breeding firms to move their research overseas.

Selectable markers of bacterial origin such as kanamycin resistance have been invaluable in plant genetic engineering, but there are worries about the spread of **antibiotic resistance** from transgenic food plants to potentially pathogenic gut bacteria in humans and farm animals. Very recently, Mentewab and Stewart (2005) described a purely plant gene from *Arabidopsis* which confers antibiotic resistance to kanamycin on transgenic plants, providing a selectable marker as an alternative to bacterial genes. There are also fears about the creation of "**superweeds**", such as herbicide-resistant weed species, from accidental pollination of wild plants by genetically engineered ones, or the engineered crops themselves becoming weeds of other crops, especially if crops grown in rotation are resistant to the same herbicide. In 2005, a wild charlock plant (related to *Brassica*s such as oil seed rape), resistant to the herbicide glyphosate, was found in Britain in a field of genetically-modified oil seed rape. The vast use of herbicides would reduce **biodiversity** of plants and animals in the agricultural environment. It would be very easy for alien genes to escape from sugar beet, oilseed rape and cereals into wild relatives, making herbicide or insect-resistant "superweeds", and perhaps disrupting the natural ecology. Herbicide-resistant oil seed rape is known to hybridise with wild turnip. Using specialised herbicide-resistant strains can greatly reduce the genetic variation and diversity of many crops.

**Genetic modification of grasses** such as cereals or forage crops is especially risky because grasses tend to cross not just with closely-related species, but also with grasses of entirely different genera, so alien genes could become widely distributed (Pain, 1999). Biodiversity can be threatened by any bred crop, whether GM or not. For example, in Switzerland, the wild sickle medic, *Medicago falcata*, has largely been replaced by escaped forage alfalfa, *M. sativa*, and by hybrids between the two species.

Watrud *et al.* (2004) studied **gene flow** from genetically modified glyphosate-resistant (RoundUp-resistant) creeping bentgrass (*Agrostis stolonifera*) in Oregon, USA. This is a wind-pollinated perennial and highly outcrossing transgenic crop in widespread commercial use, e.g., as a forage grass and in golf courses. They found gene flow by cross-pollination into wild-growing plants up to 21 km away, the furthest distance tested. They also found gene flow into a related species, *Agrostis gigantean*, up to 14 km away. Plants receiving the pollen set seed and produced glyphosate-resistant seedlings with exactly the same *CP4EPSPS* gene as in the donor GM plants. Most of the gene flow was over 2 km in the direction of the prevailing winds. The very small seeds of *A. stolonifera* (about $1.6 \times 10^7$ per kg) are very readily dispersed by wind, water and animals, so GM genes could easily spread into wild and cultivated individuals very rapidly with gene flow through pollen and seed dispersal. The distance over which pollen and seed can travel depend on wind speed and direction, and it would be very difficult to detect occasional plants affected say 100 km away from the source. Worries about genetically modified genes getting into wild plants, other cultivated stocks (including ones claimed to have no genetic modification) and even different species are therefore well-founded. Cases of glyphosate-resistant weeds, and of one glyphosate-resistant crop, e.g., oilseed rape, contaminating another, e.g., maize, are found with increasing frequency, as one would expect. Some crops, such as the brassicas, have many wild relatives they could cross with, but that is not the case in Europe for maize or potatoes, which are not cross-fertile with any wild plants there.

In November 2004, a law was passed in Germany making planters of GM crops liable for economic damages to adjacent non-GM fields, even if they followed planting instructions and other regulations. Biotechnology firms, universities and research organisations opposed the law unsuccessfully, claiming that they would have very detrimental effects on innovation, with firms leaving Germany. Syngenta, the world's largest agrochemicals group that is based in Basel, announced that public resistance had caused it to halt all GM field trials in Europe, moving them to the USA. EU laws allow non-GM crops to be contaminated with up to 0.9% of pollen from nearby GM plants, which horrifies growers of non-GM crops and seeds. In December 2004, the Royal Commission on Environmental Pollution recommended that genetically modified fish should not be released or used in British fish farms for the foreseeable future.

The possible **unexpected effects** of genetic modification were clearly demonstrated by Bergelson *et al.* (1998) with *Arabidopsis thaliana*. This is inbreeding and normally self-pollinates, but the incorporation of the dominant GM herbicide-resistance gene *Csr1-1* changed the breeding system, making the plants twenty-times more outbreeding (from 0.3% outbreeding to 6.0%) and promiscuous, and so much more likely to spread the alien gene to wild relatives. The same chlorsulphoron-resistance gene had no effect on breeding when it is not in the pBin vector. The Royal Botanic Gardens, Kew, will not store GM plants in its Millennial Seed Bank (Prance, 1999).

People allergic to Brazil nuts are also allergic to soya beans engineered to have certain Brazil nut proteins. It is possible that **new viruses** could form from transgenic RNAs combining with natural viral RNAs. The spread of Bt toxin genes to wild plants could upset ecological balances, depending on herbivorous insects and their predators. Replacement of many traditional cultivars by a few genetically-engineered ones would reduce the genetic base available for future breeding (see Sec. 15.3, gene conservation). There are also **ethical problems**, such as vegetarians objecting to plant crops containing animal-derived genes giving frost-resistance.

One interesting ethical question arose from the production of **blood clotting Factor IX** in sheep milk (for extracting to treat haemophilia B; Subsec. 12.3.2) by injecting the human gene into fertilised sheep blastocysts. Approximately 90% of the resulting female sheep had no human gene detectable; 1% had the gene but did not express it, and 9% gave factor IX in the milk. The question was whether the non-producing sheep, including all the males, could be used for meat production when they might contain human genes. The Advisory Committee for Novel Foods and Processes (which carries out the assessment of novel foods in the UK) declared that there was no health risk, but several pressure groups did not want to eat human genes, on ethical grounds.

## Suggested Reading

Anke, T., *Fungal Biotechnology*. (1997) Chapman and Hall, London.

Arora, D. K. and G. C. Khachatourians (eds.) *Applied Mycology and Biotechnology, Vol. 3, Fungal Genomics* (2003) Elsevier, Amsterdam.

Avasthi, A., Can fish factories make cheap drugs? *New Scientist* (2004) 183(2464): 8.

Bergelson, J., C. B. Purrington and G. Wichmann, Promiscuity in transgenic plants. *Nature* (1998) 395: 25.

Bishop, J., *Transgenic Mammals*. (1999) Longman, Harlow.

Delhaize, E. *et al.*, Engineering high-level aluminium tolerance in barley with the *ALMT1* gene. *Proc Natl Acad Sci USA* (2004) 101: 15249–15254.

Editorial, Reburnishing golden rice. *Nat Biotechnol* (2005) 23: 395.

Galun, E. and A. Breiman (eds.) *Transgenic Plants* (1997). Imperial College Press, London.

Grant, S. R., Dissecting the mechanisms of posttranslational gene silencing: divide and conquer. *Cell* (1999) 96: 303–306.

Griffiths, A. J. F. *et al.*, *Introduction to Genetic Analysis*. 8th ed. (2005) W. H. Freeman and Co., New York.

Halford, N. G., *Genetically Modified Crops*. (2003) Imperial College Press, London.

Hartl, D. L. and E. W. Jones, *Genetics. Analysis of Genes and Genomes*, 6th ed. (2005) Jones and Bartlett, London.

Klein, T. M. *et al.*, High velocity microprojectiles for delivering nucleic acids into living cells. *Nature* (1987) 327: 70–73.

Maijala, K., Genetic aspects of domestication, common breeds and their origin, in Piper, L. and A. Ruvinsky (eds.) *The Genetics of Sheep*. (1997), pp. 13–50, CAB International, Wallingford.

Mariani, C. *et al.*, Induction of male sterility in plants by a chimeric ribonuclease gene. *Nature* (1990) 347: 737–741.

Mariani, C. *et al.*, A chimaeric ribonucleotide inhibitor gene restores fertility to male sterile plants. *Nature* (1992) 357: 384–387.

Mentewab, A. and C. N. Stewart, Overexpression of an *Arabidopsis thaliana* ABC transporter confers kanamycin resistance to transgenic plants. *Nat Biotechnol* (2005) 23: 1177–1180.

Owen, M. R. L. and J. Pen, *Transgenic Plants: a Production System for Industrial and Pharmaceutical Proteins*. (1996) John Wiley and Sons, Chichester.

Pain, S., Selfish genes warrant caution. *Kew* (1999) Spring, 56.

Prance, G., Genetic modification and Kew. *Kew* (1999) Spring, 3.

Ruf, S. *et al.*, Stable genetic transformation of tomato plastids and expression of a foreign protein in fruit. *Nat Biotechnol* (2001) 19: 870–875.

Schillberg, S. *et al.*, Antibody-based resistance to plant pathogens. *Trans Res* (2001) 10: 1–12.

Sedivy, J. M. and A. L. Joyner, *Gene Targeting*. (1992) W. H. Freeman and Co., New York.

Shankar, P., N. Manjunath and J. Lieberman, The prospect of silencing diseases using RNA interference. *J Am Med Assoc* (2005) 293: 1367–1373.

Tang, G. and G. Galili, Using RNAi to improve plant nutritional value: from mechanism to application. *Trends Biotechnol* (2004) 22: 463–469.

Waterhouse, P. M., M. W. Graham and M.-B. Wang, Virus resistance and gene silencing in plants can be induced by simultaneous expression of sense and antisense RNA. *Proc Natl Acad Sci USA* (1998) 95: 13959–13964.

Watrud, L. S. *et al.*, Evidence for landscape-level, pollen-mediated gene flow from genetically modified creeping bentgrass with *CP4EPSPS* as a marker. *Proc Natl Acad Sci USA* (2004) 101: 14533–14538.

Weatherall, D. J. *et al.* (eds.) *Oxford Textbook of Medicine*, 3rd ed. (1995) Oxford University Press, Oxford.

Wilmut, I., K., H. S. Campbell and L. Young, Modern reproductive technologies and transgenics, in Piper, L. and A. Ruvinsky (eds.) *The Genetics of Sheep.* (1997), pp. 395–411, CAB International, Wallingford.

Wright, G. *et al.*, High-level expression of active human alpha-1-antitrypsin in the milk of transgenic sheep. *Biotechnology* (1991) 9: 830–834.

Xoconostle-Càzares, B., E. Lozoya-Gloria and H. Herrera-Estrella, Gene cloning and identification, in Haywood, M. D., N. O. Bosemark and I. Romagosa (eds.) *Plant Breeding. Principles and Prospects.* (1993), pp. 107–125, Chapman and Hall, London.

# Chapter 15

# Genetic Variation in Wild and Agricultural Populations; Genetic Conservation

## 15.1. The forces controlling the amounts of variation in a population

Chapter 1 dealt with definitions of populations (Subsec. 1.3.12) and polymorphisms (Subsec. 1.3.13), which can be for alleles at a locus or chromosome aberrations. Chapter 4 covered basic population genetics, including genetic drift and the effects of population size, and mutation, selection and migration. Chapter 5 covered various types of selection, Chap. 7 covered mutation, Chap. 8 covered recombination, and Chaps. 9 and 10 covered chromosome aberrations of structure and number respectively. We now look at the factors affecting the amount of genetic variation in a population, both in the wild and in agriculture.

### 15.1.1. *The forces or processes which increase or maintain genetic variation within a population*

The forces or processes which generally **increase** the amount of genetic variation within a population are:

- **gene mutation**. Mutation frequencies are controlled by the amount of exposure to external and internal mutagens; by the rate of spontaneous tautomerisation of the DNA bases; by the efficiency of various methods of protection from mutagens and of repair processes, such as proof-reading during DNA replication, and excision-repair of pre-mutational lesions such as thymine-thymine dimers. Mutation frequencies are affected by environmental factors such as temperature. Various mutagenic processes can cause single base-pair changes such as frame-shifts and base substitutions, and larger changes varying from deletions or additions of a few bases to changes in many bases, especially large deletions.

- **production of chromosome aberrations** by agents which break chromosomes, and where broken ends may re-anneal in different patterns. Chromosome aberrations can also be produced by unequal crossing-over (Chap. 9, Fig. 9.3). Unequal crossing-over could occur at meiosis, mitosis (more rarely), at DNA replication, or at other times, by chance breakage or from agents (including radiation and enzymes), causing double-strand breaks in DNA.

- **changes in the number of copies of all chromosomes**. **Increases** in the number of copies of whole genomes can come about (Sec. 10.2) by the failure of a cell to cleave after the chromosomes have replicated, when a diploid could give an autotetraploid, or by double fertilisations giving triploids, or unreduced gametes giving triploids or occasionally autotetraploids. Series of events can give higher ploidies. In haploid fungi, nuclei may rarely fuse by chance in vegetative hyphae, giving diploid nuclei (Sec. 18.4). **Reductions** in number can come about by development of a female gamete without fertilisation, giving a monoploid from a diploid, or a diploid from a tetraploid. This occasionally happens spontaneously and sometimes may be triggered by pollination with pollen of related varieties which do not fertilise the egg cells (Sec. 10.2). Fungi with a parasexual cycle can have spontaneous haploidisation by progressive loss of chromosomes from the diploid until a stable haploid condition (with one of each type of chromosome) is reached (Sec. 18.4).

- **changes in the number of copies of one type of chromosome**. **Non-disjunction** for a single chromosome at meiosis or mitosis in a diploid can cause monosomy and trisomy, generating aneuploid chromosome numbers such as $2X - 1$ and $2X + 1$. Non-disjunction for a pair of homologous chromosomes in a diploid can produce a $2X + 2$ cell and a $2X - 2$ cell, but the latter will almost certainly die if the chromosome contains any essential genes, as for a haploid which loses one chromosome. Aneuploids from polyploids are less affected by the abnormality than are aneuploids of diploids, and may be vigorous. Animals are more sensitive than plants to abnormal chromosome numbers, and in plants and animals, certain aberrations of number or of structure are more often passed through female gametes than male gametes. Non-disjunction frequencies are influenced by the environment, with extremes of temperature and some heavy metals increasing them.

- **immigration**. Immigration increases the number of individuals in the population, and if the immigrants have different alleles or allele frequencies from the resident population, they will increase the amount of genetic variation and/or change the allele frequencies (Subsec. 4.5.2). Migrants from different environments will often have different allele frequencies from the resident population because of selection, and even if from similar habitats, they will often have different allele frequencies by chance. On a farm, immigration may be accidental, as with another farmer's ram breaking into a field of ewes, or deliberate, by introduction of animals or semen from another farm.

- **recombination**. If there is genetic variation giving heterozygosity at two or more loci, then recombination will produce new combinations of existing genetic variation. Recombination can occur by independent assortment for non-syntenic loci, by meiotic crossing-over for syntenic loci, by mitotic crossing-over, by gene conversion at meiosis or mitosis, or by haploidisation in some fungi (Chap. 8 and Sec. 18.4). Crossing-over and gene conversion frequencies are under genetic control and are also influenced by environment, especially by temperature.

The forces or processes which tend to **maintain** the amount of genetic variation within a population are:

- **heterozygote advantage**. By favouring individuals in which the frequency of two alleles is equal, heterozygote advantage helps to maintain existing genetic variation, even if other forces tend to favour one allele (Secs. 13.1 and 15.3).

- **cyclic selection**. With cyclic selection, different forms or alleles are favoured at different times, tending to maintain both in the population (Sec. 5.4).

- having **patchy or diverse habitats** tends to maintain genetic variation, as some types are favoured in some habitats and others in other habitats.

- **selection for rarity**. If rare forms are favoured because they are rare, their rare genes will tend to be maintained in the population and may increase until they become relatively common (Sec. 15.3).

## 15.1.2. *The forces or processes which reduce genetic variation within a population*

The forces or processes which generally **reduce** the amount of genetic variation within a population are:

- **directional selection**. If the breeder or nature or the opposite sex selects for organisms at one extreme for breeding, then alleles giving average phenotypes or phenotypes towards the other extreme will be lost, whether major genes or polygenes.
- **stabilising selection**. If the breeder or nature or the opposite sex selects for phenotypes in the middle of the range, then alleles giving extreme phenotypes at either end of the range will be lost.
- **genetic drift**. Genetic drift will decrease the amount of genetic variation purely by chance, with the effect being much greater in small populations than in large ones (Sec. 4.3). The closer an allele's frequency is to zero, the more likely that allele will be lost by chance, even if it is selectively neutral or mildly advantageous.
- **emigration**. Emigration reduces population size and therefore makes drift more likely. The loss of some individuals through emigration leads to the loss or the reduction in the frequency of certain alleles, if they are more frequent in the emigrants than in the resident population.

## 15.1.3. *The interactions of forces or processes affecting the amount of variation within a population*

We have seen that there are some forces or processes creating new genetic variation, some tending to maintain existing variation, and some tending to reduce the amount of genetic variation in a population. These **forces will interact** in different ways in different organisms. For example, in haploid organisms, deleterious recessives cannot be hidden from selection in heterozygotes, and the loss of a chromosome by non-disjunction will probably be lethal. Polyploids are less sensitive than haploids to aneuploidy and to the effects of a deleterious recessive. New recessive mutations could easily show in a haploid, but would need several to many generations to become homozygous and expressed in an autotetraploid.

The **amount of phenotype variation** in a population depends on its amount of genetic variation, on the amount of environmental variation, and on the heritability of different characters. The amount of phenotype variation for particular types of genetic variation also depends on dominance, additive action, epistasis, etc. For example, if a diploid maize population is homozygous for *a a* (colourless grain) and that allele shows recessive epistasis to the *P/p* locus (purple versus recessive red grain), then genetic variation at the hypostatic *P/p* locus will not show in the phenotype (Subsec. 2.2.2 and Plate 2.3). Similarly, inducible and repressible genes and conditional mutations will only be subjected to selection in environments in which they are expressed. Selection only acts on phenotypes, not genotypes.

The amount of **migration** depends on population structure and the mechanisms for gene flow in an organism (Sec. 4.4). The amount of genetic drift will be affected by whether population numbers are fairly stable or whether they undergo big seasonal or annual fluctuations. Periodic events such as droughts, fires or floods, or farmers selling off all their stock for meat, can wipe out whole populations, but migration from surviving neighbouring populations can give recolonisation later, perhaps with different allele frequencies.

## 15.2. Using a knowledge of the origins of genetic variation to solve a practical problem

Suppose we wanted to use some land with **salt-pollution** for growing an annual seed crop. How could we utilise our knowledge of genetic variation to solve the problem? A first approach would be to use **artificial migration**, selecting possibly pre-adapted seeds from existing varieties of this crop already growing on salt-polluted land elsewhere. If that did not work, or if one wanted to adapt an existing non-tolerant variety to grow in this habitat, one could try **mutation, recombination, chromosome aberrations, or changes in chromosome number** in non-tolerant strains, using spontaneous or induced changes as convenient. One could make **polyploids** such as autotetraploids or allotetraploids to test in the new environment.

One could try **inbreeding** existing lines to expose any beneficial recessive alleles, and could make **F1 hybrids** (Subsec. 13.1.3) to exploit the power of hybrid vigour in this difficult environment. One could also try a

**composite cross with agricultural mass selection** (Secs. 13.2, 13.3), of different varieties in a semi-polluted salty habitat, trying a sample of seed each year on the fully salt-polluted land. One could also try gradual adaptation, growing seeds from a composite cross in successively more salty environments each season, or try **interspecific crosses**.

Once a salt-pollution-resistant strain had been found, a safe policy for keeping it would be to reproduce it vegetatively (Subsec. 17.1.5), in case favourable gene combinations were lost at meiosis and fertilisation. Recurrent backcrossing to high-yielding non-tolerant lines might be used to improve yield (Sec. 13.4). Various selfings and crosses to non-tolerant plants could be made to find out whether the character is qualitative or quantitative, and to study segregation ratios and the **genetics of tolerance**, such as the number of loci involved, dominance, epistasis, etc. Genes for tolerance could be crossed into other varieties or could be incorporated in polyploids or F1 hybrids, or used in DNA manipulations into other species.

## 15.3. The maintenance of polymorphism in populations

For adaptation and evolution in plants, animals, micro-organisms and humans, it is useful for populations to have a variety of genetic polymorphisms. These will however tend to disappear because of selection, if any alleles have any phenotypic advantage over the others. Even for selectively neutral alleles, genetic drift will tend to eliminate polymorphisms, especially from small populations. With selection often being strong and mutations being fairly rare per gene per generation, it might seem strange that polymorphisms for many characters are so common in most organisms. There are several **mechanisms which favour their retention**.

The first is **selection in favour of rarity**, that is, an allele is favoured by selection because it is rare. A plant example is self-incompatibility alleles in the evening primrose, *Oenothera*. If self-sterility allele ($s$) type $s^1$ has a frequency of 1% in a particular population, then plants carrying it can cross with approximately 99% of the population, subject to assumptions about dominance in pollen or style. If allele $s^2$ has a frequency of 40%, existing in heterozygotes, then plants carrying it can only cross with about 60% of the population, so much $s^2$ pollen is wasted in incompatible pollinations. Rare alleles for this character are therefore favoured until they become common, when they are at a disadvantage to rarer alleles. This leads to situations with many alleles at a locus, all of them fairly rare yet still preventing self-pollination. *Trifolium* (clovers) and *Brassica* populations typically have many self-incompatibility alleles, with up to 200 different self-incompatibility alleles in a single field of clover.

Selection for rarity can also occur for **prey/predator** and **host/parasite relations**. A bird might eat the common form of a butterfly, getting used to the taste and visual form of that common type, but might ignore a rare form, not associating it with the common form as the same favoured food source. Similarly, a fungal parasite might adapt to attacking the common form of some plant species, but not adapt to attacking a rarer form. Even a relatively low frequency of disease- or pest-resistant polymorphisms helps to curtail the spread of diseases or pests. A build-up of different alleles giving resistance to challenges by different races of a pathogen helps to control epidemics, but those alleles' frequencies only rise in response to such challenges.

A second possible mechanism for preserving polymorphisms is **an equilibrium between mutation, selection and gene conversion** (Sec. 4.5). Mutation pressures might favour one direction of change of allele frequency over the other, typically favouring wild-type to mutant changes over mutant back to wild-type changes. Selection might favour a different direction of change, and there might well be disparity in the direction of gene conversion (e.g., with conversion from mutant to wild-type being more frequent than conversions from wild-type to mutant). Depending on the parameter values for mutation, selection, gene conversion frequency and the disparity in the two directions of correction for a pair of alleles, an equilibrium may be set up between the three forces, tending to preserve the polymorphism. Gene conversion disparity can also remove polymorphisms. The equations and data on conversion frequencies and typical amounts of conversion direction disparity were given by Lamb and Helmi (1982) and Lamb (1985, 1998, 2003).

A third and very important mechanism favouring polymorphism retention is **heterozygote advantage** (Sec. 13.1). The heterozygote is sometimes fitter than either homozygote, showing overdominance for fitness. In Mendel's pea plants, one usually describes height as showing complete dominance, but $t$ $t$ is about 30 cm high, $T$ $T$ is about 182 cm and $T$ $t$ is about 213 cm, showing some overdominance. Heterozygote advantage favours the retention of polymorphism because it favours the genotype with equal numbers of the two alleles.

The classic human case of heterozygote advantage is **sickle-cell anaemia** (Subsec. 12.3.1) in a malarial area. The recessive homozygote, $a$ $a$, has anaemia which is often lethal in childhood, because at low oxygen concentrations, the red blood cells go sickle-shaped and are removed by the spleen. The bad allele, however, is maintained at much too high a frequency to be due to recurrent mutation. The explanation is that the heterozygote, $A$ $a$, is more resistant to a common form of malaria than is the dominant homozygote, $A$ $A$. In a malarial region, $A$ $a$ is fittest, so let its fitness be 1.0; let the fitness of $a$ $a$ be reduced by $t$ from anaemia, and let the fitness of $A$ $A$ be reduced by $s$ by malaria, so that the relative fitnesses are $(1 - s)$ for $A$ $A$, 1.0 for $A$ $a$, and $(1 - t)$ for $a$ $a$. The genotype frequencies before selection in a generation are the Hardy–Weinberg ones of $p^2$, $2pq$ and $q^2$ respectively, so what is eliminated by selection in a generation is $p^2 s$ by malaria and $q^2 t$ by anaemia. As the frequency of $A$ alleles is $p$ and $p^2 s$ are eliminated, the proportion of $A$ alleles eliminated is $p^2 s/p$, which is $ps$; similarly, the proportion of the $q$ of $a$ alleles eliminated is $q^2 t/q$, which is $qt$. At equilibrium, the proportions of the two alleles being eliminated must be equal, which is why the allele frequencies remain constant, so at equilibrium, $ps = qt$, and because $p + q = 1$, equilibrium $\hat{p} = \frac{t}{s+t}$ and equilibrium $\hat{q} = \frac{s}{s+t}$.

In Africa, $a$ $a$ is usually lethal, so $t = 1$, and in malarial regions, $A$ $A$ has about a 10% fitness disadvantage compared with $A$ $a$, so $s = 0.1$. At equilibrium, we therefore get $\hat{p} = \frac{1}{0.1+1} = \mathbf{0.91}$, and $\hat{q} = \frac{0.1}{0.1+1} = \mathbf{0.09}$. The genotype frequencies expected at birth, before selection, are therefore $A$ $A$, **0.826**; $A$ $a$, **0.165**, $a$ $a$, **0.008**, in a malarial area. If, however, by migration to a non-malarial area or the elimination of mosquitoes, malaria disappears, then $t$ stays 1.0 for anaemia, but $s$ becomes 0.0, so at equilibrium, we get $\hat{p} = 1.0$, $\hat{q} = 0.0$, with eventual **complete elimination** of the deleterious sickle-cell gene once the heterozygote advantage due to malaria goes. In accordance with these predictions, this polymorphism and sickle-cell anaemia are slowly being eliminated where swamp drainage and/or insecticide sprays have killed off the mosquitoes, and also where Africans have migrated to non-malarial countries such as Britain. Approximately 1 in 12 of US blacks is a carrier for sickle-cell anaemia.

In malarial regions, it is wasteful to have some *A A* dying of malaria and nearly all *a a* dying of anaemia. The **ideal population** there would be **all *A a***, but without vegetative reproduction or cloning that is not possible in humans because of Mendelian segregation, *A a* × *A a* giving $1/4$ *A A*, $1/2$ *A a*, $1/4$ *a a*. In plants with vegetative reproduction, or yeasts reproducing by budding, or animals with apomixis or parthenogenesis, successful heterozygous genotypes can be maintained once they have arisen by recombination or mutation. The loss of fitness due to segregation and recombination is called the **segregation load**, parallel to the mutation load. The segregation load reduces immediate fitness, but like mutation, segregation and recombination, it permits adaptation, long-term fitness and evolution by providing genetic variation on which selection can act.

## 15.4. The need for genetic conservation; methods of conservation

### 15.4.1. *The need for genetic conservation and the value of some old varieties*

It has been estimated by staff at Kew Gardens that in 50 years on present trends, one quarter of the world's **wild species** will have been lost, for plants and their dependent animals. Staff at Marwell Zoo, Hampshire, estimate that 41% of all mammalian species are at risk of extinction. In **agriculture**, as large numbers of mediocre varieties of crops and farm animals are replaced by a few superior modern varieties, the **gene pools available for future breeding** and for genetic engineering are being seriously diminished. Wars, natural disasters and disease can all devastate existing varieties. To counteract this, large collections of older stocks and their wild relatives should be maintained for possible future use, even from inferior material which might contain some very useful genes which have not yet been recognised and exploited [see Brush (1999)]. A Mediterranean vetch (*Vicia* sp.) provided a human blood clotting protein helpful in the detection of rare human blood disorders (Van Slageren, 2003). Some genes might become useful only in the future, through changes in climate, agricultural practices, medical drug discoveries or consumer tastes. Some might just be polygenes which would only show to a very small degree individually, but might have a large impact on performance collectively.

In Afghanistan, a mediocre **local wheat** strain which was being replaced by modern varieties was found to have a valuable rust-resistance which was then bred into the replacement wheats. A local wheat collected in Turkey in 1948 had poor agricultural characteristics but has been the source of several fungal-resistance genes for modern wheats. It has genes for resistance to *Puccinia striiformis* rust, to 35 strains of *Tilletia caries* (bunt) and to *T. foetida*, to 10 strains of *T. controversa*, and tolerance to several species of *Urocystis*, *Fusarium* and *Typhula*. In **rice**, *Oryza sativa*, resistance to the important Asian "Grassy Stunt" virus was found in just one population of the wild relative, *Oryza nivara*; this resistance was incorporated into the high-yielding variety *IR36*.

In some less developed parts of the world, **human malnutrition** is often more common than starvation. The problem is usually one of insufficient protein, especially of the amino acid lysine in cereals, and to a lesser extent, a deficiency of methionine and cysteine in legumes. In 1964, people at Purdue University, USA, reported that maize homozygous for the mutations *opaque-2* or *floury-2* had lysine at 3.7 g/100 g protein, instead of the normal 1.6 g/100 g. *Opaque-2* has a grain yield of 99.9% of normal varieties, so its flour can be useful in combating lysine-deficiency in human diets (Sec. 14.6). Both these **high-lysine mutants** had been known as morphological mutants since 1935, but their nutritional properties had not been investigated. Some endangered animal species have known **dietary problems**, such as the great sensitivity to vitamin E deficiency in Przewalski's Horse and the Black Rhino, so conservation or captive breeding programmes can take that into account.

We have seen that within a variety, a **need for uniformity** (for marketing or ease of mechanisation) results in stabilising selection being used. In Subsec. 15.1.2, we saw that stabilising selection, directional selection and genetic drift all reduced genetic variation. Although mutation and recombination could restore some variation, most mutations are harmful and would be eliminated by natural or artificial selection; many recombinants would not be as good as their parents and would be eliminated.

If genetic stocks are kept in small populations because their performance is mediocre and they cost money to maintain, then **genetic drift** will be severe and reduce genetic variation within each stock. It is particularly expensive to maintain big herds of **large farm animals** if they are inferior to current commercial stocks, as individual animals cost a lot more to maintain than do most plants. In rare breed cattle, sheep and pigs, most males are sold off for meat, as only a few are needed for breeding with the females, but the fewer the males that are used for breeding, the lower the effective population size and the more severe the genetic drift (Sec. 4.3).

If the world is undergoing a **climatic change** with global warming, then even more species will be endangered. There are already worries that the hole in the ozone layer, by allowing more UV light to reach the earth's surface, will reduce the viability of pollen in wind-pollinated plants. Plants particularly affected in experiments were maize, rye, pears, cherries, pistachios and poppies (Cuttings, 1998).

Sometimes, ideal types have been lost by **indiscriminate breeding** with other types. In South America, the guanacos are the wild, undomesticated ancestors of the llamas, alpacas and vicunas. **Llamas** were selected over 5,000 years ago by man as pack animals for use in the Andes mountains and were also used for meat and wool. **Alpacas** and **vicunas** were selected for fine wool production and the Incas are said to have had advanced breeding systems for improving them. The Spanish conquests of South America in the 16th century led to the mass slaughter of alpacas in favour of sheep and cattle, and indiscriminate breeding of the remaining animals led to the loss of the fine lines. There is now a British Alpaca Society, with a stud at Arunvale, Sussex, selecting for soundness of frame and wool quality.

## 15.4.2. *Conservation programmes and methods of genetic conservation*

Ideally, one should conserve **wild habitats** which contain *in situ* wild relatives of species used by man in case they are needed in future (Maxted *et al.*, 1997), but the pressures of human populations often make that difficult. There is a strong case for *ex situ* regional, national or **international centres** for plant, animal and microbial conservation, with public funding and national conservation bodies, especially as one cannot always predict what will be useful in future. For methods of plant conservation, see

Given (1994), Brush (1999), Maxted *et al.* (1997) and Razdan and Cocking (1997 and 1999).

The Food and Agriculture Organisation of the United Nations (**FAO**) promotes cooperation, and various International Agricultural Research Centres specialise in particular crop collections. The **numbers of accessions of plants** at international research centres, for *ex situ* conservation, research and further breeding can be enormous. They include 79,500 for **rice** at the International Rice Research Institute in the Philippines, about 12,350 for **maize**, 25,500 for **wheat** and 8,200 for **triticale** at the International Center for Maize and Wheat Development in Mexico, about 54,000 for **grasses** at the International Center for Agricultural Research in Dry Areas in the Lebanon, 31,500 for **pearl millet** at the International Crops Research Institute for the Semi-Arid Tropics in India, and about 880 for **potatoes** and 6,300 for **sweet potatoes** at the International Potato Center, Peru (information from ASSINSEL, 1997).

The FAO has a **Global Project for the Maintenance of Domestic Animal Genetic Diversity** for 14 species, including sheep. One aim is to conserve genetic uniqueness, using microsatellite markers to calculate the genetic distance between breeds. The FAO has a "world watch list" for **domestic animal diversity**, which is a catalogue of breeds at risk for 28 mammalian and avian species (Sherf, 1995). For the United Nations Environmental Programme, Heywood (1995) has listed the indigenous breeds of the main mammalian livestock species being conserved in various countries.

In Denmark, Finland, Iceland, Norway and Sweden, the "**Nordic Gene Bank Cooperation — Farm Animals**" was started in 1984, covering 103 breeds over 13 animal species. Farmers can receive a subsidy for loss of profit on maintaining a conserved breed. Flocks and herds are also kept by teaching organisations and agricultural museums. The loss of some cryo-preserved stocks from power failure in this Nordic programme shows the importance of keeping live animals as well as frozen stocks. See Ponzoni (1997) for other animal conservation programmes.

In the UK, there is the non-governmental, privately funded, **Rare Breeds Survival Trust**. Since its foundation in 1973, no farm animal breed has become extinct in Britain. About 1,000 owners maintain small groups of rare-breed animals. It owns Linga Holm island off the north coast of Scotland where **North Ronaldsay sheep** can survive on kelp (seaweeds).

**Plate 15.1.**   **A Gloucester Old Spot sow with piglets** at Habarana Lodge Farm, Sri Lanka. Note that the largest piglets suckle from teats nearest the head end.

This breed is unusually salt tolerant, with a tidal, not diurnal, grazing pattern, and can survive on a diet so low in copper as to kill other breeds.

The **Gloucester Old Spot pig** (Plate 15.1) is an example of what was an endangered breed, but enthusiasts such as Dave Overton of Exfold Farm, Dorking, Surrey, have led to their numbers reaching about 700 (529 breeding females in year 2000) in Britain, plus a high-quality herd in Sri Lanka. The pigs are very hardy and were traditionally fattened on whey from cheese and windfall apples. When crossed to white breeds, all offspring are white, giving commercial F1 pigs with hybrid vigour, suitable for pork or bacon. Pedigree pigs are a bit fatty for most commercial processors, but they have a superior flavour and sell through specialist "rare breed butchers". Gilts have 5–10 piglets and sows have 9–15; up to 18 piglets is not uncommon but is undesirable as mortality is then high. Piglets weigh about 1.6 kg, weaning at 11–27 kg at eight weeks, with slaughter at 18–24 weeks, with about 64 kg liveweight giving a 45 kg carcase. The **Tamworth** breed of pigs had only 223 breeding females in the UK in 2000, but was voted "Top taste" in a 1990s competition. Their long thin muscular shape gives very good bacon and pork, but less meat than in most commercial breeds.

Most **zoos** are involved in animal conservation projects and in breeding programmes to return endangered species to the wild, but these are mainly not for agricultural animals. For example, the Jersey Wildlife Preservation Trust has bred Antiguan racer snakes and the Mauritius pink pigeon; Marwell Zoo has bred Przwalski's horse, the scimitar-horned oryx, red pandas, peccaries, reddish buff moth and other species.

In Britain, the **National Council for the Conservation of Plants and Gardens** (NCCPG) aims to retrieve plants believed lost from commercial cultivation, but which may still be preserved in private gardens, and to conserve as many varieties as possible. For example, a double-flowered bramble, *Rubus rosifolius* "Coronarius", first described in 1816, was believed to have been lost for 100 years, but was spotted in a garden in Virginia and is now listed by about 30 nurseries. The NCCPG has established about 600 National Collections of separate genera, containing as many types as can be traced, with more than 50,000 types kept by institutions or enthusiastic amateurs. For example, amateur Veronica Read is the Holder of the National Collection of Hippeastrums (Knight's star lily), with more than 300 *Hippeastrum* plants covering 74 named cultivars and 20 species, including varieties no longer commercially available. Similar organisations have been set up in America, Australia, New Zealand and several European countries.

The £74 million **Eden Project** in Cornwall, England, has two "biomes", climate-controlled environments, one for plants, especially economic plants, from the humid tropics (e.g., Amazonia, West Africa, Malaysia), and one for Mediterranean-climate plants (southwestern USA, South Africa, and Mediterranean areas). One greenhouse for tropical rainforest trees is 1,000 m long by 120 m wide and 60 m high. This award-wining project for conservation and education is also a tourism success.

Unfortunately, there can be severe **legal and financial problems** about ownership, exchange and the use of conserved stocks. The Convention on Biodiversity agreement, from the 1992 Earth Summit in Rio de Janeiro, has been ratified by 174 countries, including the UK and European Union, but the sharing of the benefits from the use of biodiversity is the most controversial part (see Prance, 1998).

As inferior stocks will not usually be commercially viable on their own, there is an excellent case for **"added value" activities** with them, whereby members of the public will pay admission charges to see rarities or will sponsor individual animals. Thus, zoos, safari parks, game reserves, rare-breeds centres, botanical gardens, show farms, etc., are all valuable places for conserving non-commercial varieties. While some rare breeds have public appeal in their appearance or behaviour, others do not. To minimise the problems of genetic drift and inbreeding in small populations, many zoos and safari parks have scientifically planned programmes of exchange of animals between institutions, either permanently or just for mating. Sale of surplus seed, fruit or animals is one way of helping to fund genetic conservation. The restrictions on the range of plant or animal varieties which can be sold commercially within the EU are extremely unhelpful to those trying to preserve older or less productive types.

**Micro-organisms** are the easiest and cheapest to conserve. One can preserve most of them dried down on silica gel. For example, yeasts are suspended in non-fat milk and then poured into small chilled glass vials containing dehydrated sterile silica gel, before storing at 4° in an airtight jar containing moisture-indicating silica gel. Bacteria can be stored at room temperature in high-salt "stab" nutrient media. Phage can be stored in phage buffer at 4°. Many micro-organisms can be stored in freezers after suitable treatment, e.g., at liquid nitrogen temperature, as can animal sperm and mammalian embryos. All these treatments can result in stocks being stored in suspended animation for ten to many years, though viability should be checked periodically.

With **plants**, one can store pollen, seed, cell cultures, organ or tissue cultures, or whole plants, in different ways. Living plants at the National Fruit Collections at the Brogdale Horticultural Trust, Faversham, Kent, include at least two plants of each variety of 2,009 dessert and culinary apples, 75 cider apples, 272 cherries, 495 dessert and culinary pears, 20 perry pears, 4 medlars, 41 hazelnuts, 120 blackcurrants, 55 grape vines, 60 strawberries, etc. (1995 figures).

**Seed banks** are built to conserve seeds of many of the world's flora. For example, the Royal Botanic Gardens, Kew, has built a Millennial Seed Bank in Sussex at Wakehurst Place to conserve wild plants. The project had a cost of £80 million, with £30 million from the National Lottery Millennium Project. The construction costs were put at £13 million. The project will safeguard all the British flora and will eventually include 10%

of the world's flora, including endangered species, with some specialisation in plants from arid regions of Africa, India and Latin America (Van Slageren, 2003). The collected seeds are dried for about a month at 10–15% relative humidity at about 40° and are frozen at −20°. With storage in airtight 3-litre preserving jars, the **estimated survival time** is 200 years for about 80% of the species, and more than 1,000 years for some legumes. There is a very large seed bank in Colorado, USA, and another very large one, the International Seedbank at Svalbard, buried in the permafrost in Norway, so seeds will be preserved even if electricity fails for very long periods.

In seed banks, one wants to conserve variation within species, so a typical collection would be of 20,000 seeds of one species, taken from different plants in different areas, regions or countries, including different ecotypes. Desirable genes, say for resistance to a particular disease, may be present at only very low frequencies. **Seed bank storage** involves collection, verification of the species and cultivars, seed preparation (checking for pests and diseases, fumigation if necessary, sometimes X-raying to detect insects), germination tests, seed drying, packaging in moisture-proof containers, storage, periodic germination tests, and seed regeneration when required. The extent of mutation accumulation under long-term storage conditions needs researching.

Near Gannoruwa in Sri Lanka, there is the modern **Plant Genetic Resource Centre**, including a seed bank for Asian food plants. Outdoors in soil, there are "**living germplasm**" collections consisting of many small beds of different strains of local food plants such as **gotukola**, a low-growing plant (Plate 15.2) whose small leaves are used like spinach. Inside the building, there are **tissue cultures** of commercial plant cultivars including pineapples, bananas and jack fruit. The main facility consists of cold stores for the **seed bank**. For this, many seeds of a crop plant are collected from different sources, the identification is checked, and the seeds are tested for pests and diseases, and for germinability, before drying. The Base Collection can store 25,000 species, at 1° and 30–35% relative humidity, with typical projected seed longevities of 35–50 years. There is also an Active Collection for sending out as required, with a capacity of 25,000 species, kept at 5°, 35–40% relative humidity, with a projected longevity of about 20–25 years. The seeds are kept in large screw-topped plastic jars in

**Plate 15.2.**   Plates 15.2, 15.3 and 15.4 were taken at the **Plant Genetic Resource Centre, Gannoruwa, Sri Lanka**. This shows **living plant collections**, especially different collections of **gotukola** (foreground and middle ground).

**Plate 15.3.**   The **stainless steel refrigerated chambers** for the seed banks, especially of Asian food plants. See text for details.

**Plate 15.4.** **Inside the refrigerated chamber** shown in Plate 15.3. The seeds are in screw-top plastic jars, filed in drawers.

filing drawers inside large walk-in stainless steel chambers (see Plates 15.3 and 15.4).

**"Recalcitrant seeds"** are difficult to store for long periods, remaining viable for only two weeks to several months, as with important Asian crops, cocoa, coconut, mango and rubber; large seeds such as coconuts make preserving 20,000 very difficult. Recalcitrant seeds often have a high water content and may be damaged by low humidity and low temperature storage. Some plants can be **cryopreserved** *in vitro* at −196° as zygotic embryos, tissue cultures or single-cell cultures [see Withers (1991), and Razdan and Cocking (1997 and 1999)]. Some species are highly heterozygous and their genotypes are maintained by vegetative propagation, as for many fruit trees, strawberries, potatoes, artichokes, bananas and cassava. They can be kept in field gene banks as whole plants, or as cuttings, bulbs, tubers, etc., at low temperatures and controlled humidity for relatively short periods. They can also be stored as tissue cultures, meristems, or shoot tips in media giving very slow growth, with low light and low temperatures. This is used for potatoes and sweet potatoes at the Centro Internacional de la Pipa, Peru.

With **animals**, one can conserve them as living populations or by cryo-preservation of gametes (especially as sperm for AI; Subsec. 17.2.5) or embryos (Subsec. 17.2.6). DNA can be stored, but one cannot regenerate whole animals from isolated DNA at present. Semen and embryos freeze and thaw well, but not ova. Cryopreservation avoids genetic drift and dis-ease, but power failures can ruin stocks. For sheep over a 20 year period, the costs of keeping live populations and the same number of cryopreserved animals are very similar. See also Frankham *et al.* (2004).

## Suggested Reading

ASSINSEL, *Feeding the 8 Billion and Preserving the Planet.* (undated, but 1997). ASSINSEL (International Association of Plant Breeders), Nyon, Switzerland.

Brush, S. B., *Genes in the Field: Conserving Plant Diversity on Farms.* (1999) Lewis Publishers, Boca Raton, Florida.

Collins, W. W. and C. O. Qualset (eds.) *Biodiversity in Agroecosystems.* (1998) Lewis Publishers, Boca Raton, Florida.

Cuttings, Ozone loss may damage pollen. *Kew* (1998) Autumn, pp. 7–9.

Ford-Lloyd, B. and M. Jackson, *Plant Genetic Resources: An Introduction to their Conservation and Use.* (1986) Edward Arnold, London.

Frankham, R., J. D. Ballou and D. A. Briscoe, *A Primer of Conservation Genetics.* (2004) Cambridge University Press, Cambridge.

Given, D. R., *Principles and Practice of Plant Conservation.* (1994) Chapman and Hall, London.

Heywood, V. H. (ed.) *Global Biodiversity Assessment.* (1995) Cambridge University Press, Cambridge.

Lamb, B. C., The relative importance of meiotic gene conversion, selection and mutation pressure, in populations and evolution. *Genetica* (1985) 67: 39–49.

Lamb, B. C., Gene conversion disparity in yeast: its extent, multiple origins, and effects on allele frequencies. *Heredity* (1998) 80: 538–552.

Lamb, B. C., The extent, molecular origins and evolutionary effects of gene conversion disparity in fungi. *Recent Res Dev Genet* (2003) 3:143–162.

Lamb, B. C. and S. Helmi, The extent to which gene conversion can change allele frequencies in populations. *Genet Res* (1982) 39: 199–217.

Marwell, *Your Guide to Marwell and the World of Animals.* (1999) Marwell Zoological Park, Hampshire.

Maxted, N., B. V. Ford-Lloyd and J. G. Hawkes (eds.) *Plant Genetic Conservation: The In situ Approach.* (1997) Chapman and Hall, London.

Ponzoni, R. W., Genetic resources and conservation, in Piper, L. and A. Ruvinsky (eds.) *The Genetics of Sheep.* (1997) CAB International, Wallingford.

Prance, G. Questions of ownership and access. *Kew* (1998), pp. 3.

Prasad, B. N. (ed.) *Biotechnology and Biodiversity in Agriculture/Forestry.* (1999) Science Publishers, Inc., Enfield, USA.

Razdan, M. K. and E. C. Cocking, *Conservation of Plant Genetic Resources in vitro. Vol. 1: General Aspects.* (1997) Science Publishers, Inc., Enfield, USA.

Razdan, M. K. and E. C. Cocking, *Conservation of Plant Genetic Resources in vitro. Vol. 2: Applications and Limitations.* (1999) Science Publishers, Inc., Enfield, USA.

Sherf, B. D. (ed.) *World Watch List for Domestic Animal Diversity*, 2nd. ed. (1995) FAO, Rome.

Van Slageren, M. W., The Millennium Seed Bank: Building partnerships in arid regions for the conservation of wild species. *J Arid Envir* (2003) 54: 195–201.

Withers, L. A., Maintenance of plant tissue cultures, in Kirksop, B. E. and A. Doyle (eds.) *Maintenance of Microorganisms.* (1991) Academic Press, London.

# Chapter 16

---

# Genetic Methods of Insect Pest Control

## 16.1. Introduction

The control of insect pests of plants, animals and man is of enormous economic and medical importance. Insects attack animals, as in fly-strike of sheep and screw worm of cattle; they destroy or damage plants, as with locusts, or sawfly of apple, and may be vectors of disease, as with aphids transmitting viruses to plants, and mosquitoes transmitting malaria to man. The main methods of controlling insect damage (see ADAS, 1998) are **insecticides** to kill the insects, **chemosterilants** that sterilise but do not kill the insects, **biological control agents**, the **release of sterile insects** into the agricultural and natural environment, and breeding **genetically resistant** plants or animals.

Insecticides can be very effective, as shown by the use of DDT in Sri Lanka to kill the **mosquitoes** which cause malaria. With a human population of about 15 million, there were about 2.8 million cases a year of malaria in Sri Lanka in 1948. After the widespread use of DDT under a World Health Organisation programme, the number of cases fell to only 17 a year in 1963. When the use of DDT was banned on environmental grounds, the number of cases of malaria rose within five years to 2.5 million a year. This programme involved the use of the antimalarial drug chloroquin; the development of chloroquin-resistant strains of malaria also promoted the rise in cases. Hormone-mimicking insecticides can be fairly specific, e.g., to Lepidoptera, but some are broad-spectrum (Dhadialla *et al.*, 1998).

Over time, many insect pests develop **genetic resistance** to particular insecticides. For example, the housefly, *Musca domestica*, developed resistance to DDT, breaking it down with a dehydrohalogenase enzyme controlled by a major gene, with polygenic modifiers controlling the rate of uptake of the DDT (Sec. 2.3). Insecticides are generally sprayed over the

462

target areas, but can be used in **traps** with attractants such as pheromones or lights.

In theory, **chemosterilants** can be more efficient than insecticides. A chemosterilant spray which made 90% of an insect population sterile would be more effective than an insecticide giving 90% kill. Providing the chemosterilant does not affect mating fitness, many of the matings of the remaining 10% of fertile insects will be with sterile insects, and therefore be infertile, leaving only 1% of fertile mating combinations (10% × 10%). Using insecticides and chemosterilants does however mean loading the environment with potentially harmful chemicals, which is why other methods are used whenever suitable.

**Biological controls** of insects, e.g., using fungi, nematodes, bacteria and viruses, vary greatly in effectiveness and specificity. A baculovirus, nuclear polyhedrosis, is used on one million hectares in Brazil to control soybean caterpillar, *Anticarsi gemmatalis* (Moscardi, 1999), and fungi can be used to control roaches and termites.

## 16.2. The release of sterile insects, or of fertile insects giving inviable progeny

The **release of sterile insects** has been extremely successful with some pests, but is unsuitable for others. Typically, one cultures vast numbers of the pest one wishes to eliminate, e.g., rearing codling moth of apples on apples or a cheaper artificial diet. One then sterilises them with the minimum necessary dose of gamma rays from a cobalt-60 source (or X-rays, but they are more expensive to use), and releases them over the pest's natural habitat from the air or the ground, depending on economic and biological factors. In the natural habitat, the sterilised insects mate with natural populations, reducing those populations because the matings are sterile. One advantage of using sterile insects rather than insecticides or chemosterilants is that the sterile insects can move and actively seek out the natural insects, even in niches which sprays might not reach. One could use chemosterilants to sterilise the insects before release, instead of radiation.

The **success** of such programmes depends on many factors. They work best if the females mate only once rather than many times, since a combination of mating with sterile and fertile insects is usually fertile. The ratio of sterile insects to natural ones is crucial and should be as high as possible, at least 10 sterile insects to 1 natural, with 40 sterile codling moths being used for every natural one, which means high rearing costs. There must be a good geographical spread of the sterile insects, and the sterile insects must be reliably sterile; otherwise one is just spreading the pest itself.

With **codling moth**, the best results are from giving males a gamma ray dose which permits 10–15% egg hatch on mating to natural females, as those hatchlings are too feeble to do much damage. A dose sufficient to give complete male sterility reduces the mating competitiveness of those irradiated males. In codling moths, sterile males disperse further and more rapidly in search of wild females if they are released separately from sterile females than if they are released with both sexes together, when sterile males tend to mate with the sterile females.

In cattle screw worm control, one would like to, but cannot, kill off all females from the sterilised insects, because their laying of sterile eggs still damages the cows' hides and they lessen dispersal of the males. Even worse, females need a dose of 7,500 rads, while the males only need 2,500 rads to sterilise them, and giving the males the higher dose reduces their competitiveness. Another problem with mass culture is that the reared insects adapt to the culture conditions and may be less competitive than wild insects.

Despite these difficulties, the classic success story of release of sterile insects is the **screw worm of cattle**, *Cochliomyia hominivorax*. Just before 1957, it did about $120 million damage a year to cattle in the USA. The female lays eggs in skin lesions, where larvae feed for five to six days before falling off and pupating in the soil. In 1957, a campaign was begun with sterile insects to eliminate screw worm from Florida and Southeastern USA. A mass-rearing factory was started, producing 50 million flies a week. Pupae were given 7,500 rads of gamma rays, which stops female oviposition. As the sexes are not easily separated, mixed sex releases were made with 800 flies per square mile (per 259 ha) per week dropped from planes, with additional releases in badly infested areas. After 17 months, with the release of 3,500 million sterile insects, complete eradication was achieved in the Florida area. The cost was $10.6 million, giving estimated savings of $20 million a year. The Florida peninsula is fairly isolated from sources of reinfection; one must eliminate the pest from a whole zone, not just from isolated pockets. **Reinfection** from South America has been prevented by the regular release of sterile insects in a barrier zone in Panama. The same method was used successfully in 1990/91, at a cost of $82 million, to eliminate screw worm from Libya after its accidental introduction in 1988 in a sheep shipment from South America; up to 40 million sterile males were released each week.

Because of the cost of mass rearing of insects for such programmes, and the risk of **accidental escape of unsterile insects**, e.g., of 100,000 fertile medflies in the 1980 California medfly sterile release programme and of nonirradiated screw worm flies in Mexico in 2003, **alternatives to radiation** for sterilisation have been explored, as well as methods of getting rid of females. Heterozygous translocations reduce fertility, but releasing homozygotes for translocations to mate with natural non-translocated ones would be pointless if the translocation homozygotes could breed amongst themselves. Tetraploids give sterile triploids on mating with diploids, but tetraploid males usually have low fertility and are difficult to keep in culture. Other possibilities include cytoplasmic male sterility, or the release of conditional lethal mutations with stocks raised under permissive conditions where the mutations were not expressed.

Myers *et al.* (1998) considered the **cost-benefit assessment** of eradication programmes, pointing out that it costs as much to eliminate the last 1–10% of a population as it does the first 90–99%. Some eradication programmes have succeeded, such as the screw worm ones in Florida, 1958–1960, in USA and Mexico (at a cost of $750 million) by 1991, and Honduras by 1995. In Nigeria, traps and insecticide-impregnated targets were used to reduce the **tsetse fly** population before the release of 10 sterile males to each wild male, giving success by 1985 (Oladunmade *et al.*, 1986). A sterile male release programme was tried in the Okanagan Valley, British Columbia, in 1995–1996, for **codling moth** (*Cydia pomonella*) of apple, but the sterile males were not active enough in early spring to compete with wild males. The wild population was reduced but not eradicated. The Californian State Medfly Project was unsuccessful and resulted in 14,000 claims for damage to car paintwork. In 1997, millions of sterile male mosquitoes were released from aircraft and from the ground in the Southern Indian state of Tamil Nadu in an attempt to eradicate malaria.

The **Mediterranean fruit fly**, *Ceratitis capitata*, Medfly, is one of the world's most important pest insects, attacking more than 250 types of fruit, vegetable and nut. Gong *et al.* (2005) devised an experimental Medfly system using **tetracycline-repressible dominant lethality**. Strains were genetically engineered to have the tetracycline-repressible transactivator (tTA) which causes lethality in the early developmental stages of larvae on diets without tetracycline, but with little effect on the survival of parental transgenic tTA flies on diets with tetracycline. Low level expression of tTA does little harm, but high levels are deleterious in development. In the presence of tetracycline, tTA is inactivated, but without tetracycline, the basal expression of tTA leads to the synthesis of more tTA, accumulating to a harmful level. All genotypes survived on a diet with $100\mu$g tetracycline, but the engineered construct conferred dominant lethality in the absence of dietary tetracycline. Homozygous transgenic males had no reduction in mating competitiveness compared with wild-type males. Mass rearing of homozygous transgenic flies on a diet with tetracycline is possible, then releasing them to mate with wild flies; the heterozygous progeny in the wild without tetracycline should die during development. Ideally, only males would be released as females would cause oviposition damage. The system could be applied to a range of different insect pests, but has yet to be proved under agricultural conditions.

## 16.3. The breeding of insect-resistant varieties

The **breeding of pest-resistant** plants (see Maxwell, 1980) and animals is an ideal method of control. Farmers and plant and animal breeders have selected for yield for hundreds of years, perhaps involving indirect selection for insect resistance. More direct screening is easy. More than 2,000 apple trees of a number of varieties had the normal insecticidal sprays stopped for one year, with an assessment of pest damage. Fourteen apple varieties showed resistance to rosy apple aphid (*Sappaphis mali*) and three were totally resistant to green apple aphid (*Aphis pomi*).

For plant defence **mechanisms of resistance**, including avoidance (e.g., thick hairs on alfalfa against spotted alfalfa aphid, waxy surfaces, feeding repellents, high silica levels in rice tissues wearing away the mouth parts of stem borer *Chilo suppressalis*; life cycles out of synchrony with that of pests), antibiosis, resistance and tolerance, gene-for-gene interaction and specificity of resistance, see Niks *et al.* (1993). See Sec. 14.7 for genetic engineering with *Bacillus thuringensis* Bt toxin genes against specific insect pests of plants. At least 90 different genes from *Bt* have been isolated and sequenced. Their protoxins were classified into four types depending on their host range: *cry I*, Lepidoptera; *cry II*, Lepidoptera and Diptera; *cry III*, Coleoptera; *cry IV*, Diptera. For example, the potato variety "New Leaf Plus" has been engineered with the Bt CRYIIIC gene, giving resistance to the Colorado beetle, as well as with other genes giving some virus resistance. Bt cotton has been very successful in the USA.

**Proteinase-inhibitors** such as a trypsin-inhibitor gene from cowpea (*CpTi*) disrupt insect digestion of proteins and have been used to give plants resistance to lepidopteran insects. The proteinases occur naturally in many plants, associated with insecticidal activity; they are inactivated during cooking and are thought to be harmless to humans and farm animals.

One worry about breeding pest-resistant plants and animals is that success could lead to a big reduction in the **natural predators** of the pest species, and that mutations in the pest which might eventually enable it to overcome the resistance could lead to a huge outbreak of the pest, with little control by its predators. There are already insect pests which have developed resistance to Bt toxins. In general, insect-resistance breeding has been much more successful in plants than in animals.

In **sheep**, fly-strike has a heritability of 0.1 to 0.58 in Merinos, while fleece-rot has heritabilities of 0.05 to 0.8 (Raadsma *et al.*, 1995). Fly-strike increases with rainfall and if dags (a build-up of faeces in wool around the anus) are not removed by jetting with water. Increased genetic resistance to fly-strike would reduce treatment costs. As fleece-rot is easier to measure than fly-strike and has higher levels of expression and a good genetic correlation with fly-strike, Australian Merino breeders have selected just for resistance to fleece-rot, to improve the character and resistance to fly-strike [see Atkins (1987)].

Lindley in 1831 found that the **apple** variety Winter Majetin was much more resistant to **woolly aphid** than other varieties, and that could provide breeding material for the transfer of the resistance to the other cultivars. This resistance has proved to be stable ever since. Although it has been known for many years that American **grape vine** species are largely resistant to the **phylloxera** aphid-like root pest, the resistance has not yet been bred into European vines such as *Vitis vinifera*. Instead, most *vinifera* grape vines for wine are grafted onto American rootstocks to minimise phylloxera, which devastated the vineyards of Europe and other places in the second half of the 19th century. Even if genetic resistance is incomplete, as for carrots to carrot fly (*Psila rosae*) and potatoes to the blight fungus (*Phytophthora infestans*), partial resistance often permits significant reductions in the amount of pesticide or fungicide required for control.

After the breeding of **wheats** resistant to **Hessian fly** (*Mayetiola destructor*), infestation levels for this important pest of wheat in the midwest of the USA have dropped from over 90% to under 10%. Similarly, wheat stem sawfly (*Cephus cinctus*) has been controlled by breeding resistant varieties. The costs over 10 years of producing wheat cultivars resistant to Hessian fly, European corn borer (*Ostrinia nubialis*) and wheat stem sawfly were US$ 9.3 million, giving estimated savings of US$ 308 million a year, which include reduced pesticide and labour costs (see Niks *et al.*, 1993). In the search for genes for resistance to Hessian fly, 28 major genes were found in one survey, of which 8 were from common wheat, *Triticum aestivum*, 13 from durum wheat, *T. turgidum*, 5 from wild wheat, *T. tauschii*, and 2 from rye, *Secale cereale*. **Multifactorial (horizontal) resistance** is usually better than monogenic (vertical) resistance as the former is more durable, less likely to be overcome by mutations in the pest.

In 1997, scientists at Horticultural Research International, at East Malling, Kent, reported finding an apple gene, the dominant $Sd_1$, which gave resistance to two local biotypes of the damaging pest, the rosy leaf-curling aphid, *Dysaphis devecta*, which causes severe leaf curl and conspicuous red galling (HRI, 1996). It was found in Cox's Orange Pippin and can be crossed into other varieties. See also Sec. 8.7.

Unlike the action of most pesticides, insect resistance in plants is nearly always **species-specific**, but resistance to more than one pest can be incorporated, as with alfalfa resistant to spotted alfalfa aphid and to pea aphid. Insects sometimes form **different biotypes**, so that a plant may be resistant to some biotypes but not others, as with Hessian fly of wheat. Insects can mutate and/or have recombination, so that a plant's resistance may be overcome. Gene-for-gene systems controlling host-pest resistance and virulence have been found, parallel to the ones which Flor found for rust-resistance in flax. Resistance genes are often found in primitive cultivars or related species, and can be bred into cultivars. One such example is the transfer of greenbug resistance from rye to wheat.

# Suggested Reading

ADAS, *The ADAS Pest Control Manual: a Reference Manual for the Management of Pests*. (1998) ADAS, Guildford.

Atkins, K. D., Resistance to fleece rot and body strike: its role in a breeding objective for Merino sheep, in *Proceedings of the Sheep and Blowfly and Flystrike Management Workshop*. (1987) Department of Agriculture, Trangie, New South Wales.

Dhadialla, T. S., G. R. Carlson and D. P. Le, New insecticides with ecdysteroidal and juvenile hormone activity. *Ann Rev Entomol* (1998) 43: 545–569.

Gong, P. *et al.* A dominant lethal genetic system for autocidal control of the Mediterranean fruitfly. *Nat Biotechnol* (2005) 23: 453–456.

HRI (1996) New molecular markers and their use in apple breeding. [Work of P. Roche *et al.*] *Ann Rep 1995–96*. Horticulture Research International, Wellesbourne, Warwick.

Maxwell. F. G. (ed.) *Breeding Plants Resistant to Insects*. (1980) Wiley, New York.

Meara, T. J. and H. W. Raadsma, Phenotypic and genetic indicators of resistance to ectopathogens, in Gray, G. D., R. R. Woolaston and B. T. Eaton (eds.) *Breeding for Resistance to Infectious Diseases in Small Ruminants*. (1995), pp. 187–218, ACIAR Monograph No. 34, ACIAR, Canberra, Australia.

Moscardi, F., Assessment of the application of baculoviruses for control of Lepidoptera. *Ann Rev Entomol* (1999) 44: 257–289.

Myers, J. H., A. Savoie and E. van Randen, Eradication and pest management. *Ann Rev Entomol* (1998) 43: 471–491.

Niks, R. E., P. R. Ellis and J. E. Parlevliet, Resistance to parasites, in Hayward, M. D., N. O. Bosemark and I. Romagosa (eds.) *Plant Breeding. Principles and Prospects*. (1993), pp. 422–447, Chapman and Hall, London.

Oladunmade, M. A., L. Dengwat and H. U. Feldman, The eradicating of *Glossina palpalis palpalis* using traps, insecticide impregnated targets and sterile insect technique in Central Nigeria. *Bull Entomol Res* (1986) 76: 2775–2786.

Raadsma, H. W., G. D. Gray and R. R. Woolaston, Genetics of disease and vaccine response, in Piper, L. and A. Ruvinsky (eds.) *The Genetics of Sheep*. (1997), pp. 199–224, CAB International, Wallingford.

# Chapter 17

## Reproductive Physiology in Plants, Animals and Humans; Crossing Methods

### 17.1. Plants

#### 17.1.1. *Plant sexual reproduction*

Plants normally reproduced **sexually** by seed in commerce include: cauliflower, Brussels sprouts, pea, bean, beetroot, soya beans, sugar beet, celery, leek, radish, oilseed rape, cucumber, marrow, onion, carrot, turnip, tomato, lettuce, sweet corn, cereal grasses (including wheat, barley and rice), poppies, most trees including oaks, and most garden flowers. See North (1979) for details of individual plants' reproduction, including agricultural crops, and for relevant anatomy, physiology and development. For plant breeding systems, see Richards (1997). For plant propagation by seeds or vegetatively, see Hartmann *et al.* (1997).

**Asparagus** is one crop which can be reproduced sexually or vegetatively. The seed gives a sex ratio of about 1 male (XY chromosomes) : 1 female (XX), with the females usually being rogued out as they are inferior to the males in spear size and quality. Rare supermales (YY) can be obtained from anther culture (Sec. 10.3) and can be propagated vegetatively or crossed to females to produce only males. Excellent males can be clonally propagated *in vitro*, vegetatively.

To cross different varieties, one needs them to flower at the same time, unless one stores pollen. *Solanum* species have long **flowering periods**, so that if one species flowers from January until June, and another flowers from April until December, one could cross them between April and June. Cut-flower crops such as daffodils have much shorter flowering periods, and cereals average about 14 days. In cereals, normal spacing between plants allows only one or two tillers (offshoots) to develop. To get a longer flowering period, the breeder can plant the seeds further apart, allowing many tillers to develop, some of which may flower 7 to 21 days later than others. Some fruit trees have a juvenile period of about 6 years before they flower, so grafting buds of a newly-bred seedling onto a mature tree can speed the breeding process.

Flowering times can be manipulated by various treatments, including temperature, soil conditions and **light intensity and duration**. **Short day plants**, such as rice, soya beans and many tropical crops, have flowering promoted by day-lengths shorter than some critical value. Thus, with less than 6 hours light per day, it might take 11 days to induce flowering, but with 10 hours light per day, it might take 21 days to induce flowering, and with more than 12 hours a day, the plants might never flower. **Long day plants** such as wheat or sugar beet are induced to flower by light of longer duration than some critical value, and **day-neutral plants** such as groundnuts, tomatoes and some dwarf wheats need light for growth, but are largely indifferent for flowering to the length of the daily light period. Plate 17.1 shows artificial lights and black plastic sheeting to control day-length in a commercial greenhouse for the production of flowering **chrysanthemums** for a much longer period than is possible with natural daylight alone.

The **induction of flowering** in a given plant is sometimes permanent, sometimes reversible. The leaves are the main sensors of day-length, but often go through a non-sensitive juvenile period, i.e., rice leaves are insensitive to daylight length for two to five weeks after emergence.

Higher **temperatures**, unless too high, usually speed development, including flowering. Some plants, such as groundnuts and peas, need a certain **accumulated temperature** for flowering; this is measured in "**degree days**", i.e., total days × degrees C, e.g., needing 1,000 degree days to initiate flowering. **California's wine regions** are classified into five zones based on the number of degree days experienced in each zone in a typical growing season. Those degree days are measured, in Fahrenheit, as the length of time the average temperature remains over 50°F (10°C) between 1st April and 31st October, and by how much the average temperature exceeds this base value. Thus, if the mean temperature over a five-day period in summer was 70°F (21°C), the number of degree-days accumulated over that period would be 5 days × (70 − 50)°F = 100 degree days. The five wine

**Plate 17.1.   Glasshouse growing of chrysanthemums** at Headcorn, Kent. The lights and black plastic curtains are used **to control day-length** and so get flowering for a longer part of the year than would happen naturally. The plants are vegetatively propagated, with good uniformity for flowering time.

zones are: zone I, 2,500 degree days or fewer (e.g., the cool Napa Valley); II, 2,501–3,000; III, 3,001–3,500; IV, 3,501–4,000 (e.g., San Diego), and V, more than 4,000 (e.g., Fresno in the hot Central Valley).

The introduction of stone fruits to warmer areas was made possible by developing varieties without a cold requirement for fruiting. Pollination at 40° has been used in *Trifolium* (clover) to overcome self-incompatibility and between-species incompatibility.

Some plants need particular temperature and day-length sequences. A principal example is **vernalisation**, where plants need a period at low temperatures (e.g., 2 to 8°) to change the physiological condition of the growing points, so that the resulting leaves become perceptive to day-length. In barley, wheat and rye, there are separate **spring and winter varieties**, where the spring varieties do not need vernalisation, but winter varieties do. The winter varieties need low temperatures at the seedling stage, in order to be able to respond to long days subsequently. Winter varieties are therefore planted in late autumn or early winter, with increasing daylight the following spring initiating flowering. As the spring varieties do not need the cold winter temperatures, they can be planted after winter.

Brussels sprouts and pyrethrum plants (*Tanacetum* — also called *Chrysanthemum* or *Pyrethrum* — *cinerariaefolium*, and other species used for insecticide from the flowers) need **low temperature** to induce flowering. **Pyrethrum** is often grown in the tropics, but the need for low-temperature induction of flowering means that the plants have to be grown at high altitude. Three varieties of pyrethrum were grown at altitudes of 1,830, 2,130 and 2,440 metres in Kenya. Variety A yielded **1,000** units of flowers at the lowest altitude, 1,100 units at the middle altitude, and 1,200 units at the highest altitude. Variety B yielded 400, **1,200** and 1,400 units at these altitudes, and variety C yielded 50, 1,100 and **1,600** units respectively. At 1,830 m, one would therefore grow variety A; at 2,130 m, one would grow B, and at 2,440 m, one would grow C. These data show **genotype/environment interactions**, with different varieties behaving differently in different environments (Subsec. 1.3.10).

Other tropical crops which are subject to long dry periods may need **rain** to get flower opening. With **coffee**, long dry periods give ripe flower buds, then one shower of rain causes the flowers to open, with excellent synchronisation for pollination. The breeder can use the same method for synchronising different varieties for crossing.

Some **tuber crops** are difficult to get into flower for the breeder to make crosses, as in yams and some potato varieties. In potatoes, one can graft the stems onto tomato plants which suppresses tuber formation and diverts the energy into flowers. Breeders can also manipulate flowering times by controlling glasshouse temperatures, or by using hormones such as gibberellins or antigibberellins.

With **fruit trees**, seedlings may take many years before first flowering. Flowering many be greatly speeded up by **grafting** seedling shoots or buds onto adult trees, e.g., in apples or pears. To increase flowering in perennials, one can spray with anti-gibberellins to reduce vegetative growth, or girdle the stems with incisions into the layers below the bark to retard the downflow of photosynthetic products to the roots. In general, a **high carbon/low nitrogen ratio** encourages flowering rather than vegetative growth, so some fruit trees are encouraged to flower by partial root-pruning, or planting in poor soil, or even over concrete to restrict root growth, as with fig trees.

The duration of **stigma receptivity** is very variable between species, with 14 days for the "silks" on a maize plant and 10 days for grape vines, but only 3 to 4 hours for sweet potato and mango. The duration of **pollen viability** is often only 1 to 3 days, with cereals under dry conditions losing pollen viability after 6 hours. Pollen grains may be **binucleate**, with a vegetative nucleus and a generative nucleus which only divides in the pollen tube during germination on the stigma, or **trinucleate**, when the generative nucleus divides while the pollen is still inside the anther. This binucleate/trinucleate difference correlates with many other pollen properties such as the incompatibility system, storage life and ease of *in vitro* germination of pollen on artificial media. The **number of pollen germ pores** is usually specific, usually two or three in pollen from diploids, then gradually increasing with ploidy, as in clover and potato. In potato, pollen from diploids generally has three germ pores, but pollen from the tetraploid mainly has four pores.

Normally, only one pollen tube grows per grain, with germination on a compatible stigma taking only a few minutes. The time from pollination to egg fertilisation is typically 12–48 hours, but it is much longer in some shrubs and trees, with 3–4 months in hazel (*Corylus avellana*). It takes 12–14 months in oaks (*Quercus* spp.) and in *Hamamelis*, where part-grown pollen tubes overwinter.

**Pollen size** and **surface stickiness** are correlated with whether plants are **wind-pollinated** (grains are often 20–60 $\mu$m diameter, with a dry surface) or **insect-pollinated** (grains are larger and sticky). For example, wind-pollinated *Myosotis* (forget-me-not) has grains 4 to 5 $\mu$m across, and dry, while insect-pollinated *Curcurbita* (marrow, squash) species often have grains 200 $\mu$m across, and sticky. Pollen development typically includes approximately three days for meiosis, eight days for microspore interphase, and four days for mitosis 1 and 2. One maize plant may produce several million pollen grains, and many grains will carry spontaneous mutations (see Subsec. 7.2.1 for mutation frequencies).

Breeders should find out whether their crops are wind- or insect-pollinated when considering how to avoid **bastardisation**, the contamination of their seed lines with stray pollen from other crop or wild varieties.

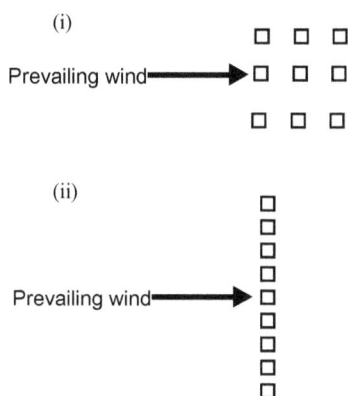

**Fig. 17.1.**   **Plot arrangements affecting bastardisation of seed lines** by each other in a wind-pollinated species. Arrangement (ii) is much better than arrangement (i) for avoiding bastardisation.

Figure 17.1 shows two arrangements of a breeder's plots in relation to the prevailing wind, with arrangement (ii) being much better than (i) for avoiding bastardisation between those plots.

**Seed production in open plots** should not be done near to other varieties of the crop, nor to wild relatives which can cross with them. Bastardisation can be reduced if there is a screen of tall trees (e.g., windbreak poplars), a hill or some buildings between an upwind source of contaminating pollen and the seed fields, so that the prevailing wind either takes the contaminating pollen right over and beyond the seed plots, or the obstruction intercepts most of the pollen. If there is an unavoidable upwind source of contaminating pollen, the breeder can discard seed from plants nearest that source, which may be pollinated by the contaminant, but the breeder can use seed from plants further downwind, as these will largely have been pollinated by the desired plants. Depending on flight patterns, insect and plant biology, and winds, the breeder of insect-pollinated plants may take a similar approach, not harvesting seed from the outer rows of plants on the seed plots, as these "guard rows" might have been pollinated from insects flying in from contaminating plants elsewhere, but using seed from more central areas of the seed plot. These plants will probably be pollinated by insects which have picked up pollen from the seed plot itself. In Holland, there are legal minimum distances between particular seed crops and the nearest known source of contaminating pollen, e.g., 50 m for *Chrysanthemum leucanthemum* and 600 m for sugar beet (see North, 1979).

**Pollen viability** can be tested by staining the grains with acetocarmine or with iodine in dilute potassium iodide solution to show the percentage of grains with cytoplasm, but this may overestimate viability. It is better to stain with fluorescein diacetate and examine under a fluorescence microscope, as tests for integrity of the plasmalemma. Especially with binucleate pollen grains, one can try *in vitro* germination in liquid or solid media (solid media for cereals). The medium usually includes sucrose, calcium nitrate and boric acid, with other requirements varying between species.

**Seedless varieties** of fruits are often preferred by consumers. They can be made by breeding types which do not set seed either on selfing or crossing, as with many seedless grapes. Breeders in Spain have recently developed seedless **clementines** called Clemenules. They are self-sterile as a variety but can set seed if they are cross-pollinated from other varieties by bees. Clemenules are therefore grown at least 10 km from other varieties. Clementines are largely seasonal citrus fruits, with main production and sales in time for Christmas in Europe. Seedless **tomatoes** are tastier than seeded ones, with increased dry matter, sugar, more soluble solids, less cellulose and lower acidity levels. Seedless tomatoes have potential for genetic engineering since this state prevents dissemination of transgenes through seed dispersal.

**Pollen preservation** (see North, 1979) works best for binucleate grains, using 0–10°C and 40–50% relative humidity. Trinucleate grains do not store well and need a higher relative humidity. Storage life can be increased by raised carbon dioxide levels, lower air pressures, storing in the dark, and mixing the pollen with casein or talcum powder. Some pollen remains viable for decades at −80°. Pollen preservation works well with Rosaceae, Primulaceae, Ranunculaceae, Liliaceae and Iridaceae.

**Seeds for commercial use** are usually dried down to a moisture content of 16% or less, and are often stored at ~5° to prolong viability. Seeds for sale must usually pass the tests for germination percentage and seedling vigour. Various post-harvest treatments may be used, such as coating with fungicide, e.g., against "damping off" (*Pythium* and other fungi), or temperature treatments related to seed dormancy, such as some stone fruits needing low-temperature treatment ("stratification"). Older seeds tend to have higher spontaneous **mutation frequencies**, with 1% in fresh *Antirrhinum* seeds rising to 5% in 9-year-old seeds. Four-year-old pea seeds have approximately five times the mutation frequency of fresh seeds. If one seeks new spontaneous mutations, using old seed would be of help.

Whether **hops** are **seeded or not** is commercially important. Hops are dioecious perennials lasting about 25 years and are normally grown from cuttings from female plants. Only two or three male plants are needed per hectare to get seeded hops. The seeds add weight but have little effect on flavour, and lagers and bitters can be made from seeded or unseeded hops. Pollination is not needed to get the bitter bracts with hop oils. British hops are often seeded, but in Germany, growing seeded hops and having male plants are illegal. Wild male hops can pollinate commercial crops. See Subsec. 13.7.8 for hop breeding.

Early ripening peaches may not form mature embryos, so they need embryo dissection and aseptic culture for propagation. Some orchid seeds are extremely small, with inadequate nutrients, and need cultivation on nutrient agar, or application of a symbiotic fungus to the roots.

## 17.1.2. *Incompatibility in higher plants and ways of overcoming it*

The haploid pollen tube usually needs to grow down a diploid style for fertilisation, but in many species there are physiological incompatibility mechanisms which prevent self-fertilisation. **Self-incompatibility** means that a plant's own pollen is rejected by its own stigma and style, and so is that of any other plant carrying the same self-incompatibility alleles. Approximately 60% of flowering plant species have self-incompatibility. Two major types are **gametophytic incompatibility**, where the pollen's own gametophytic genotype controls the pollen properties, and **sporophytic incompatibility**, where the pollen properties are determined by the genotype of the sporophyte, i.e., by proteins from the diploid anther tissue being deposited in the grain's outer layers. There can be **dominance and gene interactions** in the diploid stigma and style in either system, and in the pollen in the sporophytic system.

**Sporophytic incompatibility** occurs frequently in the Cruci-ferae (Brassicaceae), Compositae (Asteraceae), Betulaceae (birches), Caryophyllaceae and Convolvulaceae, often in species with trinuclear pollen. **Gametophytic incompatibility** is common in the Legumi-noseae and Solanaceae (monofactorial), Gramineae (Poaceae) (bifactorial), Cheniopodiaceae and Ranunculaceae (polyfactorial), often in species with binucleate pollen. For reviews of the molecular mechanisms involved, see Hiscock and Tabah (2003) for sporophytic self-incompatibility and Franklin-Tong and Franklin (2003) for gametophytic self-incompatibility. In most of those families, gametophytic self-incompatibility involves S-RNase from the style inhibiting pollen tube growth about one-third of the way down the style by degrading pollen RNA, but in the Papaveraceae, there is a signal transduction cascade giving rapid pollen tube inhibition. In sporophytic and gametophytic self-incompatibility, control is by a single polymorphic locus, *S*. The stage of pollen arrest in incompatible matings is very variable, from stopping any pollen tube growth, to slight growth and to extensive pollen tube growth before arrest.

An example of **sporophytic control** is the **primrose**, *Primula vulgaris*. This has two alleles at one locus giving two morphological flower types, pin eyed (with a tall style and low-set anthers) and thrum eyed (with a short style and anthers set high in the corolla tube). The different style and anther heights provide a **heteromorphic system** (first described by Darwin in 1877) that promotes cross-pollination by long-tongued insects. There is also an incompletely effective **physiological incompatibility** shown by hand-pollination selfings. Thrum is *S s*, giving 3 thrum: 1 pin on selfing, and pin is *s s*, giving all pin on selfing, with *S* dominant to *s*. As a result of this dominance and the sporophytic control, all pollen from thrum has thrum phenotype, although the genotypes are 1 *S*: 1 *s*. The *S s* thrum style is also thrum phenotype, so if one hand-self-pollinates thrum, the thrum phenotype pollen should not be able to grow down the thrum style, but about seven seeds per flower are set. Similarly, in a selfed pin × pin cross, the *s* pollen should not grow down the *s s* pin style, but about 35 seeds are set per flower. In pin female × thrum male and thrum female × pin male crosses, the pollen has a different phenotype from the style and fertilisation is good, producing about 61 and 57 seeds per flower respectively.

By detecting **rare recombination** of the extremely closely linked different elements, Ernst showed in *Primula viscosa* that the $S/s$ locus is a **super-gene**, made up of different parts. It contains $G$ for style length, $S$ for stigmatic surface form, $A$ for anther height, $P$ for pollen size, and $I$ for pollen incompatibility type. All these traits differ between pin and thrum, and most recombinants are at a disadvantage to the two standard types because they are less efficient at preventing either self-pollination or self-fertilisation.

In contrast to the two-allele heteromorphic sporophytic system in *Primula*, there are also **homomorphic sporophytic self-incompatibility systems** with no floral differences between compatibility types and often with multiple alleles at the $S$ locus, with typically 30 to 40 $S$-alleles in natural homomorphic populations in the Brassiceae and Asteraceae (Lawrence, 2000). There are often different dominance effects in pollen and stigmas, so that one plant may be able to pollinate and fertilise another, but not be fertilised by it when receiving pollen from it. Dominance between $S$-alleles increases the proportion of compatible crosses compared to co-dominance, as recessive alleles in common to pollen and stigmas will not be expressed in heterozygotes. Dominance can allow plants homozygous for recessive $S$-alleles to occur. In brassicas, dominance interactions are more common in pollen than in stigmas.

An example of **gametophytic** control is the **evening primrose**, *Oenothera organensis*, which has multiple alleles and no dominance at the self-incompatibility locus, so both alleles are expressed in the style and are always different because pollen with an allele the same as one allele in the style cannot grow. If we had (i), the cross $s_1$ $s_2$ female $\times$ $s_1$ $s_2$ male, the pollen genotypes and phenotypes would be 1 $s_1$: 1 $s_2$, and neither type could grow down the style expressing $s_1$ and $s_2$, so no seed is set. If we had (ii), the cross $s_1$ $s_2$ female $\times$ $s_3$ $s_4$ male, the pollen genotypes and phenotypes would be 1 $s_3$: 1 $s_4$, so both types could grow down the style expressing $s_1$ and $s_2$, so good seed is set. If we had (iii), the cross $s_1$ $s_2$ female $\times$ $s_1$ $s_3$ male, the pollen genotypes and phenotypes would be 1 $s_1$: 1 $s_3$, and only type $s_3$ could grow down the style expressing $s_1$ and $s_2$, so seed is set, giving a 1: 1 ratio of $s_1$ $s_3$ and $s_2$ $s_3$ seed.

As we saw in Sec. 15.3, selection for rarity can lead to there being very large numbers of alleles at such self-incompatibility loci, such as

45 alleles at this locus in evening primrose, *Oenothera*, and 200 alleles in clover, *Trifolium repens*. In sweet cherry, apples and potatoes, such systems prevent self-pollination. Good breeding lines can then be propagated vegetatively. In apples and cherries, fruit does not develop without fertilisation, so **compatible pollinator varieties** must be planted in the orchards. Thus a minority of English Golden Delicious trees in an orchard of Cox's Orange Pippin apples ensures fruit formation after cross-pollination of both varieties. Bramley apples are triploids with sterile pollen and need other pollinators to get fruit set. Most pear varieties are self-incompatible, requiring separate pollinator varieties to be planted with them, but Conference and Concorde are self-fertile. Some self-fertile sweet cherries have now been bred. It is clearly essential for fruit growers and breeders to know about any self-incompatibility systems in their plants.

Breeders can sometimes **overcome self-incompatibility** by: pollination at the **bud-stage** before the S gene is expressed (*Nicotiana*, *Petunia*) or of **over-mature stigmas** in which the S phenotype has deteriorated (*Lilium*, *Brassica*), or abnormally late season flowers (Solanaceae); **high temperature** (*Lolium perenne*, *Lilium*, *Trifolium*, *Lycopersicon*); **radiation** (*Lilium*, *Petunia*), **carbon dioxide** (*Brassica*) or by **chemicals** such as actinomycin D which prevent RNA or protein synthesis; **hormones** such as alpha-naphthalene acetic acid or indole acetic acid (many genera); *in vitro* **fertilisation** of cultured ovules by germinated pollen, overcoming any barriers in the stigma or style; **style grafting**; **mutilation of the stigma** with a wire brush; **electric currents** (*Brassica oleracea*). In some species such as sweet cherry, one can use radiation to induce **mutation in self-incompatibility loci** (*Oenothera*, *Prunus*, *Nicotiana*, *Petunia*, *Trifolium*), with easy selection in gametophytic systems. **Polyploidy** has also been used, where tetraploid relatives or induced tetraploids of diploids with gametophytic monofactorial self-incompatibility often have self-compatible diploid pollen (*Prunus*). If homozygotes for self-sterility alleles are required, e.g., for F1 hybrid production, a good method is to get monoploids and use colchicine on side shoots to get diploid homozygous buds and flowers (Sec. 10.3).

In poplar (*Populus*), Knox (1972) obtained large quantities of interspecific hybrids. He used **mentor pollen**, which was killed compatible pollen mixed with live incompatible pollen, when the stigma allowed the incompatible pollen to germinate and fertilise. Even just proteins from the walls of compatible pollen can sometimes act as "mentor" in sporophytic systems.

**Interspecific incompatibility** is also under sporophytic or gameto-phytic control and often occurs unilaterally, with pollen from a self-compatible species being rejected by stigmas of a self-incompatible species. When pollen from a self-incompatible species is put on stigmas from a self-compatible species, strong zygotic and postzygotic barriers usually lead to sterility even if the pollen germinates and grows. Radiation (*Nicotiana*), polyploidy and somatic hybridisation with protoplasts (*Lycopersicon*) have been used to overcome interspecific incompatibility.

## 17.1.3. *Crossing methods*

For a **controlled cross**, the stigmas of one parent must be protected from all other pollen, including its own, and pollen must be applied to the style from the anthers of the other parent. If the flowers are hermaphroditic, with male and female parts, one usually has to remove the anthers before they are ripe from the flowers to be used as females. Protection from stray pollen is provided by putting cellophane or plastic bags over individual flowers or whole inflorescences, but they must allow moisture to escape, to prevent rotting. With insect-pollinated flowers, one needs special plastic, muslin or wire cages of appropriate mesh size to exclude unwanted insects. Waterproof coverings are needed for plants grown outdoors. Details of crosses can be recorded on plastic or embossed metal tags, tied near each flower, if used outdoors, or on card tags if used indoors. Plate 17.2 shows two Brussels sprouts lines being crossed using blowflies within a pollination bag. **Blowflies** (readily available as larvae from fishing shops) are often preferred to honey bees for making breeders' pollinations because blowflies are more active than bees in dull weather and are more likely to cross-pollinate different lines, while bees often only visit one of the two lines to be crossed.

**Hand-pollination methods** vary with the species involved. Pollen can be transferred from anthers to stigmas with a small paint brush which is sterilised between pollinations with high concentrations of ethanol, industrial alcohol or methylated spirits. If they are of a convenient size, whole anthers can be detached — with alcohol-sterilised fingers or forceps — when ripe

**Plate 17.2.    Blowflies pollinating two Brassica strains,** within a pollination bag. Fruits (siliquas) can be seen above.

and rubbed on the stigmas. In maize, pollen is collected in bags from the dehiscing terminal tassels (stamens) and the pollen is then shaken over the appropriate silks (stigmas). Maize pollen is only viable for about one day under normal conditions, less if it is very dry. See North (1979) for details of hand pollination and emasculation in various species.

Clearly the male and female parts to be crossed must be ripe simultaneously. If the male parts are ready too early, part of the plant can be cut off and put in a fridge, with the stem in water, to delay ripening. Barriers to self-pollination, which is sometimes needed, include **protandry**, where the male parts ripen before the female ones as in maize, carrots and parsnips, **protogyny**, where the female parts ripen first as in walnut, **dioecism**, where there are separate male and female plants as in hops and asparagus, and **genetically determined self-incompatibility** as in many brassicas such as cabbage and sprouts, and in apples, cherries and clover.

To prevent **unwanted selfing** during breeding programmes, it is sometimes convenient to turn a monoecious, hermaphroditic plant into a dioecious one by obtaining male-sterile mutants and female-sterile mutants. In maize, the mutant "tassel seed" has the anthers converted into female parts, so is male sterile, and "barren-stalk" is female sterile.

**Emasculation** (removal of stamens) and hand pollination are slow processes and costly in labour, so are only used commercially if there is no satisfactory alternative. They are used in making **F1 hybrid tomato seed** because hybrids outyield selfed lines considerably, overcoming the disadvantage of more expensive seed. Tomato (*Lycopersicon esculentum* or *Solanum lycopersicum*) flowers are easy to emasculate and each flower pollinated yields a lot of seed. In lettuce and peas, which are naturally inbreeding, one can get higher yields from F1 hybrids but the labour costs usually make the seed too expensive. Breeders could even consider breeding varieties with flowers more convenient for emasculation and hand-pollination. Attempts have been made to use chemicals, such as 0.2% aqueous FW 450 (2,3-dichloroisobutyrate), as **selective gametocides** to give male sterility, e.g., in cabbage, lettuce and tomato, with varying success.

## 17.1.4. *Getting uniform fruit, seed or bud ripening*

The **ever-increasing mechanisation of agriculture** makes getting **uniform ripening** more important. A combine harvester going through a field of wheat, barley or rye will harvest all plants whatever their degree of ripeness. The seed-buyer does not want a mixture of ripe and unripe grains. When **Brussels sprouts** were picked by hand, the pickers picked the first-ripening, lowest on the stem, sprouts first, then picked the later-ripening upper ones a few days later. For mechanical plucking, the sprouts need to ripen simultaneously all over the stem and breeders have produced suitable varieties. In catalogues of sprouts, some varieties are marked as suitable for mechanical harvesting and others are not, depending on uniformity of ripening.

Uniform ripening is increasingly wanted for trees such as apples, oranges and olives, and bushes such as raspberries, where fruit can be harvested by mechanically by shaking the whole plant and collecting the falling fruit on sheets. **Size uniformity** is also desirable with more pre-packing of fruit, eggs and vegetables graded for size. Farmers who produce a whole range of sizes of a particular fruit have had them rejected by supermarkets because the extremes of size, too large or too small, were unwanted. In 1999, one UK supermarket started selling packets of mixed-size strawberries, as this reduced waste by 30% compared to Class One uniform fruit, enabling it to undercut the price of uniform fruit.

Breeders can collaborate with physiologists, perhaps using physiological methods to **impose phenotypic uniformity**, e.g., for ripening time, on a genetically diverse population. Especially with outbreeders or random maters, one might breed a high-yielding variety which lacked uniform ripening, then use a spray to induce uniform ripening. **Ethylene** gas and compounds which break down to ethylene within the plant are very useful for this (see Abeles *et al.*, 1992). It is interesting that growers used bonfires (which produce ethylene) and kicking plants (ethylene is produced as a wound response) to help mangoes and pineapples to ripen. One can now use ethylene-producing sprays to trigger ripening, e.g., to synchronise apple harvests, as ethylene starts formation of the abscission layer in the stalks. **Bananas** and **tomatoes** are often picked green and unripe, then after refrigerated transport are gassed with ethylene to ripen them and develop colour. This works well with bananas but ethylene-ripened tomatoes have much less flavour and aroma than vine-ripened tomatoes which are increasingly available but at higher prices.

## 17.1.5. *Somaclonal variation and vegetative propagation; grafting and rootstocks*

**Somaclonal variation** is the genetic variation arising during the vegetative propagation of plants. **Vegetative propagation** can involve specialised organs such as bulbs, corms, runners, rhizomes, stolons and tubers, or taking stem or leaf cuttings, or grafting buds or stems onto rootstocks or stemstocks. It can also involve more technological methods with meristem multiplication or callus culture, which can regenerate whole plants from single cells (which are dedifferentiated) in tissue culture.

Some ornamental plants can be propagated **sexually or vegetatively**, such as begonias, delphiniums, lilies, and some lupins, geraniums, primulas and violets. Plants normally propagated vegetatively include blackberries, black currants, blueberries, gooseberries, raspberries, red and white currants, strawberries, apples, pears, plums, potatoes, carnations, chrysanthemums, dahlias, fuchsias, gladioli, hyacinths, irises, paeonies, rhododendrons, roses and tulips. See North (1979) for details of breeding new cultivars of such plants, and Hartmann *et al.* (1997) for all aspects of plant propagation.

**Vegetative reproduction** allows the production of a large number of plants without waiting for completion of the sexual cycle, and without the risks of losing favourable genotypes through crossing, and segregation and recombination at meiosis. It avoids losing heterozygosity and avoids the segregation load (Sec. 15.3). Viruses are usually unable to replicate in meristem tissue, so regeneration of plants from isolated meristems has been used to get **virus-free** rhubarb and strawberry plants, for example. Meristem culture, heat therapy and anti-viral chemicals have all been used to get virus-free plants, but confer no resistance to infection later.

Although one might think that plants produced by mitosis from the same starting material should all be identical, one often gets **somaclonal variation**, as the origin of the variation is somatic, and not from the germline. Tissue-culture-produced variability has been found in many species, including sugar cane, tobacco, oil palms, rice, barley, oats, alfalfa, maize, wheat and soya bean. It can arise in the nuclear genome or in the mitochondrial and chloroplast genomes. It can be useful when it produces superior plants, perhaps with a higher yield of secondary metabolites or improved disease resistance, but often it results in poorer plants than the parental line. Any favourable mutations should not be accompanied by the loss of good genetic backgrounds, as there is no sexual reassortment.

Somaclonal variation also gives **heterogeneity** in what might be expected to be uniform plants. In geraniums, plants derived from stem callus were fairly stable genetically, but ones from root and leaf callus showed much more variation — perhaps there has been natural selection for more stability in those parts which normally produce the germ line tissue for sexual reproduction than in those which do not. While some somaclonal variation comes from changes in genes or chromosomes during tissue culture, it can also come from existing variation between cells in the material used in the explant. Thus, if some cells in the parental plant are diploid and some are polyploid, plantlets derived from regenerated protoplasts will include some diploids and some polyploids. Many roots cells become polyploid during normal development in diploids.

Meristem-derived cultures which have had no real dedifferentiation tend to produce normal plantlets, but regeneration from callus, after dedifferentiation, gives more somaclonal variants. Different cultivars produce different proportions of variants. The **amount of variation** obtained tends to rise with the age or number of cell divisions of the culture, as gene mutations and chromosomal changes accumulate and have longer for expression. The hormones used in tissue culture can affect the rate of variation production. In tissue culture, there is no need for cells to photosynthesise to survive and the production of albino plantlets is quite common, as in rye-grass.

The **causes of somaclonal variation** include base changes in DNA, changes in the numbers of copies of genes within a chromosome, epigenetic changes in methylation patterns of DNA, activation of transposable elements, mitotic crossing-over between homologous chromosomes, sister-chromatid exchange at non-homologous points, failure of cell cleavage after DNA replication giving autopolyploidy, aneuploidy from non-disjunction of chromosomes at mitosis, and chromosome breakage is frequent (especially in late-replicating chromosomal regions), giving translocations, inversions, deletions and duplications (see Lee and Phillips, 1988). Increases or decreases in methylation were observed in 16% of regenerated maize plants, persisting through several generations of selfing.

Some **beneficial quantitative changes** from somaclonal variation have been early maturity in maize and sorghum, increased dry matter in potatoes, increased submergence-tolerance in rice, and improved yield in oats. As most changes are unfavourable, a very large number of plants must be screened to find the few beneficial changes, as in screening sugar cane for disease resistance. **New commercial varieties** produced in this way include rice with improved chilling tolerance and potatoes with improved resistance to *Verticillium dahliae*, a serious fungal vascular wilt disease (e.g., Bertin *et al.*, 1996). One can use tissue culture conditions to select for types of **stress-resistance** at the single-cell level, e.g., tolerance to metals, low temperatures, herbicides and pathogen toxins, although the properties of the whole plants regenerated may not parallel those of single cells.

**Asexual reproduction** is widely used commercially in many species, giving a series of clones ("ramets") from one individual (the "ortet"). It is essential in triploid bananas, as the fruits do not set seed. Vegetative propagation is correctly called "cloning", e.g., layering of carnations, but confusion can arise because the term is also used for *in vitro* DNA work (Chap. 14).

In **apomixis**, seeds develop an embryo without fertilisation. A somatic cell in the ovule develops parthenogenically into an embryo without having undergone meiosis. Typically a diploid cell in the ovule, from the nucellus or a megaspore which has not undergone meiosis, develops into the embryo and the rest of the ovule becomes the seed. These seeds are viable and have been used to reproduce various citrus species, giving identical clones from the same plant, which can have heterozygosity perpetuated as the reproduction is essentially vegetative although involving seeds. Apomixis occurs mainly in polyploid species of the Gramineae, Rosaceae (e.g., *Rubus*) and Asteraceae families, and in *Citrus*. In some plants, such as bluegrass, *Poa pratensis*, both apomictic and sexual seeds are produced. A distinction is sometimes made between **agamospermy** (seed production without sexual fertilisation) and **vegetative apomixis** where no seeds are involved. Occasionally, a pollen grain may be needed to trigger apomixis although it does not fertilise the ovule; in some cases, pollination is required to get the normal triploid endosperm in the seed, even though the embryo develops without fertilisation.

Apomicts are excellent for maintaining existing good lines, but for breeding better lines one needs to find sexual plants, or use good pollen from an apomict on a female-fertile plant. Some hybrids which would be sterile in normal sexual reproduction are enabled by apomixis to set viable seed. Apomixis can occur in wheat, maize, sugar beet, apples, *Panicum* (millet) and *Poa*. When taken to include vegetative reproduction, apomixis has been suggested to occur in ~60% of the British flora. See Richards (2003) for a comprehensive review of apomixis, its consequences and possible use in plant breeding, where an ability to introduce some "apomixis gene" into F1 hybrids would be very useful so that yield could be maintained in later generations.

With **asexual reproduction**, vegetatively or by apomixis, there is no meiotic recombination or segregation, so genes on the same chromosome show complete linkage and may evolve into co-adapted sets. Many apomicts are polyploid, when accumulated recessive mutations are less likely to show than in diploids. In many plants such as *Taraxacum* (dandelions), sexual and pollen-bearing apomictic types coexist in the same populations, with apomicts sometimes fertilising the sexual types. Hybrid vigour and heterozygosity can be maintained in apomicts; in agamospermous populations of *Taraxacum*, about half the loci were usually heterozygous, much more than in related outcrossing sexual individuals (Hughes and Richards, 1988).

**Propagation by cuttings** is very widely employed, where the part taken must have a terminal or lateral bud to provide a growing point, and hormones are usually applied to stimulate new growth (see Hartmann *et al.*, 1997). It is often used for ornamental shrubs but is more labour intensive than seed production. Stem cuttings of deciduous hardwood trees are taken from matured wood when it is dormant in winter. Cuttings from semi-hardwood *Rhododendron* and *Camellia* plants are taken in summer, and cuttings from herbaceous chrysanthemums and geraniums are taken in spring. With leaf cuttings, one uses the leaf blade or the leaf blade plus petiole, e.g., in tea (Plate 17.3) and African Violet (*Saintpaulia*). One can use root cuttings in some plants, including the conifer *Picea abies* (Norway spruce).

**Grafting** is used in many woody species including grape vines, roses and apples; the rootstock chosen is very important, determining things such as phylloxera-resistance in vines and how much the apple tree is dwarfed, with some giving "trees" only one metre high. The properties of some different **apple and pear rootstocks** are shown in Table 17.1. Different rootstocks and tree heights have different staking and pruning needs, suit different soils, and suit different training methods such as dwarf pyramid, centre leader, stepover, cordon, minaret, espalier, half standard, large espalier and standard. In deciding which rootstock and training method to use, the orchard manager has to consider the necessary spacing between plants and between rows, typical yields, and age of trees at first crop. Some fruit trees tend to have alternate good and poor yields in successive years,

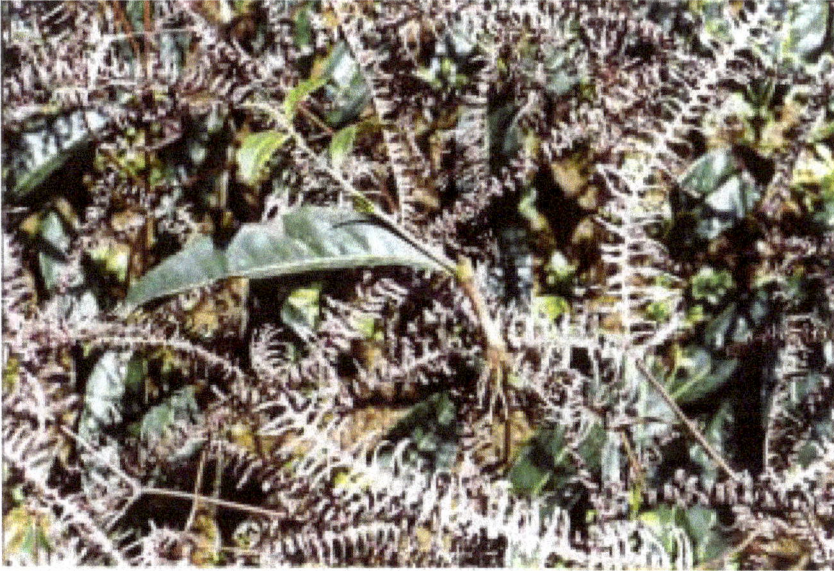

**Plate 17.3.   A rooted leaf-cutting of tea**, with a strong new axillary shoot. The fern leaves are for shade and moisture-retention. Taken at Eildon Hall Estate, Sri Lanka.

**Table 17.1.   Properties of apple and pear trees with different rootstocks**. Information from www.keepers-nursery.co.uk/rootstocks, accessed 3/11/2004.

| Rootstock name | Type | Maximum mature tree height, m | Tree spread, m | Typical mature yield per year, kg | Age at first harvest, years |
|---|---|---|---|---|---|
| Apple | | | | | |
| M27 | Very dwarfing | 1.8 | 1.5 | 6 | 2 |
| M9 | Dwarfing | 2.4 | 2.7 | 17 | 2 to 3 |
| M26 | Semi-dwarfing | 3.0 | 3.6 | 25 | 2 to 3 |
| MM106 | Semi-vigorous | 4.0 | 4.0 | 40 | 3 to 4 |
| MM111 | Vigorous | 4.5 | 4.5 | 115 | 4 to 5 |
| M25 | Very vigorous | 4.5 or more | 6.0 | 135 | 5 to 6 |
| Pear | | | | | |
| Quince C | Dwarfing | 3.0 | 2.7 | — | 3 to 4 |
| EMH | Dwarfing | 3.3 | 3.6 | — | 4 to 5 |
| Quince A | Semi-dwarfing | 3.6 | 3.6 | — | 4 to 5 |
| Pyrus | Very vigorous | 6 or more | 4.5 | — | 5 to 6 |

such as apple cultivars Laxton's Superb, Pixie and Blenheim Orange. Pruning hard after their poor years can help to restore more regular good yields. Labour and materials' costs in planting, pruning, spraying, possible netting against birds, and at harvesting, are important economic considerations, with tall trees being more costly to harvest than semi-dwarf ones.

In **grape vines**, using American rootstocks such as *Vitis labrusca* gives protection to the European scion, *Vitis vinifera*, against the phylloxera root pest insect. The **scion** (top part) may be a shoot or a bud and must be compatible with the **rootstock**. Occasionally, one has to graft the scion onto an intermediate "**interstock**" and then that onto the rootstock if the scion and rootstock are not directly compatible. Grafting is very skilled work and there are many forms of graft, such as the "omega" grape vine graft, where the stock end of the scion is cut by machine into a rounded omega-shaped protrusion, and fits into a corresponding omega-shaped hole cut in the rootstock (Plate 17.4). Wax protection is used over the graft to minimise drying out and infection during fusion of parts and rooting.

**Plate 17.4.** Top: **an omega graft in grape vines**, near the left-hand end below the buds. Middle: *Vitis vinifera* scion before shaping to an omega protrusion about one third of the way down from the left. Bottom: American rootstock before having an omega-shaped hole cut in the left-hand end below the bud. The grafted pairs are waxed over the graft before being incubated to get rooting. Taken at Geisenheim Wine School, Germany.

It is sometimes advantageous to remove the scion's mature growth and replace it with a different scion on the existing rootstock if the most desirable variety of fruit has changed, or if pathogens have developed which can attack existing scions. In South Africa, at Klein Constantia, white Sauvignon Blanc scions have been field-grafted onto rootstocks previously used for the red Shiraz grape. In California, changing the wine grape variety by grafting new scions onto existing rootstocks in the vineyard has been done extensively, and even changing the rootstock (by planting a new rootstock alongside the existing one, then changing the scion over; see Plate 17.5) when some rootstock varieties developed phylloxera in the mid 1990s.

Wild relatives of cultivated plants are often hardier than cultivars and have sometimes been used as rootstocks to extend a crop's range to areas with poorer soils or climates. For example, *Poncirus trifoliata* (trifoliate orange) is used as a cold-resistant rootstock for several citrus fruits, as well

**Plate 17.5. Inarch grafting to replace a rootstock** which had become phylloxera-susceptible. The new rootstock was planted behind and to the left of the existing one, then when it was big enough it was grafted into the existing large scion. When the graft had taken, the first rootstock was sawn off, with its upper end visible coming down to the right of the supporting pole. Taken at Domaine Carneros, Napa Valley, California.

as having been crossed with *Citrus sinensis* (sweet orange) to make the hardy "citrange".

Propagation by **bulbs, corms, tubers, rhizomes and stolons** is common in species having these reproductive or resting-stage structures, e.g., growing new plantlets from strawberry stolons or blackberry runners, after rooting. For example, many potato plants can be obtained from one seed potato, where each "eye" axillary bud can grow into a new plant. Bananas are propagated from cutting sections of rhizome, while bamboos can be propagated by laying whole aerial shoots ("culms") horizontally in trenches, when new plants grow from the nodes. See Subsec. 13.7.7 for stem cuttings in sugar cane. In bulbs, the basal plate can be scooped out, with adventitious bulblets developing from the base of exposed bulb scales when put on culture media. See Hartmann *et al.* (1997).

**Micropropagation** uses sterilised explants from apical meristems, stems, leaves, roots, seeds or pollen grains or single cells; the explants are grown in culture with nutrients and hormones. It is particularly used on plants of high individual value such as orchids and asparagus. The most common form is meristem culture to produce shoots which are then subcultured by cutting them into sections, each bearing an axillary bud. These buds come from the germ-line and are genetically stable. Adventitious buds, which come from outside the germ-line, are often less stable and are used less frequently. Shoots derived from callus cultures are even less stable due to somaclonal variation, although they may still be used in cases where the callus genome is known to be relatively stable. Valuable lines of oil palm have been successfully multiplied from callus. Callus cultures sometimes produce **somatic embryos**, which are morphologically similar to zygotic embryos in form and development, with separate root and shoot meristems.

**Tissue-culture cloning** of ideal **date palm** plants is done by Date Palm Developments of Baltonborough, Somerset, exporting about 130,000 plants a year to the Middle East. Plants are grown in Somerset greenhouses and are about 61 cm high when exported for transplanting into commercial plantations, e.g., to Kuwait. They begin to fruit at six years of age. The most popular varieties are "barhee" for fresh dates and "khalas" for stoning and exporting. Decorative palms are also supplied. No genetic engineering is involved, just tissue culture and regeneration.

Major agricultural crops for which **micropropagation** is used include sugar cane, potatoes and the Monterey pine, *Pinus radiata*. Only the very tip of the meristem is normally totally free of viruses or mycoplasmas, so meristem tips taken for culture are usually less than 1 mm long. Approximately eight viruses have been eliminated from potatoes by micropropagation. Axillary shoot cultivation is used for *Chrysanthemum*, *Dianthus*, *Anthurium*, *Fuchsia*, *Allium*, *Asparagus* and *Fragaria*.

**Anther culture** of Brussels sprouts can give about three diploid plants per anther, on average, as well as sometimes being used for getting monoploid plants for F1 hybrid production (Sec. 10.3 and Subsec. 13.1.5).

**Chimeras and variegated plants** need special consideration. Chimeras have different genotypes in different layers of the meristems and hence in the vegetative parts, and must be propagated vegetatively. For example, *Laburnocytisus adamii* is a graft chimera between *Laburnum anagyroides* and *Cytisus purpureus*, with a periclinal arrangement of tissues. There is a single outer layer of broom (*Cytisus*) cells giving a brown-purple colour to the flowers, overlying a core of *Laburnum* cells which give a laburnum-shaped inflorescence. This garden curiosity will not breed true from seed and is maintained by cuttings. Reversion to one or other parental type occasionally occurs spontaneously by vegetative segregation, such as a pure laburnum shoot with yellow flowers. See Hartmann *et al.* (1997) for the production and reproduction of chimeras.

Variegated plants have a mixture of green and white or yellow areas in their leaves, with occasional **vegetative segregation** (Plate 17.6) into pure green shoots and/or pure paler shoots which would die on their own from a lack of photosynthesis. Variegation normally results from meristem cells having two types of chloroplast genome, one wild-type, giving green chlorophyll pigments, and one mutant, giving no colour. Usually, daughter cells in variegated plants get both types of chloroplast after cell division, giving green cells, but occasionally a daughter cell may get only green chloroplasts, when its daughters will no longer give variegated areas, or only colourless chloroplasts, when its daughters will give a white or yellow leaf area. **Variegated plants** are therefore reproduced by cuttings from variegated areas. If reproduced by seed, variegation usually shows maternal inheritance, with most progeny variegated but with some all green progeny and some all white, which die. The proplastids thus go through the ovule

**Plate 17.6.** **Vegetative segregation** of an all-white branch (lower left) and some all-green leaves (centre) from a variegated garden *Euonymus*.

but not usually through the pollen (with some exceptions), with occasional segregation into pure types, by chance at cell division.

## 17.1.6. *Plant protoplast fusion*

One can use **protoplast fusion** and plant regeneration to overcome sexual incompatibility and to form new allopolyploids. Enzyme mixtures are used to break down the middle lamella and cell walls, liberating protoplasts into a medium with a suitable osmotic pressure and nutrients. Polyethylene glycol or other agents are used to promote protoplast fusion, and selective systems, such as complementation between different albino mutants or non-allelic nitrate reductase mutants, can be used to select hybrid cells. In tobacco, *Nicotiana langsdorfii* (2n = 18) leaf mesophyll protoplasts were fused with ones from *N. glauca* (2n = 24), with a selective medium for growth of fused cells. The hybrids grew and developed stems and leaves, and were grafted onto one parent. Fertile flowers gave seeds which germinated to give intact allopolyploid plants, but these two species can cross sexually anyway. Fusion of protoplasts from leaf mesophyll cells has been used to get a "pomato" from potato and tomato plants. Protoplast fusion can also be used to transfer cytoplasmic male sterility to male fertile varieties. For more information, see Pelletier (1993). To get diploids from protoplast fusions, one can use pollen culture (Sec. 10.3) of the two varieties to get haploids, then fuse those to get a diploid product.

## 17.1.7. *Gene expression and natural and artificial selection at the haploid stage*

In animals, **sperm** express some sperm-specific genes, but neither sperm nor eggs have much in the way of haploid development, nor does the plant egg cell. In contrast, **pollen** needs to germinate, penetrate the stigma, grow down the style, locate the ovule and penetrate its micropyle, and then fertilise the egg nucleus, to form the embryo, and fertilise the polar body fusion nucleus to form the endosperm. In maize, the pollen tube from a grain 90 $\mu$m diameter may have to grow down a silk 20 cm long. The male haploid generation in plants therefore needs extensive metabolic and growth capabilities, using its own stored nutrients and those from the style. It has been estimated that about 10% of plant genes are pollen-specific, 10–40% are sporophyte specific, and 50–80% are expressed in both gametophyte and sporophyte.

There can be hundreds of pollen grains deposited on one stigma, with intense **competition between pollen grains** for growth and fertilisation, at a haploid stage with no dominance. As well as **sexual selection** by incompatibility systems, there will be **natural selection** for fitness both on pollen-specific genes and on ones expressed in both gametophyte and sporophyte, including genes for growth, wall formation, mitosis and general metabolism. Any recombinants from male meiosis which prove better or worse than the parental type can be selected for or against at the haploid pollen stage, with far less resources committed to them than at the seed stage.

The breeder can try **artificial selection** at the haploid stage with *in vitro* pollen germination. This has been used to select maize and sugar beet for herbicide tolerance, and *Silene* (campion) for copper and zinc tolerance. Although the method sometimes works, with the surviving pollen tubes being used to fertilise stigmas, genes selected in the gametophyte stage often do not have reliably similar effects in the seedlings produced. See Ottaviano and Sari-Gorla (1993).

## 17.2. Animals

### 17.2.1. *Sex ratios*

The **sex ratio** is of commercial importance and is measured as the number of males born per 100 females born. It can change within a species by selection, season, maternal age and the frequency of ejaculation. It tends to be high in dogs, 110–124, 104 in cats, but is usually less than 100 in horse, sheep and chickens. In humans, cattle, sheep and pigs, more males than females abort before birth, with more male conceptions than female ones. For humans, the approximate figures are 130 males conceived per 100 females, but only 105 males per 100 females at birth. The sex ratio is economically very important in dairy cattle and in chickens for egg-laying, where most males are unwanted. In contrast, males are wanted in cattle beef herds and chickens for meat.

Although it seems wasteful of reproductive potential, sex ratios in farm animals are often adjusted by **selective slaughter after birth**. Even in the same species, the sex ratio may be adjusted by slaughter at different values according to the use of the products. For example, **deer** in Asia are raised mainly for the value of their antler velvet which is only produced on males, with a typical 60 males : 40 females (needed for breeding) sex ratio, whereas deer in New Zealand are raised for venison and velvet, with a sex ratio of about 35 males : 65 females.

Much effort has been put into experiments to separate X-bearing (female-producing) and Y-bearing (male-producing) sperm, or at least in getting **selective enrichment** for one type for artificial insemination (AI). Mechanical methods such as centrifugation, and chemicals and electrophoresis have all been tried, with mice, rabbits, sheep, cattle, pigs and humans. An analysis of boar sperm for X- or Y-specific surface proteins that might be used for immunological sorting failed to find any differences among the 1,000 proteins tested. This is not surprising as sperm proteins are largely or entirely specified by the diploid cells producing the sperm, rather than by the sperm's own genotype. There have been many claims of successful sperm separation, including a commercial system for humans, but most have been unrepeatable by others until DNA staining and quantification were used. In the late 1990s, flow cytometry for sexing sperm became a commercial reality (see Subsec. 17.2.2).

In **chickens for egg-laying**, one can use a sex-linked dominant gene for barred feathers to identify males at hatching (very young chicks are difficult to sex from their anatomy), and kill off the males at hatching to reduce rearing expenses. One could cross females from a pure-breeding barred line ($Z^B$W — females are heterogametic in birds, with ZW rather than XY sex chromosomes) to males from a pure-breeding non-barred line ($Z^b Z^b$), when all males will be barred ($Z^B Z^b$) and all females will be non-barred ($Z^b$W). In mammals, one can sex embryos, even when only 1 mm long, from the presence or absence of Barr bodies (Sec. 12.5), but it is uneconomic to do embryonic or foetal sex-determination and abortion of the unwanted sex in farm animals.

## 17.2.2. *Flow cytometry for sexing sperm*

The characteristic exploited in **flow cytometry sexing** is **the difference in amount of DNA** in X-bearing (female-producing) and Y-bearing (male producing) sperm. The amount of DNA in the X-bearing sperm exceeds that in the Y-bearing sperm by 7.5% in the chinchilla, *Chinchilla langier*, 4.2% in sheep, 4.0% in horses, 3.9% in dogs, 3.8% in cattle, 3.6% in pigs, 3.4% in elephants, 3.0% in rabbits, and only 2.8% in man (Johnson and Welch, 1999). These differences mean that one can get almost pure (98–100%) X-bearing or Y-bearing sperm in *Chinchilla*, 90% separation in bulls, and less than 80% separation in humans, where 75% Y-bearing sperm and 90% X-bearing sperm have been obtained under ideal conditions.

Flow cytometry for farm animals was pioneered in cattle by a Colorado, USA, firm, XY, Inc., in conjunction with Cogent, an English firm set up in Chester by the Duke of Westminster to improve the UK's dairy herds. About 600,000 newly born male dairy calves were being slaughtered each year in Britain and incinerated once live sale abroad for veal or beef had been stopped by the BSE scare. In 1999, the British government stopped the £28 million a year compensation for those calves. XY Inc. chose Britain for its first commercial operations, getting 93% female calves from its first trials with Cogent. Conception rates are similar for sexed and normal semen.

**Ejaculated semen samples** contain about one billion sperm per ml in cattle, 300 million per ml in pigs. They are diluted to 150 million sperm in 1 ml of a suitable buffer. A **fluorescing dye**, Hoechst 33342, penetrates living sperm and binds to AT-rich regions in the minor groove of the DNA double helix. Sperm are stained at 35° for an hour with the dye at about $7\mu$M concentration. The sperm are ejected at about 90 km/h under pressure of about 40 psi (2.8 kg per $cm^2$, 276 kPa), with vibration to break the liquid into droplets, some of which contain a single sperm. After the droplets pass through an orienting nozzle, a laser beam is used to excite the fluorescence which is measured by two detectors at 90° to each other as fluorescence from the flat side and edge of the sperm are different. After rapid signal processing by a computer, droplets containing an X-sperm or a Y-sperm are given a **positive or negative charge** respectively, so the droplets are deflected into two streams by high voltage plates. Unsorted drops are removed, and so are dead sperm and some aneuploid sperm. Separated droplets are collected in tubes containing a compound to neutralise the electrical charges, and an egg-yolk-based extender. The tubes are centrifuged and sperm pellets are diluted as required.

The **flow cytometry system is very fast** and can ideally sort about 12,000 sperm per second with about 90% accuracy in cattle, about 43 million an hour. Typical commercial separation rates are about 5,000 sperm per second per machine (personal communication, Richard Williams of Cogent). Commercial systems have been claimed to separate 6 million sperm of each sex per hour, or 18 million of X-bearing sperm. Approximately 5 million sperm are usually needed per insemination in cattle, but as this process removes dead and distorted sperm, about 2 million per straw are usually enough. Whole banks of sorters have to be used commercially, as the separation rates are still rather slow for mass production of separated semen.

The system has been used commercially with dairy cattle. In July 2000, Cogent in Britain offered "Sexed Advantage" bull semen for Holsteins, guaranteeing 90% female calves, a world first, based on a trial with 5,000 births with conception rates similar to using non-sorted sperm. In 2002, Cogent introduced a new freezing process for sperm — Harmony Freeze — which improved the robustness of sexed semen. Conventional pig AI needs 3 billion sperm per dose, so it is much harder than in cattle to get enough sperm, but more specialised insemination methods in pigs need fewer sperm. In pigs and cattle, sex-separated sperm can be used for *in vitro* fertilisation and embryo transfer, although those processes are much more expensive than normal AI.

The **advantages of using sexed semen in dairy cattle** such as Holsteins are several, although the sperm cost twice as much as unsorted sperm. As the offspring are mainly female, only half the heifers (young cows) or cows need to be inseminated with the sexed sperm, so they can be selected females, e.g., for improved udders and legs, producing replacement females. The remaining heifers or cows can be used with beef breeds such as Belgian Blue or Charolais, where the crossbred bull calves sell for about £200 a head compared with the largely unwanted Holstein bull calves fetching about £45 a head (2005 prices). Heifer calves from heifer mothers cause fewer problems at calving than bull calves, and the mother needs a shorter recovery period before being inseminated again. Using sexed sperm reduces the cost of progeny-testing dairy bulls (Sec. 5.7) as the offspring are largely female.

**Sex selection in humans** is not usually permitted, but in theory a man with a serious sex-linked recessive disorder such as haemophilia could have flow cytometry on his sperm to increase the chance of a male offspring, who would not carry his affected X-chromosome. Sex selection is practised **after birth** in some countries where sons are much more valued than daughters, with baby daughters sometimes being killed or left to die. This has been widely reported from China where the one-child per family policy has made the sex of that offspring even more important. When it leads to a marked imbalance between the sexes, as in China, social problems arise at partner choice from an excess of males over available females.

### 17.2.3. *Anatomy, progeny per pregnancy, and temperature effects*

The **anatomy of the sex organs** and accessory glands in males and females of different animals differs astonishingly widely (see Nalbandov, 1976). A major reproductive difference (with anatomical and physiological consequences) between mammalian females is whether the species is **monotocous**, usually bearing one or two offspring per pregnancy, as in women, ewes, mares and deer, or **polytocous**, litter-bearing, as in sows, cats, rabbits, bitches and mice. Within a species, the **number of offspring born per pregnancy** affects their average and total weights, and survival. In sheep, average individual weights at birth are 4 to 5 kg for a single lamb, 3 to 4 kg for twins, 2 to 3 kg for triplets, and 1.5 to 2.5 kg for quadruplets (information from the Seven Sisters Sheep Centre, East Dean, Sussex, UK). In polytocous animals, the number of young born per litter varies with the breed and the mother's age and physical condition. In cats, the litter size varies from 1 to 10, averaging 3.9, with heavier females tending to have larger litters than do lighter ones (Robinson, 1991).

Fertility shows low heritabilities in all farm animals, but in sheep a major gene (or a series of closely linked genes) for **increased lambing rate** was found in Booroola Merino sheep in 1980. In breeds of low to moderate prolificacy, one copy of this allele with additive action, $Fec^B$ on chromosome 6, increases the number of ova by 1.0 to 1.5, on average, giving an addition of 0.75 to 1.0 lambs, on average, per pregnancy. Two copies of the gene double that effect. This allele has been bred into a number of other breeds world-wide, using recurrent backcrossing (Sec. 13.4) to restore other breed characteristics. An increase of two lambs per pregnancy, from homozygous $Fec^B Fec^B$, is excessive in many breeds, as the ewe usually cannot feed more than three lambs. Major genes for increased lambing have been found in other breeds (McNatty *et al.*, 2005), with additive effects, but have usually been lethal when homozygous.

Another sheep high fertility allele, the **Inverdale gene**, was found in New Zealand in a Romney Marsh flock in the early 1980s. It is a single dominant X-linked gene, increasing the ovulation rate by 1.0 and the litter size by 0.6, but homozygous females are infertile, so care is needed in its use. Rams with the gene will pass the trait to all daughters, and a heterozygous ewe passes it on to half her sons and half her daughters. It was introduced into the UK and has been bred into British Texel sheep. Inverdale-carrying rams are crossed to hill breed ewes, improving fertility of daughters and enhanced carcase and conformation performance. Inverdale heterozygous ewes are crossed to terminal sires to give an increased number of prime carcass lambs. The gene is not used in breeds with high fertility, or too many lambs would be born per pregnancy. (Information from Inverdale Genetics, East Mains, Ormiston, East Lothian, EH35 5NG, 2005.)

**Multiple births in humans** were covered in Sec. 12.15. In contrast to sheep, twins are usually unwanted in cattle and horses as the strain on the mother's body is too great. In horses, if twins are diagnosed early in pregnancy, one is often aborted for the sake of the other. The frequency for a mare having surviving twins has been estimated at about 1 in 10,000 births. In Thoroughbred horses, twinning occurs in about 3.5% of pregnancies, usually resulting in miscarriage, and surviving foals have lowered performance compared to single births. In cattle, twinning frequencies vary with breed and environment, from 0.4% of pregnancies to over 10%, being least frequent in the first pregnancy.

In seasonally breeding males, the **descent of the testicles** from the body cavity into the scrotum is very important for fertility, reducing the **testicle temperature** by one to four °C. If the descent does not occur before the breeding season, infertility usually results. Rams can be made temporarily sterile, when not needed for breeding and when it is inconvenient to keep them separated from the ewes, by tying their testicles up against the body wall to raise their temperature. Rams may suffer reduced fertility in very hot weather, when shearing their bodies and scrotum to reduce internal temperatures may help. Fertility-restoration can be very slow because spermatogenesis takes 10 days in mice, 48 days in bulls, and 50 days in rams. Fever in man can cause temporary sterility, and infertile men are often recommended to wear loose-fitting underwear and to spray their testicles with cold water.

## 17.2.4. *Breeding seasons and oestrous cycles*

In man and most domestic animals, both sexes can breed (mate) throughout the year, though there may be **seasonal peaks of fecundity**. Thus in humans in the Northern Hemisphere, conceptions occur most frequently in May and June, with peaks of births in February and March. In North America, horses mate mainly in spring and early summer, though conceptions can occur at any season. In cats in the Northern Hemisphere, the breeding period in mainly January to July, occasionally to October (Robinson, 1991). With cattle, the farmer controls conception times to take advantage of factors such as pasture conditions and milk prices. Camels are only sexually active for a few months in the year, differing in different regions, while the related alpacas and llamas can breed throughout the year, with autumn favoured. See Nalbandov (1976), Hammond *et al.* (1983) and the various books by Gordon (1996 and 1997) for accounts of reproduction in particular farm animals.

Some animals naturally have restricted **breeding seasons**, usually controlled by day-length or sometimes by food availability. Sheep and deer are short-day breeders, mating in autumn or early winter as the days shorten, while ferrets, poultry and horses are stimulated to breed by long days. Artificial lighting in poultry sheds at night can be used to simulate long days. In sheep, there has been selection for breeding at specific times of the year, giving the **best winter survival** of the lambs. Breeds from northerly latitudes such as Iceland and Scotland have a short breeding season: lambs born too early die from cold, and those born too late do not grow enough to survive the following winter. Breeds from nearer the Equator, e.g., Merinos in Spain, have longer breeding seasons, with winter survival being less difficult, and one Asiatic breed can breed all year round. Suffolk sheep are highly seasonal breeders in England, with an anoestrous phase each year started by photoperiod, while the Romanov, Finnish Landrace and D'Man breeds are almost completely non-seasonal in their breeding. The international meat market needs lambs all the year round.

**Seasonally breeding females** go through an anoestrous period, with no oestrous cycles, when they are sexually inactive or less active. In continuous breeders, there is a succession of **oestrous cycles** (from the Latin, *oestrus*, a gadfly) throughout the year. Except in higher primates, females only allow copulation during definite periods of the oestrous cycle called "**heats**", when they are physiologically and psychologically ready for mating. Males usually show little interest in females which are not on heat and are repelled if they do. "Heat" occurs at the time of greatest development of the oestrogen-producing ovarian follicles, and this female sex hormone (the steroid $17-\beta$ oestradiol) brings about the physical and mental changes needed for optimum mating. Sometimes mares and cows have "quiet heats" with ovulation and the physiological changes, but are unwilling to mate, when they can be given extra oestrogen to overcome this. Heats can be induced by external oestrogen, sometimes with a little progesterone.

During "**heat**", there are changes in vaginal mucus viscosity and pH, changes in body temperature and increases in general physical activity: when in heat, sows and cows walk about five times as much as

**Table 17.2.** The timing of oestrous cycles, heat and ovulation in different animals. Data from Nalbandov (1976), Table 4.4.

| Species | Oestrous cycle length, days | Length of heat or sexual receptivity | Time of ovulation |
|---|---|---|---|
| Camelids | Continuous in mating season. | Continuous in mating season, with short breaks. | About 26 h after mating (induced). |
| Cat | 14 or longer. | 3 to 6 days; up to 10 days if not mated. | Induced by mating. |
| Cow | 21 | 13–17 hours. | 12–15 h after end of heat. |
| Ferret | — | — | 30 h after mating (induced), but continuous heat from March to August if unmated. |
| Goat | 19 | Around 39 hours. | 9–19 h after start of heat. |
| Horse | 19–23 | 4–7 days. | 1 day before to 1 day after heat. |
| Man | 28 | Continuous? | Days 12–15 of cycle. |
| Pig | 21 | 2–3 days. | Usually 30–40 h after start of heat. |
| Rabbit | — | — | $10^1/_2$ h after mating (induced). |
| Sheep | 16 | 30–36 hours. | 18–26 h after start of heat. |

when not in heat. These changes can be used to detect heats, and so the farmer can introduce males at the best time for producing offspring.

The length of the oestrous cycle and of "heat", and the timing of ovulation, differ greatly between species, as shown in Table 17.2 (see also Nalbandov, 1976). Thus the cycle takes 21 days in the cow and the sow, and 16 days in the ewe. These cycles are slightly shorter in heifers and gilts (i.e., young cows and young sows), and in ewes, they depend somewhat on the sex drive of the ewe or ram. In the mare, the cycle is rather variable, taking 19 to 23 days, with 4 to 7 days of heat, and ovulation varying from 1 day before to one day after heat. This vague timing of ovulation in mares is one reason why horse breeders usually use hormone treatments to induce heat and ovulation, to allow introduction of the stallion at the best time for fertilisation. In rabbits and ferrets, there is no set length to the oestrous cycle and no obvious "heat". Instead, **ovulation is induced** by mating, taking

10.5 hours after mating in rabbits and 30 hours in ferrets, so presumably the sperm stay viable for at least that period after mating. Ovulation is also induced by mating in camelids (camels, llamas, alpacas, etc.), which do not have regular oestrous cycles.

For introducing the male at the best time, it is very important to know when a female is on heat. The changes in body temperature, vaginal mucus and motor activity can all be used, or seeing if the female will mate with a vasectomised male. In cats and rats, there is a "Lordosis" reflex from a female on heat: when touched on the back she stands still; when touched in the pudendal region she adopts a mating position, especially if she smells a male of the species. In pigs, a sow in heat stands rigid if someone sits on her back or presses his or her hands on the pig's back. An experienced stockman does not need chemical tests to tell whether a cow "is bulling", but hormone tests can be used.

**Control of oestrous** in farm animals such as dairy cattle is often needed to get all of a herd reproducing at the same time, e.g., for a visit by an artificial inseminator or putting males with the females, or for the convenience of all births roughly coinciding, or to take maximum advantage of the best pasture or to boost milk production at a particular time, or to avoid exceeding milk quotas in a particular period. For a small dairy herd of fewer than 50 cows, the farmer might want an unsynchronised herd, with calves produced throughout the year and a steady small income from the sales of dairy bull calves or beef calves, while for a large herd, synchronisation could mean that extra farm labour was only needed at peak times, such as birth.

Oestrous cycles in cattle, sheep and pigs involve very similar fluctuation cycles in **reproductive hormones**, although the cycle length is shorter in ewes (see above). In pro-oestrous, a decline in progesterone levels from the previous cycle causes a rise in the reproductive hormone, follicle-stimulating hormone (FSH). This induces the growth of a follicle in the ovaries, and the follicle secretes increasing amounts of $17-\beta$ oestradiol (oestrogen), which acts on the hypothalamus in the brain, causing physiological changes such as increased blood flow to the genitals, and enlargement of the cervix due to swelling mucus cells which secrete large amounts of mucus which aid the passage of sperm to an ovum.

During oestrous, the ovarian follicle releases an unfertilised egg into the fallopian tubes because of a shift from FSH to a surge in lutenising hormone (LH), which brings about metoestrous and the development of a corpus luteum from the broken ovisac cavity. The final stage, dioestrous, has a fully developed corpus luteum secreting progesterone which prepares the uterine wall for implantation of a fertilised egg. If there is no fertilisation, the corpus luteum remains functional (19 days in cows) then degenerates, with progesterone levels falling, which stimulates the next pro-oestrous. If fertilisation occurs, the corpus luteum continues to secrete progesterone until the developing placenta takes over this function.

One way to **synchronise oestrous cycles** is to end the existing cycle by the injection of a synthetic prostaglandin (PG) such as Estrumate® (clo-prosterol) on arbitrary Day 0 to terminate the corpus luteum. In **cattle**, a second PG injection is given 12 days later, as animals in the early pro-oestrous cycle will not respond to the first PG, but they will to the second. With the termination of the corpus luteum and decreased progesterone levels, the onset of oestrous occurs within 2 to 3 days and the females can be inseminated, usually by **Artificial Insemination** (AI). Another way to synchronise females is to lengthen the cycle and delay oestrous by a controlled intravaginal drug-releasing device (CIDR) or a subcutaneous ear implant, to release progesterone or a synthetic progesterone such as Norgestomet. This delays the ripening of a new follicle until the corpus luteum has degenerated. PG is then administered as before to induce luteal breakdown and start pro-oestrous.

Hormone treatments can be used in **sheep** to increase the number of fertilised ova. Pregnant mare serum gonadotrophin (PMSG) is injected, combining FSH and LH activity. Follicular growth is induced but ovulation does not occur until the corpus luteum has regressed. Oestrous can be synchronised for AI in sheep by inserting progesterone-impregnated sponges into the vagina for 14 days, then on removal injecting the ewe with PMSG. Prostaglandins can be used in pigs, too.

Non-chemical treatments have been used to increase the number of lambs per pregnancy in sheep, where twinning is very desirable economically, but the ewe cannot usually suckle more than two or three lambs adequately. Feeding the ewes intensively (called "**flushing**") before the onset of breeding increases the ovulation rate and is widely used.

## 17.2.5. *Sperm; natural and artificial insemination*

The volume of **ejaculate** and the number of sperm per ejaculation vary enormously between species as shown in Table 17.3, which has implications for

**Table 17.3.** Average values for the volume of sperm per ejaculation and the concentration of sperm. Data adapted mainly from Nalbandov (1976). The figures differ considerably in different studies.

| Animal | Ejaculate volume, ml | Millions of sperm per ml |
|---|---|---|
| Cock | 0.2–1.5 | 4,000 |
| Turkey | 0.2–0.8 | 7,000 |
| Boar | 150–500 | 100 |
| Bull | 2–10 | 1,000-1,800 |
| Ram | 0.7–2 | 3,000 |
| Stallion | 30–300 | 100 |
| Rabbit | 0.4–6 | 700 |
| Dog | 2–14 | 3,000 |
| Man | 2–6 | 100 |

artificial insemination. A bull produces enough sperm a week to inseminate 4,000 cows, because approximately 11,400 million sperm in one ejaculate can be diluted into many doses, as only ~5 million motile sperm are needed per insemination. Thus, while the 5 ml ejaculate from a bull is enough for 500 to several thousand cows, the much lower density of sperm in the 250 ml ejaculation from a boar is only enough for ~25 sows as a dose of $2 \times 10^9$ sperm is used.

Ejaculated sperm only survive typically for 20–30 hours in the female reproductive tract, although in a bull they may live for 60 days stored in the epididymis of his testicles. Although sperm are motile, with active tails, it is mainly contractions in the female reproductive tract, especially the uterus wall, which transport sperm from the vagina to the oviducts where fertilisation normally occurs. In rats, sheep and cows, it only takes a few minutes for sperm to reach the oviducts from the vagina, and even dead sperm and Indian ink particles travel as fast as live sperm because of the wall contractions. Most sperm do not actually reach the oviducts. In rabbits, of the many million sperm introduced at ejaculation, only about 5,000 reach the oviducts and perhaps only 10 to 20 reach the ovum. Sperm need a maturation process (capacitation) in the oviduct before being able to fertilise an egg, which has implications for *in vitro* fertilisation.

**Artificial insemination** (AI) in farm animals is the process by which sperm are deposited in the female reproductive tract by mechanical means. It was first used on a large scale in Russia in 1931, on 20,000 cows. It is widely used today in dairy cattle, sheep, turkeys, goats, pigs ($\sim$70 to 75% of pigs born in the UK are from AI, 2005 figures, and nearly all turkeys) and horses. It can be used to minimise the dangers of having aggressive bulls on a farm, to avoid damage to small females from mating with large males, and to help control **sexually transmitted diseases** such as vibrosis in cattle which may be transmitted by symptomless bulls. Bulls for AI are tested before first use and annually thereafter for vibrosis; they are given antibiotics even if they test negative, and sperm diluents contain antibiotics. Although using one male many times a year through AI reduces the effective population size (Sec. 4.3), and could cause inbreeding if used for more than one generation on a given herd, AI enables genetic merit in highly tested males to be spread rapidly.

There is a range of **artificial vaginas** available for different species, designed to provide the correct temperature (from a water jacket), pressure and lubrication (e.g., Vaseline or K-Y$^{®}$ jelly) to evoke normal ejaculation from the chosen male, with up to five collections a week. Sperm are collected in a graduated glass tube at the end of the collecting tube. The artificial vagina may be held alongside a female in heat, or the male may mount a dummy containing it. A bull given a smell of a suitable cow will even mount a dummy made simply of a piece of coconut matting over a wooden trestle. Males unable to mount a dummy to serve into an artificial vagina, e.g., after injury or arthritis, or which are too small for such devices (e.g., hamsters), can be electrically stimulated to ejaculate, often by using a rectal probe, with rhythmic applications of current to achieve arousal and ejaculation. In turkeys and ganders, semen is obtained through manual massage of the ejaculatory ducts through the abdominal wall. In boars, semen may have a low sperm concentration when a dummy and artificial vagina are used, when manually deviating the penis into a flask as the boar mounts a sow gives better yields. Animals can be dangerous to their handlers and Gordon (1997) states that: "Rutting camels can turn vicious and unmanageable and collection of semen by way of the artificial vagina could be hazardous; for such reasons, an electroejaculatory approach to semen collection has been regarded as the safer alternative".

Bull sperm can be stored unfrozen for up to seven days, but chicken, boar and stallion sperm should be used within two days unless frozen. **Sperm for freezing** are diluted about 50-fold with extenders containing egg yolk or milk, buffers such as sodium citrate, antibiotics such as penicillin and streptomycin, glycerol as a protectant in freezing, glucose as an energy source and water-potential agent, and sometimes vegetable dyes to differentiate between breeds. The treated semen is cooled to just above freezing, before being packed into individual doses, usually in plastic straws, and is frozen in liquid nitrogen ($-196°$) or dry ice and alcohol ($-79°$). Bull semen remains viable frozen for over 15 years. Rapid freezing and thawing give best viability. The transport of frozen semen by air, e.g., from the UK to New Zealand and Australia, has allowed the introduction of Charolais genes when quarantine rules prohibit import of live cattle. Sheep semen can be stored fresh at $15°$ for use within 11 hours, or stored frozen at $-196°$ for several years.

For use, the straws, labelled with sire details, can be taken out frozen to the site of use, thawed, brought to body temperature, and put in an AI "gun". In cows, a dose of 0.25 to 1.0 ml is injected through the cervix, guiding the tip of the gun's tube by a hand inserted into the rectum and feeling downwards (Plate 17.7).

Females need inseminating at the **optimum time in the oestrous cycle**. In cows, ovulation occurs about 14 hours after the end of heat. Conception rates from insemination of cows at different times in the cycle have been found to be about 44% at the start of heat, 83% in the middle of heat (the optimum time), 75% at the end of heat, declining to 32% 12 hours after heat, 12% 24 hours after heat, and 0% 48 hours after heat.

In cattle, AI is widely practised in dairy herds but less so for beef cattle where "normal service" is usual. In 2005, the price of one straw of semen from a top pedigree Limousin (a beef breed) bull, "Grahams Unbeatable", was £70; his 200 day growth, 400 day growth, muscle depth and beef value scores were all in the top 1% for Limousins.

With semen dilution, one **bull** can supply 40,000 to 50,000 doses a year, so one would need fewer than 100 mature bulls for the whole of the United Kingdom. In **sheep** about 1 ml is collected per ejaculate and up to 11 collections per day can be made over short periods, with each ejaculate enough for 25 to 40 ewes. In **pigs** the semen comes in three fractions, a

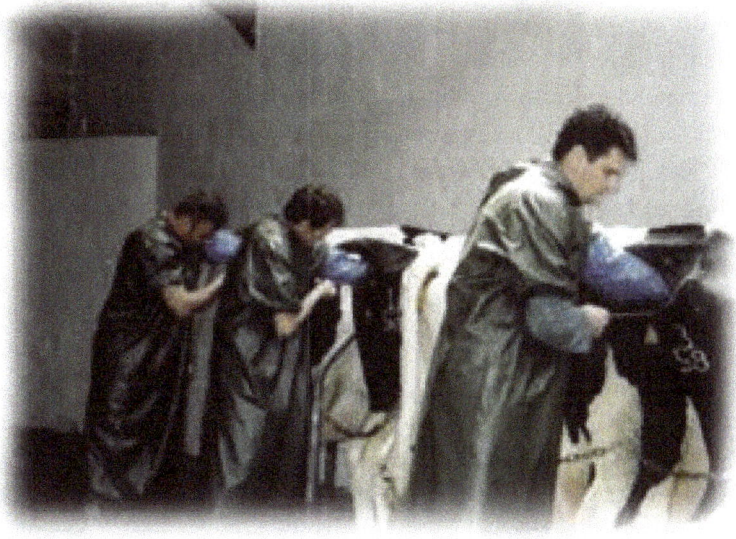

**Plate 17.7.    An artificial-insemination instruction session**. The men's left arms, in protective blue plastic, are feeling down to guide the insemination tubes held in the right hand. Courtesy of Richard Williams of Cogent.

clear pre-sperm portion, then a larger sperm-rich portion, then a gelatinous fraction amounting to half the total volume. After natural matings, the sperm is concentrated by absorption in the sow of accessory fluids. Conception rates increase from about 60% to 80% if a boar is present when the sow has AI, as the sight, sound and especially the scent of a boar helps to condition the female for pregnancy. The scent is now available in aerosol cans and tapes can be played of boar mating cries. In turkeys, natural mating gives only 50–60% fertile eggs, but AI can give 85% fertility. More than 90% of turkey hens have AI in Britain. It has been used in chickens, quail, pheasants, ducks and geese, but with less success. In horses, AI is mainly used in pedigree stock, especially Thoroughbreds and Arabs; it saves transporting mares long distances to the top sires.

A new use for AI in dogs was proposed in 2004 by the Association of Chief Police Officers to breed **police dogs** in Britain which are faster, stronger, with a stronger sense of smell, for crowd control and the detection of drugs and bombs. It will be the first national breeding programme for police dogs, with only 1 in 11 of dogs donated to the police having the right temperament. Of about 2,000 working police dogs in Britain, most are German shepherds, with spaniels and labradors for finding drugs and explosives as they have a better sense of smell. Sperm for AI would be taken from selected British and European dog donors.

**Natural insemination** in farm animals is easy to control. To prevent mating, one just keeps the sexes apart. To get mating, one puts the sexes in the same enclosure, or puts them together when the female is on heat.

## 17.2.6. *Egg transplantation and embryo freezing*

As females of high genetic merit can naturally only have a few offspring in their lifetime, especially in monotocous animals, people have tried to find ways of increasing their reproductive potential. Many procedures for the culture and manipulation of ruminant embryos result in unusually large offspring at birth and a high perinatal death rate.

While AI allows a farmer to use the best bulls, egg transplanting can make use of the best bulls and the best cows. Fertilised ova from an excellent cow can be **transplanted** into genetically inferior cows at the same stage of the oestrous cycle as the donor, i.e., post-ovulation, as the uterine lining must be ready for implantation. Embryo transfer begins with multiple injections of FSH (follicle stimulating hormone) to multiply the ovulations, insemination at oestrus three to four times, 12 hours apart. A week later, the uterus is flushed to extract embryos and ova. Unfortunately, it is difficult to extract fertilised eggs from the donor because two to three days after fertilisation the zona pellucida has been shed and the egg is easily damaged. A flexible two-channel plastic catheter is passed through the cervix to the tip of each uterine horn and an air cuff is inflated around it to seal off the tip area. Fluid is passed up through one channel and collected through the other. Embryos settle out in a dish and can then be implanted in the new recipients or frozen for storage. The method works but is expensive, so it is only used on a small scale, for the top cows. In 1983, one "super-cow" had 35 calves in a year by egg transplants, but 10 or more progeny is usual per year. See Subsec. 13.7.10 for modern Ayrshire cow figures. Frozen embryos can be transported very long distances, so founder herds have occasionally been taken to other countries by embryo transport, transplanting the embryos into local cattle.

**Sheep** were the first domestic animals used for surgical recovery of embryos, from superior donor ewes which had been induced hormonally to super-ovulate. This is used on a small scale commercially for exporting embryos but it is expensive and involves major surgery, with a mid-ventral incision. Six to ten embryos can be recovered, giving four to six lambs as the conception rate is ~60%. The operation can cause complications in the ewe from internal adhesions. Laparoscopy ("keyhole surgery") is less traumatic for the sheep but fewer embryos are recovered. The non-surgical trans-cervical method used in cows does not work in sheep as the cervical passage is convoluted and the sheep rectum is too small to allow an arm up it for guiding tubes through the cervix.

**Embryo freezing** for storage or transport works well in farm animals. In sheep, the embryo is put in 10% glycerol at 20° for a few minutes before cooling to three or four degrees below the freezing point of the freezing fluid. Rapid ice-crystal formation occurs when forceps at liquid nitrogen temperature are placed against the vessel's side. Cooling is then carried out at 0.3° per minute to −60°, with transfer to −196° for long-term storage.

## 17.2.7. Animal cloning

**Somatic cell nuclear transfer** is the basis of mammalian cloning. Genetic material from the nucleus of a donor adult cell is transferred to an egg whose nucleus has been removed. The reconstructed egg is treated with chemicals or electricity to stimulate cell division; when it is big enough, the embryo is implanted in the uterus of a suitable female at the right stage of the oestrus cycle.

The first animal to be cloned was the African clawed frog, *Xenopus laevis*, in the 1950s at Oxford University. In 1996, **sheep were cloned** by nuclear transfer, using a micromanipulator. Sheep embryo cells were tissue-cultured for 6 to 13 passages, then were fused to enucleated oocytes, using electric charges for activation and fusion. In 1997, a sheep named Dolly was cloned from an adult (six-year-old) sheep udder cell nucleus using similar technology, so the udder cell nucleus must have dedifferentiated and acted like a fertilised egg nucleus. The **dedifferentiation** from an adult tissue mammalian nucleus to a totipotent embryo nucleus was a great step forward, one which might never have worked. To get one live lamb took 277 fertilised eggs and 29 implanted embryos. Such procedures are interesting but very expensive and are unlikely to become widespread in farm life of the near future, where costs are crucial. By 1998, a Japanese team had **cloned cattle**, producing eight identical calves from one cow. Other animals which have been cloned include mice (1998), rats, rabbits, cats (2001), goats, pigs (2002), deer (2003) and horses (2003). Monkeys have been cloned by embryo splitting.

In 2004 an American firm, Genetic Savings and Clone, offered the world's first **commercial cat cloning service**. When a cat called Nicky died at the age of 17, his owner sent DNA samples to this firm. For $50,000 (£26,000), the owner received 15 months later a perfect replica, Little Nicky, which she says has Nicky's behaviour and no defects. The firm claims that their method of transferring only the donor's DNA to a surrogate egg, instead of a whole nucleus, has improved survival rates, health and looks. It hoped to offer a dog-cloning service in 2005 for $100,000. There are technical difficulties in **cloning dogs**. A still-born cloned dog was obtained in 2002, and in April 2005 two cloned Afghan puppies were born after 1,095 cloned embryos were transferred into 123 surrogate mothers by a South Korean group. That produced only three pregnancies, one of which miscarried. One of the two puppies died from pneumonia after 22 days. Cloning birds is very difficult as the eggs are too large and opaque to suck out the DNA with a pipette; mammalian eggs have a diameter of about $100 \, \mu$m.

In 2003, **a horse** was cloned for the first time and since then cells have been frozen from 30 exceptional horses, such as the world's top show jumper, ET. In 2005, after three years of attempts, cloning gave a male Arab foal, Pieraz 2. His "father", Pieraz, was the 1994 and 1996 world endurance champion, but as a gelding (castrated) he could not reproduce sexually. A Franco-Italian project at Cremona did the work for the American owner who hopes that Pieraz's merits can be preserved through cloning. Pieraz 2 has normal testicles and should be able to breed in 2007. Horse clones are not permitted to compete. Cloning of endangered species of animal has been suggested, e.g., for giant pandas and the Sumatran tiger.

Cloning of animals has **many problems**. Typically, fewer than 4% of reconstructed embryos develop to adulthood. There are high rates of spontaneous abortion or failure to implant, with high frequencies of foetal, perinatal and neonatal loss, with a high proportion of abnormal offspring. Cloned animals tend to have problems such as a more compromised immune system, higher rates of infection, tumour growth, obesity, abnormal largeness, liver failure and pneumonia. One third of cloned mice which looked normal became massively overweight a few weeks later. Microarray analysis of mouse liver and placental cells showed that about 4% of genes had abnormal function in the cloned mice. **Errors in epigenetic reprogramming** are a major problem, as well as telomere shortening (see Inui, 2005, for cloning problems in various lab and farm animals). Defects in imprinting of DNA may also cause problems in cloned embryos. Animal welfare organisations such as the RSPCA are strongly against animal cloning.

In the USA, the Food and Drug Administration's Center for Veterinary Medicine has asked companies not to introduce animal clones or their progeny or their food products such as meat or milk into the **human or animal food supply chain** (Tian *et al.*, 2005). In 2005, no country allowed cloned animals or their products to enter the food chain. Tian *et al.* (2005) tested more than 100 parameters comparing the composition of meat and milk from beef and dairy cattle from cloned animals with that of genetic- and breed-matched control animals from normal reproduction. Cloning used somatic nuclei from skin fibroblasts or cumulus cells. Eight of sixteen clones survived to adulthood. Milk and beef parameters were essentially the same for cloned and non-cloned animals, suggesting that the products from the cloned animals were safe for human consumption.

## 17.3. Humans

For basic **human reproductive physiology**, see Sherwood (1995) or other textbooks. Human reproductive advances are mainly in the field of treating infertility. **Hormone treatments** figure widely in the treatment of various sexual and reproductive problems. **Testosterone** is responsible for most male sexual development, but it is also important in females where it is produced in the ovaries and adrenal glands, giving women about five per cent of the level found in men. In women, it promotes the development of pubic hair and strengthens bone and muscle. Overproduction can cause masculinisation symptoms in women, such as excessive facial hair. In 2004, trials in Canada, USA and Australia showed that testosterone patches (trade name Intrinsa, from Proctor and Gamble) could improve the sex drive of postmenopausal women, just as Viagra helps older men. It is expected to be licensed for use in Europe in 2005.

***In vitro* fertilisation**, introduced in 1978, works fairly well, but collecting unfertilised eggs from the would-be mother or from an egg-donor involves hormone treatments and can be uncomfortable. By 1997, about 150,000 babies world-wide had been born by this method. Several fertilised eggs are usually placed in the recipient because most are lost through spontaneous abortion, but high survival rates can cause multiple births, even of eight babies in one pregnancy. Selective abortion is sometimes used to prevent excessive multiple births. The first baby resulting from a donated egg, fertilised *in vitro* and transplanted into a different woman's womb, was born in 1984 in Australia.

Where a couple carry a known serious genetic defect for which there is an appropriate DNA probe, it is possible to do *in vitro* fertilisation and use DNA analysis (from a single cell taken from each embryo after a very few cell divisions — removing one cell does little harm) by PCR, to find which embryos carry the defect, so that they are not implanted. This is **pre-implantation embryo selection**, following **pre-implantation genetic diagnosis** (PGD: see Subsec. 12.12.6).

The term "**designer babies**" is used in the media, as in the headline " 'Designer baby' cures brother in stem cell breakthrough" (*The Daily Telegraph*, 21/10/2004, p3, John Crowley). The term is often applied to selected embryos, not to babies designed by some medical scientist. In this particular case, six-year-old Charlie Whitaker was reported to have been almost cured of a rare blood disorder, Diamond Blackfan anaemia, by a transplant of his baby brother Jamie's blood stem cells from his umbilical cord. The mother received *in vitro* fertilisation treatment with pre-implantation screening in Chicago (as it was not licensed then in Britain, although it is now) to detect embryos matching Charlie for transplants. Two such three-day-old embryos were implanted in the mother in 2002 and in 2003 she gave birth to Jamie, whose umbilical cord blood stem cells (Sec. 12.18) were isolated to treat Charlie at Sheffield Children's Hospital. Charlie had previously been kept alive by frequent blood transfusions and nightly steroid injections. Three months after the transplant, his bone marrow was working so well that he no longer needed transfusions or injections, with much improved energy levels as he produced more red blood cells. Such treatments involving pre-implantation embryo screening are now available in Britain on the National Health Service in certain areas but not all.

As an example of the **economics** of such techniques, for a sufferer from severe $\beta$-thalassaemia it costs about £1,000,000 for a lifetime of treatment with transfusions and medical care. For example, a five-year-old boy in Leeds, Zain Hashmi, had already had more than 100 blood transfusions, one every three to four weeks, and lots of medication involving drugs in a drip for 12 hours, five nights a week. His parents were trying to conceive by IVF, with pre-implantation genetic diagnosis, a non-thalassaemic sibling whose umbilical cord blood could be used to cure Zain's faulty bone marrow. Four attempts at IVF and PGD cost about £20,000. A search for suitable ethnic minority bone marrow donors was unsuccessful. In the potential bone marrow donor register of the Anthony Nolan Trust in Britain, there are only 9,000 Asians and 8,500 Afro-Caribbeans out of 360,000 volunteers (*The Daily Telegraph*, 25/11/2004, David Devonshire).

Most reproductive technology in humans has a fairly low success rate and is expensive, so natural reproduction is preferred if couples are fertile. In 1999 in Britain, a single *in vitro* fertilisation treatment cost £2,000, with a success rate of only 15%. The National Health Service funded, on average, 11 single courses per 100,000 people in 1998.

**Intracytoplasmic sperm injection** (ICSI) was pioneered in Belgium in 1988. It enables men to reproduce even if they have no functioning sperm at all, or if they have had a vasectomy. Immature sperm are taken by needle directly from the testicles and are injected *in vitro* directly into the cytoplasm of an egg, as such sperm are incapable of *in vitro* fertilisation. It has improved the chance of treating male infertility from 15% to nearly 90%. Approximately 6,500 children were conceived by this method in Britain between 1992 and 1999. If genes for an inability to produce sperm are passed to the offspring, sons will be unable to reproduce naturally. Page *et al.* (1999) studied three men with very low sperm counts because of having the deletion *AZFc* on their Y chromosome. The three men fathered a total of four sons by ICSI; as expected, all sons inherited their father's Y carrying that deletion and were therefore expected to be sterile. About 2% of men have very low or zero sperm counts, with the *AZFc* deletion present in about 10% of those cases. Daughters fathered via ICSI by men who carry this Y-chromosome deletion have been normal as they inherit their father's unaffected X chromosome, not his Y.

As noted in Sec. 12.8, the frequency of meiotic non-disjunction in human females increases with age; it also increases with maternal hypothyroidism and sometimes after viral infections and irradiation. The mutation rate increases with paternal age for several autosomal dominant traits, including Apert syndrome, progressive myosotis ossificans, Marfan syndrome and achondroplastic dwarfism (see Connor and Ferguson-Smith, 1997).

## Suggested Reading

Abeles, F. B., P. W. Morgan and M. E. Saltveit, *Ethylene in Plant Biology*, 2nd ed. (1992) Academic Press, London.

Bertin, P., J. M. Kinet and J. Bouharmont, Heritable chilling tolerance improvement in rice through somaclonal variation and cell-line selection. *Aus J Botany* (1996) 44: 91–105.

Chopra, V. L. *et al.*, (eds.) *Applied Plant Biotechnology*. (1999) Science Publishers, Inc., Enfield, New Hampshire.

Connor, J. M. and M. A. Ferguson-Smith, *Essential Medical Genetics*, 5th ed. (1997) Blackwell Scientific, Oxford.

Croston, D. and G. Pollott, *Planned Sheep Production*, 2nd ed. (1994) Blackwell Scientific Publications, Oxford.

Franklin-Tong, V. E. and F. C. H. Franklin, The different mechanisms of gametophytic self-incompatibility. *Philos Trans Roy Soc (Lond) B* (2003) 358: 1025–1032.

Gordon, I., *Controlled Reproduction in Cattle and Buffaloes*. (1996) CAB International, Wallingford.

Gordon, I., *Controlled Reproduction in Pigs*. (1996) CAB International, Wallingford.

Gordon, I., *Controlled Reproduction in Sheep and Goats*. (1996) CAB International, Wallingford.

Gordon, I., *Controlled Reproduction in Horses, Deer and Camelids*. (1997) CAB International, Wallingford.

Hammond, J., J. C. Bowman and T. J. Robinson, *Hammond's Farm Animals*, 5th ed. (1983) Edward Arnold, London.

Hartmann, H. T. *et al.*, *Plant Propagation: Principles and Practices*, 6th ed. (1997) Prentice-Hall International (UK) Ltd., London.

Hiscock, S. J. and D. A. Tabah, The different mechanisms of sporophytic self-incompatibility. *Philos Trans Roy Soc (Lond) B* (2003) 358: 1037–1045.

Hughes, J. and A. J. Richards, The genetic structure of populations of sexual and asexual *Taraxacum* (dandelions). *Heredity* (1988) 60: 161–171.

Inui, A., (ed.) *Epigenetic Risks of Cloning.* (2005) CRC Press, Boca Raton.

Islam, A. S., (ed.) *Plant Tissue Culture.* (1996) Science Publishers, Inc., Enfield, New Hampshire.

Johnson, L. A., Sexing mammalian sperm for production of offspring: the state-of-the-art. *Anim Reprod Sci* (2000) 60–61: 93–107.

Johnson, L. A. and G. R. Welch, Sex preselection: high-speed flow cytometry sorting of X and Y sperm for maximum efficiency. *Theriogenology* (1999). 1323–1341.

Kinghorn, B. *et al.*, (eds.) *Animal Breeding. Use of New Technologies.* (1999) Postgraduate Foundation in Veterinary Science of the University of Sydney, Sydney.

Lawrence, M., Population genetics of the homomorphic SI polymorphism in flowering plants. *Ann Botany* (2000) 85(Suppl. A): 221–226.

Lee M. and R. L. Phillips, The chromosomal basis of somaclonal variation. *Ann Rev Plant Physiol Plant Mol Biol* (1988) 39: 413–437.

McNatty, K. P. *et al.*, Physiological effects of major genes affecting ovulation rate in sheep. *Gene Select Evol* (2005) 37(Suppl. 1): S25–S38.

Nalbandov, A. V., *Reproductive Physiology of Mammals and Birds*, 3rd ed. (1976) W. H. Freeman and Co., San Francisco.

North, C., *Plant Breeding and Genetics in Horticulture.* (1979) Macmillan, London.

Ottaviano, E. and M. Sari-Gorla, Gametophytic and sporophyic selection, in Haywood, M. D., N. O. Bosemark and I. Romagosa (eds.) *Plant Breeding. Principles and Prospects* (1993), pp. 332–352, Chapman & Hall, London.

Page, D. C., S. Silber and L. G. Brown. Men with infertility caused by *AZFc* deletion can produce sons by intracytoplasmic sperm injection, but are likely to transmit the deletion and infertility. *Hum Reprod* (1999) 14: 1722–1726.

Pelletier, G., Somatic hybridisation, in Haywood, M. D., N. O. Bosemark and I. Romagosa (eds.) *Plant Breeding. Principles and Prospects.* (1993), pp. 93–106, Chapman & Hall, London.

Prasad, B. N. (ed.) *Biotechnology and Biodiversity in Agriculture/Forestry.* (1999) Science Publishers, Inc., Enfield, New Hampshire.

Richards, A. J., *Plant Breeding Systems*, 2nd ed. (1997) Chapman and Hall, London.

Richards, A. J., Apomixis in flowering plants: an overview. *Philos Trans Roy Soc (Lond) B* (2003) 358: 1085–1093.

Robinson, R., *Genetics for Cat Breeders*, 3rd ed. (1991) Pergamon Press, Oxford.

Seidel, G. E., Jr., Sexing mammalian sperm — intertwining of commerce, technology, and biology. *Anim Reprod Sci* (2003) 79: 145–156.

Sherwood, L., *Fundamentals of Physiology. A Human Perspective*, 2nd ed. (1995) West Publishing Company, St. Paul/Minneapolis.

Tian, X. C. *et al.*, Meat and milk composition of bovine clones. *Proc Nat Acad Sci USA* (2005) 102: 6261–6266.

# Chapter 18

# Applied Fungal Genetics

## 18.1. General fungal genetics: life cycles; wild-types and mutants; spore types; control of sexual and vegetative fusions; genomics

### 18.1.1. Life cycles

The **life cycles** of fungi, including industrial fungi, are extremely varied. Some fungi have no sexual cycle, a condition which has probably arisen independently in a number of formerly sexual species or groups. Fungi with no known sexual cycle include *Aspergillus niger, Penicillium chrysogenum*, and some polyploid brewing yeasts. For example, *Saccharomyces carlsbergensis*, the lager yeast, is an allopolyploid from a hybrid thought to be between *S. cerevisiae* and *S. monacensis*. Most fungi have cell fusion between haploid cells (which may or may not be specialised gametic cells), followed sooner or later by nuclear fusion to give a diploid nucleus. There may or may not be diploid mitosis before meiosis restores the haploid condition, completing the sexual cycle which can recombine syntenic and non-syntenic loci.

Many fungi are **predominantly haploid**, including the mycelial Ascomycetes such as *Aspergillus, Penicillium, Neurospora* and *Sordaria*. In these, mitosis is mainly or exclusively in the haploid vegetative stage, although there is often mitosis in a **dikaryotic** (n + n) stage within a reproductive structure after fusion of haploids in sexual reproduction. Meiosis occurs very soon after the fusion of haploid nuclei to give a diploid nucleus from the dikaryon. There may sometimes be a dormant diploid structure, especially the zygospore in Zygomycetes such as *Mucor*, where meiosis is delayed until long after the formation of the diploid nucleus.

In the Basidiomycetes, there is a **haploid-dikaryotic** cycle in the Hymenomycetes and Gasteromycetes which are usually macrofungi with large fruiting bodies, such as the mushrooms. The haploid mycelium grows extensively but if it meets a compatible haploid, fusion occurs to give a dikaryon which can grow and can produce fruiting bodies. In these, nuclear fusion occurs in basidia to give a diploid nucleus, with meiosis rapidly occurring to give four nuclei which pass into the haploid and dispersive basidiospores. The parasitic Uredinales (rusts) also have a haploid and a dikaryotic mycelium, but usually on different hosts, e.g., barberry and wheat for wheat stem rust.

Some fungi can have prolonged **haploid and diploid** stages, both stages having vegetative reproduction after mitosis. Examples include yeasts such as *Saccharomyces cerevisiae* and the Chytridiomycete *Allomyces*. In both these types of organism, the haploid cells and diploid cells are very similar in appearance but the diploid cells have approximately twice the volume of the haploid ones. In Myxomycetes, there is an amoeboid haploid phase and a plasmodial diploid phase.

In the Oomycetes such as *Saprolegnia* and the parasitic *Pythium* and *Phytophthora*, the mycelium is **diploid**. Meiosis occurs in the male and female sex organs, the antheridia and oogonia respectively, with gametic fusion restoring the diploid state.

Two examples of life cycles will be given, both from Ascomycetes, but with very different amounts of morphological complexity; both fungi have unordered asci. ***Saccharomyces cerevisiae***, yeast, is of enormous importance in the wine, beer, spirits and baking industries, as well as being useful for food and for producing vitamins and other biochemicals, including industrial alcohol and biofuel. The asci are formed singly, not in fruit bodies, and there are only single cells, not a mycelium. The diploid vegetative cell grows then undergoes mitosis, budding off a smaller diploid bud from the mother cell. Both cells then grow and can bud, continuing the diploid vegetative cycle indefinitely under suitable conditions. Adverse growing conditions such as starvation can induce meiosis, giving an ascus, a sack containing four haploid ascospores, two of mating type *a* and two of mating type $\alpha$. The ascal sack eventually breaks down, releasing the four ascospores. They each give rise to a haploid vegetative cell which can reproduce indefinitely by budding after mitosis. There are no specialised gametic cells but a vegetative $\alpha$ cell can mate with a vegetative $a$ cell, usually but not necessarily derived from a different ascus. Cell fusion is followed quickly by nuclear fusion and a diploid bud gives rise to a new diploid vegetative cell. There is thus indefinite haploid and diploid mitosis. See Plate 1.6 for yeast crosses made for the *cis/trans* allelism test.

***Aspergillus nidulans*** is not of industrial importance but is closely related to industrial fungi such as *Aspergillus niger* (citric acid production), to the Aspergilli used in producing soya sauce (*A. sojae*), industrial enzymes and sake (*A. oryzae* and others), and to *Penicillium chrysogenum* used in penicillin production. *A. nidulans* has both a sexual and a parasexual cycle (Sec. 18.4). The vegetative mycelium is haploid, septate and multinucleate, growing indefinitely by hyphal extension. It can fuse vegetatively with other mycelia, giving a heterokaryon. It **reproduces asexually** by producing aerial conidial heads, with chains of uninucleate haploid dispersive green conidia (Plate 18.3) which form new mycelia if they land on suitable substrates.

The mycelium can also **reproduce sexually**. Hyphal sexual initials give rise to closed spherical fruiting bodies called cleistothecia. Haploid nuclei pair up to give dikaryotic hyphae within the fruit bodies. In terminal croziers, nuclear fusion gives a diploid nucleus which quickly undergoes meiosis. The four haploid nuclei in the resulting ascus undergo two mitoses, producing eight sculptured binucleate haploid ascospores per ascus. There is no discharge mechanism for the ascospores, unlike in *Neurospora*, *Sordaria* and *Ascobolus*, and the ascospores are liberated by breakdown of the cleistothecium and of the ascus. An ascospore can germinate on a suitable medium to give a new mycelium. There are no mating types in this homothallic species, so any strain can be self-fertile and can usually cross by vegetative fusion with any other strain. A selfed cleistothecium from a homokaryon will produce identical ascospores, but a heterokaryon from vegetative fusion between two genetically different haploids can, depending on whether similar or different nuclei form the dikaryotic hyphae, produce selfed cleistothecia and crossed cleistothecia, where the latter will produce a range of genotypes in the ascospores.

## 18.1.2. *Wild-types and mutants*

The **wild-type** of a fungus is the typical form found in the wild. The term is also used for the wild-type allele at a locus, with other alleles being considered mutant. Wild-types of many, but by no means all, fungi are **prototrophic**, growing on a simple defined minimal medium, e.g., water, glucose, biotin and inorganic salts (including a nitrate) for *Neurospora crassa*. Other fungi may require organic nitrogen or other vitamins. The wild-type of a given fungus has a given morphology for its vegetative and sexual parts.

One can get **spontaneous or induced mutants** where each mutant has an altered morphology (growth rate, hyphal growth pattern, fruit body form, asexual spore forms, colony form, etc.), or altered pigmentation (fruit body, sexual or asexual spores, mycelium or colony), or is **auxotrophic**, requiring an additional nutrient to minimal medium, say an amino acid, vitamin or nucleic acid base, or is resistant to a drug, metabolic inhibitor or antibiotic. One can therefore obtain a range of mutations to use in basic genetic studies of the organism, for mapping or for use in breeding better strains of useful fungi (Sec. 18.6). Mutations can be nuclear or cytoplasmic; e.g., yeast with deletions of mitochondrial DNA can give *petites*, which grow slowly by fermentation but cannot carry out aerobic respiration, giving very small colonies (Plate 18.1), while *petites* can also be due to mutations at one of at least nine nuclear loci.

**Mutagens** (see Sec. 7.2) used with fungi include:

- **UV light**, especially on conidia or hyphae, giving all kinds of gene mutation (base substitutions, frame shifts, deletions, ) and chromosome aberrations;
- **X rays** and **gamma rays**, on hyphae, sexual or asexual spores, giving all types of gene mutation and lots of chromosome aberrations;
- **EMS** (Sec. 7.2), giving mainly base substitutions but some frame shifts;
- **NMG** (Sec. 7.2), which is fairly specific for base substitutions;
- **ICR-170**, mainly giving frame shifts, but also some base substitutions;
- **intercalating agents** such as proflavin, which are fairly specific for frame shifts.

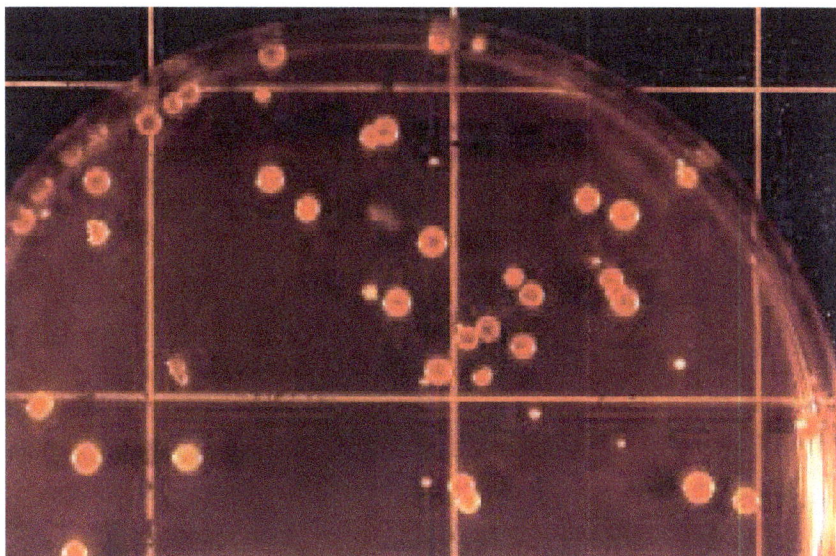

**Plate 18.1.** A red *ade-1* yeast which has been exposed to UV light. Most of the colonies are normal-sized red ones, but **induced "petites" show as small paler colonies**. "Petites" cannot respire aerobically but can grow slowly by fermentation as the medium here includes glucose. They have mutations at one of at least nine nuclear loci, or in mitochondrial DNA. Spontaneous "petite" mutations may accumulate in beer yeasts which are used in several successive fermentations, and may cause off-flavours. The dish is 9 cm diameter.

## 18.1.3. *Spores*

When working with any fungus, it is important to know what kinds of spore it produces and what their properties are, including how to germinate them. **Sexual spores** include ascospores and basidiospores. They are usually haploid but may be uninucleate, binucleate or occasionally multinucleate, depending on the group or species. They may be tough, resistant, resting and survival bodies, as well as dispersive reproductive bodies.

**Asexual spores** are absent from yeasts, but there may be several different types as in stem rust of wheat, *Puccinia graminis tritici*, which has aecidiospores, uredospores and teleutospores, as well as pycniospores involved in mating, and sexual basidiospores. Conidia are very small and efficient dispersal agents in many Ascomycetes. They are haploid and uninucleate in *Aspergillus nidulans* (Plate 18.3) and *Penicillium chrysogenum*, but have three to eight haploid nuclei in *Neurospora crassa*.

## 18.1.4. *The control of vegetative and sexual fusions*

In some fungi there are no genes which normally control **vegetative fusions**, so all mycelia can usually fuse with all other mycelia, as in *Aspergillus nidulans*. If the two strains fusing are genetically different, then **heterokaryons** are formed containing two types of nuclei. The ratio of the two types of nuclei may change drastically or slightly during growth, from natural selection for the optimum nuclear ratio. In yeast, there are no vegetative fusions, only a sexual one between haploids of opposite mating type.

In *Neurospora crassa*, only opposite mating types *A* and *a* can fuse sexually, e.g., a conidium or hypha with the trichogyne of the protoperithecium of the opposite mating type, to give a large black perithecium containing about 100 asci, each with eight black haploid ascospores in a linear order. Opposite mating types in this species will not fuse vegetatively, say two hyphae, one *A* and one *a*. The only compatible vegetative fusions are of *A* with *A*, and *a* with *a*. There are also at least 10 non-mating-type loci controlling vegetative fusions, where the fusing hyphae must carry identical alleles, so that *C,D* will fuse with *C,D* but not with *C,d*, *c,D* or *c,d*. In *Ascobolus immersus*, however, hyphae will fuse whether they are of the same or opposite mating types, (+) with (+), or (+) with (−), or (−) with (−) (Lamb and Chan, 1996). See Glass and Kuldau (1992) and Carlile *et al.* (2001).

The **control of sexual fusions** has different degrees of genetic complexity in different fungi, starting with self-fertile strains with no mating types in **homothallic** species such as *Aspergillus nidulans* and *Sordaria fimicola*. In **heterothallic species** (having more than one type of body with respect to sex), one can have two alleles at one locus as in yeast, *Neurospora crassa*, *Ascobolus immersus* and *Sordaria brevicollis*. *Coprinus comatus* has multiple alleles at one locus, $a_1, a_2, a_3, \ldots, a_{100}$, where alleles must differ for fertility, and *Coprinus fimeterius* has multiple alleles at two loci, so that $a_4$, $b_1 \times a_2$, $b_1$ is sterile because of one common allele, but $a_4$, $b_8 \times a_{23}$, $b_{98}$ is fertile. *Coprinus cinereus* has more than 12,000 mating types (Casselton and Zolan, 2002). There are even more complex systems.

Yeasts can be "homothallic" (carrying the *HO* allele) or "heterothallic", where these terms are used differently from normal usage, in relation to **mating-type switching**. In "homothallic" strains, such as most wine yeasts, a single haploid cell can switch mating type when forming a single colony, mating with unswitched cells of opposite mating type to give self-mated *a*/α **diploids** which are expected to be homozygous at all other loci. Diploids of *HO* haploids can of course be highly heterozygous when different strains of opposite mating type mate. "Heterothallic" (non-*HO*) strains such as most baker's yeasts do not switch, or do so very rarely, so a haploid cell line can only mate if it encounters a cell of opposite mating type, when diploids tend to be heterozygous at many loci.

## 18.1.5. *Fungal genomics: nuclear, mitochondrial and plasmid*

The number of fungal species has been estimated as one to two million, with fewer than 10% of them scientifically described. Only a few species have had their chromosome numbers and genome sizes determined, and only a very few have been sequenced yet. The numbers of nuclear chromosomes per haploid set and genome sizes for a number of yeast-like and filamentous fungi are shown in Table 18.1. **Chromosome numbers** and **genome sizes** are fairly low, with both attributes sometimes varying markedly even within genera, as in the *Fusarium* data. See Sec. 8.8 for general aspects of genomics.

**Table 18.1.** Chromosome numbers and genome sizes in selected fungi (data from Desjardins and Bhatnagar, 2003).

| Fungal species | Chromosomes per haploid set | Genome size, Mb |
|---|---|---|
| *Aspergillus flavus* | 6 to 8 | 33–36 |
| *Aspergillus fumigatus* | 8? | 32 |
| *Aspergillus nidulans* | 8 | 28.5 |
| *Aspergillus niger* | 8 | 37.5 |
| *Aspergillus oryzae* | 8 | 35 |
| *Aspergillus sojae* | 6 to 8 | 35.5–38.5 |
| *Candida albicans* | 8 | 16–17 |
| *Fusarium sporotrichioides* | 6 | 27.7 |
| *Fusarium verticillioides* | 12 | 46 |
| *Magnaporthe grisea* (rice blast) | 7 | 40 |
| *Neurospora crassa* | 7 | 42.9 |
| *Saccharomyces cerevisiae* | 16 | 12 |

Typically, filamentous and yeast-like fungi have about **5,000 to 15,000 functional genes**, with a gene density of about one gene per 3,000 base pairs (Desjardins and Bhatnagar, 2003). Yeast (*Saccharomyces cerevisiae*) has about 6,100 genes (Goffeau *et al.*, 1996, 1997) and has been completely sequenced. Its nuclear genome is 12,069 kb, with chromosomal lengths from 230 kb to 1,532 kb. Yeast has many duplicated chromosomal regions, with a suggested whole genome duplication some 100 million years ago (Sugino and Innan, 2005), followed by large-scale rearrangements by reciprocal translocations. Compared to higher animals and plants, fungal genomes have fewer introns and fewer repetitive sequences. Due to the small genome sizes, genes can usually be cloned easily and directly by mutant complementation with plasmid or cosmid vectors. Fungal sequence data are available on a number of websites, including http://www.apsnet.org; http://www-genome.wi.mit.edu/seq/fgi/candidates.htlm; http://genome-www.stanford.edu.Saccharomyces/.

Fungal **mitochondrial DNA** (mtDNA) is usually circular but is linear in some species. Its length varies from 19.4 kb in *Schizosaccharomyces pombe* to 86 kb in baker's yeast and to 175 kb in *Agaricus bisporus*, largely because of different extents of optional introns and in intergenic spacer regions. Some introns catalyse their own removal from precursor mRNA. In baker's yeast, there are about 20–50 copies of the mtDNA per mitochondrion, with replication throughout the cell cycle. Mitochondrial genes code for ribosomal RNAs and tRNAs, proteins in the respiratory chain such as cytochrome oxidase subunits, parts of ATP synthase, and some ribosomal proteins. Mitochondria divide by fission and can fuse, so that recombination between different mitochondrial genomes in a heteroplasm, e.g. following hyphal anastomosis between different strains (as in heterokaryon formation), can occur by chance. In most fungi where it has been investigated, mitochondrial DNA is inherited sexually largely or exclusively from one parent in a cross: the initial heteroplasmon in a dikaryotic mycelium becomes a homoplasmon by elimination of one parental type. For more details of mtDNA and plasmids, see Hausner (2003).

**Plasmids** are often present in fungi, usually with little phenotypic effect, mostly in the cytoplasm but some are associated with mitochondria. The major types are: circular, encoding a DNA polymerase; linear with terminal inverted repeats encoding a DNA polymerase and an

RNA polymerase; circular or linear retroplasmids which encode a reverse transcriptase.

## 18.2. The commercial importance of fungi

Carlile and Watkinson (1994) gave the following figures from the 1980s for the annual value in billions of US$ for fungal products, but modern values would be considerably higher: alcoholic beverages, 37; cheese (only some use fungi, especially blue-veined cheeses and some soft ones), 22; fungal antibiotics, 5 (penicillins, 3; cephalosporins, 2); mushrooms, 8; industrial alcohol from yeast, 4; fungal enzymes, 1; high fructose syrup, 2; yeast biomass, 1; citric acid, 0.5; steroids, 0.5. Since then there has been a big rise in **alcohol production** by yeast fermentation of cornstarch derivatives in America (more than a billion gallons a year) and of sugar cane in South America. Much of this alcohol is used as a fuel for cars with modified engines, or as "**gasahol**" for normal engines, with 10% ethanol added to petroleum to give a cleaner-burning fuel.

Different strains of **yeast**, usually *Saccharomyces cerevisiae*, are used for baking, wine, beer, whisky, gin, industrial alcohol, and food yeast, and for the production of vitamins, amino acids and other biochemicals. Both *Saccharomyces cerevisiae* and *Torulopsis utilis* are used as food yeasts, being added to many foods to improve flavour and nutritional quality.

*Penicillium chrysogenum* is used for **penicillin** production and other species such as *P. camembertii* and *P. roquefortii* are used for soft cheeses (e.g., Camembert) and blue-veined cheeses (e.g., Roquefort). Antibiotics are produced in many fungi, including *Penicillium*, *Aspergillus*, *Trichoderma viride* and *Cephalosporium*.

*Aspergillus niger* is important in producing citric acid from molasses, starch hydrolysates and other cheap substrates. *A. terreus* produces itaconic acid, used for pharmaceuticals, synthetic fibres, resins and surfactants, from beet and cane molasses. *A. oryzae* is used on a large scale in Japan for making soy sauce and rice wine (sake). Aspergilli also produce amyloglucosidases (for converting starch to glucose), amylases (starch and dextrins to maltose etc.), pectinases, proteases, glucose oxidase, catalases, tannases (to break down tannins), lipases to break down fats, lactases, cellulases, etc.

Various Basidiomycetes are well known for **food**, e.g., *Agaricus bisporus*, the cultivated mushroom, *Lentinus edodes*, the shiitake, and some Ascomycetes, such as *Tuber*, the truffle. *Fusarium graminearum* or *F. venenatum* mycelium (with thin-walled mutants to reduce wall material in relation to protein) is used to make fungal protein products such as the meat-substitute Quorn®.

Fungi are used for **organic acids** such as citric, gallic, gluconic and itaconic acids, for amino acids such as lysine and threonine, a vast range of enzymes, vitamins such as riboflavin (many are from yeast hydrolysates), antibiotics, alkaloids and many other biochemicals. Fungi can be used to transform metabolites in very specific ways, e.g., *Rhizopus nigricans* does one stereospecific step in steroid biosynthesis.

Fungi are also extremely important as **parasites** on plants, animals and humans, e.g., rusts and mildews on crop plants, as **saprophytes** rotting down plant remains and animal dung, and as **spoilage organisms** on stored foods and drinks, timber, leather, optical surfaces, fuels, etc. See the textbooks at the end of the chapter for more details.

Another use of fungi is for **biological control**, e.g., of roaches and termites, and of parasitic fungi on plants, such as applying *Trichoderma harzianum* spores to seeds to control seed rots and damping off of seedlings caused by the fungus *Pythium*. Fungi are used to control plant nematodes either by fungal products, e.g., DiTera from Abbott Labs, or directly, as with nematode-trapping fungi such as *Dactylaria*. *Paecilomyces lilacinus* destroys the egg masses of the nematode *Meloidogyne arenaria*, increasing yield in oats. Fungi can also be used as **mycoherbicides**, with host-specific pathogens of weeds of crops. For example, sprays of spores of *Phytophthera palmivora* have been used to control milkweeed vine, *Morrenia odorata* (see Carlile and Watkinson, 1994). A recent development is of **mycopesticides** to control locusts and grasshoppers. Lomer *et al.* (2001) stated that preparations of *Metarhizum anisopliae* spores in oil can be used to control locust plagues, killing about 80% of the locusts in two to three weeks with no impact on non-target organisms.

## 18.3. Recombination and sexual mapping

Basic aspects of **recombination** were considered in Subsec. 1.3.8. For syntenic genes in fungi, recombination frequencies depend on the distance between the loci. For non-syntenic loci or syntenic loci far apart, there is

50% recombination. For n heterozygous loci, there are $2^n$ possible combinations, with 1,024 combinations for 10 loci, 1,048,576 for 20 loci and more than a billion for 30 loci.

For reasons of space, not all the background theory will be given in this section (see Fincham *et al.*, 1979, for details). Chapter 8 dealt with general aspects of mapping, recombination and genomics, including the physical DNA mapping methods now widely used. For a summary of recombination mechanisms in fungi, see Lamb (1996, 2003). In fungi with different mating types such as yeast, it is easy to make defined crosses for mapping and for getting desired recombinants, using sexual spores from such a cross. If the fungus is homothallic with no mating types, as in *Aspergillus nidulans*, one makes a heterokaryon between the strains to be crossed, or inoculates the two strains on opposite sides of a Petri dish, but the fruit bodies formed could be selfed or crossed. Suppose in *A. nidulans* one crosses a multiply-marked strain 1, with white conidia, with a multiply-marked strain 2, with yellow conidia (Plate 18.3), with $w$ and $y$ being unlinked. One plates only some of the ascospores from specific cleistothecia onto complete medium. If a cleistothecium produces colonies only with white conidia, it was from strain 1 selfed. A cleistothecium with ascospores giving only yellow conidia is from strain 2 selfed, while a cleistothecium giving some colonies with white conidia, some with yellow conidia, and some with recombinant green conidia, must be a crossed one, and recombination frequencies can be obtained for pairs of markers by scoring the colonies growing from its remaining ascospores when their colonies are tested on suitable media.

Sexual spores such as ascospores or basidiospores can be used as isolated, separate spores for mapping, using recombination frequencies for pairs of loci, or for obtaining desired recombinant types, e.g., for industrial or agricultural use. This is **random spore mapping**. One can do more sophisticated mapping when the spores can be isolated as intact groups from a single meiosis, such as the unordered tetrads of ascospores in yeast, or the linear octads of ascospores in *Sordaria fimicola* (Plate 18.2).

With **ordered tetrads or octads**, one can find the distance in cM between a locus and its centromere from the formula,

$$\text{centromere distance} = \frac{\text{\% second division segregation}}{2}.$$

The factor of two arises because map units are based on the percentage of recombinant chromosomes from meiosis. In a second division segregation ascus, only half the chromosomes show recombination between the centromere and the locus.

Ordered asci with **first division segregation** for the locus $m$, in a $m^+ \times m^-$ cross of two haploid strains, have the two different spore-types confined to half the ascus, e.g., $m^+$ $m^+$ $m^+$ $m^+$ in one half and $m^-$ $m^-$ $m^-$ $m^-$ in the other half. Asci with **second division segregation** for that locus would have both different types of spore in each half ascus, e.g. $m^+$ $m^+$ $m^-$ $m^-$ $m^+$ $m^+$ $m^-$ $m^-$ (see Lamb, 1996, for diagrams, photos and theory, and Plate 18.2 here). If the mutation is for a character scorable in ascospores, such as for black versus mutant white pigmentation, the scoring of intact asci can be done directly down a microscope after displaying the asci from say a dissected perithecium. If the character is not scorable visually from ascospores, say being a requirement for some metabolite, then analysis is much more lengthy: ascospores have to be dissected out

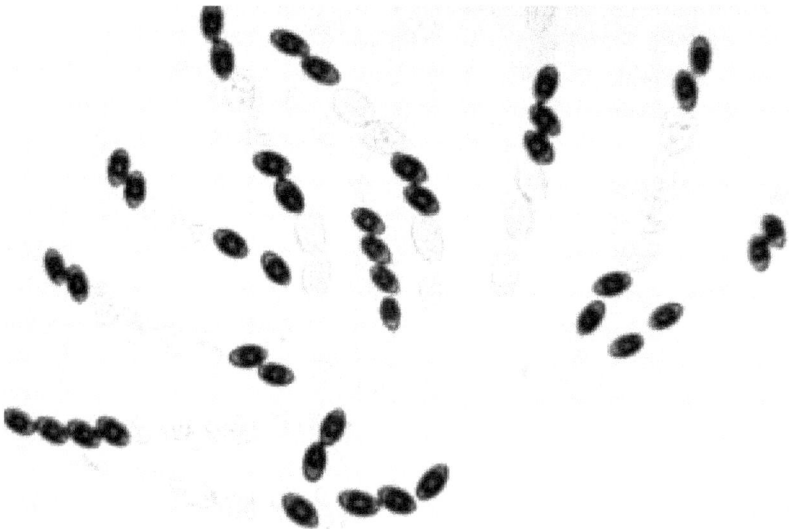

**Plate 18.2.  Ordered asci** from a monohybrid cross, + (black ascospores) × *hyaline* (colourless ascospores) in *Sordaria fimicola*, from a crossed perithecium. Clockwise from the bottom left, the segregation classes are 2, 1, 3, 6, 5, 2, 5, gene conversion to *hyaline* with postmeiotic segregation giving 3+:5$h$, 3, 6, showing first division (classes 1, 2) and second division (3, 5, 6) segregation classes.

of each ascus in their original order, then have to be germinated, then the resulting colonies need biochemical tests, e.g., for growth on minimal or complete medium, or on minimal medium with various supplements.

Centromeres can also be mapped from unordered octads, as was done extensively in yeast using the **centromere marker technique**. If one has a cross of two haploids, $A$, $B \times a$, $b$, then one can identify three **types of unordered tetrad**. A **parental ditype** (PD) has two types of spore, with a parental arrangement of alleles, which here would be two spores $A$, $B$ and two spores $a$, $b$. A **non-parental ditype** (NDP) has two types of spores, with a non-parental (recombinant) arrangement of alleles, two spores $A$, $b$ and two spores $a$, $B$. A **tetratype tetrad** (T) has four different types of spore, two parental and two recombinant, one spore each of $A$, $B$; $a$, $b$; $A$, $b$; $a$, $B$. For **non-syntenic loci**, one expects half the asci to be PD and half to be NPD when both loci show first division segregation, and all the asci to be T when one locus shows first division segregation and the other locus shows second division segregation. When both loci show second division segregation, one expects a ratio of one PD to one NPD to two T (see Fincham *et al.*, 1979, for the theory). For **linked loci**, one expects PDs (from no crossover between the loci) to be more frequent than twice the number of Ts (from a single crossover) which in turn will be more frequent than NPDs (from a four-strand double crossover). A simple test is that loci are linked if PDs are significantly more frequent than NPDs.

The **centromere marker test** is based on the fact just described, that to get a tetratype, one or both loci must show second division segregation. One chooses a marker (mutant) which is very closely linked to its centromere, so that it hardly ever shows second division segregation. In yeast, *trp1* (linkage group IV) is such a centromere marker, so one crosses it to an unlinked marker whose centromere distance one wishes to determine, say *ade1* (linkage group I), getting a tetratype frequency of ~24%. As *trp1* rarely shows second division segregation, and tetratypes show that at least one locus has segregated at the second division, the tetratype frequency will represent the second division segregation frequency of the other marker. One can therefore deduce that *ade1* has a second division segregation frequency of 24%, which we halve, using the above formula, to get a centromere distance of 12 cM for *ade1*.

If a fungus does not have ordered asci, one needs a method to identify centromere markers for the above technique. This can be done using **three unlinked markers**, in a method put into equation form by Whitehouse (1957). Let the loci be $a$, $b$ and $c$, respectively with second division fractions of $x$, $y$ and $z$, and first division fractions of $(1-x)$, $(1-y)$ and $(1-z)$. For unlinked loci, tetratype frequencies depend solely on the second division segregation frequencies of the two loci.

The **first way to get tetratypes** is with first division segregation at one locus and second division segregation at the other, which for loci $a$ and $b$ has a frequency of $x(1-y)+y(1-x)$. The **second way to get tetratypes** is with second division at both loci, when for unlinked loci half the tetrads are Ts, which will have a frequency of **1/2 $xy$**.

We therefore get **a tetratype frequency for loci $a$ and $b$** of

$$\mathbf{T_{ab}} = x(1-y)+y(1-x)+1/2\ xy,\text{ which simplifies to } x+y-3/2\ xy.$$

With $\mathrm{T_{ab}}$ obtained as data from a cross, one cannot solve for the two unknowns, $x$ and $y$. However, we have parallel equations for the other tetratype frequencies, determined from the crosses: $\mathbf{T_{ac}} = x+z-3/2\ xz$ and $\mathbf{T_{bc}} = y+z-3/2\ yz$.

With three values and three unknowns, one can solve for $x$, $y$ and $z$ by simultaneous equations. Alternatively, one can solve the following equation, $x = \frac{2}{3}\left(1 \pm \sqrt{\frac{4-6Tab-6Tac+9TabxTac}{4-6Tbc}}\right)$, and substitute the value of $x$ obtained from it in the simpler equations above. In practice it is easy to choose whether to use the positive or negative value of the square root, as one gives a sensible answer and the other gives an impossible answer, such as more than 100% second division segregation.

As all spores in an NPD are recombinant and half the spores are recombinant in a T ascus, we can calculate **recombination frequencies** from unordered tetrads from $\mathbf{RF\%} = \frac{1/2\,T+NPD}{PD+T+NPD}$.

Considerations of the effects of single, double and multiple crossovers give a maximum expected value of 50% recombination between two syntenic loci, however far apart they are, if there is no interference between crossovers (Sec. 8.2). With no interference there is a maximum expected value of 66.67% tetratypes for two loci, and of 66.67% second division segregation frequency for one locus, however far it is from its centromere.

There are two major types of **interference**, chromosome and chromatid interference. **Chromosome interference** (chiasma position interference) is concerned with whether one crossover in an interval tends to inhibit (positive chromosome interference), tends to promote (negative chromosome interference), or has no effect on the chance of a further crossover in that interval. In some organisms, e.g., *Drosophila*, there is localised positive interference, with a crossover tending to inhibit another one from occurring close by, but with no effect on crossovers further away, while the fungus *Aspergillus nidulans* generally has no chromosome interference. With no chromosome interference the chance of getting $x$ crossovers in an interval can be calculated from the Poisson distribution, $(m^x \cdot e^{-m})/x!$, where $m$ is the average number of crossovers in the interval. The main departure from this expectation is that almost all chromosomes have at least one crossover, which is probably essential for regular chromosome segregation in meiosis. Short chromosomes tend to have one crossover, with more in longer chromosomes, but there are seldom more than two or three visible chiasmata per chromosome arm. Foss *et al.* (1993) stated that localised positive chromosome interference extends over different physical distances in different organisms: $10^4$ kb in *Drosophila*, and in fungi, $10^3$ kb in *Neurospora* and $10^2$ kb in yeast. For the different findings of negative, positive and no interference in different fungi, see Lamb (1996).

**Chromatid interference** (strand interference) is concerned with whether different chromatids are involved at random in double crossovers. Random involvement would give a ratio of one two-strand double to two three-strand doubles to one four-strand double. This ratio has been found, approximately, showing no or little chromatid interference in *Neurospora*, *Aspergillus* and yeast (see Lamb, 1996). Fungi are particularly suited for studying chromatid interference because it requires tetrad analysis.

Section 8.2 mentioned **mapping functions** for allowing for undetected crossovers in multiple crossovers, e.g., from the two crossovers cancelling each other out in a two-strand double crossover. The mapping functions needed for tetrad mapping are different from those used in random-spore mapping: see Barratt *et al.* (1954). They depend on the amounts of chromosome and chromatid interference.

One further use of tetrad analysis in fungi is that in favourable circumstances it enables one to tell whether two loci giving about 50%

recombination are non-syntenic or are syntenic but far apart, which cannot be done just from the recombination frequency. For example, in *Neurospora crassa*, tetrad analysis of a cross of two mutants gave 7 PD, 11 NPD and 2 T. With the recombination being by chance a bit over 50% and with PDs approximately equal to NPDs, we deduce that the loci are unlinked and either non-syntenic or syntenic but far apart. The low tetratype frequency tells us that both loci usually segregate at the first division (because second division segregations often give tetratypes), and so both loci must be near their centromeres. They cannot both be near to the same centromere and simultaneously be far away from each other, so they must be close to different centromeres and be non-syntenic.

## 18.4. The parasexual cycle and parasexual mapping

A **parasexual cycle** is a series of events which make possible the recombination of non-syntenic loci and/or of syntenic loci, independently of the sexual cycle and meiosis. A parasexual cycle can recombine non-syntenic loci during **haploidisation**, by chance chromosome losses as a diploid nucleus becomes haploid by a series of non-disjunctions, and it can recombine syntenic loci in a diploid nucleus by **mitotic crossovers** or, less often, by mitotic gene conversion.

**Full parasexual cycles** have been demonstrated in a range of fungi, including ones of industrial importance and parasites of major crops, in Ascomycetes, Basidiomycetes and Fungi Imperfecti. They include *Penicillium chrysogenum* (penicillin production), *Aspergillus niger* (citric acid), *Cephalosporium* (the antibiotic cephalosporin C), *Coprinus lagopus, Verticillium* and *Fusarium*. *Saccharomyces cerevisiae* (yeast) and *Ustilago maydis* (maize smut) have **partial parasexual cycles**, with mitotic crossovers but no haploidisation by non-disjunction. Parasexual cycles are particularly important in nature and for the breeder where the fungus has no sexual cycle, as in *A. niger* and *P. chrysogenum*.

*Aspergillus nidulans* has a full sexual cycle and a full parasexual cycle, so is excellent for comparing the two cycles. It is the organism in which Roper, Käfer, Pontecorvo and others worked out the parasexual cycle during the 1950s. This account includes practical details of how it is used. The wild-type has green conidia ($y^+$, $w^+$), is haploid and prototrophic, growing on a defined minimal medium. To map the markers involved and to get recombinants, one can take two strains with complementary auxotrophic markers and with a different conidial colour marker each, for example, strain 1 with $a$, $b$, $c$, $d$, $e^+$, $f^+$, $y$, $w^+$, which is auxotrophic for $a$, $b$, $c$ and $d$, and has yellow conidia, and strain 2, $a^+$, $b^+$, $c^+$, $d^+$, $e$, $f$, $y^+$, $w$, which is auxotrophic for $e$ and $f$, and has white conidia (*white* is epistatic to *yellow*).

Conidia of the two strains are mixed in water over solid complete medium so that conidia can germinate and hyphae can fuse, giving a **strain 1 + strain 2 heterokaryon** which is prototrophic and can outgrow both homokaryons as the nutrients get depleted. The mycelial mat which grows overnight is allowed to drain and is washed free of nutrients. Small bits of mat are placed well apart on minimal medium, so that only the heterokaryon can grow. It produces a mixture of white and yellow conidia because the conidia are haploid and uninucleate, so each conidium is either of strain 1 or of strain 2 type. Within the growing 1 + 2 heterokaryon, one gets **rare chance nuclear fusions**, 1 + 1, 2 + 2, and 1 + 2. The first two types of diploid nuclei give only auxotrophs. The 1 + 2 diploid nuclei, however, give **prototrophic green conidia**. As the nuclear fusions are rare (being non-sexual), one plates millions of conidia from the heterokaryon on minimal medium, to select the very few 1 + 2 diploid nuclei in conidia.

These particular nuclei give green prototrophic colonies which are isolated as the diploids, and their conidial size is checked against that of the parental haploids: diploid conidia have about twice the volume of the haploid conidia. The diploids are then grown on complete medium, so that any segregants can grow, even if auxotrophic. Well-spaced inocula are used, and after about three days, colonies are checked visually for the presence of **sectors with white or yellow conidia**. The sectors will be large if they arose early in colony development or very small if they arose late. Conidia from the sectors are isolated onto complete medium, with subculturing to get rid of any wild-type green contaminants.

The purified segregants are checked for **conidial size**, to find which are diploids and which are haploids. They (and the two parental strains and the diploid as controls) are then inoculated onto particular positions on a master plate of complete medium (see Plate 18.3). After the growth of new conidia, they are then **replica-plated** (using a pin replicator or a damp

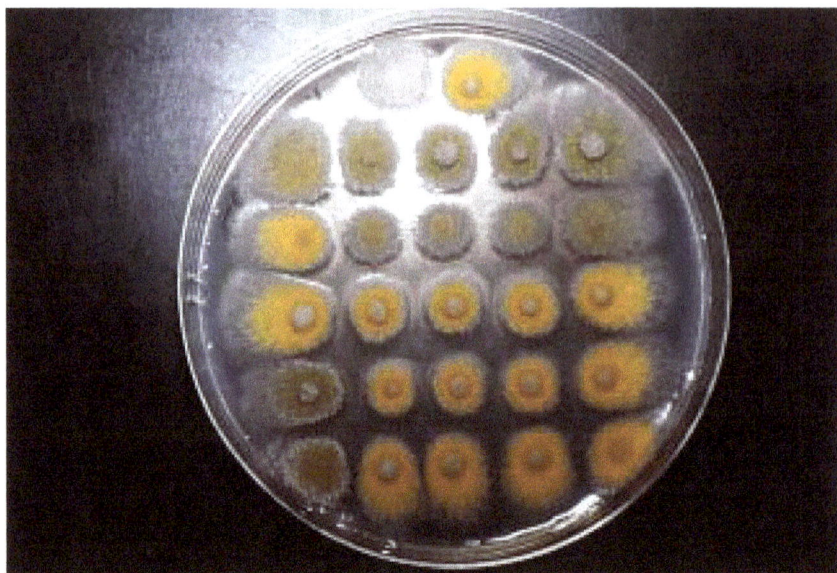

**Plate 18.3.** *Aspergillus nidulans* **colonies inoculated onto a master plate of complete medium for replica-plating** with a 27-pin pin-inoculator onto a series of diagnostic plates to determine which auxotrophic mutants are expressed by which colonies. This is an experiment on the parasexual cycle and mapping. The **two haploid parental complementing auxotrophs are at the top**, one with yellow conidia and one with white conidia. The green colonies include the prototrophic diploid obtained from them, and a haploid wild-type. The other yellow and white colonies are **diploid and haploid segregants** from the green diploid; a few are not pure, having more than one colour of conidia.

sterile velvet pad) onto a series of plates, e.g., minimal medium, complete medium, minimal supplemented with all requirements except that required by *a*, minimal supplemented with all requirements except that required by *b*, etc. That enables the genotypes of the yellow haploid segregants to be worked out, and that of the white haploids except for the *y* locus (because *white* is epistatic to *yellow*). In the diploid segregants, those loci which show a recessive phenotype have become homozygous. One cannot directly tell whether loci showing the dominant phenotype are still heterozygous or whether they have become homozygous dominant.

Suppose that maps for the two parental strains are as shown in Fig. 18.1, with *a, b, c, y* and *d* syntenic, and the other three loci non-syntenic with any other of these loci. The diploid between strains 1 and 2 is

Strain 1

$$\underset{o}{\rule{0pt}{0pt}}\underline{\text{a b c y d}} \qquad \underline{e}^{+} \qquad \underline{f}^{+} \qquad \underline{w}^{+}$$

Strain 2

$$\underset{o}{\rule{0pt}{0pt}}\underline{\overset{+\ +\ +\ +\ +}{\text{a b c y d}}} \qquad \underline{e} \qquad \underline{f} \qquad \underline{w}$$

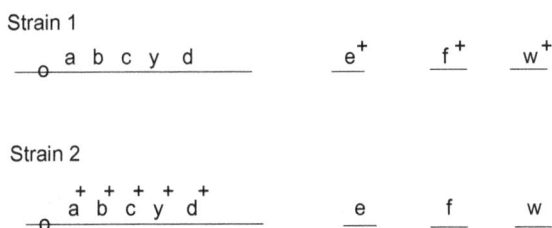

**Fig. 18.1.** **Possible maps for strains 1 and 2,** with syntenic and non-syntenic loci.

heterozygous for all these loci. The three causes of yellow or white segregants from that diploid are haploidisation, giving haploid segregants;
mitotic crossovers, giving diploid segregants, and non-disjunction diploids,
also giving diploid segregants. The commonest cause of colour segregants
is mitotic crossovers, then haploidisation, with non-disjunction diploids the
rarest.

In **haploidisation,** a diploid nucleus occasionally has chance mitotic
non-disjunction, giving one $2X + 1$ product and one $2X - 1$ product. The
latter monosomic nucleus is often less stable than the diploid nucleus and
tends to lose more chromosomes at later mitoses, only reaching stability
when it has the haploid chromosome number, with a complete genome, one
of each type of chromosome. The diploid only shows the dominant markers
in the phenotype, but in haploids any retained recessive markers show in the
phenotype, e.g., $y$ or $w$, which respectively give yellow or white conidia,
in contrast to the green conidia of the diploid. As whole chromosomes are
retained or lost, syntenic loci stay in their parental arrangements, with 0%
recombination but non-syntenic loci recombine freely, with 50% recombination. One can see from Fig. 18.1 that if the chromosome carrying $y$ is
retained and the homologue carrying $y^+$ is lost, then the haploid yellow
segregant will carry and express those markers on the $y$ chromosome, $a, b,$
$c, d,$ but it is a matter of chance whether this yellow haploid carries $e$ or $e^+$,
$f$ or $f^+$; because of the epistasis of white over yellow, it must carry $w^+$,
not $w$, or it would be a white segregant, mutant at both conidial colour loci.
Syntenic loci $y, a, b, c, d$ therefore show 0% recombination with any of
each other, but 50% recombination with non-syntenic loci $e/e^+$ and $f/f^+$;
we cannot map $y$ relative to $w$ because of the epistasis. Looking for such
a very big difference as 0% versus 50% is very easy. Using haploids is

therefore excellent for finding out whether loci are syntenic or not, but does not give any indication of distances between syntenic loci. Approximately 1 in 200 haploids has had a mitotic crossover before haploidisation, so in practice there is very rare recombination for syntenic loci in the haploid products, rather than none.

For distances, we use the **diploids from mitotic crossing-over**, for loci which the haploids have shown to be syntenic, which in this case are $y$, $a$, $b$, $c$, $d$. Mitotic crossovers occur by chance during mitosis, after replication, when each chromosome is present as two chromatids, with no zygotene pairing or synaptonemal complex holding non-sister homologues together, unlike in meiosis. Double crossovers are so rare in mitosis that they can be ignored.

Consider one crossover at the four-stand stage of mitosis in the $y$-bearing chromosomes, between $a$ and $b$, as in Fig. 18.2, with chromatids 1 and 3 going to one pole and chromatids 2 and 4 going to the other pole at anaphase. One daughter nucleus, the lower left-hand one in the figure, will express the alleles $a^+$, $b$, $c$, $y$, $d$, so it will be a yellow diploid not requiring the compound needed by $a$, but needing the compounds required by $b$, $c$ and $d$. The other nucleus, the lower right-hand one in the figure, will give

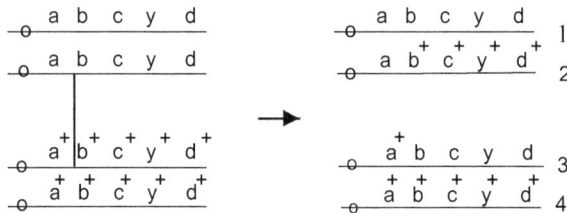

If chromatids 1 and 3 segregate to one pole and chromatids 2 and 4 go to the other pole, it then gives:

The alleles expressed in the phenotypes will be:
left set,
right set,

**Fig. 18.2.   The effect of a mitotic crossover on loci proximal and distal to the crossover.**

rise to a phenotypically normal colony which is not isolated as a colour segregant.

The important point to note is that, given the right centromere segregations, any loci **proximal** to the crossover (between the crossover and the centromere) stay heterozygous and express their dominant allele, like locus $a$ in Fig. 18.2. Loci **distal** to the crossover (further away from the centromere than the crossover) become homozygous and the recessive alleles can show in the phenotype, like $b$, $c$, $y$ and $d$ in the figure.

We could have crossovers in other places along the chromosome. One between the centromere and $a$ would make all the loci in that chromosome arm homozygous, so $a$ would be expressed as well as the other loci. A crossover between $c$ and $y$ would make only $y$ and $d$ homozygous, so the colony would express $a^+$, $b^+$, $c^+$, $y$, $d$. Suppose our yellow diploid segregants were phenotypically 20 $a, b, c, y, d$; 10 $a^+, b, c, y, d$; 30 $a^+, b^+, c, y, d$; 15 $a^+, b^+, c^+, y, d$, and no others for these syntenic loci. The fact that $d$ is always homozygous whenever $y$ is homozygous recessive shows that $d$ is **distal** to $y$ (or extremely closely linked and proximal to $y$). Alleles $a$, $b$ and $c$ must be proximal to $y$ because they can stay heterozygous, expressing the dominant allele, while $y$ and $d$ are homozygous and expressed. The example in Fig. 18.2 shows that $a$ must be proximal to $b$ and $c$, and the $a^+$, $b^+$, $c$, $y$, $d$ class shows that $b$ is proximal to $c$. We can therefore deduce the order of the syntenic loci within one arm of a chromosome: centromere, $a$, $b$, $c$, y and $d$. The frequencies of the different diploid yellow segregants give the relative distances, since the larger physical distances between loci are expected to have the most crossovers. We then get the **order and relative distances** of centromere - 20/75 - $a$ - 10/75 - $b$ - 30/75 - $c$ - 15/75 - $y$, and an unknown distance to $d$.

We only detect half the mitotic crossovers because the alternative segregation of centromeres, as in Fig. 18.2, of chromatid 1 with 4, and chromatid 2 with 3, results in daughter nuclei heterozygous at all loci, so they will not be detected as colour segregants.

In **non-disjunction diploids**, the diploid has a first non-disjunction, giving $2X + 1$ and $2X - 1$. Suppose the extra chromosome in the $2X + 1$ is the one carrying the $y$ allele, so the trisomic has $y^+$, $y$, $y$. It may later have a second non-disjunction in which the $y^+$ chromosome is lost, giving a stable diploid homozygous for $a$, $b$, $c$, $y$ and $d$, and expressing all those alleles in the phenotype as a yellow sector. It will differ from mitotic crossover diploids in that it will be homozygous for all markers in **both arms of the chromosome**, not just those distal to a crossover in one arm.

The analysis has been explained for sectors for conidial colour mutants. One can also use some other types of markers such as **acriflavin resistance** and sensitivity. On medium with the right concentration of acriflavin, the heterozygous diploid $acr^R/acr^S$ grows slowly. By eye, one can identify fast-growing sectors which have become $acr^R/acr^R$ by mitotic crossovers or $acr^R$ by haploidisation. If one needs to increase the frequency of haploidisation, say to analyse a diploid fully, one can use 60 $\mu$g/ml of paraflu-orophenylalanine to increase non-disjunction.

**Mitotic crossovers** are much rarer than meiotic crossovers, with roughly 0.1 to 0.3% of crossovers per chromosome arm per mitosis, compared with 100 to 300% of crossovers per arm in meiosis. In *Aspergillus nidulans*, we can compare the maps from meiotic and parasexual mapping. The order of loci is the same in both types of map, but the relative distances are sometimes different, showing different "hot spots" for recombination at meiosis and mitosis (see Pritchard, 1963, quoted in Rédei, 1982).

The **complete parasexual cycle** in *Aspergillus nidulans* can be summed up as a fusion of unlike haploid hyphae to give an n + n heterokaryon, which can reproduce by growth but not from conidia which only contain a haploid single nucleus; in the heterokaryon one gets rare nuclear fusion to give rare diploid nuclei, 2n, which may sometimes be fusions of the unlike nuclei; the diploid nuclei may get into uninucleate conidia to give diploid colonies capable of indefinite asexual reproduction by conidia and vegetative growth. Rare mitotic crossovers within diploid nuclei give recombination of syntenic loci, and haploidisation by progressive non-disjunctions gives haploid nuclei, which can get into conidia, be dispersed, and start normal homokaryotic haploid colonies, reproducing by growth and conidia.

One of the fascinating developments in the 1970s was the application by Pontecorvo and others of ideas and methods from fungal parasexual analysis to **human genetics**, especially the fusion of say a human skin fibroblast or a peripheral blood leukocyte with a tissue-culture adapted mouse cell to give a heterokaryon, then with nuclear fusion giving a synkaryon. Studies of gene and chromosome losses in proliferating hybrid cell populations greatly helped with mapping and with assigning linkage groups to visible chromosomes (Subsec. 8.6.2). Cell fusion methods in humans also helped with diagnosis of genetic defects in hereditary diseases through locus assignment of mutations by *cis/trans* tests (Subsec. 1.3.9) in tissue culture.

## 18.5. The induction and isolation of mutants, including auxotrophs

If a fungus has asexual spores such as conidia, they are the best material for mutation, although ascospores, basidiospores and hyphae can be used if necessary. If the conidium is haploid and uninucleate, as in *Penicillium* and *Aspergillus*, a typical procedure would be to **assess survival** at various dosages of mutagen and to choose a dose giving 99% kill, 1% survival. That is a good compromise between getting enough mutants and not making all survivors too weak to be of use. If the conidia are multinucleate, a higher percentage kill is used, e.g., 99.9% for *Neurospora crassa*. That fungus has three to eight haploid nuclei per conidium, so if one were isolating *his⁻* auxotrophs, one would want the mutagen to destroy all but one nucleus in a conidium, and to mutate that remaining nucleus, because a conidium with a mixture of mutant and wild-type nuclei has the wild-type phenotype, prototrophic, so could not be used for selecting *his⁻* mutants. One could grow the conidia on media giving low numbers of nuclei per conidium, or use microconidial mutants of *Neurospora* which usually have only one nucleus per conidium.

Some types of mutant can be identified by **direct selection**. One can select conidial colour mutants by eye, subculturing any growth with mutant colour. Using the naked eye or the microscope, one can identify non-conidiating mutants, morphological mutants with altered colony form or colour, growth rate, growth pattern (e.g., growth in waves, or clock mutants with periodic growth patterns), branching habit, failure to produce fruit bodies, etc. One can select directly for resistance to drugs or metabolic inhibitors by plating spores on medium containing the appropriate chemical at a suitable concentration, e.g., acriflavin.

Although **selection for auxotrophs** is not often needed directly for industrial strains, it is very useful for getting mutants for mapping and general genetic studies of a fungus. Auxotrophs are invaluable for many selective techniques, such as using two non-allelic adenine-requiring (red) mutations in repulsion crosses in yeast, so that the prototrophic zygotes (white) can be selected on minimal medium, to separate them from the two auxotrophic parental strains, with a colour confirmation.

For selecting auxotrophs, one can use the **filtration enrichment** method with filamentous fungi and highly flocculent yeasts. The suitably mutagenised fungus, say conidia or yeast cells, is put into liquid minimal medium and allowed to grow say for four to six hours, with mechanical agitation to reduce aggregation and possible fusions. Prototrophs can grow hyphae, or yeast cells can form flocculating groups, but auxotrophs cannot grow in the minimal medium, so remain as conidia or separate yeast cells. The suspension is poured through a filter with an appropriate pore size, so that auxotrophs can pass through, but grown prototrophs are retained. Approximately four to six periods of growth and filtration are used, so that nearly all prototrophs are removed. The number of auxotrophs should remain nearly constant, but become a much higher proportion of the remaining organisms, hence the term "filtration enrichment". Finally, the filtered suspension is plated out on complete medium, and all colonies growing are isolated into individual tubes of complete medium. They are then tested on minimal and complete medium to check which are auxotrophs, and then their particular requirements are investigated.

If one wants **particular types of auxotrophs**, e.g., histidine-requirers, the method can be made selective. For *his*⁻ mutants, one would replace the liquid minimal medium by liquid minimal medium supplemented with all the common vitamins, DNA bases and amino acids except histidine. All auxotrophs except *his*⁻ ones should therefore be able to grow and would be filtered out along with the prototrophs. The final plating out should be on minimal medium plus histidine, not complete medium.

Another method to get auxotrophs in fungi and bacteria is **killing enrichment**. It can be used with filamentous or single-cell fungi such as yeasts. For fungi, one uses an anti-fungal antibiotic such as nystatin, but for bacteria one uses an appropriate anti-bacterial antibiotic such as penicillin. The mutagenised organism is placed in liquid minimal medium plus the antibiotic. Suitable antibiotics kill growing cells much more quickly than non-growing cells, so prototrophs grow in the minimal medium and are rapidly killed, while auxotrophs do not grow, and they survive. After a period of growth, the suspension is centrifuged down; the organisms are washed free of the antibiotic and are plated at a number of dilutions on complete medium without antibiotic for the isolation of individual survivors, which can then be tested for auxotrophy. This method can also be made specific for particular types of auxotrophs by using liquid minimal medium plus all except one nutrient.

A third method is **double-mutant survival**, using the fact that certain double auxotrophs survive longer in minimal medium than do single auxotrophs, which tend to starve to death more quickly. For example in *Aspergillus nidulans*, biotin-requiring conidia were mutagenised with UV and were spread on minimal medium, and covered with additional solid minimal medium. After 96 hours at 37°, 99% of *bi⁻* conidia had died. A layer of complete medium was poured over the top, with nutrients diffusing down to the conidia. Colonies which grew were isolated and tested, with up to 60% having a second auxotrophic mutation, e.g., being *bi⁻*, *arg⁻*. In *Neurospora crassa*, an inositol-requiring mutation was used, with minimal medium plus sorbose to slow down growth.

Any auxotrophs obtained by any method can be tested on minimal medium with a series of different supplements to find what compound is required. A quicker method is to use a **combinatorial approach** developed by Holliday in 1957 for 36 common requirements (see Clowes and Hayes, 1968). This involves having 12 pools of nutrients, each pool containing six different nutrients, with each nutrient occurring in only two different pools. Pool 7 might contain adenine, biotin, phenyl alanine, alanine, arginine and leucine, while pool 1 might contain adenine, uracil, and four other compounds; pool 8 might contain serine, glycine and four other compounds, and pool 12 might contain uracil, valine and four other compounds.

A mutant which grew only on medium supplemented with pool 7 and on medium supplemented with pool 1 could then be identified as an adenine-requirer, because adenine only occurs in pools 7 and 1. A mutant growing on only one pool could have a double requirement, so that a uracil/valine double mutant could grow on pool 12, which contains both nutrients, but could not grow on pool 1, containing uracil but not valine, nor on pool 6, containing valine but not uracil. Growth on more than two pools indicates an alternative requirement. For example, a mutant requiring either serine or glycine could grow on pools 3, 6 or 8.

In species such as *Aspergillus nidulans* which can use inorganic nitrogen, **chlorate resistance** can be used to select mutants defective in nitrate reductase, and fluoroacetate resistance selects for mutants that lack acetyl CoA synthetase activity, when they can no longer use acetate as sole carbon source. **Two-way selection** is then possible, as revertants which have regained the enzyme function can be selected on acetate. Selection for **amino-acid overproducers** can be done by growth in the presence of toxic amino-acid analogues, where increased amounts of the amino acid dilute the toxic effects of the analogue. The next section contains information on obtaining other types of mutant, especially industrially useful overproducers of extracellular enzymes and antibiotics.

## 18.6. Obtaining improved strains for industry

### 18.6.1. *Aims and methods*

**The aims** are to increase the yield and quality of desirable products and the rate of production, to decrease the yield of unwanted products, to reduce the costs of nutrients, equipment, fuel, labour, transport and purification, and to increase the reliability of the yield. Depending on the fungus, the main techniques to use are selection, mutation (spontaneous or induced), recombination and segregation through the sexual or parasexual cycles, genetic engineering, or trying different ploidies, chromosome aberrations, heterokaryons to utilise complementation, hybrid vigour in diploids or polyploids, and even mixtures of genotypes or of species. For details of fungal genetic engineering, see Sec. 14.9, and for fungal biotechnology, see Anke (1996), Peberdy *et al.* (1991), Bennett and Lasure (1991), Bos (1996), Attfield and Bell (2003), and Nevalainen and Te'o (2003).

**Selection** only works if there is heritable variation present, which can come from mutation, recombination, immigration, or genetic engineering. Selection procedures can be very simple, such as selecting wine yeasts with higher alcohol tolerance from setting the yeasts to grow at increasing alcohol concentrations. Selection for high yields of diffusible excreted products, such as amino acids, extracellular enzymes or antibiotics, can be done very easily by using the **overlay method**. Suppose one wants overproducers of tryptophan from a fungus. In a Petri dish one puts fungal minimal medium, then the mutagenised fungus at low concentration, followed by an overlay of *tryp⁻* bacteria in a bacterial minimal medium. After a period of growth at a temperature suiting both organisms, one examines the size of the bacterial colony over each fungal colony. Fungal colonies which overproduce tryptophan will have larger bacterial colonies above them than other fungal colonies, and under-producers will have smaller bacterial colonies.

For **antibiotic production**, e.g., penicillin by a fungus, one would have fungal medium (not minimal), then over that the mutagenised fungus, then a penicillin-sensitive bacterium in bacterial medium. The size of the zone of killing of the bacterium would indicate the amount of penicillin excreted and diffusing through the bacterial medium.

For **extracellular enzymes**, one would have fungal medium, the mutagenised fungus, then a layer containing the enzyme's substrate. After growth, one would look for breakdown of the substrate or for the presence of the breakdown products. This could be used for proteases, cellulases, amylases, lipases, pectinases, tannases, etc. For enzymes breaking down starch or cellulose, one could look for visible zones of clearing of the starch or cellulose powder, or one could stain for starch with dilute iodine in potassium iodide solution.

For bacterial antibiotics, one can use **indirect selection**, and the method could be applied in fungi. Where the antibiotic has amino acid precursors, one can select for overproduction of those amino acids, and this often results in higher antibiotic titres; e.g., penicillin has cysteine and valine precursors (Fig. 18.4). One uses a chemical analogue of the amino acid to act as a competitive inhibitor of the normal amino acid in protein synthesis. By selecting for resistance to the analogue, one can often get overproducers of the normal amino acid, although some mutations might have other modes of action, such as permease changes affecting access of the analogue to the cells.

**Types of mutation** which one might get or want include the following. Mutations in **structural genes** coding for enzymes might give no enzyme or an enzyme with altered catalytic properties, giving faster or slower rates of action or altered substrate specificity or different end products, or altered regulation. Many biochemical pathways in synthetic metabolism have end-product-inhibition of the first enzyme by the final product, as shown in Fig. 18.3 where the end-product might be an amino acid such as alanine. If a mutation made enzyme 1 no longer sensitive to allosteric binding and

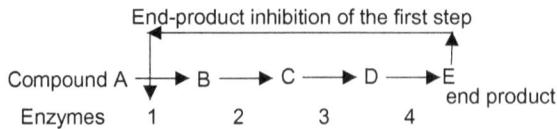

**Fig. 18.3.   End-product inhibition of the first enzyme in an anabolic pathway**.

end-product inhibition by alanine, one could get overproduction of alanine because all enzymes in pathway could work even in the presence of a high concentration of the end-product.

One could get mutations in **regulator genes** controlling transcription, if there were bacterial-type repressor proteins and operator genes controlling transcription, changing inducible enzyme production to constitutive production. Stronger promoters or ones with altered regulation could be induced by mutation or introduced by genetic engineering (Chap. 14). Mutations in **permease genes** could affect the rate at which substrates or products got into or out of cells, especially where active transport is involved.

Mutation can be nuclear or cytoplasmic, where nuclear genes show more reliable segregation. Mitochondrial DNA mutations can cause "petites", with slow growth (Plate 18.1) and an inability to respire aerobically. Such mutations occur spontaneously in beer yeasts and can give off-flavours. Mutations can also be in plasmids, which are used in fungal genetic engineering (Sec. 14.9) and in bacterial starter cultures for dairy products.

Mutations can result in metabolic gains in ability, or more often, in losses of ability. An example of a **desirable loss mutation** occurred in *Penicillium chrysogenum* strain Q176. A UV-induced mutation gave strain BL3-D10 which lacks chrysogenin, the wild-type's unwanted yellow pigment which has to be purified away in penicillin production, so the mutation reduced purification costs.

The most common fungal **mutagen** is UV light (Casselton and Zolan, 2002), which was used to improve itaconic acid production from *Aspergillus terreus*. Chemical mutagens are also in frequent use. N-methyl-N-nitrosoguanidine was used to improve $\beta$-galactosidase production by 28% in *A. niger*, and then UV gave a further 118% increase.

**Recombination** through the parasexual cycle has been used to increase glucoamylase production in *A. niger*, and the secretion of kojic acid in

*A. oryzae* and *A. sojae*. **Protoplast fusion** has been used to make interspecific hybrids in *Aspergillus*, recombining the fast growth of one strain with the better production of glucosamylase of another strain. As well as changing fungal strains, **culture conditions** also need optimising for each new strain, e.g., the initial temperature, sugar concentration and pH for citric acid production by *A. niger.*

Protoplasts are often subjected to **transformation** for **genetic engineering**, with polyethylene glycol to aid DNA uptake. Approximately 100 transformants per $\mu$g of DNA can be obtained, followed by selection for transformants, e.g., using prototrophic markers, drug- or antibiotic-resistance. Genetic engineering can also use *Agrobacterium tumefaciens* Ti-DNA to transform filamentous fungi without first making protoplasts (Casselton and Zoltan, 2002). For species without well-developed genetics, dominant mutations giving resistance to the fungicide benomyl can be incorporated into the vector for transformation, to select successful transformants on medium containing benomyl. The main methods to improve **protein yields** through genetic engineering include using strong promoters, overexpression of transcriptional regulators, raising gene copy number and reducing proteolytic activity. For example, by using a stronger promoter and a higher gene dosage, the yield of the protein sweetener thaumatin was more than doubled in *Aspergillus awamori*.

**Gene identification** in fungi can be done by random insertional mutagenesis with tagged DNA, then using the inserted tag to recover adjacent DNA from the mutant of known phenotype. Restriction enzyme-mediated integration with a promoter on the DNA tag allows the identification of genes activated by the promoter on insertion near to them.

## 18.6.2. *Different aims in different yeasts*

**Yeast** is an example of a very valuable industrial fungus. The most common yeast used is *Saccharomyces cerevisiae*, as a vegetative diploid, which can therefore show hybrid vigour. Most strains are sexually fertile and can form ascospores while others cannot. One can use the sexual cycle for getting recombinants, and can select recessives using haploid cells. Yeast species sometimes hybridise naturally, e.g., *S. cerevisiae* and *S. bayanus* (Masneuf *et al.*, 1998). The breeding aims are quite different in different industries.

For **baker's yeast** one wants a rapid production of $CO_2$ to get dough to rise, a good flavour and aroma, alcohol tolerance only up to 5%, and an ability to work at 37° during leavening and at even higher temperatures during the early stages of baking. Bread yeasts can grow up to 42° and ferment up to 55°.

**Wine yeasts** must produce 8 to 15% alcohol, depending on the type of wine; they must work at pH 2.9 to 3.6, give a good flavour and bouquet, and must not autolyse rapidly as that can give off-flavours. They must resist the $SO_2$ used to control bacteria and wild yeasts, and should settle out well at the end of fermentation. Many wineries used carefully selected cultivated yeasts, and those which use "natural" yeasts are usually using ones which have been through the winery many times, with "**natural oenological selection**" for the right properties. One very widely used wine yeast is Prisse de Mousse, an Institute Pasteur *bayanus* race, ideal for crisp, dry wines, including sparkling and still white wines and fruit wines. It has low foaming, is excellent for barrel fermentation, has good flocculation and ferments well at low temperatures (13–15°). It can also be used for high-alcohol beers such as barley wine.

**Genetic engineering** has been used to transfer the alpha-L-rhamnosidase gene from *Aspergillus aculeatus* to a wine yeast to increase production of the aromatic monoterpene linalool (a bouquet compound) from Muscat grape juice, when combined with another transgenic wine yeast strain expressing a *Candida molischiana* β-D-glucosidase gene (Manzanares *et al.*, 2003). The authors comment that "Problems related to the social acceptance of genetically modified foods, mainly in the European Union, make the use of such products an alternative for the future but not for the present".

**Beer yeasts**: for bitters and stouts, one uses a top-fermenting yeast, *S. cerevisiae*, but a bottom-fermenting yeast, *S. carlsbergensis*, is used for lagers. These two species confer quite different flavours. Beer yeasts are selected for flavour and aroma characteristics, flocculence and speed of fermentation. They work at higher pHs, 4.5–5.5, than do wine yeasts and do not need such high alcohol tolerance or $SO_2$ tolerance. They must resist hop oils. Different beer yeasts are needed for different types of beer, and these descriptions modified from a WYEAST Laboratories (Mt. Hood, Oregon, USA) Brewer's Choice™ leaflet illustrate that. Note the ability of the lager yeast to ferment at the low temperatures traditionally used for producing and maturing lagers.

**Ale yeasts**: (i), German ale yeast. Ferments dry and crisp, leaving a complex but mild flavour. Produces an extremely rocky head and ferments well down to 13°. Flocculation — low. (ii), London ale yeast. Rich, minerally profile, bold and crisp, with some diacetyl production. Flocculation — medium. 16–22°. (iii), London ESB ale yeast. Highly flocculent top-fermenting strain with rich, malty character and balanced fruitiness. This strain is so flocculent that additional aeration and agitation are needed. An excellent strain for cask-conditioned ales. 18–22°. **Lager yeast**: (i), Danish lager yeast II. Clean dry flavour profile often used in aggressively hopped pilsner. Clean, very mild flavour, slight sulphur production, dry finish. Flocculation low. 8–13°.

Beer yeast **flocculence** is controlled by three main loci, with alleles *Flo-1*, *Flo-2* (these are 8 cM apart) and *flo-3* (unlinked to the other two) increasing flocculence. Of these three alleles, one recessive and two dominant, only *Flo-1* is common in commercial strains. Even without knowing genetics, brewers have controlled flocculence for centuries by the time in fermentation when they skim off yeast (from the top, for top-fermenters, from the bottom for lager yeasts) to use in the next fermentation. The most flocculent yeasts come to the surface early and less flocculent ones rise later, for ale yeasts. Beer yeasts are often a mixture of strains, especially if a brewery keeps reusing yeasts for many successive fermentations. In a batch fermentation with a top-fermenting ale yeast, taking about seven days to complete, the brewers might skim off for reuse the head formed after 48 hours for use in the next fermentation as that might have the right mixture of strains for the desired degree of flocculence. As beer yeasts multiply five to ten times during a fermentation, one fifth to one tenth needs to be retained for reuse.

Some breweries reuse yeasts indefinitely, but most keep lab stocks of highly selected yeasts which they grow up, test, use sequentially for four to six fermentations, then they sell off the surplus yeast (e.g., for food after de-bittering) because mutations tend to build up and give off-flavours. Adnam's Brewery in Suffolk, England, uses a mixture of four different yeast strains in their bitter, as they find that this gives better results than any single yeast. They have had the same yeasts since 1943 and need to reisolate and retest them periodically to get rid of contaminating wild yeasts.

**Adaptations to change:** Breweries used to use batch fermentations in open-topped rectangular vats with yeasts which formed creamy heads that could be skimmed off as required. **Modern breweries** now usually use large enclosed cylindro-conical stainless steel fermenters from which skimming would be very difficult and in which the heads would take up a lot of space. It was therefore necessary to select **non-head-forming strains** of *S. cerevisiae* with the same flavour profile as the head-formers. For ales, they could not use bottom-fermenting lager yeasts as their flavour profiles are quite different. Many breweries now use high-gravity brewing to make optimum use of equipment. They use double-strength wort to make double-strength beer, then dilute it to make different weaker beers for sale. This obviously requires yeasts with more osmotic tolerance to the stronger wort, and an ability to produce more alcohol (Campbell, 2000).

For a **food yeast**, one wants maximum production from very cheap media, good protein and vitamin content, and no toxins or off-flavours. For details of the preparation of **microbial starters** for the food industry, including mixtures of bacteria for dairy-products, and the use of plasmids in bacteria, see Frazier and Westhoff (1988). For details of fermented beverages, see Lea and Piggott (1995).

If one has a highly selected yeast, sending it through the sexual cycle to get better recombinants often loses good gene combinations, so mutation or genetic engineering (e.g., using the $2\,\mu$m plasmid, Chap. 14) is often preferred.

### 18.6.3. *Improving baker's yeast*

**Baker's yeast**, *Saccharomyces cerevisiae*, has well understood genetics, with many classical and molecular techniques available for strain improvement, but the use of molecular techniques has been somewhat hampered by public opposition to genetically modified organisms. In their comprehensive review of improving baker's yeast, Attfield and Bell (2003) state that *"Genetic modification of baker's yeast can be achieved by classical or molecular procedures, or a combination of both approaches. However, given the general negativity surrounding GMO's, we contend that classical strategies remain the most practical approach to developing strains for commercial applications. Nevertheless, genomics and molecular techniques* *remain important for determining key genes, pathways and associated phys-iological functions that need to be enhanced in novel strains of baker's yeast. ... [There is] the need to avoid the "Genetically Modified" tag in many food markets".*

Baker's yeasts are diploid, aneuploid or polyploid, with high levels of genomic diversity, and often having chromosome-length polymorphisms. There is much variation in the number and type of **Ty transposable elements**; for example, lab strain S288c (the one with the sequenced genome) has 33 copies of Ty1, 12 of Ty2, 2 of Ty3, 3 of Ty4, and none of Ty5 (quoted by Attfield and Bell, 2003). Changes in industrial yeast strains occur independently of the sexual cycle through transposition of genetic elements such as Ty, and by mitotic crossing over and mitotic gene conversion, plus some translocations.

The diploid or polyploid nature of industrial strains means that attempts to improve them by **mutation** by using standard mutagens on vegetative cells can work if the mutants are dominant or show additive action, but recessive mutations will not show unless made homozygous. Mutating haploid cells would allow recessive mutations to show, but industrial strains do not always produce haploids through the sexual cycle (see below). Even if they do, the highly heterozygous nature of most industrial strains means that many desirable characteristics could be lost by going through the sexual cycle and mating to reconstitute diploids after mutation and selection among haploids. Mutation has been used to speed up maltose utilisation in industrial strains. Mutation has also been used to get yeasts with very low activity at refrigeration temperatures but which have normal activity when raised above 14°, so they can be used with a longer storage life in refrigerated compressed blocks or suspensions, e.g., for bread, wine or beer.

A classic method of **improving yeast through recombination in the sexual cycle** is to sporulate diploid or polyploid strains, select amongst the haploid progeny, and cross haploids of opposite mating type to make new diploids, some of which may show hybrid vigour. In lab strains that is easy. Industrial yeasts are diploid or polyploid, often aneuploid, frequently giving non-Mendelian ratios, and often having genetic abnormalities. Wine yeasts typically give sporulation (production of asci and ascospores) frequencies of 0 to 80%, while brewer's yeast strains often fail to sporulate or have very low sporulation frequencies. Baker's yeast strains typically have 0% to over

50% sporulation. Because of their complex ploidies, industrial yeast strains often produce fewer than four ascospores per ascus, with spore viabilities varying from 0 to 95% (Johnston *et al.*, 2000) in both naturally occurring and commercial cultivated wine yeasts. Baker's yeast colonies from ascospores sometimes have a clear mating type, sometimes an unstable mating type, and often no mating type, being *a/α* **diploids** from a tetraploid parent.

In lab crosses to get recombinants via diploids, one often uses auxotrophy, especially between complementing adenine-requiring red mutant haploids, to select white prototrophic diploids, as in Plate 1.6. One can also induce different antibiotic markers in the two haploids, then select diploids on medium with both antibiotics, but neither method is easy to use directly on haploids from industrial strains where the auxotrophy and antibiotic genes are unwanted. Flow-cell cytometry can also be used in which say one haploid is dyed green and the other is dyed orange with fluorescent dyes, and only dual-fluorescing cells are selected by the machine. Without using any of those methods for identifying diploids from accompanying haploids in the mating mix, one can often simply streak for colonies of single-cell origin and then recognise diploids by their larger colony size and cell size.

**Commercial baker's yeasts** are grown aerobically in batch culture on low-cost substrates such as sugar beet or cane molasses. They are harvested, processed, and supplied as a cream (concentrated suspension), compressed blocks, or as an active dried yeast (as used in home baking). **Different types of bread-making** may need different yeast strains. In sourdough bread, lactic acid and other bacteria give lactic acid, acetic acid and other organic acids, so the yeasts have to work at unusually low pHs. In some procedures, yeasts need to work quickly, fermenting and leavening (raising by $CO_2$ production) dough in a few minutes; in others, fermentation for several hours is normal. Bread doughs are usually made by mixing all ingredients together but sponges may be made in stages, with the yeast exposed to different flour and water concentrations. **Ready-mixed doughs** are now often frozen for weeks or months before being thawed and baked, so yeasts must tolerate those conditions and work rapidly later. Bread doughs may contain no free sugar (needing the yeast to adapt to utilising maltose), while sweetened doughs may have 30 kg sugar added per 100 kg flour, causing osmotic stresses. As mentioned earlier, resistance to anti-mould preservatives such as sorbic acid, may be needed.

Some **qualities selected in particular bread yeasts** include: better freeze-thaw tolerance for frozen doughs; improved tolerance to drying and rehydration in dried yeasts, as these tend to lose activity with time especially if stored above 18°; tolerance to preservatives used in foods; improved $CO_2$ production. Selection has been successful for the following properties (Attfield and Bell, 2003): melibiose utilisation for full benefit from raffinose in beet molasses; improved maltose utilisation in plain doughs; higher fermentation rate; improved osmotic stress adaptation in sugar doughs and frozen doughs; improved flavour profile. Shelf-life and drying and freezing tolerance are strongly influenced by intracellular levels of glycogen and trehalose, which have multigenic control and are affected by environmental conditions (see Francois and Parrou, 2001).

There is a vast amount of work in progress, studying gene function in yeast through deletions in each of the 6,100 known genes, analysis of mRNA transcripts, knock-out cassettes, protein studies, transposon insertions, gene fusions, etc. (Attfield and Bell, 2003). Some systems are well understood. For example, in plain (unsugared) doughs, maltase activity is most important for liberating fermentable sugars from starchy food reserves. There are five unlinked maltose loci, and at least one of these *MAL* genes needs to be active. Each locus codes for a maltose permease, a maltase ($\alpha$-glucosidase), and a positive regulatory protein. These loci are not constitutive but are induced by maltose and usually repressed by glucose. Flour contains low levels of glucose and fructose (1 to 2% w/w of free sugars), so in unsugared doughs, rapid maltose utilisation is essential to maintain $CO_2$ production once they are used up; the better industrial bread yeasts have been selected for higher expression of maltase and maltose permease, and for their production not to be repressed by low levels of glucose and fructose. Invertase works very fast on any free sucrose, and the fermentation enzymes are constitutive, not inducible.

With molecular **genetic engineering techniques**, desired gene combinations can be achieved with more certainty than through sexual recombination or random mutation, but Attfield and Bell (2003) state that to their knowledge, no recombinant engineered yeast strains are currently used or sold in baking. Precise genetic changes can be made and expression levels can be changed by altering or adding promoters. Unfortunately for that approach, many industrially relevant characters are multigenic and incompletely understood, and in diploids or polyploids, all copies of a gene need altering unless expression of an engineered change is dominant.

**Vectors** available include autonomous and integrating plasmids, plasmids with centromeric sequences giving stability through mitosis, and yeast artificial chromosomes with a very high gene-carrying capacity. Transformation with engineered DNA can be done by biolistic methods, electroporation, chemical treatments, e.g., with lithium salts, or with using protoplasts. Detection of gene uptake by looking for uptake of a wild-type gene into an auxotroph is difficult as the industrial strains are not auxotrophic, so uptake of an antibiotic resistance gene (e.g., to geneticin, canavanine, hygromycin) is simpler, although that gene should be removed later by recombination. If one is introducing the ability to utilise something like a new carbohydrate source, one can select for that gene uptake on a medium with that as the sole carbon source, for example. Homologous recombination can be used to introduce novel gene activity into yeasts, or to modify or delete genes, as its requirement for sequence homology ensures accurate targeting of the DNA introduced by transformation to the desired gene on the yeast chromosome.

## 18.6.4. *Improving enzymes in industrial fungi*

Unlike the situation for yeasts used in the food and drinks industries, users of **industrial enzymes** do not mind genetically manipulated organisms' products, which they usually are now. **Filamentous fungi "Generally Regarded As Safe"**, which are massively used for producing excreted industrial enzymes, are selected strains of *Aspergillus niger* var. *awamori*, *Aspergillus oryzae*, *Trichoderma reesei*, *Rhizomucor miehi* and *Humicola lanuginosa*. Nealainen and Te'o (2003) give the world market for industrial enzymes as US$ 2 billion in 2000, and rising rapidly. The major uses are for laundry detergents, wood pulp, paper and textile processes, food and drink process such as pectolases for clarifying fruit juices and wine musts, amylases for starch breakdown, enzymes for animal feed production, with medical and diagnostic enzymes being most important by cash value. These fungi can generally be grown on very cheap undefined substrates, often in submerged culture in volumes up to 500,000 cubic metres. Most of these enzymes are extracellular, making for relatively easy processing to obtain the desired product in sufficiently pure form.

The various **aspergilli** produce a large and important diversity of industrial enzymes, including glucoamylase for the saccharification of bread and soft drinks, naringase for debittering of citrus juices and pentosanase for dehazing wines. They also produce about 350,000 tonnes a year of citric acid.

Products are generally classified as **homologous** if intrinsic to the fungus or **heterologous** if specified by an engineered transgene which has been transformed in from another organism. One cannot just put a gene into a fungus and expect good results. A powerful appropriate promoter is required, preferably easily inducible, and appropriate gene regulation, while transcription factors, chaperone molecules and protein foldase genes may be required. Correct processing of mRNA is important, as is protein stability, correct cleavage, and the correct direction of the protein through the cell for excretion.

Two important **fungal promoters** are the cellobiohydrolase 1 promoter (*cbh1*) from *Trichoderma reesei*, and the glucoamylase promoter (*glaA*) from *Aspergillus niger* var. *awamori*. They both suffer from catabolite repression from the CRE protein if glucose is present. The strain RUT-C30 of *Trichoderma reesei*, which is an important over-producer of cellulolytic enzymes, was selected by classic mutagenesis and screening. It was found to be mutated in the *cre 1* gene, relieving that repression. In unmutated promoters, the difference in expression levels can vary more than one thousand-fold depending on induction versus glucose repression. One successful method of greatly improving the production of heterologous products is in-frame fusion with a homologous gene, giving better mRNA stability, better translocation of the protein in the secretory pathway, and protection of the heterologous part from enzymatic degradation. This has been especially effective for heterologous genes from non-fungal organisms, with 5 to 1,000-fold improvements (see Nealainen and Te'o, 2003).

**Deleting unwanted genes** can also lead to big yield improvements, for example, deleting the aspergillopepsin protease gene in *Aspergillus* strains used for heterologous proteins which would otherwise be degraded by that enzyme. Both intracellular and extracellular fungal proteases may attack the desired protein product; mutation plus screening can identify strains with reduced production of fungal proteases such as aspartyl protease. The correct **post-translational glycosylation** of proteins can also be crucial to function, and *Aspergillus* and *Trichoderma* differ in glycosylation properties. **Protein processing** before secretion, especially by **cleavage**, is needed by various lipases and proteases. One approach is to engineer a cleavage site recognised by fungal cleavage systems such as Kex-2p (often active in the trans-Golgi in fungi) into fusion proteins, to get cleavage at the correct site, such as between the fungal carrier protein and the foreign protein, or in the correct part of the foreign pre-protein.

In an industrial environment, enzymes from fungi may have to work in totally different conditions from the natural ones, for example, at different pHs, temperatures, osmotic conditions, and in the presence of detergents, proteases, unusual substrates, oxygen and heavy metals. If the enzyme structure and action are well characterised, one can use *in vitro* **protein-engineering techniques** of **site-directed mutagenesis** to alter thermal stability, substrate specificity, rate of action, etc. For example, a protein-engineered Lipolase (a lipase gene from *Humicola lanuginosa* engineered into *A. oryzae*) is used in detergents in household and industrial cleaners to remove fats; in leather processing and the food industry it is used to interesterify fats and oils to produce modified acylglycerols. Its washing performance was enhanced by mutation to put mainly hydrophobic or positively charged amino acids in place of negatively charged residues in the lipid contact zone. If there is insufficient information for protein engineering, random mutation and screening are often very useful. For example, production of a thermophilic proteinase engineered into *Trichoderma reesei* from a thermophile *Thermus* species was improved by mutating the transformants with UV and screening on plates with a protease-substrate (skimmed milk) at 85°.

Archer (2000) lists fungally-derived **recombinant enzymes approved for food use**: catalase, cellulase, chymosin, $\alpha$-galactosidase, $\beta$-glucanase, glucose oxidase, lipase, phytase, protease and xylanase. **Chymosin**, an acid protease, cleaves kappa-casein and is widely used in the cheese industry for coagulating milk. It is produced in precursor form in the fourth stomach of a young calf, and was provided in rennet from dried stomachs of calves slaughtered for veal. Supplies became limited as veal production lessened, and the chymosin gene from cows was cloned into fungi and bacteria, producing large quantities of more stable chymosin more cheaply. The US Food and Drug Administration approved the use of chymosin from genetically modified microorganisms in 1990. It is often produced in the dairy yeast, *Kluyveromyces lactis* (e.g., Maxiren™), *Mucor* spp., *Aspergillus niger* var. *awamori*, *A. nidulans* or various bacteria. Cheeses labelled as "Suitable for vegetarians" are made with genetically engineered chymosin; the labels do not mention genetic modification.

### 18.6.5. *Penicillin production*

Another extremely valuable industrial fungus is *Penicillium chrysogenum* for the production of penicillin, an antibiotic effective against gram positive bacteria. Sir Alexander Fleming's original strain (in 1928) produced 2 units/ml, compared with more than 6,000 units/ml for modern strains. The antibiotic was in use during the Second World War, after the work of Lord Florey, Chain, Abraham and others.

A **1944 survey** of 241 wild isolates showed penicillin yields varying from zero to 100+ units/ml in surface culture and from one to 80 in submerged culture. NRRL 1951 B25, a culture from a single haploid uninucleate conidium from a rotten cantaloupe melon, gave yields of 100–200 units/ml in submerged culture and is the ancestor of all industrial strains which therefore had no genetic variability. Yields were increased purely by mutations, including spontaneous ones, and ones induced by X rays, UV, nitrogen mustard, NMG, diepoxybutane, and other agents. Interestingly, the best dose of UV for improving penicillin yield was one giving 25–30% survivors, rather than a higher dose. The mutation giving a loss of chrysogenin pigment was mentioned in Subsec. 18.6.1. Selection was made for adaptation to particular culture conditions and for the production of **different penicillins**, such as I, F, G, K or X, which have different effectiveness against different bacteria, e.g., penicillin X is less effective against *Staphylococcus aureus* than against *Bacillus subtilis*.

The **genetics of penicillin production** were studied using the parasexual cycle, e.g., by Ball, Normansell and others in the 1970s. It was often difficult to obtain heterokaryons, diploids were often unstable, and some high-yielding strains had poor conidiation. Ball found from haploidisation that there were three linkage groups, while wild-type *Penicillium notatum*

(*P. chrysogenum* is in this group) has n = 5. The many mutagenic treatments may have caused translocations. High yield was generally recessive to low yield in diploids. Many genes not directly connected with penicillin acted as modifiers of penicillin yield, e.g., various auxotrophs and conidial colour mutations.

Normansell and others studied the genetics by using *npe* **mutations**, giving **no penicillin** or very low levels. Complementation *cis/trans* tests, using protoplast fusion, showed that the 12 mutants were in five complementation groups, A to E, suggesting that there were five main loci for penicillin production. Work was also done on penicillin production in *Aspergillus nidulans* because it was easier technically, and where yields were increased by recombination as well as by repeated mutation. 52 different wild-type strains had yields of 0 to 14 units per ml.

Figure 18.4 shows a modern **pathway for penicillin synthesis**, with identical steps in *Aspergillus nidulans* and *Penicillium chrysogenum*, and with the first steps also in common to the production of penicillin N

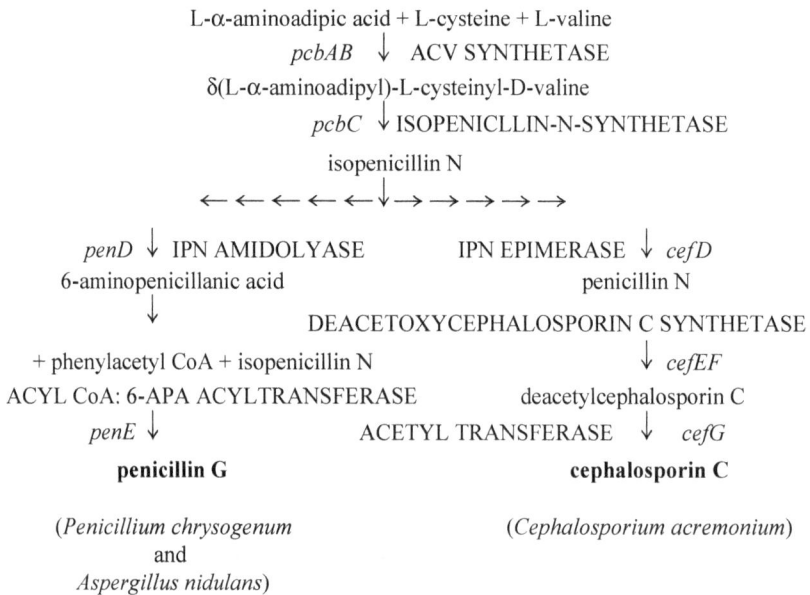

L-α-aminoadipic acid + L-cysteine + L-valine

*pcbAB*  ↓   ACV SYNTHETASE

δ(L-α-aminoadipyl)-L-cysteinyl-D-valine

*pcbC* ↓ ISOPENICLLIN-N-SYNTHETASE

isopenicillin N

← ← ← ← ←↓→ → → → →

*penD* ↓ IPN AMIDOLYASE          IPN EPIMERASE ↓ *cefD*

6-aminopenicillanic acid                    penicillin N

↓

+ phenylacetyl CoA + isopenicillin N

DEACETOXYCEPHALOSPORIN C SYNTHETASE

ACYL CoA: 6-APA ACYLTRANSFERASE          ↓ *cefEF*

*penE* ↓                    deacetylcephalosporin C

ACETYL TRANSFERASE ↓ *cefG*

**penicillin G**                              **cephalosporin C**

(*Penicillium chrysogenum*            (*Cephalosporium acremonium*)
          and
*Aspergillus nidulans*)

**Fig. 18.4.   Compounds, enzymes and genes in the penicillin and cephalosporin C pathways.** Simplified from Chiang and Elander (1992).

and cephalosporin C in *Cephalosporium acremonium*. The genes *pcbAB* and *pcbC* are common to both pathways, *penD* and *penE* are specific to penicillin G synthesis, and *cefD*, *cefEF* and *cefG* are specific to the cephalosporin C pathway. When one looks at the structure of the genes and transcripts in *P. chrysogenum*, *A. nidulans* and *C. acremonium*, there are great similarities, e.g., with *pcbAB* and *pcbC* being adjacent but transcribed in opposite directions, with *penDE* next to *pcbC* in the first two fungi (see Martin and Gutierrey, 1995). The penicillin story is remarkable for the starting material with no genetic variation and for the major improvements coming from repeated mutagenesis with a variety of mutagens. Improvements are possible from genetic engineering (see Chiang and Elander, 1992) but there are worries that physiological constraints may mean that yields cannot be improved dramatically any further.

## Suggested Reading

Anke, T. (ed.) *Fungal Biotechnology*. (1996) Chapman and Hall, London.

Archer, D. B., Filamentous fungi as microbial cell factories for food use. *Curr Opin Biotechnol* (2000) 11: 478–483.

Attfield, P. V. and P. J. L. Bell, Genetic improvement of baker's yeast, in D. K. Arora and G. C. Khachatourians (eds.) *Applied Mycology and Biotechnology*, Vol. 3, *Fungal Genomics*. (2003), pp. 213–240, Elsevier, Amsterdam.

Ball, C. (ed.) *Genetics and Breeding of Industrial Microorganisms*. (1984) CRC Press, Boca Raton, Florida.

Barratt, R. W. *et al.*, Map construction in *Neurospora crassa*. *Adv Genet* (1954) 6: 1–93.

Bennett, J. W. and L. L. Lasure (eds.) *More Genetic Manipulation in Fungi*. (1991) Academic Press, New York.

Bos, C. J. (ed.) *Fungal Genetics, Principles and Practice*. (1996) Marcel Dekker, New York.

Campbell, I., Brewing yeast selection. *Microbiol Today* (2000) 27: 122–124.

Carlile, M. J. and S. C. Watkinson, *The Fungi*. (1994) Academic Press, London.

Carlile, M. J., S. C. Watkinson and G. W. Gooday, *The Fungi*, 2nd ed. (2001) Academic Press, London.

Casselton, L. and M. Zolan, The art and design of genetic screens: filamentous fungi. *Nat Rev Genet* (2002) 3: 683–697.

Chiang, S. D. and R. P. Elander, The application of genetic engineering to strain improvement in b-lactam-producing filamentous fungi, in Akora, D. K. *et al.*

(eds.) *Handbook of Applied Mycology* Vol. 4, *Fungal Biotechnology*. (1992), pp. 197–211, M. Dekker, New York.

Clowes, R. C. and W. Hayes (eds.) *Experiments in Microbial Genetics*. (1968) Blackwell Scientific, Oxford.

Desjardins, A. E. and D. Bhatnagar, Fungal genomics: an overview, in D. K. Arora and G. C. Khachatourians (eds.) *Applied Mycology and Biotechnology, Vol. 3, Fungal Genomics*. (2003), pp. 1–13, Elsevier, Amsterdam.

Elliot, C. G., *Reproduction in the Fungi: Genetical and Physiological Aspects*. (1993) Edward Arnold, London.

Fincham, J. R. S., P. R. Day and A. Radford, *Fungal Genetics*, 4th ed. (1979) Blackwell, Oxford.

Foss, E. *et al.*, Chiasma interference as a function of gene distance. *Genetics* (1993) 133: 681–691.

Francois, J. and J. L. Parrou, Reserve carbohydrate metabolism in the yeast *Saccharomyces cerevisiae*. *FEMS Microbiol Rev* (2001) 25: 125–145.

Frazier, W. C. and D. C. Westhoff, *Food Microbiology*, 4th ed. (1988) McGraw-Hill, New York.

Glass, N. L. and G. A. Kuldau, Mating type: vegetative incompatibility in filamentous ascomycetes. *Ann Rev Phytopathol* (1992) 30: 201–224.

Goffeau, A. *et al.*, Life with 6000 genes. *Science* (1996) 274: 546–567.

Goffeau, A. *et al.*, The yeast genome directory. *Nature* (1997) 387(Suppl.): 1–105.

Gow, N. A. R. and G. M. Gadd (eds.) *The Growing Fungus*. (1995) Chapman and Hall, London.

Hausner, G., Fungal mitochondrial genomes, plasmids and introns, in D. K. Arora and G. C. Khachatourians (eds.) *Applied Mycology and Biotechnology, Vol. 3, Fungal Genomics*. (2003), pp. 101–131, Elsevier, Amsterdom.

Johnston, J. R., C. Baccari and R. K. Mortimer, Genotypic characterisation of strains of commercial wine yeasts by tetrad analysis. *Res Microbiol* (2000) 151: 583–590.

Kirsop, B. E. and A. Doyle (eds.) *Maintenance of Microorganisms and Cultured Cells*, 2nd ed. (1991) Academic Press, New York.

Lamb, B. C., Ascomycete genetics: the part played by ascus segregation phenomena in our understanding of the mechanisms of recombination. *Mycol Res* (1996) 100: 1025–1059.

Lamb, B. C., Meiotic recombination in fungi: mechanisms and control of crossing-over and gene conversion, in D. K. Arora and G. C. Khachatourians (eds.) *Applied Mycology and Biotechnology, Vol. 3, Fungal Genomics*. (2003), pp. 15–41, Elsevier, Amsterdam.

Lamb, B. C. and W. M. Chan, Heterokaryon formation in *Ascobolus immersus* is not affected by mating type. *Fung Genet News* (1996) 43: 33–34.

Lea, A. G. H. and J. R. Piggott, *Fermented Beverage Production*. (1995) Blackie Academic and Professional, London.

Lomer, C. J. *et al.*, Biological control of locusts and grasshoppers. *Annu Rev Entomol* (2001) 46: 667–702.

Manners, J. G., *Principles of Plant Pathology*, 2nd ed. (1993) Cambridge University Press, Cambridge.

Manzanares, P. *et al.*, Construction of a genetically modified wine yeast strain expressing the *Aspergillus aculeatus rhaA* gene, encoding an alpha-L-rhamnosidase of enological interest. *Appl Environ Microbiol* (2003) 69: 7558–7562.

Martin, J. F. and S. Gutierrey, Genes for b-lactam antibiotic biosynthesis. *Antonie van Leeuwenhoek Int J Gen Mol Microbiol* (1995) 67: 181–200.

Masneuf, I. *et al.*, New hybrids between *Saccharomyces* sensu stricto yeast species found among wine and cider production strains. *Appl Environ Microbiol* (1998) 64: 3887–3892.

Nevalainen, K. M. H. and V. S. J. Te'o, Enzyme production in industrial fungi: molecular genetic strategies for integrated strain improvement, in D. K. Arora and G. C. Khachatourians (eds.) *Applied Mycology and Biotechnology, Vol. 3, Fungal Genomics*. (2003), pp. 241–259, Elsevier, Amsterdam.

Peberdy, J. F. *et al.*, (eds.) *Applied Molecular Genetics of Fungi.* (1991) *18th Symposium of the British Mycological Society*. Cambridge University Press, Cambridge.

Rédei, G. P., *Genetics*. (1982) Macmillan, New York.

Sugino, R. P. and H. Innan, Estimating the time to the whole-genome duplication and the duration of concerted evolution via gene conversion in yeast. *Genetics* (2005) 171: 63–69.

Tudzynski, P. and A. Sharon, Fungal pathogenicity genes, in D. K. Arora and G. C. Khachatourians (eds.) *Applied Mycology and Biotechnology, Vol. 3, Fungal Genomics*. (2003), pp. 187–212, Elsevier, Amsterdam.

Webster, J., *Introduction to Fungi*. (1980) Cambridge University Press, Cambridge.

Whitehouse, H. L. K., Mapping chromosome centromeres from tetratype frequencies. *J Genet* (1957) 348–360.

# Chapter 19

## The Economics of Agricultural Products and Breeding Programmes

### 19.1. Basic economics: economic systems; price theory; factors affecting supply and demand; perfect and imperfect competition; monopolies; inflation

#### 19.1.1. *Factors of production; types of economic system*

The **production of most goods and services** requires labour for manpower; enterprise to organise and bear the risks of production; land, e.g., for crops, shops, factories or offices; capital equipment, such as machines, factories, vehicles and roads, and working capital in the financial sense. There is a limited willingness of people to provide labour, land and finance, but an almost unlimited potential demand for goods and services. An **economic system** determines who produces what and who consumes what. The two main systems are the command system and the money-price system, and most countries have a mixture of the two, although often in very different proportions.

The **command system** may be autocratic, where the rulers or planners do not consult the public as to their wishes, but the rulers dictate who will produce what and who can consume what. Alternatively, it can be consultative, where the planners do consult the public about their willingness to supply labour, land, etc., and what they want to consume. A complete consultative system for Britain or America for every commodity would be administratively impossible, and most people are keener to consume than to produce. Command systems are difficult to operate, slow to respond to changes in demand, and may involve a large loss of individual freedom of choice in jobs and products. They characteristically have occasional large gluts of certain products, and more frequently, severe shortages, often with a private "black market" supplying more choices than the state system, and at higher prices.

The **money-price system** has individual firms producing what they think the consumers want; if the firms produce goods or services at a price people are prepared to pay, their products sell. The **preferences of the public** in their own buying, and in their willingness to hire out their own labour, land and savings, determine what is produced and at what prices. There is a circular functioning, as shown in Fig. 19.1. The households' expenditure becomes the firms' income; what the firms pay out in wages and rents become the households' income, etc. If people decide to spend more on vegetables and less on meat, then vegetable suppliers can expand, but meat producers would have to cut back. The sum of decisions of individual purchasers, and people's willingness to do different jobs or give others use of their land or savings, determine such things as what is produced and the relative wage rates for different jobs. The money-price system is highly democratic in that each purchase is a vote for the continued production of that item or service.

One factor missing from Fig. 19.1 is the **role of governments**, local and national. For administration, legal services, armed services, education, health, roads, etc., governments consume resources which could otherwise have gone to consumer spending. The allocation of resources between governments and citizens is sometimes called the "guns or butter" problem.

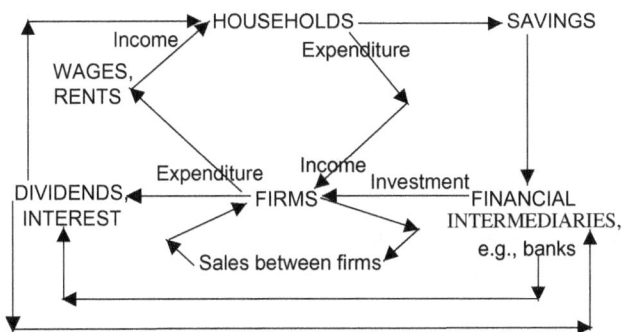

**Fig. 19.1.** The circular functioning of a money-price economic system.

Governments curtail the spending power of citizens by direct and indirect taxation, and alter the distribution of incomes by differential taxation, taxing higher incomes at proportionately higher rates. During and after wars, or at other times of shortages, governments may regulate consumption by **rationing**, issuing ration coupon books. Governments also affect consumption by **differential taxes** on different items, such as different rates of tax on fruit juices, beer, cider, wine and spirits. Governments also work through the command system, e.g., banning cyclamate sweeteners or the sale of beef on the bone. National and local governments together consume a large proportion of the gross national product in Britain. In 2004 in Britain, taxes and excise duties amounted to £439 billion, i.e., 42% of the national income of £1,057 billion. In 1999, 85% of the pump price of petrol was government tax. Under the Labour Government, total public spending rose from £316 billion in 1996 to a planned £580 billion by 2007 (equal to 42% of GDP and £10,000 per household).

Governments are made of humans who naturally like to see their power and job opportunities expand, and they can pass the necessary laws. There is thus a dangerous tendency for governments to expand, even though it is not announced as being a partial nationalisation of jobs and resources. An editorial (*The Daily Telegraph*, 22/1/2004) stated that: "*Britain's growth in recent years has been based on an unprecedented expansion of the government payroll: two out of every three jobs created since Mr Brown* [*Chancellor of the Exchequer since 1997*] *took office have been in the state sector. Taxing the productive part of the economy in order to inflate the unproductive part was always bound, sooner or later, to end in tears*".

## 19.1.2. *Price theory*

A **price** is a value agreed upon between producers and consumers for the exchange of goods or services. The main **factors affecting demand** are: income; the prices of substitutes and complements; tastes, and the availability and convenience of the use of products.

**Income**: for most goods, demand rises as income rises. A 19th century economist, Engels (not the Engels who co-wrote the Communist Manifesto) made an empirical finding that as incomes increase, proportionately less is spent on food. In Britain, between 1955 and 1992, average real earnings (i.e., earnings in relation to the index of retail price) rose by 210%; expenditure on food fell from 35% to 15% of household spending, while expenditure on cars and transport increased from 7% to 16%. There is a different **elasticity of demand** for different products. "**Inferior goods**" are those on which spending decreases in absolute terms as incomes rise; this happens when cheap items are replaced by more expensive ones as people get richer, so potatoes and margarine have sometimes had reduced sales when people are able to afford more varied vegetables or butter, as incomes have risen.

**Availability and price of substitutes**: demand for beef is affected by the price and availability of lamb, pork, chicken and fish, as people can substitute one product for another, depending on price, availability and quality. The demand for one type of product might be fairly consistent, e.g., for green vegetables, but the demand for a particular type of green vegetable might be elastic, depending on price and quality. In the 1990s, health scares over beef from BSE (bovine spongiform encephalopathy) greatly increased the demand for other meats as beef substitutes in the UK.

**Availability and price of complements**: the demand for cranberry jelly, for example, is influenced by the price of turkey and of game birds, as it is usually bought as a complement to those meats.

**Tastes**: people's tastes can change. Certain foods and drinks come into or go out of fashion, irrespective of price. In 1999, a major UK supermarket chain asked its melon growers in Spain to produce small melons (up to 540 g) rather than large ones (around 965 g), as the small ones had started selling very much better than the large ones. When breeders first hybridised different species of *Hippeastrum* lily, they aimed for large showy flowers, but gardening tastes have changed, with smaller-flowered types coming back into fashion (Mathew, 1999). Tastes are influenced by advertising, press comments, consumer organisations, people's upbringing and fashionable trend-setters. In October 2004, it was reported that Christian shops and jewellers in Britain were being overwhelmed by demand from teenagers for chains of rosary beads after footballer David Beckham and singer Britney Spears began wearing them as necklaces and bracelets, instead of using them as aids to prayer. For factors influencing consumers' choice in meat, see Subsec. 13.7.13.

**Availability and convenience of use**: people usually prefer to buy what is locally available, rather than to go far to obtain something else.

Convenience foods are very popular as they take less time to prepare, even though they cost more than starting with the basic ingredients. There is a trend for more seedless varieties of fruit, such as seedless oranges and grapes, while fruits such as Russett apples, with thick corky skins which usually need to be peeled off, are less popular than thinner skinned varieties which do not need to be peeled. "Easy-peel" citrus fruits are favoured.

A **demand curve** shows the relation between price and demand for a particular product, and the term is used even if the relation is linear. Fig. 19.2 shows two possible demand curves for one product. In curve 2, more is demanded at the same price than in curve 1; for example, at price OA, amount OC is demanded for curve 2, compared with only OB in curve 1. Demand curve 1 could change to curve 2 if incomes rise, or if the price of substitutes rises, or if the price of complements falls, or with favourable changes in tastes, or wider availability or increased convenience. For almost all types of goods or service, the demand curves slope upwards to the left (less demand at higher prices), but the degree of slope varies with the **price-elasticity of demand** for that product, as shown in Fig. 19.3. Some "**snob goods**" actually have increasing demand as their prices rise, as some people value exclusivity of possession, e.g., for very expensive cars or watches.

With elastic demand, one gets large changes in demand with changing price, which is true of many manufactured goods such as cameras. With

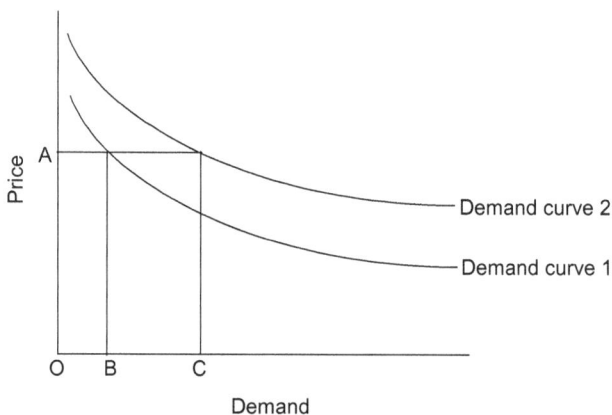

**Fig. 19.2.**   Different demand curves in relation to price.

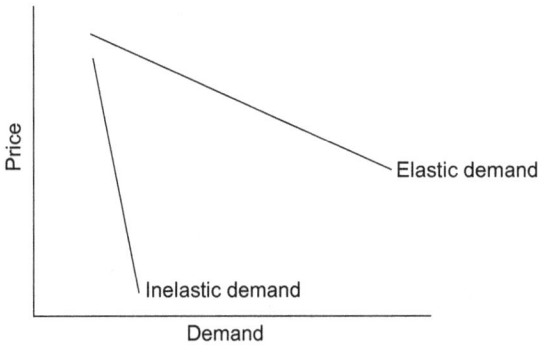

**Fig. 19.3.** Differences in price-elasticity of demand.

inelastic demand, one only gets relatively small changes in demand with changing price, which is true of most agricultural produce and other primary products such as metal ores.

The **price-elasticity of demand**
$$= \frac{\textbf{the change (e.g., rise) in the quantity demanded}}{\textbf{the opposite change (e.g., fall) in price}}.$$

For example, if demand goes down 20% when the price rises 10%, the price-elasticity of demand is 2.0. Elastic demand has values of 1.0 or over, and values less than that are inelastic demand.

One must distinguish between price and **total expenditure**, the relation between which depends on the shape of the demand curve. In Fig. 19.4, the total expenditure at the higher price OA is price × demand = OA × OB,

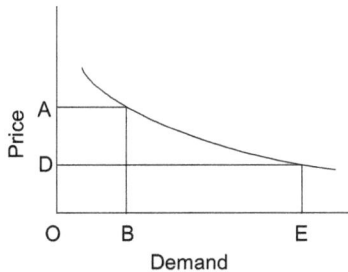

**Fig. 19.4.** The relation between total expenditure and price.

which, depending on the exact shape of the curve, may be greater than, equal to, or less than the total expenditure at the lower price OD, i.e., OD × OE. In agriculture, one must also distinguish between expenditure on food and **farmers' personal income**, because there are many costs in the food chain. It has been estimated that a Danish farmer in the 1990s only received on average 27% of the market price for his or her produce. Of the 27%, only 2% is for the farmer personally, on average, and 98% goes to employees, suppliers, government, transport, debt-servicing, and other items.

One of the main problems with agriculture is that of inelastic demand: falling prices only expand demand a little, so falling prices greatly diminish the total expenditure on food. If there is a food surplus, cutting prices will not help much in getting rid of the surplus, but it will cut farmers' income greatly.

**The determination of supply**. Profit = revenue - costs. Most firms aim to maximise long-term profits, and this may well mean not taking the maximum **profit per item**, if a smaller profit per item results in proportionately larger sales and a greater total profit. Let P be the market price, then the supply curve for one firm might be approximately shown in Fig. 19.5.

As supply expands, between O and A in Fig. 19.5, costs per item usually decrease initially, with economies of scale, but eventually, costs per item increase again, as in the region to the right of A, due to the limits of managerial efficiency in very large firms; inefficiencies which might be spotted in a small firm may go unnoticed in larger ones. The biggest firms are not

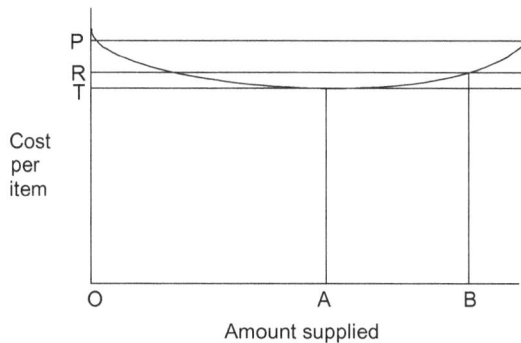

**Fig. 19.5.** Curve for cost per item for different amounts supplied. See text for symbols.

always the most efficient nor the most profitable. The firm's total profit, at the maximum profit per item, would be PT × OA, but by taking a smaller profit per item, PR at greater output OB, the total profit, PR × OB, may be greater than PT × OA. In general, firms will want to supply more when market prices are high, but firms differ in efficiency, so different supply curves for different firms are obtained and shown in Fig. 19.6. At market price OA, only firms 2 and 3 could operate.

Putting the demand and supply curves together, we get an equilibrium point where the demand and supply curves intersect, as shown in Fig. 19.7. Quantity OB would be exchanged, at price OA. The intersection point is actually an equilibrium point and might take time to reach. For example, firms might have to build or sell factories, hire or shed labour, to cope with increased or decreased demand.

An example of **reduced supply increasing prices** occurred for **hazelnuts**, an important ingredient of many chocolates. Turkey supplies approximately 80% of the world's hazelnuts and frosts severely reduced production in 2004, with other producers such as Italy, Spain, the USA, Georgia and Azerbaijan being unable to increase supplies significantly at short notice. The price went from ~£1,100 a ton to over £6,000 a ton in 2005, so chocolatiers tried switching to substitutes such as almonds, walnuts and brazil nuts, reducing demand for hazelnuts.

**Speculators** are routinely damned as harmful by political parties of the left and of the right, but they are useful if their operations are not large in relation to the market. A successful speculator buys when prices are low and sells when there is a shortage, with high prices. Speculators therefore

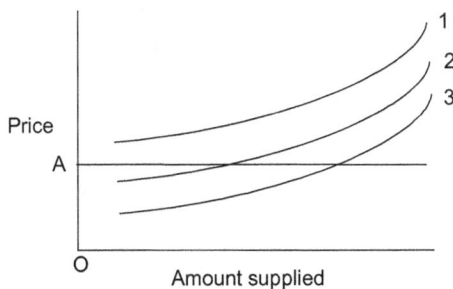

**Fig. 19.6.** Different supply curves for firms of different efficiency.

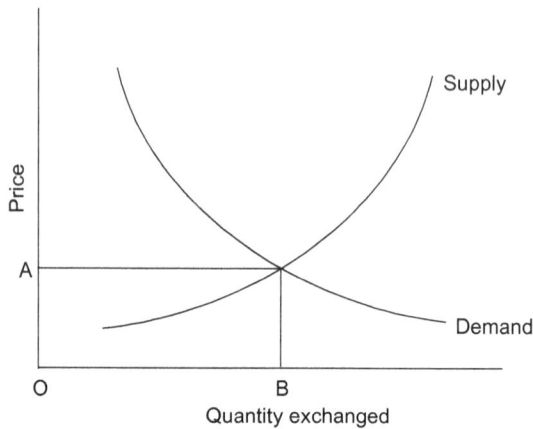

**Fig. 19.7.**  **The determination of price and quantity exchanged by the intersection of supply and demand curves.**

improve demand when it is slack, and increase supplies when they are short, driving down high prices. Speculators can therefore help to reduce peaks and troughs in demand and supply.

### 19.1.3. *Types of competition*

The commonest types of **economic competition** are called "perfect", "imperfect", oligopoly, and monopoly, in order of decreasing numbers of competing firms. With **"perfect competition"**, no firm is large in relation to the total market; no one acts as a price leader. There is free entry into the market and free exit from it, with no great costs of entering or leaving the market. Many firms produce almost identical products, and the price is determined by market conditions of supply and demand, and not by individual firms. "Perfection" is rarely found but is approximated by the Chicago wheat, maize and meat markets where many farmers supply almost identical products. If a farmer raised his prices, no one would buy his produce, while if he lowered his prices, he could expand sales but would not get an adequate return.

With **"imperfect competition"**, the products of different firms are not identical. They may differ in quality, advertisement, packaging, service, availability, etc., e.g., say in the pleasantness of sales staff or in the location of sales outlets. For example, someone might go 50 metres for better bread, but not 50 km. With this difference between products, people are willing to pay different prices for them. Each firm can fix its own prices and try to maximise long-term profits, but usually they do not know what the demand curve is. Many firms charge a fixed "mark-up"; a trendy dress shop in London charged a 300% mark up, so if the shop paid £200 for a dress, they sold it for £200 + (£200 × 3) = £800. A firm may charge different mark-ups for different items. For example, in 1999, a computer firm charged £19 retail each for a keyboard and a mouse, with cost prices of £5–91 and £3–54 respectively.

With an **oligopoly**, there are few sellers, mostly large in relation to the market, as with petrol companies in the UK. The action of one firm affects all the others. For example, if Shell lowered its petrol prices, it could expand sales, but that would diminish sales by Texaco, BP and Esso. The other firms could regain sales by lowering their prices, but all firms' profit margins would be lower, including Shell's. **Price wars** can harm all firms in a market, as evident in freight shipping rates in the Indian Ocean in the 1950s. Firms kept reducing their prices in order to retain custom, in the hope that other firms would abandon the market as unprofitable, so that rates could rise again. Prices actually fell to covering only 10% of costs, so all firms made losses and many went bankrupt. As a result of the dangers to all firms from a price war, there is often an informal but usually illegal "gentlemen's agreement" between the firms not to start price wars, to compete by advertising and services rather than by price, but even a big advertising war can be harmful by putting up costs, unless it expands the total market. In 1998, the European Commission fined four British sugar companies, including British Sugar and Tate & Lyle, £36 million for attempting to fix sugar prices in the late 1980s. The Commission criticised British Sugar for instigating the price-fixing cartel in 1986 at the end of a price war.

**Cartels** are combinations of firms which cooperate to keep up prices and to dominate markets by a collective monopoly, instead of competing. A famous example is OPEC, the **Organisation of Oil-Exporting Countries**, involving 11 major producers such as Saudi Arabia. In December 2004, they agreed to cut production by a million barrels of oil a day to keep oil prices high when Iraq restored oil exports. Their spokesman claimed to have helped calm market prices, by OPEC countries pumping more oil when political troubles in various countries reduced oil production and increased world prices to record levels. In 1973, OPEC was much more Arab-country-dominated; after the Israeli success in the Yom Kippur war, the Arabs reduced production severely. Oil prices multiplied 10-fold, to the equivalent of $70 a barrel today, showing a very sharp relation between inelastic demand, cartel-controlled supply and oil price.

With a complete **monopoly**, one firm has all the market, with no competitors. In some countries, a monopoly may be defined legally as having some lesser share of the market, say 1/4 or 1/3 for a particular product. In a complete monopoly, there is no need for a firm to worry about the action of the competitors, or of being undercut on price. The monopoly firm can usually make more profits than under competition, making excess monopoly profits. In Britain, supermarkets control ~70% of the fruit and vegetable market, but with no single chain having a monopoly position.

In Britain, the Fair Trading Act 1973 defines a monopoly as a situation where a company, or a linked group of companies, supplies or purchases 25% or more of all the goods or services of a particular type in the UK or in a defined part of it, so there can be national or local monopolies. There is no assumption that monopolies are wrong in themselves. The invention of a new device will make the inventor a monopolist, or a firm may gain a monopoly from its efficiency. The Fair Trade Act just defines situations where it is possible that excessive market power could exist and could be misused, contrary to the public interest. The Monopolies and Mergers Commission (MMC), the Director General of Fair Trading and the Restrictive Practices Court all have responsibilities corresponding to their names. Under the Competition Act 1980, an anti-competitive practice is defined as one which restricts, distorts or prevents competition in some market in the UK. Practices which are acceptable in a market where competition is strong, may not be acceptable in a market where competition is weak. Companies are excluded from provisions of the act if they have a turnover of less than £10 million a year, or if they have less than 25% of a relevant UK market. In 1992, the MMC found that Bryant and May had 78% of the UK market in matches and disposable lighters, and were making monopoly profits, so price controls were imposed. In 1991, the MMC found that Nestlé had 56% of the UK soluble coffee market, and higher profits than other firms, but MMC attributed this to a good performance in the face of strong competition, and no action was taken. Rules on monopolies vary widely between countries.

There are several **origins of monopolies**:

- **Legal**. This is conferred by a government licence or legal charter, as for the Royal Mail having a monopoly in Britain for most types of letter delivery.
- **Too high a cost** of entering a market. For example, few firms could afford to lay out a competing London underground railway system, or a national railway system.
- One firm may "**corner the market**", e.g., buying up all sources of supply of a commodity, e.g., nutmeg.
- **Patents** on new processes and inventions. This applies to plant breeders' rights on a new variety (Subsec. 19.2.6), or to a new drug. After World War II, Mr Biro had a monopoly on the sale of his invention, of ball-point pens. Such monopolies are usually broken eventually by better new inventions.

Monopolies are often contrary to the consumers' interests, as there is no other source to buy from if they do not like that firm's prices or products. Monopolies can have some benefits such as big economies of scale of production and more efficient planning in the absence of uncertainty about competitors' plans. Big firms often do more research and development than small firms. Sometimes, there has been no reduction in prices on breaking up a monopoly, but the break-up of many state monopolies (nationalised industries) in Britain under Mrs Thatcher's Conservative Government's privatisation programme from 1979 into the 1990s generally resulted in lower prices (e.g., for telephones and gas) and better services. For example, by 1999, household electricity bills had fallen by 26% since privatisation in 1990. Many governments abroad also instituted programmes of privatisation.

It is not only firms which try for monopolies. The **Trades Unions** also try to get a monopoly control over the supply of labour in particular firms or industries, to try to drive wages above the free-market price of labour.

## 19.1.4. *Inflation*

**Inflation** is the reduction with time in the purchasing power of a unit of money, similar to a general rise in prices. For example, the retail price index rose 26% in 1976 under Mr Wilson's Labour Government in Britain. A London suburban semi-detached house which cost £6,500 in 1967 was, through inflation, worth £500,000 in 2004. Inflation is most damaging to the standard of living of people on fixed incomes, or who live on savings, or who are in weak bargaining positions, such as university lecturers.

**Demand inflation** occurs when spending power is greater than the availability of goods, so that excess demand drives up prices. This is fairly rare except during and after wars when people have not had much to spend money on, as resources have gone into war-related products and overseas supplies may have been cut off by blockades. At the end of a war, consumers want to buy more than the suppliers can provide, until war factories have switched to producing consumer goods and imports have been resumed. Possible remedies include increased production, reduced government consumption, or for governments to increase taxes "to mop up surplus spending power". The British Government restricted consumption during and after World War II by **rationing** many items, so that one had to exchange ration-book coupons, as well as money, to buy meat, sugar, butter, margarine and cooking fats, tea, cheese, chocolate, sweets, coal, clothes, petrol, furniture and other items. For example, the weekly allowances per person in 1941 were: butter, 113 g; sugar 340 g; tea 57 g; cooking fats and margarine, 57 g; bacon and ham, 113 g; other meats were rationed by price, not weight, to 1s 10d (9p) a week. Meat rationing only ended in 1954, nine years after the war ended.

**Cost inflation** is much more common and more serious than demand inflation. After cost rises, firms raise their prices; higher prices lead to demands for higher wages, leading to cost increases if these demands are met. These increased costs lead to price rises, then to demands for increased wages, etc., in an inflationary spiral. This is usually associated with wages rising faster than productivity, often under Trades Unions' use of monopoly control over the labour supply. Inflation can continue until high unemployment weakens the Trades Unions' bargaining position. There are no easy, painless remedies for cost inflation, although there are many different actions that governments can take. They are too complicated and controversial to discuss here; they include control of the money supply, interest rates, and international exchange rates, and government-imposed price and/or wage controls.

## 19.2. Economics applied to agriculture

### 19.2.1. *Gluts and shortages; how governments intervene in agriculture; European Union policies*

Although plant and animal breeders breed to increase yields, if all farmers have a high yield in one season, it can lead to a **glut** and **decrease farmers' income** because of inelastic demand. With a glut, farmers try to reduce the amount of surplus produce coming onto the market, to stop it from driving down the prices. Farmers' organisations agree to plough crops under, or dump them at sea, or render them inedible by sprays, or leave them to rot. This arouses great journalistic fury and public resentment about waste, but it may be the only way for farmers to stay in business. In 1995, the EU taxpayers paid for 720,000 tons of surplus peaches to be destroyed. In a glut year in Britain, the Potato Marketing Board can order that surplus potatoes be dyed blue and sold for stock-feeding, making it an offence to sell blue-dyed potatoes for domestic consumption.

Consider the difference between a **year of glut** and a **year of shortage**, as shown in Fig. 19.8. Due to inelastic demand, the farmers' income is lower with supply curve 2, with a glut, than in a poor harvest year, supply curve 1, because reducing prices substantially only results in a slightly increased demand: the quantity demanded at low price OC, is OD, which is only slightly more than demand OB obtained at substantially higher price OA in the poor year.

It is not in anyone's interests to have the farmers ruined, so most **governments intervene** in farming and agricultural supply. As Devon farmer Richard Yeomans put it (*Freedom Today*, 1999, February/March, pp. 15):

"*Agriculture cannot be left to the market. The goods are perishable, production lines can take years to stop, and over-production of one per cent can drive down prices 10 per cent. So over-production has to be controlled*

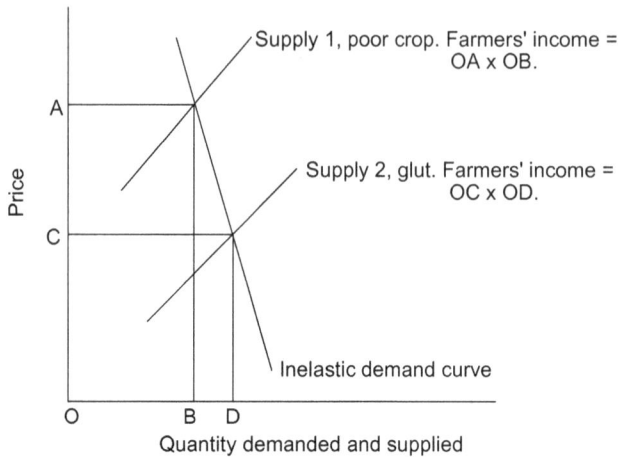

**Fig. 19.8.** Farmers' income in a free market, in a glut year and in a shortage year.

*in some way, because agriculture must be kept alert, modern, prosperous, vibrant and 'ready to go'".*

Agriculture is heavily controlled by governments in all the major economies, including the USA, Japan and the European Union (EU). A study of seven major countries recently found that government assistance to agriculture exceeded one third of the value of the agricultural produce, with approximately half of the money coming from taxpayers (from subsidies) and approximately half from consumers (from higher-than-free-market prices).

A common government policy is to try to avoid overproduction by restricting supply via **licensing**. A government agency issues licences for a particular crop, restricting the area planted, as with potatoes in the UK, or restricting yields by means of quotas as with milk yield in the EU where there are fines for exceeding one's quota. Spotter planes are sent out in Britain to look for unlicensed commercial-scale potato growing. This method of restricting supply has the obvious and serious drawback that one cannot predict yields in advance. If farmers in 1999 are licensed to grow only a restricted area of potatoes in the following year, and that year turns out to be a bad year, then shortages will result. The EU's method of controlling fish stocks and sales, by making fishermen throw back into the sea those species for which they do not have a quota, is the worst possible conservation strategy, as 40% of the catch just dies.

The second type of government policy to control production is to adopt **financial measures** such as subsidies or intervention buying. The system of **subsidies** (deficiency payments) was used in Britain before Britain adopted EU methods. It ensured a good supply of food at fairly low prices to the consumer, but with a high cost to the taxpayer. It is illustrated in Fig. 19.9.

OA is the price with no intervention, with supply OB, but the government agency (then the Ministry of Food) agreed on price OC after discussions with farmers' representatives (the National Farmers' Union). With agreed price OC, the farmers supplied OD, but with supply OD, the market price obtained was only OE. The government (the taxpayer) then had to pay for the amount by which the actual market price fell short of the agreed price, $(OC - OE = CE)$, with total payment $CE \times OD$. The amount of subsidy varied widely, but was often about one third of the shop price for many food products.

The American system of **intervention buying** is different. The farmers and the government agree on a market price, with government agencies buying up any surplus in order to maintain that agreed market price. This is shown in Fig. 19.10. With the agreed price OC, farmers supply OD, but at that price, consumers only buy OF, so the difference, FD, has to be bought up by the intervention agency. This gives higher prices than free-market

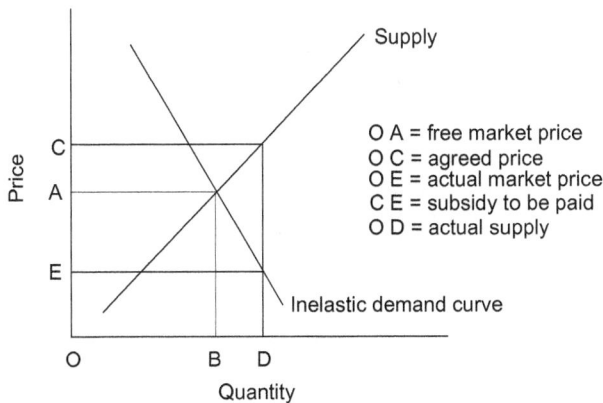

**Fig. 19.9.   The former British system of subsidies.**

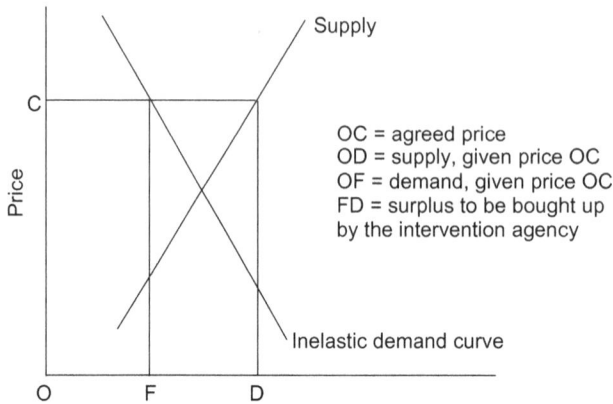

**Fig. 19.10.**    **The traditional American intervention-buying system.**

prices, a bill OC × FD for the taxpayer, and leads to **food mountains** from the intervention buying. In America, the produce bought has been used for strategic stockpiles (e.g., for use in the event of war), for foreign aid programmes, and for selling abroad. In the EU, intervention buying has at various times led to so-called butter mountains, beef mountains, wine lakes, etc.

**The EU method** is similar to the USA system. It aims to protect EU farmers from overproduction within the EU and from competition from overseas countries. A **target price** is agreed for various foods but is not guaranteed. If EU production is large enough to cause the market price to drop much below the target price, intervention agencies buy up enough to restore the market price to the target level. Imports from outside the EU which reach the EU at prices below the target price are subject to **import levies** to bring their price up to EU levels, so that the EU consumer does not benefit from cheaper world prices.

In order to get rid of the mountains of intervention surpluses, the EU pays a **subsidy** to lower the price of exports to the world price, although the World Trade Organisation tries to limit the amounts involved. For example, the EU placed a massive subsidy on the sale of butter to Russia to help get rid of a butter mountain, so that Russians were able to buy EU butter at much lower prices than EU consumers, with a large cost to the EU citizens. The cost of the **EU's Common Agriculture Policy** was £32 billion a year in 2004, approximately half the EU total budget, with high bills to EU members and distorted competition with other countries because of the import levies and export subsidies. British farmers receive about £1.8 billion a year from this.

To avoid surpluses of major crops, the EU often subsidises minor crops, and even subsidises the growing of no crops on agricultural land under the **set-aside scheme** (see Ansell and Tranter, 1992). In 1998, the EU decreed that 10% of arable land in Britain, ~650,000 ha, must be "set-aside" to reduce grain production by 1.6 million tons. See Sec. 13.9 for a linseed flax example of the absurdities to which EU policies can lead. The EU suffers from over-regulation, even issuing a directive on the amount of permitted curvature in bananas, which is surely a matter of consumer purchasing choice, not bureaucrats.

The EU policies, like those of many areas, are a political compromise between producers and consumers, politicians and voters. The various EU countries agree that the present system is cumbersome, expensive and unfair, but they cannot agree on how it should be reformed, especially countries such as France making massive gains from the existing system. The EU Common Agricultural Policy costs an average of £550 a year per UK taxpayer, plus £300 extra per head on food prices (1998 figures), with ~48% of farmers' incomes coming from various subsidies. In recent years, New Zealand has greatly reduced government intervention in agriculture, reducing subsidies from 60% to ~4%, with promising results for farmers and consumers, and less overgrazing. 1998 figures (*The Daily Telegraph*, 16/10/98) for EU subsidies per hectare in England included £242 for cereals, £349 for proteins, £427 for oilseeds, £568 for non-cannabis-producing mutant hemp for fibre, £305 for set-aside, £112 for suckler cows, and £105 bull premium.

Even with all these subsidies, there is a crisis in many branches of agriculture, especially livestock farming. The Meat and Livestock Commission found in 1998 that the UK shop price of lamb had risen slightly over a year to £4.58 a kilo, while the price paid to farmers had dropped from £2.42 to £1.69 a kilo. Of the shop price of £4.58 a kilo, the farmer typically gets £1.69, slaughter costs 22p, cutting and packing cost 33p, inspection and offal disposal cost 18p, transport costs 11p, and shops' overheads were £2.05.

In May 2004, 10 countries joined the EU and many, such as Poland with its multitude of peasant farmers, were appalled at the complexity of the agricultural subsidy forms and the amount of bureaucracy they became subject to. The Maltese were horrified at having to pay the protectionist EU sugar price of €840 a tonne instead of the world price of around €200. An extra sugar subsidy for Malta was hurriedly arranged, with much of it being paid for by the Maltese taxpayers. The EU produces 17 million tonnes of sugar a year but consumes only 12 million tonnes, with the surplus being dumped on world markets with a subsidy of €525 a tonne. The price of sugar in the EU is three to four times the world market price, and Third World producers such as Mozambique, which can produce sugar much more cheaply, are inhibited from exporting to the EU by tariffs of about 324% (figures from B. Johnson, *The Daily Telegraph*, 7/7/2005, pp. 28). In 2003–2004, the EU paid Tate & Lyle £127 million to subsidise its sugar exports. In November 2005, the European Court of Auditors refused to approve the EU accounts for the 11th successive year "because the vast majority of the payment budget was again materially affected by errors of legality and regularity." Most of the spending was unsafe or riddled with errors, with widespread fraud in farm aid payments in Italy, Greece, Spain and France. In 2004, Britain had a current account deficit with the EU of £22 billion.

In January 2005, various **changes to the EU's Common Agricultural Policy** took place, exposing farmers more to market forces. Previously, subsidies to British beef farmers, among others, were coupled to production and were paid on every animal, amounting to 40 to 45% of its sale value, i.e., ~£350 out of £800, usually making the difference between a profit and a loss on the animal. Subsidies are now being decoupled from production so that the EU payment will no longer depend on the number of animals kept or the area of cereals grown. Instead, farmers now receive a Single Farm Payment lump sum, based on individual historic subsidy levels, farm size and good environmental management. This means that farmers are better off producing nothing if they cannot sell at a profit. Beef market prices fell in July 2005 to 190p a kilo, when some farmers break even only when the price is 220 to 300p a kilo, depending on the type of production and farm efficiency. Increased imports of beef from South America depressed prices. Many beef farmers in Britain feared having to go out of business, especially in hill areas, unless prices or subsidies rise substantially.

## 19.2.2. *Seasonal and perishable produce*

In a free market, prices for the same commodity can show sharp **seasonal variations**, with higher prices at times of scarcity and the lowest prices during peak supplies. Many agricultural products are seasonal, with the lowest prices just when most farmers are selling their crops. There is therefore an economic advantage in breeding early-maturing varieties, with peak production when prices are high, as with early tomatoes and strawberries, and new potatoes. According to Dunnett (2000), Scottish potato variety Swift is the earliest variety on the British market each year and must be sold in the first days of the new season when prices are highest, with a niche market of not much longer than a week.

One can also supply goods out of season by **cold-storage**, as for many British apples harvested in autumn for sale during the winter and spring. One grower in Kent, England, puts his apples into cold store at 2° for just 10 to 14 days, to avoid supplying the market just when the other growers are selling their crop. There are losses from the costs of moving goods into and out of the store, from pests and diseases, hiring storage space, and from bruising from extra handling, which have to be offset against any gain in price obtained.

Many foods which used to be seasonal in the UK are now **available year-round**, from growing early varieties, from cold storage (e.g., frozen peas and beans), and from imports (e.g., of South African grapes, apples and citrus fruits, Plate 19.1; Spanish and American strawberries; Kenyan beans). The prices and qualities may still show strong seasonal variation, e.g., for strawberries, where UK-grown ones in summer have the best flavour. In 2005, using heated greenhouses and only natural light on the Isle of Wight (extreme south of England), Wights Salads were able to market English cherry tomatoes in late January, the earliest ever. Many imported fruits are picked before optimum ripeness and flavour to reduce losses during transport and handling, and to improve shelf-life. Tomatoes sold in Britain in winter are usually imported from Spain or Israel, picked green and with

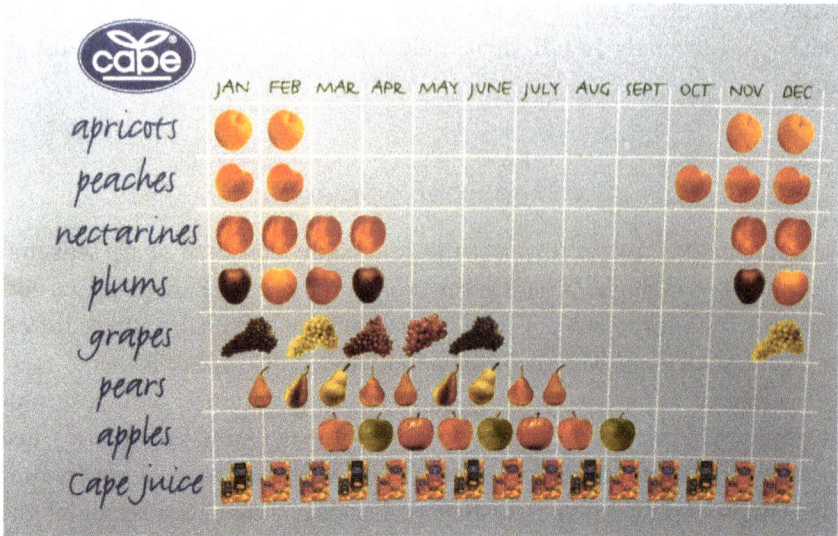

**Plate 19.1.** **A chart of the seasonal availability of fruit from the Cape, South Africa,** for export to the UK. Within one type of fruit, say grapes, different varieties are available at different times.

thicker, tougher skins, while these English-grown ones have better flavours from ripening on the vine, and thinner skins.

The **perishable** nature of many foodstuffs means that breeders, farmers and food technologists try to produce varieties, foods or processed food products with a longer shelf-life, to reduce wastage. Chilled storage and preservatives such as sulphur dioxide are widely used. Waxes and medicinal paraffin are often used on citrus fruits and partly dried grape products such as raisins to delay spoilage by micro-organisms.

## 19.2.3. *The value of rarities*

Some **rarities** are valuable, some are not, depending to a large extent on whether they catch the public's interest and on how well marketed they are. **Orchids** are generally perceived as exotic, mysterious, rare and interesting, with a £6 billion a year international trade, including rarities fetching more than £16,000 a plant.

A classic case of rarities fetching high prices was that of **tulips** in Holland, where the first bulbs were planted in 1593. They became extremely popular and in 1610–1637, and especially in 1634–1637, Holland suffered from "Tulip Mania" with a huge rise in the price of bulbs, particularly for rarities, even though only the rich could afford them. At the peak of demand, the variety Semper Augustus fetched from 5,000 to 13,000 florins a bulb, equivalent to the price of a grand house on the Amsterdam canals, when the average annual income was 150 florins (see Pavord, 1999). In April 1637, the Dutch government intervened, cancelled all speculative agreements, and fixed the maximum price at 50 florins a bulb. There are now more than 3,000 tulip cultivars, classified as early-flowering, medium-early flowering, and late-flowering, and subdivided into types such as single, double, Darwin, lily-flowered, fringed (Plate 19.2), parrot, viridiflora, Rembrant and Botanical tulips.

A later case is the **potato** (Dunnett, 2000). During the "great potato boom" in 1903 and 1904, a Scotsman, Archibald Findlay (1841–1921) was able to sell tubers of his latest-bred varieties for £25 each, equivalent to

**Plate 19.2.   Fringed tulips, variety Johann Gutenberg**. Photographed at Pashley Manor, Ticehurst, Sussex. There is an enormous range of different types of tulip, for flower form and colour.

~£1,100 now. As there were no Plant Breeders' Rights then, anyone buying a tuber could reproduce it and sell the seed potatoes themselves, even under another variety name.

A well-documented animal rarity is the **white tiger**. In 1951, a white tiger aged about nine months, Mohan, was caught in India from the wild. He was later mated to a normal tigress, Begum, and all 10 cubs were normal. Mohan (Plate 19.3) was mated to one of his daughters, Radha, and this eventually produced 11 white cubs and three normal ones. All white x white matings give all white offspring, because the condition is due to an autosomal recessive allele. White tigers are not albinos. They have icy-blue eyes, not pink, and the off-white coat has brown or chocolate stripes. The animals are larger and heavier than normal tigers. Selection within zoos has now made the coat a brighter white than Mohan's. In 1983, there were 65 white tigers in the world, 21 in India, 40 in USA, 4 in the UK. Their extreme attractiveness to zoo visitors and their rarity then made them worth about 1.1 million Indian rupees (at about 17 rupees to the pound sterling) a

**Plate 19.3. Mohan, a male Indian white tiger from the wild,** homozygous for an autosomal recessive allele. He is not an albino and has blue eyes and some coloured stripes. Such rarities can have very high values: see text. Photographed at the International Congress of Genetics, New Delhi.

breeding pair. When Bristol Zoo in England first had white tiger cubs, the number of paying visitors increased dramatically.

### 19.2.4. *"Health foods" and "organic" products*

Many people in developed countries are becoming more **health conscious**, helped by an avalanche of articles and programmes in the media, so that it is worth breeding and manufacturing to cater for their wishes, whether soundly based or not. Meat with reduced fat and cholesterol commands a good premium, and there is a huge market for reduced-fat milk, yoghurt, spreads, sauces, manufactured foods, and non-dairy "milks". See Table 5.3, in Chap. 5, for the calories, protein, fat and cholesterol content of different meats, where ostrich meat has the least fat and by far the least cholesterol. Ostrich meat is regarded as healthy and is increasing sales in Europe and elsewhere. Another low-fat meat is from the camel. There are about 700,000 feral camels in Australia, where dromedaries were introduced in the 19th century for transport in the Outback. It was announced in 2004 that Aborigines in the Northern Territory were planning to catch camels, with a carcass value of about £300 each, and export meat to the Middle East and Asia.

"**Organic**" foods are increasingly popular, commanding higher prices than normal food, e.g., 29p a litre for organic milk against 18p for normal milk (1999 figures), and see Murphy (1992). According to the Soil Association in 2005, organic food sales in Britain are worth £1.2 billion a year and are growing by £2.3 million a week, twice the rate of the total food market. The fastest developing product is organic meat, with sales rising 30% a year, especially for poultry. Approximately 56% of organic food in Britain is imported, including potatoes, carrots, onions, apples and pears. Varieties used for organic sales should be bred for disease- and pest-resistance, as few chemicals are approved for use on such produce. An example would be the potato Sarpo Mira, released in 2004, and claimed to be resistant to blight, viruses, slugs and wireworms (see Subsec. 13.7.5).

An example which combines animal conservation (Sec. 15.4) with health food is the **bison** (buffalo). In North America in 1830, there were about 60 million wild bison grazing on the prairies. By 1880, hunting had reduced them to a few hundreds. A small number of ranchers saved them from extinction and some established breeding stocks for meat production. The meat is high in flavour, has a quarter the fat of most other red meats and is low in calories, so it can command prices up to twice those of beef. Bison-rearing is growing in importance in Canada, the USA and Scotland. The animals need no shelter even in Edmonton, Canada, where winter temperatures fall to −40°, and the animals tend to be healthy and easy to manage. **Water buffalo** meat is similarly beef-like and low in fat and cholesterol. By 2005, there were about 2,500 water buffalo in Britain reared for milk, cheese and meat. According to the Napton Water Buffalo company (www.waterbuffalo.co.uk), water buffalo meat has 44% less fat, 40% less cholesterol and 55% fewer calories than traditional beef. Water buffalo are related to Asian and African buffalo, not to the American bison, and the Asian breed used in the UK is very docile, unlike many Asian breeds.

See Subsec. 13.7.11 for a new breed of Peruvian guinea pigs, claimed to produce meat low in cholesterol and high in protein. **Venison** from deer is low in fat, cholesterol, sodium and saturated fatty acids, and high in copper, iron and zinc (http://www.maf.govt.nz/mafnet/rural-nz, accessed 22/10/2004). See Sec. 13.10 for more on venison production. **Velvet** is the young antlers (skin and horn) of male deer and of male and female reindeer, removed under anaesthetic within 55–60 days' growth. It is used in Chinese medicine and is claimed to prevent or reduce osteoporosis and arthritis, to promote healing after surgery and stamina in athletes. In New Zealand, which exports 98% of the products from its one million farmed deer, the velvet is worth as much as the venison. Most antler velvet is produced in Russia which has 3.5 million reindeer and 258,000 elk and red deer, followed by New Zealand with 200,000 producing male red deer, China with 180,000 producing deer, and lesser amounts from the Republic of Korea, North America and Australia, with world production of about 10,000 tonnes a year. European countries generally do not produce antler velvet; in most of them, it is illegal to harvest it, presumably on animal welfare grounds.

As well as good sales for fat-reduced milk products on possible health grounds, there is also a market for **non-dairy milk-substitutes** for vegetarians and people unable to take dairy products, because of lactose-intolerance. A Swedish company markets Mill Milk, made from rapeseed oil and oats, in natural and mango flavours. Soya milk is good for lactose-intolerant people, and there are oat-based alternatives such as Oatly, rice-based ones such as Rice Dream and Provamel Rice Drink. Coconut milk is rich in iron and selenium, and almond milk has a delicate flavour. With cholesterol consciousness ever increasing, preparations containing plant sterols to lower blood cholesterol, by reducing their passage through the gut wall into the bloodstream, are being marketed, such as Flora pro-activ®, a butter-substitute spread and also a yoghourt drink. "Probiotic" yoghourts and drinks containing so-called "good bacteria" are also strongly marketed.

A surprising development in dairy products is the increased popularity of **horse milk** in Europe, even when a litre costs as much as €11 (£7). It is rich in minerals and vitamins, is easily digested and has only 1.5% fat compared with 3.7% in cow's milk. In 10 years, production has changed from isolated small German units to dozens of large-scale units in Germany, Austria, France, Belgium and Holland. A Belgian producer has 57 mares and exports more than 75,000 litres to America and European countries. A mare only produces about 3 litres of milk a day, compared with 40 litres per cow. Horse milk is sweeter than cow's milk and has been claimed to help with skin conditions such as eczema and dermatitis if drunk or rubbed into the skin. It has also been used by hospitals for premature babies to increase strength and immunity to diseases. Selected placid breeds of horse are used, such as the South Tyrolean Haflinger. They give birth each year and are stabled with their foals for months, whereas cows are separated from their calf after about a week. **Other milks** are increasingly available. For people with allergies to cows' milk, goats' milk is sometimes a useful substitute, and may help with eczema and some intestinal problems. Buffalo milk is richer than cows' milk in calcium (good for bones), proteins and fats.

Some foods are promoted as being good for health, e.g., having **anti-cancer properties**. Resveratrol from red grapes and red wines is said to "mop up harmful free radicals which damage DNA", as is lycopene red pigment from tomatoes. A report (Roger Highfield, *The Daily Telegraph*, 7/12/2005, pp. 10) was headed "Super broccoli may protect millions from cancer". The Institute of Food Research in Norwich has bred broccoli with 3.4 times the usual amount of sulphoraphane, with no change in taste. This isothiocyanate has powerful anti-cancer properties. About half the human population has a gene, GSTM1, which enables them to retain this anti-cancer compound, while the half lacking this gene excretes it rapidly. A number of other cruciferous plants, such as cabbage, cauliflowers, sprouts and turnips also produce isothiocyanates.

## 19.2.5. *Discounted cashflow assessment of breeding programmes*

**Different breeding programmes** give different rates of response and cost different amounts to run. Some result in an immediate improvement, while others may take years, as with the progeny testing of bulls for milk yield genes (Sec. 5.7). Breeders are interested in the rate of response per unit time, and the times of investment and benefit. To make the best decisions in the long term between alternative breeding programmes, one needs to work out their economics over time, e.g., a 20-year period. The **discounted cashflow** method is the standard approach.

One predicts some realistic annual rate of compound interest that an investment might receive, e.g., 8%, in order to compare the effects of investment in the breeding programmes at different times. The earlier the money is invested in the breeding programme, rather than in a bank, the greater its relative cost. £1 invested in year 1 would have produced £1.08 by year 2 if invested at 8% interest, and $£(1.08)^2 = £1.166$ by year 3. The cost of investing £1 in the breeding programme in year 1 is £1, compared with $£1/1.08 = £0.92$ in year 2, $£1/1.166 = £0.857$ in year 3, while investing £1 in year 20 only costs £0.232, in relation to year 1. Thus, we can calculate a discount factor for each year, multiplying costs and returns in each year by the discount factor to assess the net benefits or losses at different stages of the programmes.

This is shown in Table 19.1 for **two possible breeding programmes**. Programme 1 has consistently fairly heavy costs each year and no commercial benefit until year 6, perhaps from the progeny testing of bulls. Programme 2 has a very heavy initial cost, then low costs thereafter, with benefits from year 7. Programme 1 shows a net profit after 17 years, compared with only 14 years for programme 2. Programme 2 has a net discounted sum advantage over programme 1 after only 4 years, so with this interest rate, programme 2 would be preferred to programme 1. One

**Table 19.1.**  A discounted cashflow assessment of alternative cattle breeding programmes.

| | | Programme 1 | | | | | Programme 2 | | | | |
|---|---|---|---|---|---|---|---|---|---|---|---|
| | | Actual | | | Discounted | | Actual | | | Discounted | |
| Year | Discount factor | Costs that year, £k | Returns that year, £k | Net balance that year, £k | Net balance that year, £k | Sum over all years, £k | Costs that year, £k | Returns that year, £k | Net balance that year, £k | Net balance that year, £k | Sum over all years, £k |
| 1 | | 250 | 0 | −250 | −250 | −250 | 600 | 0 | −600 | −600 | −600 |
| 2 | 0.926 | 250 | 0 | −250 | −232 | −482 | 100 | 0 | −100 | −93 | −693 |
| 3 | 0.857 | 250 | 0 | −250 | −214 | −696 | 100 | 0 | −100 | −86 | −779 |
| 4 | 0.794 | 250 | 0 | −250 | −199 | −895 | 100 | 0 | −100 | −79 | −858 |
| 5 | 0.735 | 250 | 0 | −250 | −184 | −1,079 | 100 | 0 | −100 | −74 | −932 |
| 6 | 0.681 | 250 | 100 | −150 | −102 | −1,181 | 100 | 0 | −100 | −68 | −1,000 |
| 7 | 0.630 | 250 | 150 | −100 | −63 | −1,244 | 100 | 50 | −50 | −32 | −1,032 |
| 8 | 0.584 | 250 | 200 | −50 | −29 | −1,273 | 100 | 100 | 0 | 0 | −1,032 |
| 9 | 0.540 | 250 | 300 | 50 | 27 | −1,246 | 100 | 200 | 100 | 54 | −978 |
| 10 | 0.500 | 250 | 400 | 150 | 75 | −1,171 | 100 | 300 | 200 | 100 | −878 |
| 11 | 0.463 | 250 | 500 | 250 | 116 | −1,055 | 100 | 400 | 300 | 139 | −739 |
| 12 | 0.429 | 250 | 600 | 350 | 150 | −905 | 100 | 600 | 500 | 215 | −524 |
| 13 | 0.397 | 250 | 700 | 450 | 179 | −726 | 80 | 800 | 720 | 286 | −238 |
| 14 | 0.368 | 250 | 800 | 550 | 202 | −524 | 80 | 1,000 | 920 | 339 | **101** |
| 15 | 0.340 | 250 | 900 | 650 | 221 | −303 | 80 | 1,200 | 1,120 | 381 | 482 |
| 16 | 0.315 | 250 | 1,000 | 750 | 236 | −67 | 80 | 1,400 | 1,320 | 416 | 898 |
| 17 | 0.292 | 250 | 1,100 | 850 | 248 | **181** | 80 | 1,500 | 1,420 | 415 | 1,313 |
| 18 | 0.270 | 250 | 1,200 | 950 | 257 | 438 | 80 | 1,600 | 1,520 | 410 | 1,723 |
| 19 | 0.250 | 250 | 1,300 | 1,050 | 263 | 701 | 80 | 1,700 | 1,620 | 405 | 2,128 |
| 20 | 0.232 | 250 | 1,500 | 1,250 | 290 | 991 | 80 | 1,800 | 1,720 | 399 | 2,527 |

could repeat the calculations with different interest rates. **Inflation** would affect both costs and returns, though perhaps not equally, so the calculations are partly balanced against the effects of inflation. Even a very well-planned breeding programme can have its success seriously affected by changes in government policies, regulations and price-support, or by unforeseen circumstances such as the BSE beef crisis in the 1990s in the UK.

## 19.2.6. *Breeders' rights*

The British Society of Plant Breeders provided the following information. In the UK, a plant breeder's right is a form of intellectual property granted to protect plant varieties. It is only granted to varieties that comply with certain criteria laid down by statute. The varieties are tested during two years of National List trials carried out independently by an organisation contracted by the relevant Agricultural Ministry, depending on whether the trials are undertaken in Scotland, England & Wales or Northern Ireland. The right can either give protection in the UK only (National right) or across Europe (EU right). The Plant Variety Office in Cambridge, England, and the Community Plant Variety Office in Angers, France, grant the rights for UK and EU protection respectively. The right entitles the holder to control the following specific acts undertaken with propagating material of the protected variety: production or reproduction (multiplication); conditioning for the purpose of propagation; offering for sale; selling or other marketing; exporting from the Community; importing to the Community; stocking for any of the purposes above mentioned. The holder of those rights can authorise others to carry out these acts under whatever terms and conditions they wish to impose. The plant variety right can therefore be licensed and a royalty collected in exchange for the use of that intellectual property.

In 1987, the **British Society of Plant Breeders** (BSPB) was established by combining the Plant Royalty Bureau and the British Association of Plant Breeders. Its members comprise nearly all public and private sector plant breeders in the UK. The BSPB represents the political, commercial and technical interests of the plant breeding industry. It promotes plant breeding's role in maintaining the competitiveness of UK agriculture and encourages the education of plant breeders.

Its other function is to collect and distribute **royalties** on behalf of the breeders, originally under the 1964 UK Plant Varieties and Seeds Act. This extremely important Act gave plant breeders the exclusive right to charge a royalty for a licence to produce and sell a variety. The Plant Varieties Act 1997 brings UK law into line with the EU legislation agreed in 1994. A major change was to enable the BSPB to charge royalties (at a reduced rate compared to seed sales) on **farm-saved** seed from 8th May 1998, for reuse on the same farm, which had been exempted from previous legislation, which only referred to sale of seed. Popular old varieties are often free of this charge, but all recent varieties come under it. In 1997/98, the BSPB received about £2.3 million from farm-saved seed royalties. Payments for particular crops were agreed between the BSPB and the farmers' unions. Unlike the royalties of certified seed, farm-saved seed payments are fixed for each crop species, regardless of variety. They are charged on an area-sown basis, e.g., for 1998/99 per ha, £4.76 for wheat, £8.57 for oilseed rape, or per tonne if cleaned by a BSPB-registered processor, e.g., for 1998/99 per tonne, £25.03 for wheat, £27.42 for beans.

Licensing schemes now operate in the UK for cereals, fodder plants and legumes, tree fruits, linseed, potatoes, soft fruit, cane fruits, tree fruit, rootstocks, oilseed rape, field peas and beans, bush fruits and triticale. The breeder grants a Head Licence to the BSPB which can issue sub-licences to seed merchants for the production and sale of seeds of that variety. Royalties are on the sale of the certified seed (e.g., cereals) and/or the hectarage sown (e.g., vining peas). The BSPB collects the royalties and passes them on to the breeder after deductions for the BSPB running costs. The royalty is about £8 a tonne on seed potatoes, equivalent to less than 50p on each tonne of potatoes sold. On cereals, it is just over £6 a hectare. For certified seed for sale (i.e., not farm-saved), the royalties differ between varieties of a crop, and are fixed by the breeding firm owning the rights. Thus, in 1999, royalty rates per tonne on Otira spring barley were £70 for first generation seed and £50 for second generation seed, while for Saloon spring barley, they were £91 and £53.50 respectively. Breeders' rights last 30 years for potatoes, raspberries, apples and rootstock crops, and 20 years for other crops such as cereals. Wheat charges average about £42 a ton, perhaps £70 when first released, decreasing to perhaps £30 subsequently. This information was provided by the BSPB in 1999. Similar schemes operate for plant crops in the rest of the EU, but arrangements elsewhere are very variable.

There is no equivalent organisation or system of breeders' rights for **animals** in the UK. For beef cattle, sheep and pigs, there is a **slaughter levy** to fund the Meat and Livestock Commission, which does recording of animal performance. In recent years, the Commission has formed an organisation named Signet to provide performance records and genetic evaluations on beef cattle and sheep. For dairy cattle, there is a **milk levy** for the Milk Development Council, which funds an Animal Data Centre for genetic evaluation. Animal breeders just get payment for the animals or semen they have developed, with no continuing royalties on subsequent sales.

## 19.2.7. *Breeding or using for niche markets*

Modern wheat varieties are often bred for specific **niche markets** such as bread or biscuits. Modern oilseed rape is bred to suit a lot of different industrial needs, with different chemical products in specific strains, as well as for rapeseed oil for cooking. Potato variety Dell was bred for blight resistance, but the blight fungus developed strains that are able to overcome its major genes for blight-resistance. However, it remains a leading variety because it is so suitable for producing frozen chips. Potato variety Russett Burbank was bred in 1876 in the USA and has little or no resistance to blight, viruses or eelworm. It gives long and often bent tubers, but is on the UK National List, because the McDonalds international restaurant chain uses it in enormous quantities for long thin chips or French fries which do not break in the middle (Dunnett, 2000). It is even being grown in Russia for McDonalds in Moscow. It was not bred for making par-fried frozen chips even though it was found to be very suitable.

It must not be supposed that the most-widely used varieties are necessarily recently bred. Dunnett (2000) went into a branch of Marks and Spencer's in Inverness, Scotland, in 1998, and found only the following potato varieties on sale (their number of years since being bred is given in brackets): Jersey Royal (120), King Edward (97), Maris Peer (37), Desirée (37), Maris Piper (36), Nicola (25) and Marfona (16). Some of the more modern varieties, including the second earlies white Nadine and red Stroma, both bred by Dunnett (2000), are very widely cultivated in Britain and abroad. Nadine and Stroma both are high-yielding with a lot of disease resistances, and because they produce fairly even-sized tubers, they are excellent for the supermarket prepacked potato market. As well as having good virus resistance, Stroma has very good tolerance to, or resistance to, various eelworm pathotypes, which would become even more valuable if current nematicides were banned on health grounds.

## 19.2.8. *Who does the breeding?*

Breeding plants, animals and microbes may be done by isolated individuals, family businesses, small or large firms, government research stations, or by international consortia or several of these simultaneously. Much depends on the organism's reproductive potential and its economics. Dunnett (2000) is an example of a very successful independent potato breeder. His variety Nadine had sales of **seed potatoes** of more than a million pounds in its third year of sale, and is now more widely grown than any UK state-bred variety that entered the British market over the past 16 years in the case of the Cambridge Plant Breeding Institute, and over the past 20 years for the Scottish Plant Breeding Station, even though he could only assess 5,000 lines a year. In contrast, the state-funded Plant Breeding Stations in Scotland, England and Northern Ireland could raise 200,000 seedlings a year, while a million seedlings a year were raised in the Netherlands, where much work was done at Wageningen University and International Agricultural Centre.

**Apple breeding** has often been done by individuals, selecting seedlings from deliberate or even accidental crosses, and multiplying up successful variants by grafting onto suitable rootstocks since apples do not breed true. Raspberry breeding in the UK has largely been done by small family firms, while breeding wheat, barley and other cereals has usually been done by research stations or large companies, with millions of seeds being germinated in trials each year. With pigs and sheep, farmers often make their own selections among their own animals, but may buy in selected semen or hire superior males when needed. Cattle breeding tends to be done by fairly large organisations. There are often national animal breeding programmes with elite herds.

## Suggested Reading

*An Outline of United Kingdom Competition Policy*. (1998) The Office of Fair Trading/The Stationery Office, London.

Ansell, D. J. and R. B. Tranter, *Set-aside: in Theory and in Practice*. (1992) University of Reading, Reading.

Dunnett, J., *A Scottish Potato Breeder's Harvest*. (2000) North of Scotland Newspapers, Inverness.

Epp, D. J. and J. W. Malone, *Introduction to Agricultural Economics*. (1981) Macmillan, New York.

Mathew, B., The secret of the knights star lily. *Kew* (1999) 34–37.

*Monopolies and Anti-competitive Practices*. (1995) The Office of Fair Trading/The Stationery Office, London.

Murphy, M. C., *Organic Farming as a Business in Great Britain*. (1992) Agricultural Economics Unit, University of Cambridge, Cambridge.

Pavord, A., *The Tulip*. (1999) Bloomsbury, London.

Turner, J. and M. Taylor, *Applied Farm Management*. (1988) Blackwell Scientific, Oxford.

*Whitaker's Almanack*. (1999) J. Whitaker & Sons Ltd, London.

# Index

$\alpha$1-antitrypsin, 431
$\alpha$-thalassaemia, 262, 263
$\alpha$-thalassaemia trait, 263
2-micron plasmid, 433, 550
46,XX, 297
46,XY, 297
47,XXX, 297
47,XYY, 297
£PLI, 131

abnormalities of chromosome number, 299
abortion, 514
abortuses, 298
accessions, 453
achondroplasia, 256
achondroplastic dwarfism, 258, 299, 516
acriflavin, 541
acriflavin resistance, 540
adaptation, 101, 114, 120, 161, 245, 447, 450
adapting, 176
adaptive gene combinations, 101
addition lines, 236
additive action, 19, 20, 22, 38, 41, 52, 56, 59, 66, 67, 80, 81, 154, 446
additive effects, 20, 267, 500
adeno-associated virus, 340
adenosine deaminase deficiency (ADA), 342, 343
adenovirus, 340, 341, 343, 344
adult or adolescent screening, 313
adult stem cells, 345
adultery, 248
*Aegilops*, 234, 236

aflatoxins, 279
*Agaricus bisporus*, 526, 527
age, 274, 276
agricultural mass selection, 364, 366, 377, 447
*Agrobacterium*, 423–425
*Agrobacterium tumefaciens*, 547
*Agropyron*, 234, 235
*Agrostis*, 437
AI, 124, 132, 141, 171, 391, 398, 498, 499, 508–510 (*see* also Artificial Insemination)
airline pilots, 280
albinism, 261, 275, 317, 318, 325
albinos, 39, 225, 260, 584
alcohol, 520, 527, 548
alcohol tolerance, 548
alcoholism, 273
ale yeasts, 549
alfalfa, 145, 224, 437, 466, 468, 485
alkaptonuria, 260, 261, 318
alkylating agents, 169
allele frequency, 86, 88, 92, 109, 145, 272, 444, 449
allele ratios, 184
alleles, 14, 15, 26
allohexaploids, 231
allopolyploids, 221, 231, 232, 238, 371–375, 494, 519
allosteric binding, 545
allotetraploid, 221, 228, 230, 231, 242, 371, 372, 376
alpaca, 370, 452, 502
alpha-fetoprotein (AFP), 309, 322

aluminium tolerance, 427
Alzheimer's disease, 251, 268, 290, 422
Ames test for mutagenicity, 171, 279
amniocentesis, 287, 307, 311, 312, 314
amorph, 17
anaemia, 263, 265, 303, 514
anencephaly, 309
aneuploid, 17, 18, 220, 230, 232, 385, 551
aneuploidy, 445, 486
anhidrotic ectodermal dysplasia, 270
animal breeders, 592
animal breeding programmes, 377, 593
animal cloning, 511, 513
ankylosing spondylitis, 306
anther culture, 225–227, 362, 470, 493
anti-bacterial antibiotic, 542
anti-fungal antibiotic, 542
anti-sense RNA, 421
antibiotic, 434, 527, 528, 544, 545, 557
antibiotic production, 545
antibiotic resistance, 437, 547, 554
antigen, 299
*Antirrhinum*, 39, 229, 477
aphid, 115, 462, 466, 467
apomixis, 450, 487, 488
apples, 115, 178, 179, 196, 197, 228, 367,
    371, 374, 426, 456, 462, 463, 465–467,
    480, 483–485, 487, 488, 490, 566, 581,
    585, 591
apricots, 179
areca nut, 278
*Arabidopsis*, 437, 438
Argentinean beef, 393
arthritis, 260, 273, 307, 315
artichokes, 460
Artificial Insemination (AI), 389, 391,
    398, 496, 505, 507, 509,
artificial selection, 114, 495
artificial vaginas, 507
*Ascobolus*, 521
*Ascobolus immersus*, 524
Ascomycetes, 519, 520, 528, 534
ascospores, 520, 521, 523, 541, 547
asexual reproduction, 487, 488, 540
asexual spores, 523
Ashkenazi Jews, 313, 314

asparagus, 40, 470, 224, 424, 492
*Aspergillus*, 519, 533, 541, 547
*Aspergillus aculeatus*, 548
*Aspergillus awamori*, 547
*Aspergillus nidulans*, 521, 523, 524, 529,
    533, 534, 536, 540, 543, 544, 557, 558
*Aspergillus niger*, 519, 521, 525, 527, 534,
    546, 554, 555
*Aspergillus oryzae*, 527, 547, 554
*Aspergillus sojae*, 547
*Aspergillus terreus*, 527, 546
association studies, 202
assortative mating, 34, 86, 140–142
asthma, 255
atopic disease, 254
autism, 273
autoimmune, 347
autoimmune conditions, 253
autoimmune disease, 268, 306
autoimmune-type diseases, 337
autopolyploidy, 486
autosomal abnormalities, 289, 294
autosomal dominant allele, 245, 255, 256,
    268, 313, 320, 321, 323, 516
autosomal genes, 256
autosomal loci, 38
autosomal recessive allele, 245, 248, 255,
    257, 278, 323, 325
autosomal recessive conditions, 323
autosomal recessive disorder, 259, 321,
    342
autotetraploid, 221, 223, 228–231, 242,
    376, 380, 443, 445
autozygous, 152
autumn crocus, 222
auxotrophy, 522, 535, 536, 542, 552, 558
average number of recessive lethal alleles
    per person, 316
average pair differences, 249

$\beta$-galactosidase, 546
$\beta$-glucuronidase gene (*Gus*), 424
$\beta$-thalassaemia, 262–264, 310, 311, 314,
    315, 339
$\beta$-thalassaemia intermedia, 264
$\beta$-thalassaemia major, 264

$\beta^+$-thalassaemia, 264
$\beta^0$-thalassaemia, 264
*Bacteroides fragilis*, 213
B centromeres, 193
B chromosome, 194, 238–242
back-mutation, 23, 111
bacteria, 456, 542
bacterial antibiotics, 545
baculovirus, 463
baker's yeast, 525, 526, 543, 552
baking, 520
balanced lethal recessives, 217
balanced lethal system, 219
balanced non-reciprocal translocation, 284
balanced translocation, 295, 299, 311
bamboos, 492
banana, 178, 228, 427, 460, 484, 487, 492, 579
barley, 17, 56, 101, 117, 133, 173, 179, 192, 197, 226, 366, 377, 404, 427, 428, 470, 472, 483, 485, 591, 593
Barr bodies, 270, 307, 497
barriers to self-pollination, 483
barstar, 426
base analogue, 168, 277
base substitution, 264, 312, 522
basic risk, 321, 322
Basidiomycetes, 520, 534
basidiospores, 523, 541
bastardisation, 362, 474, 475
batch fermentations, 550
bean, 32, 265, 363, 364, 470, 581, 591
beech, 23
beef, 395, 565, 578, 580, 581
beef calves, 504
beef cattle, 158, 396
    beef breeds, 392, 402
    beef herds, 363
beer, 520, 523, 527, 546, 548
beer yeasts, 548
bees, 223
beet, 10, 376
beetles, 238
beetroot, 470
behavioural traits, 13
behaviours, 401

bentgrass, 104
benzypyrenes, 278
Best Linear Unbiased Predictions (BLUP), 83
beta thalassaemia, 259
betel, 278, 279
binucleate, 474
binucleate grains, 476
biodiversity, 437, 455
biolistic method, 415, 554
biolistics, 424
biological control, 463, 528
biotypes, 468
bird, 120, 406, 448
birth weight, 117, 332
bison, 585, 586
bivalents, 221, 230, 232, 235, 242
black market, 563
blackberry, 20, 223, 369, 492
blastocyst, 429
blindness, 325
blood, 336, 430
blood cancers, 347
blood chimera, 337
blood clotting, 334
blood donor, 336
blood group, 16, 21, 33, 34, 89, 91, 141, 299–303, 305, 335, 337
    ABO, 299–304, 330, 346
    Duffy, 300
    Lewis, 300
    M/N, 253, 299, 300, 330
    Rhesus, 299–301, 303–305, 310, 346
    S, 300
blood transfusions, 263, 264, 299, 301, 302, 337, 515
blowflies, 481, 482
BLUP, 395
bone marrow, 337, 339, 343, 345–347, 515
bone marrow transplants, 264, 347
Booroola high fecundity allele, 21, 119
bottom-fermenting yeast, 548
bovine spongiform encephalopathy (BSE), 565
*Brassica*, 226, 354, 361, 405, 448, 480, 482

brassicas, 426, 437, 438, 479, 482, 483
Brazil, 439
Brazil nuts, 439
bread, 402, 555
bread wheat, 232, 234, 235
bread yeasts, 548, 553
bread-making, 552
breast cancer, 315, 326
breed handbook, 400
breed society, 400, 408
breeder's rights, 364
breeding, 547
breeding for shows, 399
breeding programmes, 377, 588
breeding seasons, 501, 502
breeding system, 438
breeding value, 80, 82, 83, 125, 389
brewing yeast, 519, 551
bridge species, 236
bridging species, 382
British National Improvement Scheme for
     Pigs, 133
British Society of Plant Breeders, 402,
     426, 590
British Texel, 500
brittle bone disease, 322
broad beans, 379
broad sense heritability, 74, 77
broccoli, 587
brother-sister and half-brother-half sister
     marriages, 316
brother-sister mating, 144
Brussels sprouts, 227, 470, 473, 481, 483,
     493
Bt cotton, 467
Bt toxins, 425, 435, 439, 466, 467
bud-pollination, 362
buffalo milk, 587
bulb, 492
bulk population breeding, 364
bull, 399, 506
bull semen, 498
bull sperm, 508
bullfighting, 396
bulls, *see* cattle
*Burkholderia*, 262

butter, 578
butterfly, 120, 448
buyer's specification, 397, 398

cabbage, 229, 426, 483, 588
Caesarian section, 333
caffeine, 277, 278, 422
calico cats, 269, 270
callus, 492
callus culture, 485
camel, 585
camel-guanaco hybrid, 370
camelids, 370, 371, 503, 504
camels, 9, 370, 371, 501, 507
cancer, 120, 162, 175, 200, 268, 273, 274,
     276, 278–281, 314, 315, 320, 342, 344,
     345, 588
cancer genes, 282
carcase characters, 125, 136, 137, 394
     carcase characteristics, 353
carcinogen, 274, 278
carcinogenicity, 279
carrier detection, 266, 314
carrots, 483, 585
cartels, 571
cascade screening, 313
cats, 15–17, 22, 34, 53, 54, 71, 121, 122,
     129, 130, 151, 152, 169, 170, 269, 370,
     400–402, 496, 500–504, 512
cat cloning, 512
cattle, 17, 58, 61, 63, 78, 83, 119, 121, 126,
     132, 133, 137, 146, 248, 305, 353, 363,
     377, 389, 400, 408, 433, 452, 462, 464,
     496–499, 501, 504, 505, 507, 508, 511
     Aberdeen-Angus, 58, 89, 92, 363,
          389, 389, 392–394
     Angus, 390, 392
     Ayrshire, 389–392, 400, 510
     beef, 136
     beef bulls, 59
     beef cattle, 5, 9, 127, 137, 157, 158
     Belgian Blue, 395, 499
     Bonsmara cattle, 371
     Brahmin, 363
     bull, 123, 124, 132, 501, 510, 579,
          588

bull fighting, 396
calves, 556
Charolais, 59, 390, 392, 395, 499, 508
Chillingham Wild Cattle, 147
cow, 503, 504, 506, 507, 510, 579, 587
dairy cattle, 9, 121, 123, 130, 135, 499
dairy cows, 9, 14, 124
Friesian, 395, 396
Guernsey, 389, 390
Hereford, 58, 59, 363, 390, 392, 400
Hereford/Shorthorn, 371
Holstein-Friesians, 131, 132, 396
Holsteins, 131, 396, 498, 499
Indian Brahman, 371
Limousin, 83, 392–394, 508
native cattle, 371
Polled Herefords, 395
Saler, 392, 394
Scottish Highland, 395
Shorthorn, 363
Simmental, 390, 392
super-cow, 510
Zebu 363
cattle breeding, 593
cattle breeding programmes, 589
cattle/cows/bulls, *see* cattle
cell fusion methods, 541
centric fusions, 295
centromere, 539
centromere distance, 529, 531
centromere marker technique, 531
cephalosporin C, 558, 559
*Cephalosporium*, 527, 534
*Cephalosporium acremonium*, 559
cereals, 591
certification, 402
chance, 87, 97, 105
cheese, 527, 557, 574, 586
cheetah, 100, 149, 370
chemical analogue, 545
chemical mutagens, 168, 178, 278, 546
chemicals, 277
chemosterilants, 462, 463

chemotherapy, 345
cherries, 179, 452, 456, 480, 483
chiasmata, 242, 294
chickens, 14, 17, 76, 115, 136, 198, 433, 496, 497, 508, 509, 565
chicken genome, 198
chicken polymorphism, 197
chickpea, 179
Chihuahua, 400
chimera, 178, 335–338, 346, 493
chimeric, 178
chimpanzee, 205, 295
chips, 592
chlorate resistance, 544
chloroplast genes, 105
chloroplast genome, 493
chloroplasts, 424
chorionic villus sampling, 308, 311
chromatid interference, 186, 533
chromosomal non-disjunction, 334
chromosome aberration, 24, 135, 186, 205, 280, 282, 283, 334, 443, 446, 522, 544
chromosome abnormalities, 178, 281, 297, 298, 307, 311
chromosome addition lines, 234
chromosome additions, 234
chromosome breakage, 280, 282
chromosome imbalance, 290
chromosome interference, 186
chromosome number, 446, 525
chromosome number abnormalities, 283
chromosome prints, 286
chromosome rearrangements, 272, 290
chromosome substitutions, 234
chromosomes, 198
chronic myelogenous leukaemia, 347
chronic myeloid leukaemia, 216, 282
*Chrysanthemum*, 476
chrysanthemums, 471, 472, 485, 488
chymosin, 436, 556
cirrhosis, 273
*cis/trans* allelism test, 26, 27, 195, 521, 541
citrus, 487, 491, 555, 566, 581, 582
citrus fruits, 476
cleft lip, 63, 250, 251, 255, 267, 275, 326

cleft palate, 63
cleistothecia, 521, 529
climatic change, 452
clinical cytogenetics, 288
cloned cattle, 512
cloned mice, 513
cloning, 450, 487
cloning of animals, 513
cloning vectors, 433
close inbreeding, 159
closed stud method, 150, 151
clover, 14, 472, 474, 480, 483
club foot, 63, 250, 251, 267, 309, 411
co-adapted gene complexes, 11
co-dominance, 19, 21
co-transformation, 425
coca, 380
cocaine, 380
coconut, 9, 103, 124, 356, 357, 459, 587
codling moth, 463–465
coeliac disease, 307
coffee, 2, 277, 278, 473, 573
coffee drinking, 252
Cogent, 497, 498
cointegrate plasmid, 423
cola, 277
colchicine, 222, 223, 226, 242, 283, 371, 376, 380, 384
cold-storage, 581
colorectal cancer, 313
colour blind, 272
colour blindness, 15, 40, 41, 265, 297, 329
combinatorial approach, 543
combine harvesting, 379
command system, 562, 564
commercial importance of fungi, 527
commercial selection, 31
Common Agricultural Policy, 578–580
common ancestor, 152, 156, 158
comparative genomic hybridisation, 288, 291, 292
competition, 135, 495, 570, 573
complementary dominant alleles, 47
complementary dominant genes, 48
complementation, 355, 544
complementation *cis/trans* tests, 558

complementation groups, 558
complements, 565
complete parasexual cycle, 540
composite cross, 366, 447
computer-aided tomography, 136
conception rates, 508, 509
concordance score, 250, 251, 255
conditional lethal mutations, 465
conditional mutations, 70, 71, 446
congenital, 18
congenital hypothyroidism, 313
congenital malformations, 315, 332
conidia, 523
conidial colour mutants, 541
conifers, 371, 374
consanguineous, 316
consanguineous marriages, 317, 318
conservation, 455, 577, 585
conservation programmes, 452
consumer acceptability, 383
continuous variation, 49, 52, 61
contraceptive, 281
control of oestrous, 504
convenience foods, 566
*Coprinus cinereus*, 524
*Coprinus comatus*, 524
*Coprinus fimeterius*, 524
*Coprinus lagopus*, 534
cord blood, 345, 347, 411
cord blood stem cells, 347
cord-blood transplantations, 346
corms, 492
coronary disease, 313
coronary sclerosis, 273
correlated characters, 126
correlation, 9, 75, 79, 80, 151, 388
correlations between characters, 62
cosmic rays, 276, 280, 281
cost-benefit assessment, 465
cotton, 179, 192, 425, 426, 436
coupling, 25
coupling and repulsion, 24
coupling phase, 43
cousin-mating, 358
cows, *see* cattle
cows' milk, 587

*Crepis*, 240
Cri-du-Chat syndrome, 210, 295
criminal record, 250, 251
crocodiles, 397
Crohn's disease, 199
cross-breeding, 115, 380, 385, 387
cross-pollination, 140, 141, 375
crossing methods, 481
crossing over, 147, 183, 234, 245, 444
crossover, 236, 295, 539
cryo-preservation, 461
cucumbers, 19
*Cucurbita*, 45
*Cucurbita pepo*, marrow, 45, 50
customer choice, 396
cut-flower crops, 470
cuttings, 460, 485, 488, 489, 493
cyclic selection, 119, 120, 211, 377, 444
cysteine, 451
cystic fibrosis, 256, 261, 268, 299, 311,
    312, 314, 339, 343, 418, 430
cytoplasmic factors, 360
cytoplasmic male sterility, 360, 361, 465,
    495

*Dactylis glomerata*, 229
dahlia, 221, 369
dairy products, 546
dark-repair, 175
date palm, 492
*Datura*, 224
day length, 381, 386, 502
deafness, 142, 318, 322, 325
deer, 9, 17, 174, 406, 496, 500, 502, 512,
    586
degree days, 471
deleterious dominant allele, 108, 111
deleterious mutations, 178
deleterious rare recessive, 111
deleterious recessive allele, 108, 145, 147,
    224, 359, 445
deletion, 193, 165, 205, 207–211, 214,
    261, 294, 298, 302, 419, 434, 486, 515,
    516, 522, 553
deletion mutations, 266
demand, 566

demand curve, 566, 570
departure from equilibrium, 90, 92
*Delphinium*, 101
descent of the testicles, 501
designer babies, 514
determination of supply, 568
diabetes, 268, 315, 334, 347, 431, 432
diabetes mellitus type 1 (insulin
    dependent), 251, 267, 268, 375
diabetes mellitus type 2 (non-insulin
    dependent), 250, 268, 375
diet, 277, 313, 325
dietary mutagens, 279
dietary problems, 451
dikaryotic $(n + n)$ stage, 519
dioecism, 483
diploid, 220, 222, 223, 228, 232, 236, 371,
    465, 535
direct injection of DNA, 341
direct selection, 541
directional dominance, 53
directional gene flow, 102
directional selection, 118, 445
discounted cashflow, 588, 589
disease resistance, 382, 383, 403, 404,
    426, 593
diseases, 273
disequilibrium, 80, 202
disomic, 235
disruptive selection, 120
dizygotic twins, 249, 251, 255
DNA analysis, 305, 307, 308, 314,
    322
DNA fingerprinting, 200, 407, 408
DNA methylase, 412, 413
DNA methylation, 269, 331
DNA probe, 307, 514
DNA screening, 314
DNA testing, 330
DNA tests, 305
dogs, 4, 17, 19, 121, 122, 150, 151, 400,
    401, 496, 497, 506, 509, 510
Dolly, 511
dominance, 19, 65, 66, 68, 446, 477
    additive action, 187
    complete dominance, 154, 187

incomplete dominance, 187
  overdominance, 38, 41, 187
dominance hypothesis, 354, 355, 359
dominant alleles, 106
dominant autosomal gene, 267
donkey, 353, 369, 370
double allohexaploid, 232
double crossovers, 538
double first cousins, 319
double-mutant survival, 543
double-strand breaks, 206, 280
doughs, 552
Down syndrome, 18, 205, 233, 250, 251,
  287, 289, 291, 293–295, 298, 308, 309,
  334, 335
drift, 408
drive, 239, 240
*Drosophila*, 26, 70, 73, 140, 147, 169, 180,
  182, 185, 211, 214, 281, 213, 240, 376
Duchenne muscular dystrophy, 265, 268,
  270, 274, 275, 299, 311, 322, 329, 433
duplication, 206, 213, 215, 294, 486, 526
dwarfing, 489
dwarfism, 256, 257, 275, 309, 318
dyslexia, 315

early amniocentesis, 308
early-maturing varieties, 581
earthworms, 148
EBVs, 395
economic system, 562
economics, 8, 593
eczema, 254, 587
Eden Project, 455
Edwards syndrome, 294, 298
effective population size, 98, 452
egg, 135, 497
egg transplantation, 510
elasticity of demand, 565
electroporation, 339, 415, 419, 554
elephants, 497
elite flock, 387
elite herds, 593
emasculation, 483
embryo dissection, 477
embryo freezing, 510, 511

embryo rescue, 362, 374
embryo transfer, 391, 510
embryonic stem (ES) cell lines, 419
embryonic stem cells, 345
embryos, 460
emigration, 445
emotions, 324
emphysema, 430, 431
EMS, 522
encapsulated enzymes, 262
end-product inhibition, 546
endangered breed, 454
endangered species, 455, 457, 512
endonucleases, 412
endosperm, 222
environment, 114, 266, 323, 444, 556
environmental deviation, 82
environmental effects, 65, 66, 70, 71,
  80–83, 250
environmental factors, 253, 267
environmental influences, 273
environmental mutagenesis, 279
environmental mutagens, 273
environmental tolerance, 120, 377
environmental variation, 52, 56, 60, 446
environmentally-sensitive genetic male
  sterile line, 362
epigenetic changes, 331, 486
epigenetic markers, 309
epigenetic reprogramming, 513
epigenetic signals, 252
episome, 340
epistasis, 45–47, 65, 66, 446, 537
equation for quantitative characters, 133
equilibrium between mutation, selection
  and gene conversion, 111, 448
*Escherichia coli*, 26, 412, 414–416, 419,
  425, 430
Estimated Breeding Values (EBV), 83,
  389, 398
ethylene, 421, 484
eucalyptus, 354
*Euonymus*, 494
European Union (EU), 389, 402, 404, 405,
  436, 438, 456, 576, 579, 580, 591
EU method, 578

EU Set-Aside scheme, 405
euploid, 17
European corn borer, 425, 468
evolution, 202, 213, 214, 221, 447, 450
evolutionary, 242
excision repair, 277, 279
exonucleases, 412
export subsidies, 579
exposures to radiation, 276
expressed sequence tags, 385
expressivity, 323
extinction, 450
extracellular enzymes, 544, 545
eye colour, 19, 22, 30, 33, 251, 311, 336
eye diseases, 345

F, 151
F1 hybrid, 224, 353–355, 356, 357,
    359–363, 366, 426, 446, 447, 493
F1 hybrid breeding programmes, 355
F1 hybrid Brussels sprouts, 361
F1 hybrid coconut, 357
F1 hybrid rice, 362
F1 hybrid tomato, 483
Factor VII, 432
Factor VIII, 265, 266
Factor IX, 431, 439
factors affecting demand, 564
Fair Trading Act 1973, 572
familial hyper-cholesterolaemia, 313
familial incidence, 253
familiality, 254, 255, 275
family balancing, 311
family selection, 125, 135
Fanconi's anaemia, 346, 347
farm-saved seed, 591
feed conversion efficiency, 126
female-sterile mutants, 483
feral herds, 406
fermentation, 549
fermented beverages, 550
ferret, 503
fertility, 78, 241, 363, 500
fertility restoration, 361
fertility restorers, 180
fertility treatments, 327, 329

*Festuca*, 236
filtration enrichment, 542
fingerprint ridges, 266, 330
first cousins, 157, 158, 315, 316
first division segregation, 530
first-cousin marriages, 317–319
first-cousin mating, 144
first-degree relatives, 326
fish, 103, 220, 239, 565
fixation, 97, 99
flax, 468
flocculation, 548, 549
flocculence, 549
*floury-2*, 451
flow, 146
flow cytometry, 135, 497–499
flow-cell cytometry, 552
flowering periods, 470
flowering times, 471
flowers, 565
fluorescent *in situ* hybridisation (FISH),
    10, 285, 287, 290, 307, 335
flushing, 505
foetal DNA, 309, 310
foetal DNA screening, 311
foetal stem cells, 345
Food and Agriculture Organisation of the
    United Nations (FAO), 453
food mountains, 578
food yeast, 527, 550
foot blistering, 247
forage grass, 437
forensic applications, 407
forest trees, 61, 224
founder effect, 99
four-way cross, 356
fowls, 362
fox, 34
foxhounds, 121
fragile-X syndrome, 165, 255, 265, 311
frame shift mutations, 279
frame shifts, 226, 264, 522
free radicals, 280
free-market, 577
frequencies, 188
frequency of twins births, 327

frequency-dependent selection, 108
frog, 103, 114, 511
frozen embryos, 510
frozen semen, 508
fruit trees, 224, 460
full-sib mating, 159, 160
fungal biotechnology, 544
fungal genetic engineering, 544
fungal genetics, 519
fungal genomics, 201
fungal sequence data, 526
fungicide, 547
*Fusarium*, 525

G-banding, 284
galactosaemia, 313
gametocides, 483
gametophytic incompatibility, 477, 478
gamma rays, 172, 177, 276, 280, 463, 464, 522
Garrod, Archibald, 261
gasahol, 527
Gaucher disease, 314
gendicine, 344
gene, 14, 15
gene activators, 412
gene chips, 200
gene conversion, 26, 111, 112, 183, 184, 444, 448, 530
gene density, 526
gene doses, 270
gene expression, 200, 340
gene flow, 33, 100–105, 147, 437, 438
gene function, 553
gene gun, 424
gene identification, 547
gene interactions, 45, 65, 477
gene pool, 391, 450
gene targeting, 419, 420
gene therapy, 4, 326, 338, 339, 341–344, 411, 416
gene transfer, 368
gene-for-gene systems, 468
gene-product interaction, 61
gene-silencing, 422
general combining ability, 358

general formulae, 187
generation intervals, 124
generation times, 132
genetic conservation, 450, 452
genetic counselling, 4, 10, 86, 263, 299, 313, 314, 319, 322, 324
genetic defect, 324
genetic disorders, 310
genetic diversity, 101, 105, 202, 453
genetic drift, 97, 99, 105, 445–447, 451, 452, 456, 460
genetic engineering, 338, 378, 380, 411, 426, 450, 466, 544, 546–548, 553, 559
genetic fingerprinting, 403, 407
genetic gain per year, 132, 133
genetic liability, 325, 326
genetic mapping, 182
genetic maps, 138, 182, 188, 191
genetic ratios, 322
genetic susceptibilities, 277
genetic symbols, 8
genetic testing, 315
genetic variation, 99, 120, 123, 199, 408, 442, 446, 452, 484
genetically determined predisposition, 267
genetically modified (GM) crops, 13, 104
genome, 220, 338, 340
genome analysis, 199
genome sizes, 525
genomes, 198, 424
genomics, 198, 525, 529
genotype, 18
genotype frequency, 86, 88, 94, 144
genotype value, 80, 81
genotype/environment interactions, 74, 473
Genus, 124, 131
germplasm, 457
Giemsa banding, 285
glucose-6-phosphate-dehydrogenase deficiency, 259, 265, 270
gluts, 14, 563, 575, 576
glyphosate, 425, 437
glyphosate-resistant, 438
GM crops, 435
goat, 133, 369, 405, 503, 507, 430, 512

goats' milk, 587
Golden Rice, 428
gorilla, 17
government, 563, 564, 573–577, 590, 593
governments intervene, 575
grading up, 123
graft, 490
graft-versus-host disease, 347
graft-versus-host reaction, 336
grafting, 485, 488, 593
grapes, 371, 456, 467, 488, 490, 491, 566, 581, 582, 587
grasses, 437
grasshoppers, 238
great tits, 147
Great Wall of China, 102
Green Revolution, 377, 378
growth hormone, 296, 432
growth rate, 126–128
guanaco, 370, 371, 452
guinea pig, 62, 396

haemoglobin, 263
haemoglobin H disease, 263
haemoglobin-based oxygen carriers (HBOCs), 302
haemolytic disease of the newborn, 303–305
haemophilia, 15, 23, 271, 272, 274, 275, 297, 310, 311, 329, 344, 432, 499
haemophilia A, 265, 266, 430
haemophilia B, 266, 439
hair, 260, 331
hair colour, 251
Haldane's mapping function, 186
Haldane–Muller principle, 112
half-sib mating, 159, 160
half-sib selection, 125
half-sibs, 156
hamster, 195
hand-pollination, 101
hand-pollination methods, 481
haploid, 220, 223, 225, 232, 234, 236, 445, 535, 547
haploid stage, 495
haploid-dikaryotic cycle, 520

haploidisation, 443, 444, 534, 537, 538, 540, 557
haplotype, 202, 306
HapMap Project, 201, 202
Hardy–Weinberg, 253, 304, 449
Hardy–Weinberg analysis, 89
Hardy–Weinberg equilibrium, 87, 89, 91, 108, 110, 140, 151, 256, 272
Hardy–Weinberg equilibrium genotype frequencies, 152
harmful recessives, 318
harvest index, 8, 14
hazelnuts, 569
head, 162
head width, 249, 250
health, 586
health foods, 585
heart disease, 268, 314, 315
heat, 502, 503, 504, 507
height, 141, 249, 296
hemizygous, 18, 109, 193, 265, 271
herbicide resistance, 424, 436, 438
herbicide tolerance, 495
herd books, 121
hereditary diseases, 541
hereditary early baldness, 70
heritability, 29, 63, 74–78, 114, 121–125, 132, 134, 136, 248, 250, 255, 389, 446
hermaphrodites, 338
hermaphroditism, 335, 337, 338
hessian fly, 468
heterochromatic, 240
heterokaryon, 194, 524, 535, 540, 544, 557
heteromorphic system, 478
heterothallic species, 524, 525
heterozygosity, 152–154, 359, 368
heterozygote advantage, 211, 218, 259, 262, 263, 314, 354, 444, 448, 449
heterozygous, 18
heterozygous advantage, 123
heterozygous carriers, 323
hexaploid, 17, 217, 219, 231, 233, 234, 236
hexaploid loganberry, 223
high-gravity brewing, 550
high-value compounds, 426

Hippeastrums, 455
histone acetylation, 331
HO allele, 524
Home-Grown Cereals Authority, 402
homeologous pairing, 232, 234, 242
homocystinuria, 313
homogentisic acid, 260
homologous recombination, 420, 424
homology-dependent gene silencing
    (cosuppression), 421
homothallic species, 521, 524, 529
homozygosity, 152
homozygous, 18
hops, 386, 477, 548
hormone treatments, 374, 505, 513
horse, 121, 122, 133, 353, 369, 370, 408,
    451, 455, 496, 503
        Cleveland Bays, 122
        Quarter Horses, 122
        Thoroughbred breed, 122
horse clones, 512
horse milk, 587
host/parasite relations, 448
house fly, 55
human, 141, 206
Human Artificial Chromosomes, 341
human chromosome methods, 283
human gene sequencing, 201
human genetics, 540
human genome, 195, 201
human genome project, 4
human growth hormone, 432
human height, 244
human inbreeding, 145, 315
human karyotype, 284
Human Leucocyte Antigen (HLA) system,
    305, 337, 346, 347
human mutation, 274
human reproductive physiology, 513
human/animal chimeras, 338
Huntington disease, 255, 268, 311, 315,
    320
hyacinths, 233
hybrid animals, 362
hybrid maize, 358, 359
hybrid rice, 362

hybrid sprouts, 361
hybrid vigour, 21, 58, 61, 78, 178, 213,
    218, 223, 353, 354, 356, 358, 362, 363,
    387, 389, 446, 488, 544, 547
hybrid vigour in eucalyptus, 354
hybrid wheats, 359
hybridisation, 101
hybridisation probes, 195
hydrops fetalis, 263
*Hymenoptera*, 223
hypertension, 267, 268
hypomorphs, 17

ICR-170, 522
ideal type, 400
identical alleles, 156
identical by recent descent, 152
identical twins, 319, 328, 330–332
illegitimate cross-overs, 207, 215
immigrants, 142, 317
immigration, 91, 444, 544
immune responses, 305, 307, 341
immunoglobulin G, 304
immunoglobulin M, 304
imperfect competition, 570
import levies, 578, 579
imprinting of DNA, 513
in vitro fertilisation (IVF), 310, 311, 333,
    337, 506
*in vitro* selection, 134
inactive X, 297
inarch grafting, 491
inbred line, 224, 353, 358, 361, 362, 426
inbreeders, 363, 365
inbreeding, 34, 86, 123, 143–155, 160,
    244, 315, 358, 363, 365, 386, 438, 446,
    456, 483, 507
inbreeding coefficient F, 153, 157, 159
inbreeding depression, 145, 146, 150, 151,
    358, 365
inbreeding diagram, 157, 158
inbreeding pathway, 157
incest register, 316
incestuous matings, 316
income, 564
incompatibility, 374

incomplete dominance, 19, 20, 38, 39, 41, 60, 106
incomplete penetrance, 72, 74, 266, 320, 323
independent assortment, 25, 183, 444
independent culling levels, 127–129
index scores, 125
index selection, 128, 129, 131
indirect selection, 545
individual selection, 124
induced mutations, 167, 170, 177, 179, 522
inducible enzyme production, 546
induction of flowering, 471
industrial alcohol, 527
industrial enzymes, 554, 555
industrial yeast, 552
inelastic demand, 568, 575
infantile Krabbe's disease, 347
inferior goods, 565
infertility, 311
inflammation, 199
inflation, 574
influenza, 273, 340
insect pests, 462
insect resistance, 426, 468
insect-pollinated, 474
insect-resistant varieties, 466
insecticides, 462, 463
insertion, 419
insertion sequences, 169, 207
insertional mutagenesis, 341, 345
insulin, 431
intellectual property rights, 427
intelligence, 69, 70, 77, 141, 290, 296, 297
    IQ, 318
intercalating agents, 169, 522
interchanges, 216
interference, 183, 186, 533
intergeneric hybrid, 370, 371, 374, 375, 376
intermediate vector, 423
International Agricultural Centre, 593
interspecific cross, 371
interspecific hybrids, 221, 362, 369, 371, 373, 374, 376, 385

interspecific hybridisation, 231, 238, 371
interspecific incompatibility, 481
intervention buying, 577, 578
intracytoplasmic sperm injection (ICSI), 333, 515, 516
inversion, 201, 202, 205, 210–213, 266, 486
ionising radiations, 168, 280, 281
IQ, 142, 260, 290, 297, 318, 332
IQ scores, 141
iron-chelating agents, 264
island population structure, 103
isochromosome, 239, 296
isolation by distance, 102
ITEM (index of total economic merit), 131

Japanese quail, 147
jaundice, 303

kale, 353, 356
karotyped, 307
karyotype, 283, 284, 288, 289, 298
karyotyping, 239
killing enrichment, 542
kindred, 246
Klinefelter syndrome, 270, 296
knockouts, 422

lactose intolerance, 431
lager yeast, 549
lagers, 549
lamb, *see* sheep
Lamb, 16, 112, 163, 174, 176, 183, 186, 524, 529, 530, 533
larch, 374
large deletion, 164, 177
lavender, 373
Leber's optic atrophy, 256, 323, 324
leek, 17
legumes, 457
length polymorphisms, 190, 196
lentiviral vectors, 340
lethal recessives, 318
lettuce, 79, 483
leukaemia, 273, 280, 282, 290, 344–347
Leyland Cypress, 374, 375

licensing, 576
life cycles, 519
ligase, 412, 414
liger, 369
light, 523
*Lilium*, 374
line mixture, 364
linkage, 24, 44, 61, 80, 87, 147, 187, 188, 202, 488
linkage between any loci, 188
linkage disequilibrium, 80, 87, 90, 96, 192, 202, 306
linkage equilibrium, 90, 96
linkage group, 193
linkage in maize, 44
linked genes, 24
linked loci, 531
linseed, 405, 591
lion, 149, 369, 370
liposomes, 341, 343
llama, 370, 452, 502
localised positive chromosome interference, 533
loci, 14
locusts, 462
loganberries, 226, 369
*Lolium*, 232, 236, 242, 372
lordosis, 504
loss mutation, 546
loss of effective recombination, 149
loss of heterozygote advantage, 147
*Lotus corniculatus*, 232
lymphoma, 346
lysine, 451

macaroni wheats, 231
maize, 1, 2, 17, 33, 42, 46, 49, 57, 71, 80, 88, 107, 117, 133, 166, 178, 179, 181, 192–194, 198, 227, 239–242, 353, 358–361, 366, 373, 376, 378, 422, 425, 426, 428, 435, 436, 438, 446, 451–453, 482, 483, 485–487, 495, 534, 570
maize F1 hybrid, 359
major genes, 54, 407
major histocompatibility complex (MHC), 305

malaria, 256, 259, 263, 314, 449, 462, 465
male, 179
male sterility, 360, 361, 426, 483
male-sterile mutants, 483
malignant, 344
malignant melanoma, 277
malnutrition, 451
man, 17, 63, 198, 497, 503, 506
mango, 459, 587
Manx cat, 170, 402
map distances, 186
maple syrup urine disease, 313
mapping, 188, 192, 195, 197, 245, 379, 529, 541
mapping functions, 533
maps, 536
Marfan syndrome, 275
mark up, 571
marker-assisted selection, 5, 138, 197, 368, 403
market price, 577, 578
marriages between relatives, 316
marrow, 45
mass agricultural selection, 115
mate-choice, 317
maternal age, 294, 296
maternal blood sampling, 309
maternal inheritance, 360
maternally inherited, 256
mating type, 552
mating-type switching, 524
maturity-onset diabetes of youth, 268
meat, 2, 127, 397, 439, 452, 565, 570, 574, 585
Meat and Livestock Commission, 579, 592
meat carcase characters, 124
meat characters, 135, 137
meat composition, 136
meat quality, 119, 138
meat quality factors, 397
mechanical harvesting, 484
mechanisms of resistance, 466
Meckel syndrome, 256
medflies, 465
*Medicago*, 437
medical diagnostic methods, 281

medical X-rays, 280
melanin, 259
melanoma, 344
melons, 565
Mendel's First Law, of Segregation, 38
mental illnesses, 315
mental retardation, 295
mentor pollen, 480
meristem culture, 492
meristem multiplication, 485
meristems, 460
metallothionein-I, 432
methionine, 451
mice, 63, 103, 194, 331, 429, 500, 501, 512
micro-injection, 339, 419, 423, 429, 431
micro-organisms, 456
microbial starters, 550
microchimeras, 337
micropropagation, 492, 493
microRNAs, 342
microsatellite, 407, 408, 453
migration, 87, 88, 109, 110, 446
milk, 14, 78, 79, 83, 118, 121, 124, 130, 131, 363, 388–390, 392, 395, 396, 430, 439, 556, 585–587
Milk Development Council, 592
milk levy, 592
milk-substitutes, 586
Millennial Seed Bank, 439, 456
millet, 487
minimal medium, 536
minisatellite, 407, 408
mirror imaging, 330
miscarriage, 298, 308
mismatch repair system, 419
missense mutation, 259
mite, 146
mitochondrial disorders, 324
mitochondrial DNA (mtDNA), 526, 546
mitochondrial genes, 105
mitochondrial genome, 323
mitochondrial inheritance, 323
mitochondrial mutation, 256, 324
mitotic crossing-over, 185, 444, 486
mitotic crossover, 534, 537–540, 551

mitotic gene conversion, 185, 534, 551
mitotic recombination, 26, 185, 186
mixture of lines, 364
mixture of strains, 549
MN blood group, 253, 330
modified risk, 321
modifier polygenes, 55
modifiers, 49, 54, 55, 401
molecular mapping, 199
molecular markers, 197, 367, 398, 403, 408
Moloney leukaemia virus (MLV), 340
money-price system, 562, 563
mongrels, 151
monoploid, 135, 221, 224, 227, 362, 443
monoploid Brussels sprouts, 226
monoploid plants, 493
monopoly, 572, 573
monosomic, 232, 235, 236, 294, 298
monosomy, 443
monotocous, 499, 510
monozygotic twins, 249, 251, 255
morphological mutants, 541
mosaics, 270, 294, 298, 320, 334, 335, 338
mother's age, 329
mountain ibex, 149
mouse, 17, 180, 195, 198, 201, 341, 416, 420, 432, 513, 540
mouse leukaemia virus, 342
mule, 353, 370
multifactorial, 250, 254, 411
multifactorial character, 250
multifactorial conditions, 321
multifactorial diseases, 267, 268
multifactorial disorders, 63, 266, 267, 325
multifactorial (horizontal) resistance, 468
multifactorial traits, 73, 255, 273
multiple alleles, 16, 89, 305, 479
multiple births, 245, 249, 309, 327–329, 332, 333, 337, 514
multiple cloning site, 416
multiple conceptions, 337
multiple crossovers, 532
multiple sclerosis, 251, 306
multiple translocation stocks, 219
multiple translocations, 217

multiplicative action, 22
mungbean, 179
muscular dystrophy, 19, 99, 312, 315, 320, 321
mushrooms, 520, 527
mutagenesis, 339, 340, 416, 555
mutagenicity, 171, 279
mutagens, 169, 171, 174, 178, 274, 278, 522, 551, 559
mutant, 14, 15, 264, 338, 383, 541, 544, 579, 558
mutation, 23, 87, 88, 111, 112, 161, 162, 165, 167, 173, 175, 177, 179, 195, 221, 257, 274–277, 280, 282, 323, 334, 343, 420, 434, 442, 446, 447, 467, 451, 480, 486, 522, 530, 541, 544–546, 551, 553, 556, 557
mutation accumulation, 457
mutation breeding, 369
mutation frequencies, 161, 276, 442, 477
mutation load, 450
mutation rate, 274, 281, 516
mutational hot-spots, 274, 276
mutational load, 112
mycoherbicides, 528
mycopesticides, 528
myotonic dystrophy, 255, 321

N-methyl-N-nitrosoguanidine, 546
narrow sense heritability, 74, 75
National Council for the Conservation of Plants and Gardens, 455
National Fruit Collections at the Brogdale Horticultural Trust, 456
National Institute of Agricultural Botany, 402
National List, 381, 402, 592
National List Trials, 381, 590
National Recommended Lists, 403
national trials, 403, 404
natural insemination, 510
natural oenological selection, 548
natural selection, 114, 298, 486, 495
neonatal screening, 312
neuroblastoma, 200, 347
neurofibromatosis, 274, 275

*Neurospora*, 17, 22, 24, 71, 182, 214, 219, 519, 521, 533
*Neurospora crassa*, 522–525, 534, 541, 543
newborn children, 298
niche markets, 592
nitrogen mustard, 557
NMG, 522, 557
non-disjunction, 194, 240, 251, 294, 296, 443–445, 486, 516, 534, 539, 540
non-disjunction diploids, 537, 539
non-head-forming strains, 550
non-identical twin formation, 329
non-identical twins, 328–330, 332, 337
non-Mendelian ratios, 551
non-parental ditype, 531
non-reciprocal translocations, 216
non-syntenic loci, 531
nonrandom mating, 142, 149
nonsense base substitution mutations, 279
nonsense mutations, 264, 266
Nordic Gene Bank Cooperation — Farm Animals, 453
nuclear fertility restoration, 426
nuclear fusions, 535, 540
nuclear ratio, 524
nuclear restorer systems, 360
nuclear transfer, 511
number of offspring born per pregnancy, 500
number of progeny to raise, 181

oats, 178, 485, 486, 587
octaploid, 17, 235, 373, 385
*Oenothera*, 217–219, 447, 479, 480
oestrogen, 504
oestrous, 504
oestrous cycle, 508
oestrus, 510
oil palm, 122, 124, 485, 492
oilseed rape, 591, 592
oligopoly, 571
olives, 179, 484
oncogenes, 282
one-child per family policy, 499
onion, 17, 360, 470, 585

oomycetes, 520
*opaque-2*, 451
opposite-sex character testing, 124
opposite-sex characters, 122, 125
optimum mutation rates, 176
oranges, 178, 491
orchids, 371, 374, 376, 477, 492
ordered asci, 530
ordered tetrads, 529
organelle genes, 148
organic acids, 528
organic foods, 585
organically modified silica nanoparticles (ORMOSIL), 341
origins of replication, 415
ORMOSIL nanoparticles, 342
osteogenesis imperfecta, 322
osteoporosis, 334
ostrich, 136, 585
other milks, 587
outbreeders, 365
outbreeding, 34, 86, 145, 146, 148
outbreeding depression, 101, 149
overdominance, 19, 21, 38, 41, 449
overdominance hypothesis, 354, 355, 359
overlay method, 544
overproducers, 544
overproduction, 545, 546

palmyrah palms, 14
pancreas, 261
paracentric inversion, 210, 211, 212
parasexual analysis, 540
parasexual cycle, 226, 443, 534, 540, 544, 546, 557
parasexual mapping, 540
parasexual methods, 194
parasites, 528
parent-offspring, 158, 358
parent-offspring mating, 158
parental ditype, 531
parthenogenesis, 223, 224, 450
parthenogenically, 223
partial parasexual cycles, 534
Pascal's triangle, 50
Patau syndrome, 291, 294, 298

patents, 573
paternity tests, 305
pathogen-derived resistance, 428
pBluescript plasmid, 415
pea, 15–19, 38, 240, 404, 470, 477
peach, 575
peacocks, 114
pear rootstocks, 488
pearl millet, 453
pears, 374, 452, 456, 480, 485, 585
pedigree analysis, 408
pedigree breeding, 121, 365, 388
pedigree diagrams, 155, 157, 158, 160, 246
pedigree information, 321
pedigree selection, 115, 121–124, 403
pedigree studies, 246
pedigrees, 37, 157
penetrance, 320
penicillin, 9, 30, 54, 508, 521, 527, 534, 542, 545, 558, 559
*Penicillium*, 519, 541
*Penicillium chrysogenum*, 519, 521, 523, 527, 534, 546, 557, 558
perfect competition, 570
pericentric inversion, 210, 211
perishable nature, 582
perithecium, 524
permease, 545
permease genes, 546
Peruvian guinea pigs, 586
pest-resistance, 585
petites, 522, 523, 546
petunia, 421
phage, 456
pharaohs, 146
phenocopies, 70
phenotype, 18, 144
phenotype frequencies, 86, 144
phenotype plasticity, 71, 72
phenotype ratio, 187
phenotype value, 80–82
phenotype variation, 267, 446
phenotypic uniformity, 484
phenylalanine hydroxylase, 259
phenylketonuria (PKU), 259

Philadelphia chromosome, 282
photo-repair, 175
photoperiod, 381
phylloxera, 491
phylloxera-resistance, 488
physical mapping, 188, 195
physical maps, 191
physiological incompatibility, 478
pig genome, 199
pig heritabilities, 137
pigs, 75, 118, 119, 125–129, 132–134,
    136, 138, 145, 146, 150, 200, 305, 353,
    358, 397–399, 452, 454, 496–499, 504,
    507, 508, 512, 593
    Gloucester Old Spot, 454
    Tamworth, 454
pine, 17, 145, 493
plant breeders' rights, 573, 584
plant breeding stations, 593
plant breeding systems, 470
Plant Genetic Resource Centre, 457, 458
plant propagation , 485
plant sexual reproduction, 470
plant vaccines, 427
Plant Varieties Act 1997, 591
plant variety right, 590
*Plantago*, 238
plasmids, 279, 342, 415, 423, 433, 526,
    546, 554
plasticity, 71
pleiotropy, 22, 80
ploidies,17, 220, 222, 544
plum, 231, 485
*Poa*, 239
polar body, 222
police dogs, 509, 510
polio, 434
pollen, 104, 452, 495
pollen germ pores, 474
pollen preservation, 476
pollen size, 474
pollen viability, 474, 476
pollination, 104
pollinator varieties, 480
polycystic kidney disease, 320

polydactyly, 19, 62, 72, 73, 255–257, 267,
    309, 320
polygenes, 49, 52–54, 191, 250, 267, 401,
    407, 451
polygenic, 61, 62
polygenic disorders, 266
polymerase chain reaction, 311, 407
polymorphism, 37, 140, 310, 448, 551
polyploid, 17, 220, 232, 233, 376, 385,
    486, 488, 551, 554
polyploidy, 480, 481
polytene chromosomes, 213
polytocous, 500
poplar, 480
population, 31, 32, 86, 255
population crashes, 149
population genetics, 86, 87
population mixing, 87
population size, 97, 100, 109, 316, 507
population structure, 33, 100, 102, 103
pork, 136, 565
position effects, 207, 216
positive assortative mating, 318, 325
positive-negative selection, 419
post-harvest treatments, 476
post-translational glycosylation, 556
post-translational modifications, 430
post-translational processing, 429
potato, 178, 223–225, 229, 357, 380–384,
    402, 426, 427, 436, 438, 473, 474, 492,
    495, 575, 576, 581, 583, 585,
    591–593
potato breeder, 593
potato breeding, 380
pre-implantation embryo screening, 515
pre-implantation embryo selection, 514
pre-implantation genetic diagnosis (PGD),
    310, 311, 514, 515
pre-implantation selection, 345
pre-mutational lesion, 174
predators, 467
Predicted Transmitting Ability (PTA), 130
predisposition, 267
premature babies, 587
premeiotic deletions, 214
prenatal diagnosis, 166, 266, 307, 310, 322

prepotent, 122, 388
preservatives, 582
prey/predator, 448
price, 566–572, 574, 575, 577, 581
price controls, 573
price theory, 564
price wars, 571
price-elasticity of demand, 567
price-fixing, 571
prime symbol, 20
primrose, 141
*Primula*, 231, 369, 478, 479
privatisation, 573
proband, 246, 247
probes, 286, 293, 311, 312, 314
Profit Index (PIN), 131
proflavin, 522
progeny testing, 82, 115, 123, 124, 391
prognosis, 320
prolificacy, 500
promoter, 546, 547, 553, 555
propagation, 590
prostaglandin, 505
protandry, 483
protein processing, 556
protein yields, 547
protein-engineering, 556
protogyny, 483
protoperithecium, 524
protoplast, 227, 415, 423, 425
protoplast fusion, 494, 547, 558
prototrophs, 224, 481, 486, 494, 542, 547, 554
*Prunus*, 480
pseudodominance, 193, 209
pseudodominant, 40, 209
pteridophytes, 232
*Puccinia graminis tritici*, 523
pumpkins, 46
Punnett chequer board, 42, 43, 89
pure breeding, 21
pure lines, 361
pure-breeding lines, 366
pyloric stenosis, 251, 326
pyrethrum, 29, 473

quadrivalent, 216, 221, 230, 231, 242
quadruplets, 327, 500
qualitative characters, 37, 401
qualitative traits, 118
quantitative traits, 155
quantitative analysis, 59
quantitative characters, 49, 61, 65, 68, 117, 154
quantitative inheritance, 49, 52, 61
quantitative trait loci (QTL), 49, 53–55, 61, 62, 124, 138, 191, 192, 197–199
quantitative trait locus (QTL) mapping, 4
quintuplets, 327, 329
Quorn, 528
quota, 576, 577

rabbit, 71, 401, 497, 500, 503, 506, 512
rabies, 434
race, 328
race horses, 121
racial differences, 277
racial groups, 306
radiation, 276, 480, 481
radioisotopes, 276
radiotherapy, 280
radon gas, 174, 276, 280, 281
random amplified polymorphic DNA (RAPD), 407
random mating, 34, 87, 88, 89, 91, 95, 105, 140, 317
random spore mapping, 529
rape, 405, 426, 436–438, 470
rare breeds, 399, 456
Rare Breeds Survival Trust, 453
rare recombination, 479
rarities, 582
raspberry, 222, 369, 591, 593
rate of fixation, 98
rationing, 564, 574
ratios
    1:1, 42
    1:1:1:1, 42–44
    1:2:1, 39, 50, 55
    1:4:6:4:1, 50, 55
    1:6:15:20:15:6:1, 55
    3:1, 42, 50

5:1, 231
9:3:3:1, 42–44, 47, 50
9:3:4, 45, 46
9:7, 47, 48
12:3:1, 46
13:3, 47
15:1, 47
27:9:9:9:3:3:3:1, 50, 59
35:1, 231
81:108:36:12:1, 53
81:27:27:27:27:9:9:9:9:3:3:3:3:1, 53
dihybrid ratio, 42
modified dihybrid ratio, 45, 46
trihybrid ratio, 59
rats, 145, 239, 512
realised heritability, 134
rearrangements, 286
recalcitrant seeds, 459
recessive alleles, 106
recessive lethal, 106, 258, 316
recessiveness, 19
reciprocal crosses, 41, 374, 375
reciprocal translocation, 11, 217, 236, 286
recognition sequence, 416, 417
recombinant enzymes, 556
recombination frequencies, 183
recombinants, 550
recombination, 11, 24, 60, 87, 148, 149,
    181–184, 198, 202, 213, 234, 366, 419,
    424, 444, 446, 450, 451, 468, 485, 488,
    528–530, 538, 540, 544, 546, 551, 554,
    558
recombination frequency, 24, 95, 183,
    191, 529, 532
recombination hotspots, 202
recombination mechanisms, 529
recombine, 537
recurrence rate, 267, 323
recurrence risk, 320
recurrent backcrossing, 234, 368, 369,
    378, 447
recurrent parent, 368
red hair, 142
reflex, 504
regression, 62, 65, 66, 68, 69
regulator genes, 546

relative risk, 306
release of sterile insects, 462, 463
repair systems, 174
repeated induced point mutations, 214
replica-plated, 535, 536
reporter genes, 424
reproductive hormones, 504
reproductive physiology, 470
reproductive potential, 122, 132
repulsion, 25
repulsion phase, 42
restriction endonuclease, 411, 412
restriction enzyme, 190, 412–414, 416,
    417, 547
restriction fragment, 190, 196
restriction fragment length
    polymorphisms (RFLPs), 312, 407
restriction site, 312, 423
resveratrol, 587
retinoblastoma, 275, 283
retroviruses, 339–341
reverse genetics, 419
reverse transcriptase, 339, 417, 429, 527
reversion, 23, 161, 162, 164, 279
revertants, 544
Rhesus, 91
rhino, 451
rhizome, 492
*Rhododendron*, 488
rhododendrons, 369, 384, 385
ribosomal RNA genes, 295
rice, 1, 2, 14, 118, 119, 178, 179, 197,
    198, 201, 361, 362, 377, 379, 436, 451,
    453, 466, 470, 471, 485, 486
rice genomics, 201
ring chromosome complex, 218
risk factors, 326
RNA interference (RNAi), 421, 422
Robertsonian translocations, 295
role of governments, 563
rootstock, 385, 489, 490, 491, 591, 593
roses, 167, 369, 371, 485, 488
Roundup, 104, 436
royalty, 404
rubber, 459
*Rumex*, 232

rust-resistance, 451
rusts, 520
rye, 55, 118, 222, 235, 238, 241, 242, 360, 372, 376, 452, 468, 472, 483
rye-grass, 229, 236, 242, 372

*S-alleles*, 479
*Saccharomyces bayanus*, 547
*Saccharomyces carlsbergensis*, 519, 548
*Saccharomyces cerevisiae*, 519, 520, 525, 527, 534, 547, 548, 550
*Salmonella*, 279
Saint Bernard dog, 400
salt-pollution, 446
satellite DNA, 295
saxifrage, 228
schizophrenia, 30, 63, 73, 74, 250–252, 267, 311
*Schizosaccharomyces pombe*, 526
scrapie, 200, 201, 389
screening, 556
screw worm, 465
seasonal availability, 582
seasonal peaks of fecundity, 501
seasonal variation, 581
seasonally breeding, 501
second division segregation, 530
secretor allele, 303
seed banks, 456, 457
seed potatoes, 402, 584, 591, 593
seedless varieties, 476, 566
seedlings, 593
segmental allotetraploids, 231, 232
segregant, 535–537, 539
segregation, 11, 60, 324, 366, 450, 485, 488, 494, 544
segregation load, 450
segregation ratios, 230, 447
selectable markers, 437
selection, 13, 29, 30, 63, 77–79, 82, 87–89, 106–108, 115, 124, 133, 134, 138, 150, 151, 182, 202, 213, 235, 297–299, 315, 353, 358, 364, 365, 367, 378, 385, 387, 396, 398, 403, 408, 417, 445–449, 544, 551
selection and gene conversion, 448

selection at the haploid stage, 135, 495
selection coefficient, 30, 107
selection differential, 134
selection for auxotrophs, 541
selection for rarity, 444, 447–479
selection for uniformity, 117
selection in haploids, 109
selection intensity, 132
selective slaughter, 496
self-fertilisation, 143
self-incompatibility, 140, 141, 354, 356, 361, 367, 447, 448, 472, 483
self-incompatibility systems, 479, 480
self-pollination, 140, 448, 480
selfing, 144, 159, 160
semen, 460, 498, 507, 592, 593
semi-dwarf rice, 2, 378
semi-dwarf wheat, 2, 14, 377, 378
semi-leafless combining pea, 379
sensory properties of meat, 137
Set-Aside Scheme, 405, 579
severe combined immunodeficiency disease (SCID), 342, 347
sex chromosome abnormalities, 296
sex chromosomes, 38, 40
sex determination, 310
sex differences in disease susceptibility, 271
sex drive, 514
sex hormone, 296
sex linkage, 40
sex organs, 499
sex ratio, 135, 470, 496
sex selection in humans, 499
sex-limited characters, 70
sex-linked, 401, 497
sex-linked recessive disorder, 499
sex-separated sperm, 499
sexed advantage, 498
sexing sperm, 497
sextuplets, 327
sexual cycle, 547, 550, 551
sexual fusions, 524
sexual reproduction, 486
sexual selection, 114, 120, 407, 495
sexual spores, 522, 529

sexually transmitted diseases, 507
sheep, 5, 48, 75, 80, 121, 127, 132, 133,
    136, 137, 146, 152, 154, 195, 200, 201,
    305, 338, 353, 368, 369, 387, 389, 406,
    408, 412, 429–431, 439, 452, 453, 460,
    462, 465, 467, 496, 497, 500, 501–505,
    507, 508, 511
        Blackface, 39, 387
        Booroola Merino, 368, 500
        Border Leicester, 121, 387
        Cheviots, 121, 387
        D'Man, 502
        Finnish Landrace, 502
        Hamshire Down, 39, 387
        lambs, 136, 505, 512, 579
        Leicester Longwools, 121, 387
        Merino, 20, 115, 467, 502
        North Ronaldsay, 453
        rams, 501
        Romanov, 502
        Romney Marsh, 500
        Scottish Greyface, 387
        Southdown, 115, 387, 388, 391, 392
        Suffolk, 387, 388, 502
        Welsh mountain, 387
sheep genome, 199
shoat, 369
short-day breeders, 502
shortage, 563, 576
show rules, 400
shows, 121, 123, 402
shuttle vectors, 415
Siamese twins (conjoined twins), 331
sib-mating, 358
sickle-cell anaemia, 256, 257, 259, 264,
    312, 314, 315, 347, 449, 450
sickle-cell trait, 258, 259
Signet, 592
similar, 152
simultaneous selection for several
    characters, 133
Single Farm Payment, 580
single line selection, 363
single nucleotide polymorphisms (SNPs),
    198
single-strand beaks, 280

site-directed mutagenesis, 418, 434, 556
size uniformity, 484
skin, 260, 277, 336
skin cancer, 277, 281
skin colour, 277, 336
skin pigmentation, 119
slaughter levy, 592
Sleeping Beauty, 341
smoking, 278
snails, 102, 148, 216, 220
social factors, 273
sockeye salmon, 432
somaclonal variation, 173, 484, 485, 486,
    492
*Solanum phureja*, 223, 380, 382
*Solanum tuberosum*, 223, 382
somatic cell nuclear transfer, 511
somatic embryos, 492
somatic hybridisation, 481
somatic mutation, 162, 252, 270, 277
*Sordaria*, 519, 521
*Sordaria brevicollis*, 524
*Sordaria fimicola*, 524
*Sorghum*, 239
sorghum, 10, 13, 14, 360, 378, 486
Southern Corn Blight, 361
soya bean, 147, 179, 425, 436, 439, 463,
    470, 471, 485
soya milk, 587
sparrow, 140
specific combining ability, 358
specific markets, 403, 404
Spectral Karyotyping (SKY), 287
speculators, 569, 570
sperm, 456, 495, 498, 504, 507, 515
sperm count, 281, 516
sperm separation, 496
spina bifida, 251, 309
spinach, 355
spoilage organisms, 528
spontaneous abortion, 221, 296, 298
spontaneous identical twin formation, 329
spontaneous mutation, 167, 248, 274
spontaneous mutation frequencies, 274,
    275, 281
sporadic conditions, 323

sporophytic incompatibility, 477, 478
spring and winter varieties, 472
sprouts, 353
spruce, 239
S-RNase, 478
stabilising selection, 117, 445, 451
staggered-break, 413, 414
standard deviation of a proportion, 97
stem cell, 345, 346, 411, 514
stepping-stone population structure, 103
sterility, 179, 374, 501
sticky ends, 413, 414
stigma receptivity, 474
strawberry, 17, 373, 456, 460, 484, 485,
  492, 581
Streptomycetes, 434
stress-resistance, 487
stroke, 341
structural abnormalities, 298, 299
structural genes, 545
stud, 122, 123
stud books, 121
subsidy, 405, 577, 578, 580
substitutes, 565
substitution line, 235
sugar, 10, 571, 574, 580
sugar beet, 229, 360, 426, 436, 470, 471,
  487, 495, 552
sugar cane, 1, 17, 135, 385, 386, 485, 486,
  492, 493, 527
Sugar Cane EST Genome Project, 385
sunflower, 101, 179, 360
super-gene, 479
supermales, 470
supernumerary("B") chromosomes, 238,
  282
superweeds, 437
supply, 568
supply curve, 568, 569
suppression, 421
suppression of crossing over, 147
suppressor mutations, 175, 176
survival, 541
survival time, 457
sweet pea, 47, 374
sweet potatoes, 453, 460
synkaryon, 194

syntenic, 24, 538
Syntgenta, 436
systems, 562

tetraploid, 230
tetraploidy, 232
trivalents, 228, 230
Table of Kindred and Affinity, 319
tandem duplication, 214
tandem selection, 127
target price, 578
tastes, 565
tautomeric shifts, 167
taxation, 564
Tay-Sach's disease, 4, 313, 315
tea, 2, 9, 75, 277, 488, 489, 574
temperature, 281
teosinte, 373
terminal deletion, 195, 207
termination, 307, 308
terminator technology, 426
test crosses, 42, 61
test-crossing, 368
testicular feminisation, 266, 299
testosterone, 513, 514
tests, 195
tetracycline, 412, 419
tetracycline-repressible transactivator
  (tTA), 466
tetrad analysis, 533, 534
tetraploid, 17, 220–223, 225, 228–230,
  298, 443, 465, 480, 552
tetrasomic inheritance, 230, 232
tetratype frequencies, 532
tetratype tetrad, 531
thalassaemia, 4, 256, 259, 262, 315, 347
theoretical correlation between relatives,
  75
therapeutic proteins, 432
therapy, 342
three unlinked markers, 532
three-way cross, 356
threshold character, 49, 62, 63
threshold of liability, 326
thyroxine growth hormone, 259
Ti plasmid, 423–426
tiger, 370, 584

tiger cubs, 585
tigress, 369
tigron, 369
tilapia, 432
tissue culture, 457, 460, 487, 492
tissue plasminogen activator, 430
tissue-specific expression, 426
tissue-specific vector, 339
tobacco, 60, 79, 229, 231, 279, 425, 485, 494
tomato, 61, 62, 118,, 172, 178, 179, 193, 224, 229, 353, 371, 373, 421, 425–427, 436, 470, 471, 473, 476, 495, 581, 587
top-fermenting yeast, 548
*Torulopsis utilis*, 527
totipotent, 512
Trades Unions, 574
transcriptional silencers, 412
transfected, 342
transfection, 341, 342
transferring genes, 373
transformation, 415, 419, 424, 547, 554
transfusions, 299, 300, 303, 304, 342
transgene, 339, 341, 411, 421, 424, 427
transgenic animals, 389
transgenic chickens, 432
transgenic crop, 437
transgenic farm animals, 430
transgenic plants, 424, 427
transgression, 56, 62, 65, 66, 68, 69
transgressive segregation, 101, 105
translation, 342
translocation, 205, 215, 216, 234, 237, 286–288, 292, 294, 298, 299, 308, 338, 465, 486, 551, 558
translocation Down's, 292, 294, 295
transmissible spongiform encephalopathies, 200
transplant, 306, 338, 344, 345, 432, 515
transplantation, 348
transposable elements, 282, 341, 486, 551
transposon, 169, 207, 341, 553
*Trichoderma reesei*, 554–556
*Trichoderma viride*, 527
*Trifolium*, 448
trinuclear pollen, 478
trincleate, 474

trinucleate grains, 476
triple marker (Bart's) test, 309
triplet, 327, 329, 332, 500
triploid, 17, 220, 221, 225, 228, 230, 298, 236, 443, 465, 480, 487
trisomic, 298, 332, 233
trisomy, 256, 292, 294, 310, 443
triticale, 235, 372, 376, 453, 591
*Triticum*, 371
trivalents, 228
true hermaphroditism, 337
truffle, 528
tsetse fly, 465
tuber crops, 473
tulip, 485, 582, 583
Tulip Mania, 583
tumour, 279, 282, 339, 340, 344, 423, 513
tumour-suppressor genes, 282
turkey, 136, 506, 509, 507, 565
Turner syndrome, 270, 296, 298
turnip, 229, 437, 588
twins, 39, 63, 74, 76, 77, 135, 249, 327, 328, 332, 337, 500, 501
twin registers, 249, 334
twin studies, 248
two-way cross, 356
two-way selection, 544
type 1 and type 2 diabetes, *see* diabetes mellitus
types of mutation, 545

UK Recommended List, 381
ultrasonics, 136
ultrasound screening, 309
umbilical cord, 514
umbilical cord blood, 346
umbilical cord blood stem cells, 346
unbalanced non-reciprocal translocation, 206
unbalanced translocations, 299
uncle/niece marriages, 317
unequal crossing-over, 214, 443
unequal gene doses, 270
uniform ripening, 483, 484
uniformity, 117, 356, 451
univalent, 221, 224, 235, 239, 242
unordered octads, 531

unordered tetrads, 529
unstable length mutation diseases, 323
unstable length mutations, 165, 255
*Ustilago maydis*, 534
UV light, 168, 174, 175, 177, 277, 334, 522, 523, 543, 546, 557

vaccines, 434
variable age of onset, 320
variable expressivity, 72, 256, 266, 320, 323
variation, 451, 544
variegated plants, 493
varietal purity, 405
vector, 339–341, 414–416, 418, 420, 424, 554
vegetable, 565, 572
vegetative fusions, 524
vegetative propagation, 460, 484
vegetative reproduction, 173, 450, 485
vegetative segregation, 493, 494
velvet, 586
venison, 586
vernalisation, 472
*Vicia* sp., 450
vicuña, 370, 452
virus resistance, 382
virus-induced gene silencing (VIGS), 422
viruses, 282
viscous gene flow, 102
vitamin A, 428
vitiligo, 253, 254, 260

water buffalo, 586
water melons, 228
weight, 249, 250, 332
wheat, 1, 17, 49, 101, 120, 124, 127, 128, 133, 135, 147, 173, 179, 197, 231, 234, 236, 359, 360, 364–366, 371, 372, 376–379, 402–405, 426, 427, 436, 451, 453, 468, 470–472, 483, 485, 487, 520, 570, 591
whisky, 527
white tiger, 584
wild-type, 14, 15, 522, 541
wind-pollinated, 474, 475

wine, 471, 491, 520, 524, 527, 544, 554, 555, 578, 587
wine yeasts, 548, 551, 552
wool, 49, 115, 127, 388, 389, 431, 452
Wright's coefficient of inbreeding, F, 145, 151
Wright's equilibrium, 153
Wright's equilibrium frequencies, 154, 157
Wright Sewell, 99, 151

xenia, 222, 230
*Xenopus laevis*, 511
X-inactivation, 252, 269–271, 329
X-inactivation and Barr bodies, 268
X-linked, 256, 271, 297
X-linked characters and genes, 248
X-linked conditions, 270
X-linked diseases, 271
X-linked dominants, 321
X-linked genes, 109, 265, 270
X-linked recessive, 265, 266, 278, 310, 320–323
X-linked severe combined immunodeficiency (SCID-X1), 343, 344
X-rays, 177, 193, 276, 280, 281, 463, 522, 557
xeroderma pigmentosum, 175, 277, 318
XXX, 297, 298
XXY, 298
XYY, 298

yeast, 17, 23, 24, 27, 80, 132, 182, 198, 222, 415, 434, 456, 522–524, 527, 529, 531, 533, 544, 546, 547, 549, 554 (*see* also *Saccharomyces*)
yeast artificial chromosomes (YACs), 433, 554
yeast crosses, 521
yield, 544
yoghurt, 585

Z, 44
zebra, 369, 370
zebroid, 369
zeedonk, 369, 370
zoo, 455, 456, 585
Zygomycetes, 520

www.ingramcontent.com/pod-product-compliance
Lightning Source LLC
Chambersburg PA
CBHW060416220326
41598CB00021BA/2198